Applied Ecology

Applied Ecology

monitoring, managing, and conserving

Anne E. Goodenough & Adam G. Hart

University of Gloucestershire

Great Clarendon Street, Oxford, OX2 6DP,
United Kingdom

Oxford University Press is a department of the University of Oxford. It furthers the University's objective of excellence in research, scholarship, and education by publishing worldwide. Oxford is a registered trade mark of Oxford University Press in the UK and in certain other countries

Impression: 1

Published in the United States of America by Oxford University Press
198 Madison Avenue, New York, NY 10016, United States of America

British Library Cataloguing in Publication Data
Data available

Library of Congress Control Number: 2016953951

ISBN 978–0–19–872328–8

Printed in Great Britain by
Bell & Bain Ltd., Glasgow

Chapter opening images

Chapters in Part 1: Photograph © istock.com/ amenic181

Chapters in Parts 2, 3, 4: Photographs © Anne Goodenough

He who loves practice without theory is like the sailor who boards ship without a rudder and compass and never knows where he may cast

Leonardo da Vinci (1452–1519)

OVERVIEW OF CONTENTS

Part 1 **Overview** 1

1 **Introducing Applied Ecology** 3

2 **Fundamentals of Ecology** 8

Part 2 **Monitoring** 27

3 **Ecological Surveying and Monitoring** 29

4 **Ecological Indicators** 69

Part 3 **Managing** 97

5 **Ecological Impact Assessment** 99

6 **Remediation Ecology** 134

7 **Landscape Ecology and Management** 156

8 **Non-native Species Management** 193

9 **Pest Management** 224

Part 4 **Conserving** 251

10 **Principles of Conservation** 253

11 ***In Situ* Conservation** 289

12 ***Ex Situ* Conservation** 331

13 **Reintroduction and Rewilding** 357

Glossary 393

Index 399

FULL CONTENTS

Part 1 **Overview** 1

1 **Introducing Applied Ecology** 3

1.1 **Introduction** 3

1.1.1 Applied Ecology: not a new idea 4

1.2 **About this book** 5

1.2.1 Our approach to Applied Ecology 5

1.2.2 The topics covered in this book 6

2 **Fundamentals of Ecology** 8

2.1 **Untangling the 'tangled bank' of ecology** 8

2.2 **Feeding relationships** 11

2.2.1 Herbivores and predators 11

2.2.2 Parasitism 12

2.2.3 Parasitoids 14

2.2.4 Decomposition 14

2.3 **Beneficial relationships between organisms** 16

2.3.1 Mutualisms: where both parties benefit 16

2.3.2 Commensalism: 'sharing the same table' 17

2.4 **The concept of the niche** 18

2.5 **Competition** 20

2.5.1 Intraspecific competition 20

2.5.2 Interspecific competition 21

2.6 **The importance of evolution in ecology** 21

2.6.1 Life history theory and r/K strategists 23

2.7 **Patterns in ecology** 24

2.7.1 Biogeography 24

2.7.2 Ecological succession 25

2.8 **Conclusions** 26

Part 2 **Monitoring** 27

3 **Ecological Surveying and Monitoring** 29

3.1 **Introduction** 29

3.2 **What is ecological surveying and monitoring?** 30

3.3 **Purposes of surveying and monitoring** 31

3.3.1 Baseline surveying 32

3.3.2 Spatial surveying 33

3.3.3 Temporal monitoring 38

3.3.4 Alert monitoring 41

3.3.5 Compliance monitoring 41

3.3.6 Impact, mitigation, and compensation monitoring 44

3.4 **Approaches to monitoring** 44

3.4.1 Use of secondary data for monitoring 44

3.4.2 Use of primary data for monitoring 45

3.4.3 Proxy monitoring 47

3.4.4 Using citizen science for monitoring 47

3.5 **Monitoring habitats** 50

3.5.1 Classifying and quantifying habitats 50

3.5.2 Assessing habitat condition 51

3.6 **Monitoring species** 54

3.6.1 Collection and use of direct data 55

3.6.2 Collection and use of indirect data 58

3.6.3 Collection and use of remote data 59

3.6.4 Key surveying principles and links back to theory 64

3.7 **From monitoring to management** 65

3.8 **Conclusions** 65

4 **Ecological Indicators** 69

4.1 **Introduction** 69

4.2 **Types of ecological indicator** 70

4.2.1 What can ecological indicators be used to indicate? 71

4.2.2 Single- and multispecies approaches 71

4.3 **Environmental indicators** 73

4.3.1 Theory underpinning environmental indicators: niches and tolerance ranges 73

4.3.2 Exemplar environmental indicators 75

4.3.3 General lessons from environmental indicator schemes 83

4.3.4 Species that make good indicators 85

4.4 **Biological indicators** 88

4.4.1 Exemplar biological indicators 88

4.5 **Biodiversity indicators** 91

4.5.1 Exemplar biodiversity indicators 91

4.5.2 Caveats when using biodiversity indicators 92

4.6 **Reconstructing historic landscapes** 93

4.7 **Conclusions** 94

Part 3 **Managing** 97

5 **Ecological Impact Assessment** 99

5.1 **Introduction** 99

5.2 **What is Environmental Impact Assessment?** 100

5.2.1 Aim and purpose of EIA 100

5.2.2 Legislative history and current context of EIA 100

5.2.3 What developments need an EIA? 101

5.2.4 Key components and processes of EIA 103

5.3 **What is Ecological Impact Assessment?** 104

5.3.1 Aim and purpose of EcIA 104

5.3.2 Ecological receptors used in EcIA 105
5.3.3 Collecting ecological data: scoping and follow-up 107

5.4 **Defining ecological value** **112**
5.4.1 Site value 112
5.4.2 Habitat value 113
5.4.3 Species value 114
5.4.4 Individual organism value 116
5.4.5 Ecosystem service value 116

5.5 **Assessing likely impacts of development** **116**
5.5.1 Types of impact 117
5.5.2 The potential for positive impacts 118
5.5.3 Impact magnitude and importance 118
5.5.4 Predicting impact 119

5.6 **Mitigation and compensation strategies** **121**
5.6.1 Mitigation 122
5.6.2 Compensation 127
5.6.3 Enhancement 128

5.7 **Recommendations and outcomes of EcIA** **129**
5.7.1 Forming recommendations following EcIA 129
5.7.2 Outcomes of EcIA and what happens next 129
5.7.3 Post-development monitoring 129

5.8 **Limitations and challenges of the EcIA process** **130**
5.8.1 Strategic Environmental Assessment 130

5.9 **Conclusions** **131**

6 Remediation Ecology 134

6.1 **Introduction** **134**

6.2 **Pollution: an overview** **134**
6.2.1 Types and sources of pollution 135
6.2.2 Scale of pollution 136
6.2.3 Tackling polluted sites: remediation and bioremediation 136

6.3 **The general principles of bioremediation** **137**
6.3.1 Bioremediation through metabolic breakdown 137
6.3.2 Bioremediation through hyperaccumulation 137

6.4 **Bioremediation using microorganisms** **138**
6.4.1 Bioremediation using indigenous microorganisms 138
6.4.2 Bioremediation by stimulating indigenous microbial growth 139
6.4.3 The role of bio-augmentation in bioremediation 142
6.4.4 Techniques for bioremediation using microorganisms 143

6.5 **Phytoremediation: bioremediation using plants** **147**
6.5.1 Rhizofiltration: reed beds and wetland systems 150

6.6 **Conclusions** **153**

7 Landscape Ecology and Management 156

7.1 **Introduction** **156**

7.2 **Landscape elements** **156**
7.2.1 Background matrix 157
7.2.2 Patches 158
7.2.3 Linear features 159
7.2.4 Boundaries 160

7.3 **Studying landscape ecology** **162**
7.3.1 Spatial patterns of species: quantifying range and distribution 162
7.3.2 Spatial patterns of individuals: quantifying home range and movement 165
7.3.3 Spatial analysis framework: Geographical Information Systems 169

7.4 **Effects of landscape processes: habitat loss and fragmentation** **172**
7.4.1 Direct effects on individual species 174
7.4.2 Effects on wider ecology 179
7.4.3 Fragmentation metrics and mapping 179

7.5 **Effects of non-landscape processes on landscape ecology** **181**
7.5.1 Climate change 181

7.6 **Landscape ecology management** **184**
7.6.1 Heterogeneity 184
7.6.2 Connectivity 185
7.6.3 Management to prevent species movement 187
7.6.4 Ecoregions and 'living landscapes' 188

7.7 **Conclusions** **190**

8 Non-native Species Management 193

8.1 **Introduction** **193**

8.2 **Non-native species: key concepts and questions** **194**
8.2.1 'Nativeness': an academic debate with applied implications 194
8.2.2 The importance of non-native species in Applied Ecology 197

8.3 **Translocation** **199**
8.3.1 Invasion pathways 199
8.3.2 Likelihood of establishment 200
8.3.3 Invasion risk 203
8.3.4 Understanding translocation: preventing species introduction 206

8.4 **Impacts of non-native species** **207**
8.4.1 Ecological impacts: negative and positive 208
8.4.2 Impact complexity 208
8.4.3 Responses of native species to non-native species 210
8.4.4 Expansive native species 210

8.5 **Importance of monitoring** **211**
8.5.1 The role of citizen science 211
8.5.2 Lag effects 212

8.6 **Management** **212**
8.6.1 Non-native species databases 214
8.6.2 Aims of non-native species management 214
8.6.3 Controlling non-native species can be counter-productive 218
8.7 **Conclusions** **221**

9 **Pest Management** **224**
9.1 **Introduction** **224**
9.2 **What are pests?** **224**
9.2.1 The definition of 'pest' 226
9.2.2 The problem of defining 'pests' in practice 226
9.2.3 Pest management 227
9.3 **The theory of pest management** **228**
9.3.1 The Economic Injury Level 228
9.3.2 Problems with the EIL approach 231
9.3.3 The economic threshold 232
9.4 **Pest management in practice** **232**
9.4.1 Physical pest management 233
9.4.2 Chemical pest management 234
9.4.3 Biological pest management 241
9.4.4 Problems with pest management 244
9.5 **Integrated Pest Management** **247**
9.5.1 IPM control of purple loosestrife 247
9.5.2 IPM control of pocket gophers 248
9.6 **Conclusions** **249**

Part 4 **Conserving** **251**

10 **Principles of Conservation** **253**
10.1 **Introduction** **253**
10.2 **What is conservation and why is it necessary?** **254**
10.2.1 Conservation versus preservation 254
10.2.2 The focus of conservation 255
10.2.3 Threats to populations and communities 256
10.3 **Extinction risk** **258**
10.3.1 Studying extinction risk 260
10.3.2 International species-at-risk classification systems 264
10.3.3 National species-at-risk classification systems 265
10.3.4 Considering evolutionary distinctiveness 268
10.4 **Conservation strategies** **270**
10.5 **Conservation decisions** **272**
10.5.1 The need for conservation triage 272
10.5.2 Species-focused priorities in conservation triage 275
10.5.3 Site-focused priorities in conservation triage 280
10.5.4 Setting priorities within *ex situ* scenarios 285
10.6 **Conclusions** **286**

11 ***In Situ* Conservation** **289**
11.1 **Introduction** **289**
11.2 **An overview of *in situ* management** **289**
11.2.1 Aims of *in situ* management for conservation 290
11.2.2 Active versus custodial management 291
11.3 ***In situ* conservation through active management** **291**
11.3.1 Managing habitat 292
11.3.2 Managing species 303
11.4 ***In situ* conservation through custodial management** **310**
11.4.1 Legislation 310
11.4.2 Policy 313
11.5 **Reserve-based conservation versus wider countryside management** **314**
11.5.1 Creation of protected areas and nature reserves 314
11.5.2 Limitations of protected areas and nature reserves 316
11.5.3 Wider countryside initiatives 319
11.6 **The public and *in situ* conservation** **320**
11.6.1 Management of visitor pressure 320
11.6.2 Education and community engagement 320
11.6.3 Community-linked conservation 321
11.7 **The future: evidence-based initiatives** **324**
11.7.1 Researching species–habitat interactions 324
11.8 **Conclusions** **328**

12 ***Ex Situ* Conservation** **331**
12.1 **Introduction** **331**
12.2 ***Ex situ* conservation: an overview** **332**
12.3 **The Lord Howe Island stick insect: *ex situ* conservation in action** **335**
12.4 **The stages of *ex situ* conservation** **337**
12.4.1 Collection 337
12.4.2 Transport 342
12.4.3 Captive breeding 343
12.4.4 Reintroduction, supplementation, and reinforcement 348
12.5 **Gene banking** **351**
12.6 **Conclusions** **354**

13 **Reintroduction and Rewilding** **357**

13.1 **Introduction** **357**

13.2 **Basic principles of species reintroduction** **358**

13.2.1 Aims of species reintroduction 358

13.2.2 Reasons for reintroduction 359

13.2.3 Taxonomic bias in reintroduced species 364

13.3 **Reintroduction regulatory frameworks** **365**

13.3.1 Reintroduction policies 365

13.3.2 Codes of practice for reintroduction 366

13.3.3 Legislation for reintroduction 367

13.4 **The process of species reintroduction** **368**

13.4.1 Initial feasibility study 368

13.4.2 Founder individuals 370

13.4.3 Choice of release site 374

13.4.4 Reintroduction methods 376

13.4.5 Post-release monitoring 380

13.5 **Factors affecting the success of reintroductions** **381**

13.6 **Rewilding** **383**

13.6.1 Is rewilding just 'big and ambitious' *in situ* conservation? 384

13.6.2 The challenges faced by rewilding schemes 385

13.6.3 Rewilding successes 387

13.7 **Conclusions** **390**

Glossary 393

Index 399

COMMON ABBREVIATIONS USED IN APPLIED ECOLOGY

AONB: Area of Outstanding Natural Beauty [UK designation].

AWVP: Ancient Woodland Vascular Plants.

AZA: Association of Zoos and Aquariums.

BAP: Biodiversity Action Plan.

BIAZA: British and Irish Association of Zoos and Aquariums.

BMWP: British Monitoring Working Party.

CBD: Convention of Biological Diversity.

CEM: Climate Envelope Modelling.

CITES: The Convention on International Trade in Endangered Species of Wild Fauna and Flora.

CMR: Capture–mark–recapture.

DAFOR: Dominant, Abundant, Frequent, Occasional, Rare.

DEFRA: Department for Environment, Food and Rural Affairs [UK government department].

DNA: Deoxyribonucleic acid.

EcIA: Ecological Impact Assessment.

EDGE: Evolutionarily Distinct and Globally Endangered.

eDNA: Environmental deoxyribonucleic acid.

EIA: Environmental Impact Assessment.

EIL: Economic Injury Level.

ET: Economic Threshold (also known as the **CAT:** Control Action Threshold).

GEM: Genetically engineered microorganism.

GIS: Geographical Information Systems.

GMO: Genetically modified organism.

IBA: Important bird area.

IPM: Integrated Pest Management.

IUCN: International Union for Conservation of Nature.

JNCC: Joint Nature Conservation Committee [UK public body].

LNR: Local nature reserve.

MEE: Mass extinction event.

MPA: Marine protection area.

MVP: Minimum viable population.

NBSAP: National Biodiversity Strategy Action Plan.

NNR: National Nature Reserve [UK designation].

NVC: National Vegetation Classification.

PCR: Polymerase chain reaction.

PIP: Plant-incorporated protectants.

PPE: Personal protective equipment.

PVA: Population viability analysis.

RNA: Ribonucleic Acid.

SAC: Special Area of Conservation [European designation].

SEA: Strategic Environmental Assessment.

SLOSS: Single large or several small.

SPA: Special Protection Area [European designation].

SSC: Species Survival Commission.

SSSI: Site of Special Scientific Interest [UK designation].

VOC: Volatile organic compounds.

WAZA: World Association of Zoos and Aquaria.

PART 1

Overview

At heart, Applied Ecology is an evidence-based discipline that takes our understanding of ecological theory and applies that theory to monitor, manage, and conserve species and habitats in the diverse ecosystems found around the world. In **Chapter 1** we introduce Applied Ecology as a discipline, and set out the approach taken to the subject in the remainder of the book. **Chapter 2** then reviews some of the fundamental ecological concepts that Applied Ecology draws on most heavily.

1 **Introducing Applied Ecology**

2 **Fundamentals of Ecology**

Introducing Applied Ecology

1.1 Introduction.

Planet Earth is a complex place. An astonishing diversity of animals, plants, bacteria, fungi, and protists are involved in tangled webs of **relationships** with themselves and with the physical environment around them. Plants use atmospheric carbon dioxide, water in the soil, and sunlight to photosynthesize; animals eat plants and other animals; molecules are exchanged and transformed; habitats change; species evolve; and species become extinct. **Patterns** of diversity and relationships are played out across a dynamic planet of oceans, rock, soil, air, mountains, deserts, forests, rivers, and lakes. Humans, of course, also add to the complexity. Our industry, agriculture, and other activities dramatically alter the living (**biotic**) and non-living (**abiotic**) environment at local and global scale.

It is the **interactions** between organisms and between the biotic and abiotic environment that underpin the complexity of the natural world. Interactions between organisms and the abiotic environment can have major, global-level consequences. Photosynthesizing plants, for example, interact with the atmosphere, removing carbon dioxide and producing oxygen, while coral polyps can build immense barrier reefs that can have a profound effect on coastal environments. Interactions between organisms often play out at a smaller scale, but can still have important consequences. For example, novel predators finding their way onto coastal islands (perhaps because of an especially low tide) can eradicate ground-breeding birds through predation.

The study of interactions in the natural world is the science of **ecology**. A broad subject, ecology bridges the divide between the study of life (biology) and the study of the physical world (including environmental science and physical geography). Ecology is also concerned with patterns and processes. An ecologist might be interested, for example, in the altered distribution of a species as the climate changes. To understand how climate change would affect a species' distribution requires an understanding of the physical consequences of climate change, as well as an appreciation of the biological requirements and interactions of the species in question.

Increasingly, there is interest in answering ecological **questions** that arise as a consequence of human activities. There is no doubt that the 21st century is a period of unprecedented change on a global scale. These changes produce **problems** for humans and for the rest of the biota on Earth. Some of these anthropogenic (human-induced) impacts are shown in Figure 1.1.

As the human population continues to increase, increasing demands are being made on the environment for **resources,** whilst we continue to degrade the environment as a consequence of our activities. We need food, and to produce it requires intensive

Figure 1.1. Key anthropogenic (human-induced) changes affecting ecology include: (top left) climate change accelerated by our production of greenhouse gases; (top right) contamination of land, air, and water by chemicals; (bottom left) biodiversity losses sufficient for many to now consider us living through a sixth mass extinction event; and (bottom right) global homogenization of species and the impact of non-native species.

Source: Top-left image courtesy of Juan-Vidal Díaz/CC BY-ND 2.0; top-right image courtesy of Indi Samarajiva/CC BY 2.0; bottom-left image from the Smithsonian Institution archives c. 1904; bottom-right image courtesy of Paul Peters, Ribble Rivers Trust/CC BY-ND 2.0

agriculture and pest management. To fuel human development requires a staggering diversity of industry, such as disposal of waste, diverting rivers, building on biodiverse habitat, and travelling around the globe. All of these activities have the potential to affect the interactions that underpin the complexity of the natural world.

In order to solve the problems caused by human activities then it is necessary to understand their **effects**. For example, the degradation of a habitat, such as a salt marsh by a coastal reclamation project, might grab some attention because of a charismatic bird species that will be affected, but in reality it is the entire suite of potentially affected biotic and abiotic interactions that need to be understood for the consequence of the project to be fully appreciated. We need ecological insight and ecologically informed management to solve such problems and find ways to lessen the impact of human interaction with the natural world. This approach is the discipline of **Applied Ecology** and it is Applied Ecology that is the focus of this book.

1.1.1 Applied Ecology: not a new idea.

The importance of the link between theory and application is a recurring theme in this book, but the relationship between applied and what is sometimes called 'pure' ecology has not always been clear. In 1904, one of the fathers of ecology, the British plant ecologist George Arthur Tansley outlined what he considered to be the 'problems of ecology'. One of the major problems he identified was that the subject was, at that time, too descriptive and lacking in **systematic methodology**. In other words, Tansley considered that ecology was too applied. His concern, shared by the American ecologist Frederic Edward Clements, was that ecology lacked an **experimental approach** and was not based on **scientific analysis**.

Pure ecology is the scientific study of ecology for its own sake. In other words, the study of the relationships and patterns in the natural world for no reason other than intellectual curiosity driven by a desire to understand how the world works. Thus, the pure ecologist would study the interactions between fungi and plants because of scientific interest, not because a greater understanding of those interactions might lead to, say, an improvement in crop yields. However, the application of scientifically derived ecological knowledge to specific ecological problems (such as crop yield) is what defines Applied Ecology.

The concern of Tansley, Clements, and others was that ecology, at that time, was predominately applied and was merely building up a store of disconnected knowledge uninformed by the scientific method. They were concerned that ecology was developing as a series of case studies and anecdotes, rather than as a coherent body of theory and knowledge. The problem was that without a robust **scientific foundation** it is not possible to meaningfully compare the phenomena and patterns being investigated. You might generate great knowledge on the growing and harvesting of certain crops, or the treatment of certain pests or diseases, but it will not generate more general solutions to related problems since the generation of that knowledge has not been undertaken in a systematic manner.

The 20th century saw this issue resolved, with the modern science of ecology emerging. While the pure ecologist may be interested in species coexistence and niche differentiation from intellectual curiosity, such knowledge may be useful to the Applied Ecologist evaluating the potential impact of an invasive plant species. Likewise, the pure ecologist may be interested in developing sampling techniques to determine population changes as part of a study on intraspecific competition, but such techniques are invaluable to the Applied Ecologist monitoring species indicating ecosystem health. Insect pests are a major problem to agriculture, but understanding the ecology of insect parasites might lead to novel pest management strategies. Even larger-scale and seemingly quite theoretical concepts like biogeography can be useful when insights gained are applied to landscape-scale conservation projects. It is easy to think of the pure–applied relationship as a one-way street, with knowledge going solely to Applied Ecology, but the links between the two are such that pure ecology benefits considerably from observations, data, and insight gained from putting ecological principles into practice.

1.2 About this book.

It is sometimes easy, when reading about Applied Ecology, to feel rather like Tansley must have felt when he evaluated the field back in 1904. There is a wealth of case studies, anecdotes, and examples of particular types of project or approach available, and there are many specialized texts devoted to the various subdisciplines of Applied Ecology (such as pest control or conservation). However, making sense of all those studies, linking them to theory and appreciating how the different subdisciplines fit together is often difficult. This book aims to provide that framework of understanding, allowing those new to the field of Applied Ecology to make sense of what is a very **broad discipline**.

1.2.1 Our approach to Applied Ecology.

Ecology is inherently complex, encompassing many interlinked concepts and subdisciplines. In order to reflect real-world issues and practice in this book, an **overarching** approach will be taken, bringing together many of the themes of Applied Ecology in one place. We also, at the start, outline some of the most important concepts in ecology and provide the theoretical background that is necessary to appreciate many of the applications introduced in the rest of the book. We have divided the book into three broad topics that encompass the main aspects of Applied Ecology:

- **Monitoring:** studying species (individuals, populations, and communities), interactions between species, and the abiotic environment over time to answer a specific question, record change, or check compliance with legislation or policy.
- **Managing:** devising and implementing processes and procedures that enable humans to manage the natural world for its benefit, for human benefit, and in many cases, often for the benefit of both.
- **Conserving:** a component of management, conservation seeks to prevent the over-exploitation, degradation, and destruction of environments and species.

Applied Ecology operates within a constantly changing and complex economic, political, social, and increasingly globalized landscape. There are multiple 'grey areas' where different approaches might be used and the 'right answer' depends upon individual circumstances, stakeholder perspectives, ethical considerations, and political motivations. We have ensured that this complexity is reflected throughout by taking a **real-world** approach and discussing such situations openly to develop a questioning, critical, and evidence-based attitude among readers. All real-world applications are **underpinned by theory** and, accordingly, this book links back to ecological theory throughout as is relevant and appropriate. Perhaps most importantly, this book is fundamentally **application-focused,** using a variety of case studies and examples to demonstrate how ecology can be monitored, managed, and conserved within a range of different environments.

It is important to note that this book provides a detailed overview of the approaches and underpinning theory used in Applied Ecology to solve specific problems. It does not provide a manual of applied ecological techniques. It will not, for instance, explain species survey protocols, nor guide you through the intricacies of calculating diversity indices. Many other books do this, and do so well, as highlighted in the suggested further reading following most sections.

There is one deliberate omission from this book—there is no chapter on policy and legislation. Such a chapter would be extremely tedious to read, would necessarily be focused on specific countries or regions, and would quickly become outdated. Instead, examples of policy and legislation are included within specific chapters and online resources as necessary. Anyone undertaking applied ecological projects is advised to consider carefully any legislation, policy, or guidelines before they start.

1.2.2 The topics covered in this book.

Part 1: Overview

Chapter 1 Introducing Applied Ecology: introduces Applied Ecology as a subject, and considers the purpose and scope of this book.

Chapter 2 Fundamentals of Ecology: provides a short introduction to the key concepts and terminology of ecology. This chapter provides a broad introduction for readers with little ecological background and a refresher for those that have studied the topic before. Focusing mainly on biological concepts, it covers fundamental topics, such as feeding relationships, symbioses, niches, and competition, as well as the importance of scale and evolution. The concepts covered are those that are important for understanding the material in the rest of the book.

Part 2: Monitoring

Chapter 3 Ecological Surveying and Monitoring: considers why it is important to monitor and survey ecological systems, and specific ecological features (species and habitats).

Chapter 4 Ecological Indicators: leads on from surveying and monitoring, considering what the presence and abundance of certain species can tell ecologists about environmental conditions.

Part 3: Managing

Chapter 5 Ecological Impact Assessment: examines why it is important to consider possible changes to ecology as a result of proposed building, infrastructure, and other anthropogenic developments. It follows on from the Monitoring section (Chapters 3 and 4), both of which are important in assessing the baseline conditions of a site.

Chapter 6 Remediation Ecology: considers how species can be used to remediate (clean up) specific environments. This chapter follows on from Chapter 5,

where measures are considered that attempt to prevent such problems from occurring in the first place.
Chapter 7 Landscape Ecology and Management: provides details of how species interact with their environment at a landscape scale and how such interactions can be managed.
Chapter 8 Non-native Species Management: the movement of species is having profound effects in many ecosystems and this chapter examines the impact and management of such species.
Chapter 9 Pest Management: following on from non-native species, which may often be considered undesirable and therefore pests, this chapter considers how pest organisms can be controlled and managed.

Part 4: Conserving

Chapter 10 Principles of Conservation: this chapter considers the ethical and moral underpinnings of conservation, as well as the scientific background, highlighting the concept that there is often no 'right' answer. It sets conservation in a theoretical context, but also debates why and under what circumstances species should be conserved and how conservation resources should be allocated.
Chapter 11 *In Situ* Conservation: develops ideas considered in Chapter 10 and considers approaches to managing species *in situ* and the habitats that support them.
Chapter 12 *Ex Situ* Conservation: examines in detail the role of *ex situ* conservation, which often occurs because of changes in landscape ecology (Chapter 7), alongside *in situ* conservation (Chapter 11), and prior to species reintroduction (Chapter 13),
Chapter 13 Reintroduction and Rewilding: considers the important concept of species reintroduction and looks to the future by examining the concept of 'rewilding' (habitat and species restoration at a landscape level).

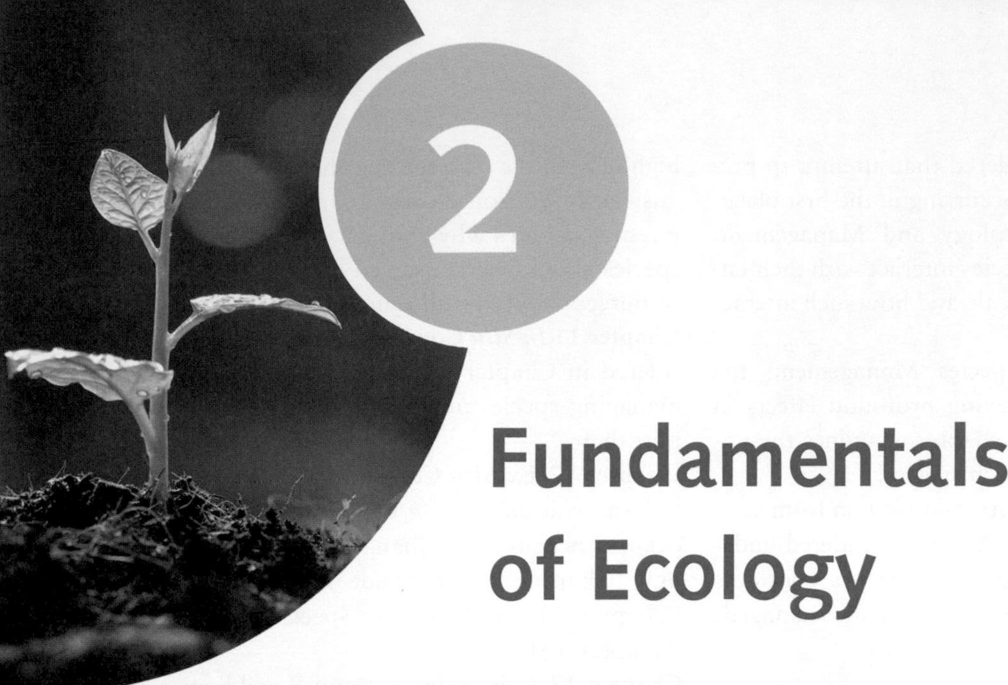

2 Fundamentals of Ecology

2.1 Untangling the 'tangled bank' of ecology.

'It is interesting to contemplate a tangled bank, clothed with many plants of many kinds, with birds singing on the bushes, with various insects flitting about, and with worms crawling through the damp earth, and to reflect that these elaborately constructed forms, so different from each other, and dependent upon each other in so complex a manner, have all been produced by laws acting around us.'
(Opening sentence of the final paragraph of *On the Origin of Species* by Charles Darwin.)

This famous quote comes at the end of one of the most important scientific books ever written, Charles Darwin's *On the Origin of Species*, published in 1859. Encapsulated within Darwin's metaphor of the 'tangled bank' (Figure 2.1) is the notion that the natural world is an elaborate and complex network of **interactions** that combine both **biotic** (the living component of the natural world) and **abiotic** (non-living) components. It is this interrelatedness that is fundamental to the science of ecology, although Darwin would need to wait seven years for the term to be coined and defined by the German biologist Ernst Haeckel (1866) (see 'Meet the Early Ecologists').

Originally '*Ökologie*', ecology was initially viewed as the science of the **relationships** of organisms to the environment and the definition of ecology has not changed much in the 160 years since then. Thus, from its earliest beginnings, ecology has been about understanding the patterns and processes of how species interact with one another and the natural world. This book focuses on **Applied Ecology**, which deals with the application of scientifically derived ecological knowledge to specific situations and problems. In other words, Applied Ecology is about using ecological knowledge to **monitor**, **manage**, or **conserve** species and habitats. Before examining practical applications, however, it is necessary to understand some of the fundamental principles of ecology.

Figure 2.1 Darwin's 'tangled bank' is a metaphor for the complexity and interrelatedness of the natural world.
Source: Photograph by Anne Goodenough.

Meet the early ecologists

As with the development of most scientific disciplines, many individuals can rightfully lay claim to be among the founders of ecology. A number of notable early ecologists, including Frederic Clements, who published the first American ecology book, and Sir George Arthur Tansley, who was the first President of the British Ecological Society and a founding editor of the *Journal of Ecology,* are rightfully regarded as important pioneers in the subject, but the three ecologists featured here have a special place in the early years of ecology.

Ernst Haekel

Ernst Heinrich Phillipp August Haeckel (Figure A) (1834–1919) was the first to coin the term Ecology, or *Ökologie* in his native German (Haeckel, 1866), although he was not the first scientist, or the only one of his contemporaries, to be thinking 'ecologically'. A prominent zoologist and biologist, as well as philosopher, physician, and artist, Haeckel coined other important biological terms, including phylum and stem cell. He worked extensively with invertebrates, and was fascinated with embryology and development. He was a contemporary of Charles Darwin and the two scientists met in 1866 on a trip that Haeckel paid to England, which also saw him meeting with the geologist Charles Lyell, whose writings were influential in the development of Darwin's theory of evolution. It is notable that Haeckel, the first to coin the term to describe such an inclusive and interconnected science as ecology, was himself an interdisciplinary scientist.

Eugen Warming

Johannes Eugenius Bülow Warming (Figure B) (1841–1924), known as Eugen Warming, was a Danish botanist who wrote the first textbook on plant ecology (Warming, 1895, 1896) and developed the first university course on ecology. Warming was well-travelled, taking botanical expeditions to Trinidad, Brazil, the West Indies, and many European destinations in search of plants and to study plant communities, a topic that was central both to his pioneering ecological work and his ecology textbook. His book, *Plantesamfund*, was based on his university lectures, and he was highly regarded as a teacher, lecturing and writing for university students, as well producing textbooks for schools. Perhaps a reflection of the inspiration Warming gained through expeditions, he was a strong proponent of 'field courses'—opportunities for hands-on learning outside the lecture theatre. Warming himself took students all over Denmark for field excursions and his notes from these trips were published as highly useful guides to those regions and habitats. Warming's enthusiasm for field trips continues in the teaching of ecology today.

Figure A Dr Ernst Haeckel, who first coined the term 'ecology' and defined it in 1866.

Figure B Eugen Warming, Danish botanist and the first to write a textbook on ecology and to teach a university course on the topic.

Charles Sutherland Elton

Many early ecologists, like Warming, Frederic Clements, and Arthur Tansley, were botanists-turned-ecologists. Charles Sutherland Elton (Figure C) (1900–1991) in contrast started life as a zoologist and went on to be a founding father of the subdiscipline of animal ecology. Elton published what has become a classic book, *Animal Ecology* (1927), in which he outlined many of the underpinning concepts of modern ecology. These include trophic relationships expressed as food chains, ecological studies of animal behaviour (itself still developing as a discipline at that time), life histories, quantitative representation of ecosystems, and the concept of the niche. He was also the first editor of the *Journal of Animal Ecology*, founded in 1932. In later years, Elton went on to develop niche theory, developing what become known as the Eltonian Niche, which emphasized the functional attributes of animals (later popularized by another prominent ecologist, Eugene Odum, as an animal's 'profession'), rather than its physical requirements (its habitat, or as Odum put it, its 'address'). Elton was the first ecologist to make use of long-term data sets from the Hudson's Bay Company on the trapping records of Canadian lynx *Lynx canadensis* and snowshoe hare *Lepus americanus*, investigating and developing an understanding of the population fluctuations those records revealed. As well as a scientist, Elton was an important figure in conservation, especially in the UK, where he helped to establish the Nature Conservancy Council in 1949.

Figure C Charles Sutherland Elton FRS, first editor of the *Journal of Animal Ecology* and pioneering animal ecologist.

Source: http://people.wku.edu/charles.smith/chronob/ELTO1900.htm Used with permission from *Journal of Animal Ecology*.

REFERENCES

Elton, C.S. (1927) *Animal Ecology*. New York, NY: Macmillan Company.

Haeckel, E.H.P.A. (1866) *Generelle Morphologie der Organismen. Allgemeine Grundzüge der organischen Formen-Wissenschaft, mechanische Begründet durch die von Charles Darwin reformirte Descendenz-Theorie*. Volume I: *Allgemeine Anatomie der Organismen*. Volume II: *Alllgemeine Entwickelungsgeschichte der Organismen*. Berlin: Georg Reimer.

Warming, E. (1895) *Plantesamfund—Grundtræk af den økologiske Plantegeografi*. Copenhagen: P.G. Philipsens Forlag. Note: the German translation of 1896 became a far more widely read book.

Warming, E. (1896) *Lehrbuch der ökologischen Pflanzengeographie—Eine Einführung in die Kenntnis der Pflanzenvereine*. Transl. Emil Knoblauch. Berlin: Gebrüder Borntraeger.

This chapter examines some of the main themes of ecology, and acts as a refresher for those readers already familiar with ecology and as a broad introduction for those that are new to the field. It is not possible to cover all the topics that fall under the umbrella of ecology, but ecological terms and concepts that are important in later chapters dealing with applied topics are included. In particular, this chapter emphasizes the concepts that ecology is about interactions and that interactions produce patterns. Readers interested in pursuing these ideas further, or wishing to find out more other ecological concepts, are advised to read one of the ecological texts recommended as further reading throughout the chapter.

2.2 Feeding relationships.

Plants can photosynthesize, making use of sunlight, water, and carbon dioxide to build carbohydrates. Through the addition of nitrogen, sulphur, and other elements from the soil, plants can also build the amino acids required to make proteins. These activities mean that plants create biomass from the abiotic environment around them, and it is this activity that makes them (and some bacteria) 'producers' or **autotrophs** (self-feeders). Organisms that cannot fix carbon and other elements from the abiotic environment must feed on organisms to gain the molecules they need for growth, repair, and reproduction. Such organisms are called **heterotrophs** and include the animals, fungi, many protists, and most bacteria.

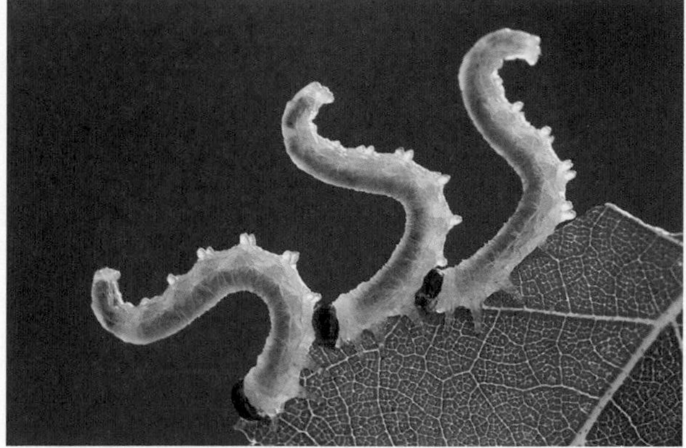

Figure 2.2 Insects, such as these sawfly larvae, are important herbivores that can easily be overlooked. Many insect larvae are economically important pests.

Source: Photograph by Anthony H Cooper, used with kind permission.

2.2.1 Herbivores and predators.

Many animals eat plants. This direct animal–plant **trophic interaction** is called **herbivory**. Some animals can only eat plants and are therefore **obligate herbivores,** while other animals incorporate plant material as part of a wider diet that also includes eating other organisms, such as animals and fungi.

It is tempting to think of herbivores as large animals, such as cows, sheep, or elephants, but a considerable amount of herbivory is carried out by insects, particularly the larval forms of some insect orders like Lepidoptera (butterflies and moths) and Hymenoptera (such as the sawfly larvae shown in Figure 2.2). Indeed, while some mammal herbivores may be a nuisance to farmers, it is insects that most commonly cause sufficient damage to crops for them to become **pests** (see Chapter 9).

Some trophic (or feeding) interactions result in the death of the consumed organism. Such interactions involve what are often termed the **true predators**. These animals attack, kill, and eat their prey in quick succession, kill a number of prey over their lifetime, and eat some or all of each prey item. It is easy to think of large animals such as lions, *Panthera leo*, as true predators, but many true predators are much less obvious. Seed-eating animals, including many rodents and the seed-harvesting ants (such as those belonging to the genera *Messor* and *Pogonomyrmex*) are, ecologically speaking, true predators because by consuming a seed they are, in effect, consuming an entire (potential) organism. Also included are filter-feeding animals like barnacles that feed on planktonic organisms, and larger consumers of plankton, such as the baleen whales (including the largest of all living animals, the blue whale *Balaenoptera musculus* and the whale shark *Rhincodon typus*). It is also tempting to think of all predators as being **apex predators** at the top of the food chain. However, many predators can become prey for larger predators (Figure 2.3).

By contrast, most herbivores tend to nibble at their 'prey', attacking potentially large numbers of individual plants during their lifetime, but consuming only a part of each plant. This type of herbivory is termed **grazing** if it involves ground-based plants or **browsing** if it involves trees and shrubs. These feeding interactions occur in both the large vertebrates commonly thought of as grazers (such as antelope, cattle, and sheep) and smaller micrograzers (such as butterfly and moth caterpillars, aphids consuming plant sap, sawfly larvae, and certain beetles).

Grazing as a trophic interaction can also include animals consuming parts of other animals, such as mosquitoes (e.g. *Culex*), leeches (e.g. *Hirudo*

Figure 2.3 Predators can become prey—here a predatory fish becomes prey for an osprey *Pandion haliaetus*.

Source: Courtesy of Andy Morffew/CC BY-ND 2.0.

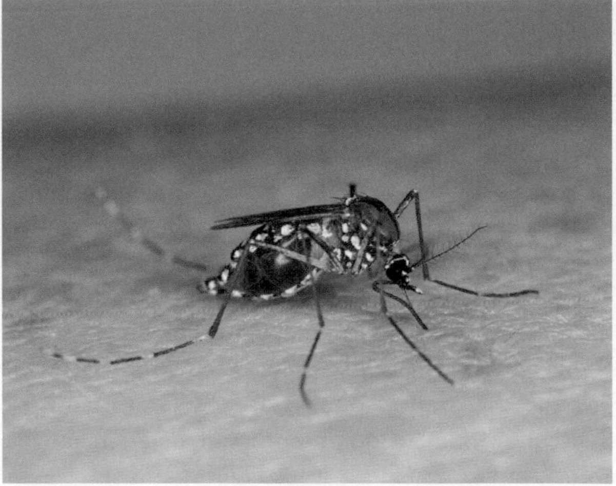

Figure 2.4 A female mosquito grazes on a human, taking a blood meal and potentially transmitting parasites, such as *Plasmodium* (causing malaria) or the dengue virus (causing dengue fever).

Source: Courtesy of the U.S. Department of Agriculture/CC BY 2.0.

medicinalis), and vampire bats *Desmodus rotundus* taking blood meals. From a functional perspective, an antelope feeding on grass or a mosquito feeding on a human are identical—one organism is consuming a part of another organism, and the individual being consumed remains alive after the consumption has occurred. While such grazing clearly has an effect (for example, leaves are damaged and this reduces leaf area for photosynthesis), the effect is not lethal. Similarly, when blood is taken the animal is often little affected. Exceptions to this include if the animal is in poor condition, if there are a great many individuals feeding upon it, or if the act of grazing transmits disease-causing parasites, as can happen when certain mosquitoes feed on humans (Figure 2.4).

2.2.2 Parasitism.

In contrast to grazers, some organisms consume part of their prey, but target only one or a very few individuals in their lifetime, causing appreciable harm as they do so. In such cases the 'grazed' individuals are known as **hosts**, the 'grazers' are called **parasites,** and the relationship is called **parasitism**.

Some grazers should properly be called parasites. Caterpillars emerging on a single oak tree, for example, are unlikely to feed on any other individual trees and do cause harm, even if that harm is not especially great. Functional definitions in ecology can have blurred boundaries that reflect the complex relationships they are attempting to define. For example, an aphid might spend its entire life on a single plant and may cause considerable harm. It is functionally a parasite, but it may be more useful to describe it as a parasitic herbivore.

Just as in other areas of ecology, scale is important. Some parasites, macroparasites, are large (relatively speaking), while many microparasites are very small, even microscopic. Parasites can also be grouped functionally as ectoparasites that live on the surface of their host, such as fleas and lice, and endoparasites, like liver flukes and thread worms that live inside their host. Both forms of parasite result in a close association of parasites and hosts. This usually results in an **evolutionary arms race**, in which parasites and hosts continually evolve to counter the effect each has on the other.

A consequence of the evolutionary arms race between parasites and hosts is that parasites are frequently highly specialized to their host. With each host species potentially able to evolve different countermeasures it becomes progressively more difficult for a generalist parasite to out-evolve these measures or, to use a human analogy, to fight the war on many fronts. Thus, evolution has tended to push parasitism towards species-limited, but tightly co-evolved, parasite–host associations. One such association is described in Case Study 2.1.

CASE STUDY 2.1  Guinea worm

Parasites and hosts live in intimate association and this sets the stage for an evolutionary arms race between them, as the host evolves to counter the parasite, and the parasite evolves to counter those counter-measures. This tends to select for highly specialized adaptations, including elaborate life cycles and life histories.

The guinea worm *Dracunculus* is a nematode parasite of humans that causes dracunculiasis, also called guinea worm disease (GWD) (Figure A). As well being an interesting example of parasitism it also demonstrates other ecological principles, including life histories, adult–juvenile niche differentiation, and behavioural modification by parasites.

GWD is caused by drinking water that contains copepods (small crustaceans commonly found in freshwater) that host guinea worm stage 2 larvae. Copepods ingested in drinking water (or by accidentally ingesting water not intended for drinking) are digested in the stomach and release the next stage of larvae that penetrate the stomach or intestinal wall. From here the stage 3 larvae enter the abdominal cavity where they mature for approximately three months, developing into adult worms, at which point they mate and the males die (Figure B).

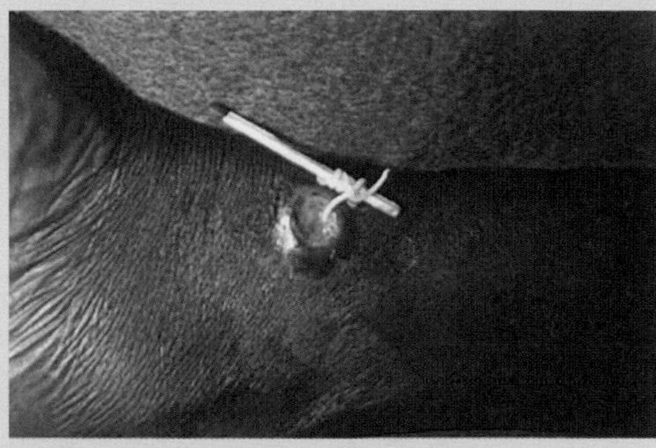

Figure A A guinea worm (*Dracunculus*) being removed from a person's leg.

Source: Image from the Centers for Disease Control and Prevention.

The females remain in the abdominal cavity for around a year before migrating into subcutaneous tissues of the limbs. Once near joints or extremities the adult female worms move towards the surface where they cause a painful blister on the skin. The burning sensation that the blister causes results in infected individuals seeking relief by immersing the affected limb into water. At this point the female worm

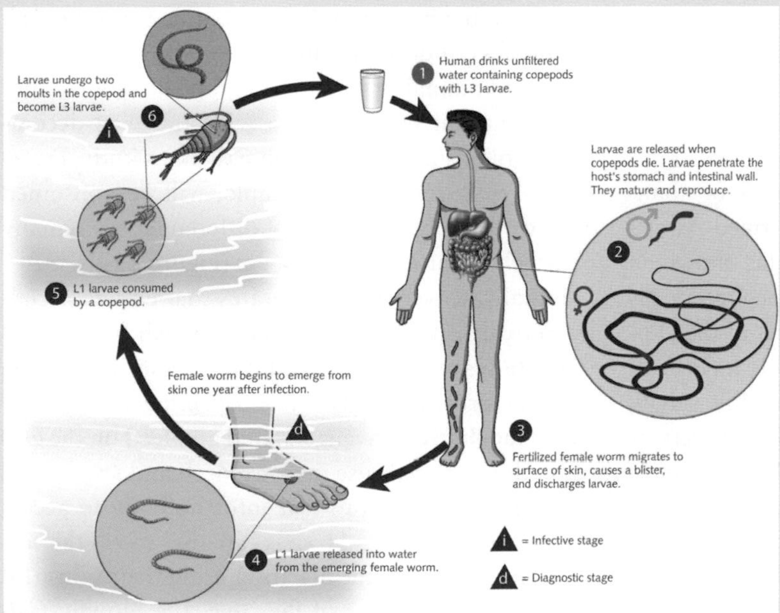

Figure B The life cycle of the guinea worm, a nematode, illustrates the complexity that can evolve when organisms live in intimate association. In this case, the guinea worm is involved in a parasitic relationship with two hosts, the second being humans.

Source: Adapted from: http://www.cdc.gov/parasites/guineaworm/biology.html

releases hundreds of stage 1 larvae into the water that can go on to infect copepods, thereby beginning the cycle again.

The female guinea worm remains in the human host, releasing more larvae whenever the affected limb is in contact with water. She must be removed carefully, in her entirety, to allow the blister (which may now have turned into an open sore) to heal. Infection can cause localized oedema and rashes, as well as nausea and diarrhoea. If adult worms die in joints they can cause arthritis and even paralysis. The guinea worm clearly fulfils all the criteria of a parasite.

FURTHER READING

Further details on Guinea worm: World Health Organization Dracunculiasis: http://www.who.int/dracunculiasis/en/

Figure 2.5 A tarantula hawk wasp (family Pompilidae) attacks an orange-kneed tarantula *Brachypelma smithi* injecting paralysing venom before dragging the spider to a burrow, where the wasp's eggs hatch into larvae that will consume the living spider, avoiding vital organs to keep its host 'buffet' alive as long as possible.

Source: Courtesy of Charles Sharp/CC BY 2.0.

2.2.3 Parasitoids.

A large number of insects (both in terms of abundance and diversity) are ecologically termed parasitoids. Parasitoids are typically wasps (as shown in Figure 2.5), sawflies in the order Hymenoptera (as shown in Figure 2.2), and some members of the Diptera (the so-called true flies). Parasitoids have free-living (i.e. not associated with a host) adult stages, but females lay their eggs in, on, or near insects or other arthropods such as spiders. The larval form of the parasitoid then develops inside the host, consuming it from within. Initially, parasitoid larvae cause little harm, but eventually they kill the host. This lifestyle combines a number of ecological features that straddle some of the functional definitions of predation. A parasitoid has an intimate association with a single host (like a parasite), it kills its host (making it like a true predator), but unlike a true predator it does not attack, kill, and eat its host in quick succession, making it functionally similar to a grazer at least for the first part of its larval stage.

2.2.4 Decomposition.

When primary producers convert inorganic components like carbon, nitrogen, oxygen, and hydrogen into organic molecules like sugars and amino acids they immobilize those components, incorporating them inside their bodies. Other organisms can access those organic components by feeding on the producers when they are alive. However, when organisms die they enter into a trophic relationship very distinct from herbivory and predation. Immobilized nutrients are converted back into their inorganic form through the process of decomposition.

A dead animal or plant is a complex resource that offers nutrient potential to a wide range of **decomposer** organisms. A large mammal, like a kudu *Tragelaphus strepsiceros*, for example, will be consumed after its death by a number of different species, many of which occur in high numbers (Figure 2.6).

Initially, scavenging vertebrates that either specialize on carrion (such as vultures, which are termed **obligate** scavengers) or will take carrion if available (such as brown hyenas *Hyaena brunnea*, which are called **facultative** scavengers) find and feed on the carcass. These animals use their strength and specialized physical adaptations to tear into the body to access the soft tissues inside. Hyenas can also access the soft tissue inside bones (the marrow) using their powerful jaws. A large number of flies (e.g. the blowfly *Calliphora latifrons*) use the carcass as a nursery, laying eggs that will develop into larvae that will consume much of the remaining flesh and organs. Different

Figure 2.6 A before and after shot of a greater kudu *Tragelaphus strepsiceros* decomposing in the African bush, photographs taken 40 hours apart. Note that the grass has been worn away in the 'after' photograph due to the activity of scavengers that included black-backed jackal *Canis mesomelas* and brown hyena *Hyaena brunnea*.

Source: Photographs by Anne Goodenough.

species of flies then come in waves to take advantage of the carcass as it moves through different stages of decomposition, which is enhanced by the action of bacteria that are now active throughout the carcass exploiting the range of different food types it presents (e.g. bone, horn, hair, skin, soft tissue, and faeces within the gut).

Burying beetles (sometimes called sextant beetles) belong to the genus *Nicrophorus*, and feed on the flesh of smaller carcasses (such as rodents and birds) or on fly larvae. If they discover a fresh carcass, they will set about the task of burying it and converting it into a ball of flesh that acts as a nursery for larvae developing from eggs laid on it by a female. They are followed by further waves of arthropods, such as dermestid beetles (e.g. *Dermestes maculatus*, sometimes called the hide beetle), exploiting the now dried hide of the animal. The final stage of decomposition will be achieved by a natural **succession** of different decomposing bacteria and fungi that will decompose whatever remains into smaller organic and inorganic molecules, which eventually leach into the soil to re-enter the biotic environment. Earthworms and other soil-dwelling invertebrates will aid in this process by pulling material down into the soil.

Within a very short period of time the animal will have been converted to recycled molecules within the soil, atmosphere, and the biota, as well as heat dissipated into the atmosphere as a consequence of the metabolic processes involved in feeding on the carcass. This process involves a large number of different types of organisms forming a complex network of relationships.

Decomposers also feed on animal faeces, which in some habitats can be a plentiful resource as Figure 2.7 shows. Yet more decomposers feed on the parts of organisms that are shed during life. A diversity of organic matter is shed by plants and

Figure 2.7 Dung is a highly available and potentially nutritious resource for those able to exploit it, such as dung beetles, Scarabidae.

Source: Photograph by Anne Goodenough.

animals during their lives. Leaves are shed by deciduous trees during the autumn fall. Crustaceans (like crabs, lobsters, and prawns), insects, and other arthropods have rigid exoskeletons, and can only grow by shedding this hard external 'shell', expanding when soft and unconstrained, and then forming a new hard exoskeleton. In insects, this process takes place in the juvenile or larval stages, and in some groups during the process of **metamorphosis**. Other arthropods, like spiders and some crustaceans, can continue to grow when adult. This shedding, or **ecdysis**, produces larval, pupal, and adult 'skins' or exuviae, which, given the abundance of arthropods in most ecosystems, represents a considerable potential resource for those organisms able to exploit it.

Lizards and snakes also grow by shedding their hard external layer and again these shed 'skins' contain nutrients. Birds shed feathers, both as part of an annual or bi-annual **moult,** and as a consequence of damage or growth. Mammals generally shed small amounts of hair continually and, for some species, there may be one or two annual moults, when large amounts of hair are shed. Additionally, mammals that grow horns and antlers may shed them when they are no longer required. Both mammals and birds also shed copious quantities of skin cells, a fact well known in humans, where dust mites that have evolved to feed on shed human skin cells can cause serious allergies in some people. In the case of birds, there are also feather mites and feather lice that are adapted for exploiting shed skin and feather scales.

FURTHER READING FOR SECTION

Overview of trophic relationships: Begon, M., Townsend, C.R., & Harper, J.L. (2005) *Ecology: from Individuals to Ecosystems*, 4th Edition. Oxford: Wiley Blackwell.

Parasitoid diversity and biology: Godfray, H.C.J. (1994) *Parasitoids: Behavioral and Evolutionary Ecology*. Princeton, NJ: Princeton University Press.

Parasite biology: Zimmer, C. (2002) *Parasite Rex: Inside the Bizarre World of Nature's Most Dangerous Creatures*. New York, NY: Atria Paperbacks.

The important role of predators in ecosystems: Terborgh, J. & Estes, J.A. (Eds) (2010) *Trophic Cascades: Predators, Prey, and the Changing Dynamics of Nature*. Washington, DC: Island Press.

2.3 Beneficial relationships between organisms.

The trophic relationships outlined in Section 2.2 are fundamentally one-sided. One individual benefits, while another is disadvantaged. A grazing antelope gains, whereas the grass being grazed is damaged. In true predation, or interactions with parasitoids and some parasites, the cost is the loss of the consumed individual's life.

2.3.1 Mutualisms: where both parties benefit.

Trophic relationships are not inevitably imbalanced. An elegant example of this is the case provided by the fungus-growing ants (tribe Attini) of Central and South America. These ants have a fungus that they grow within their underground nest chambers as a 'crop', which is shown in Figure 2.8. The ants feed their developing larvae on the fungus as it grows. In the largest and most sophisticated of the fungus-growing ant species, the leafcutting ants of the genera *Atta* and *Acromyrmex*, adult ants cut leaf fragments from fresh vegetation and process these in 'fungal gardens' back at their large nests, which may contain millions of ant workers.

Although the fungus is being eaten, this relationship is more complex. The fungus of leafcutting ants is only found in leafcutting ant nests and it is highly successful, being taken by new queens when they leave their home colony to start a new nest. As well as being dispersed far and wide, it is carefully nurtured in a scrupulously hygienic environment. In response, the leafcutting ants' fungus has evolved special swollen tips at the end of the hyphal strands that make up the fungus, these little tips providing nutritious 'apples' for the ants to harvest. Without the fungus the

Figure 2.8 Leafcutting ants of the genera *Atta* and *Acromyrmex* cultivate a fungus in their nest that they consume. The large individual shown here is the queen and the smaller individuals are the workers. They are resting on a substrate composed of the fungal mutualist. Neither the ants nor the fungus can survive without each other.

Source: Photograph by Ayub Khan, used with kind permission.

ants would starve; without the ants tending it the fungus dies.

Organisms that live together in close and long-term relationships are said to be in **symbiosis.** In everyday language, symbiosis is usually assumed to be to the mutual benefit of the interacting organisms. However, strictly speaking symbiosis is any close, long-term relationship between organisms, regardless of the costs and benefits of that relationship to each of the organisms involved. Relationships that are to the mutual benefit of both are termed mutualisms**.** In the case of the leafcutting ants the two **mutualists** have evolved an **obligate** mutualism; neither party can survive without the other. Mutualisms where one or both parties can survive without the other are termed **facultative** mutualisms.

There are a great many mutualistic relationships within the natural world. Well-studied examples include:

- The gut biota of animals, sometimes known as gut flora and consisting mainly of bacteria.
- The association between mycorrhyzal fungi and the roots of many plants, which provides extra nutrients through **nitrogen fixation** (converting atmospheric nitrogen to nitrogen that is available for plants to use).
- The relationships between cleaner wrasse (a reef fish) and their 'clients', who attend cleaning stations to have ectoparasites removed, thereby providing food for the cleaner wrasse.

It may be that one party benefits to a greater extent than the other, but the crucial factor in mutualisms is that both parties gain overall. Relationships where one gains and one loses, and where both live in a close and long-term situation, are parasitic relationships, where the losing party is termed the host.

2.3.2 Commensalism: 'sharing the same table'.

Not all relationships are clear-cut. Consider the decomposing kudu carcass described in Figure 2.6. When large carcass breakers like brown hyenas rip the carcass apart, they provide scraps of meat, as well as openings into the carcass that smaller scavengers (such as jackals *Canis mesomelas*) struggle to make on their own. The association that exists between hyenas and jackals at carcasses is not one from which the hyena gains, but neither does it lose, since a hyena is unable to consume the entire carcass itself. However, jackals and other scavengers gain a great deal.

Relationships where one party gains and the other is unaffected are termed **commensal relationships**, which literally (and appropriately, given the example above) means 'at the same table'. A problem with defining and quantifying commensalism is that while proving benefit is often straightforward it can be very difficult indeed to prove that one party is unaffected.

An example of a relationship usually termed a commensalism is that of **phoresy.** Phoretic organisms are hitch-hikers, attaching to another animal and thereby transported to new places. Many mite species have a phoretic stage in their life cycle that allows them to disperse, and phoretic mites can be readily observed on beetles and other larger insects. However, while phoretic mites are typically very small compared with their transporters they are not weightless, and for any animal, especially those that fly, additional weight has an energetic cost. It may also be that the attached mites affect aerodynamics and could make the animal less manoeuvrable and more likely to become prey (Figure 2.9).

Figure 2.9 A ladybird barely visible under a 'cloak' of phoretic mites. It is unlikely that this animal could fly (or indeed open its wings) and, in this case, the assumed commensal nature of the relationship would seem more appropriately termed parasitic.
Source: Photograph by Roger Key, used with kind permission.

Another potential commensal relationship occurs when birds nest in tree holes. This might seem to be commensal (birds gaining and the tree being unaffected), but adult birds forage over comparatively large distances and may pick up fungal spores of disease-causing species that they take inside the tree. If their nesting behaviour introduces fungal parasites, then the tree is far from unaffected by the presence of the birds. Problems with definitions and issues with the complexities that can develop in real-world scenarios are part of the fabric of ecology and something of which any Applied Ecologist must be aware.

FURTHER READING FOR SECTION

Relationships between organisms: Cain, M.L., Bowman, W.D., & Hacker, S.D. (2014) *Ecology*, 3rd Edition. Sunderland, MA: Sinauer Associates Inc.

Leafcutting ants and their mutualistic fungus: Hölldobler, B. & Wilson, E.O. (2010) *The Leafcutter Ants: Civilization by Instinct*. New York, NY: W.W. Norton and Company.

Mutualisms in general: Bronstein, J.L. (2015) *Mutualism*. Oxford: Oxford University Press.

2.4 The concept of the niche.

In everyday language, a niche is 'a shallow place, especially in a wall to display a statue or other ornament', but an ecological niche is not a physical place. An animal's life cannot be summed up by its habitat and range alone; both are useful to know, but they do not describe the animal's niche in ecological terms. To do that requires knowledge of its range of **requirements** and **limitations.**

No organism can live everywhere on Earth. Humans have a remarkable 'natural' range and with the use of advanced technology can live in extreme environments, but without considerable external support humans cannot live under the sea or in extremes of temperature. If modern technology is removed and humans in a natural state are considered, then we can start to build up a picture of their requirements and tolerances:

- Humans need to breathe oxygen and expel carbon dioxide as gases, not as gases dissolved in water. This constrains human life to the land.
- They are heterotrophs and so are constrained to living in regions where the presence of plant life (and, ideally, other consumers) means that they have something to eat.
- Humans cannot live in extreme cold and, although they cope well with higher temperatures, they can do so only when there is sufficient drinkable water available.
- They are endothermic, large, and have a relatively high metabolic rate (partly because of their large brains) and so need a relatively high food intake, which tends to constrain them to more productive regions of the world where energy and nutrients are sufficiently available to sustain them. Human dentition and digestive biochemistry mean that they do best with a mixed diet, and cannot survive or reproduce unless their dietary intake exceeds an energetic and a nutritive threshold.
- Humans are susceptible to parasites, and to grazers like mosquitoes that spread parasites, and they are of interest to a number of large predators.
- They are a diurnal species, active during the day, when their predominately visual senses are most

efficient for finding food and avoiding predators. At night they are vulnerable and need to seek shelter from the physical and biological environment.
- Their size and social behaviour mean that they are highly visible and unable to take refuge in the array of nooks and crevices that smaller animals can use.

The biology of humans acts to constrain their options and to begin to define their niche. It should be noted that 'niche', 'geographical range', and 'habitat' are often confused. Indeed, these terms are sometimes used interchangeably when, in fact, there are important differences:

1. **A niche** describes the conditions needed for a species to occur. Niches are, to some extent, a theoretical construct, being the result of evolutionary adaptations to abiotic parameters. Those adaptations can be **physical** (size, shape, or structure), **physiological** (functional processes, such as respiration or photosynthesis) or **behavioural.** Niche is often defined as an '*n*-dimensional hyperspace', a theoretical space in multiple dimensions constrained and defined by a species' tolerances along multiple axes. This is discussed from an applied perspective in Section 4.3.1 and is illustrated in Figure 4.4.
2. By contrast, a **habitat** is the physical location in which a species occurs. There is a complex two-way interaction between species and habitat: habitat is partly abiotic (influenced by parameters such as temperature and rainfall, which affect what species can colonize successfully) and partly biotic (being influenced by what species have colonized successfully). This is exemplified by the notion of a tropical rainforest. When broken down, this simple term—which is instantly recognizable as a 'habitat'—is somewhere warm (tropical) and wet (rain), and dominated by trees (forest).
3. The sum of all the habitats that a species can colonize, and the location of those habitats, constitutes a species' **geographical range.**

The niche space that a species is theoretically capable of occupying under ideal conditions is termed its **fundamental niche.** This is likely to be constrained because of competition (Section 2.5), predation (Section 2.2), or some other interaction to a smaller niche space, known as its **realized niche.**

Some organisms have more constrained niches than others, especially when it comes to feeding. Some of the animals feeding on the kudu carcass discussed in section 2.2.4 can be described as **generalists.** Vultures (e.g. Cape vultures *Gyps coprotheres*) can feed on the carcass of just about any medium-to-large species and although fresh carcasses are preferred they will feed on carcasses in quite advanced states of decay. Likewise many blowfly species (e.g. *Calliphora vomitoria*) are indiscriminate in their choice of nurseries and will lay eggs on virtually any dead vertebrate. Although they are generalists in one respect, both vultures and blowflies are highly specialized to their decomposer mode of life. Their sensory systems are attuned to finding carcasses. Vultures use highly acute eyesight to find carcasses from great distances, while conserving energy by gliding on thermals high above the plains. Blowflies use their antennae to detect airborne molecules associated with death and decay, often arriving on a body within minutes of death. Fly larvae, commonly called maggots, are also highly adapted to their mode of life, with breathing 'snorkels' allowing them to feed deep within almost liquid soft tissue with their simple but effective mouthparts.

Other visitors to the carcass are ecological **specialists** with a highly constrained niche. Insects feeding on horn, hair or skin for example require a specialized set of digestive enzymes, and these biochemical adaptations may not be compatible with other diets. Such materials might also require physical adaptations, specifically mouthparts capable of processing tough items like horn.

FURTHER READING FOR SECTION

Introduction to the concept of niche: Smith, R.L. & Smith, T.M. (2001) *Ecology and Field Biology*, 6th Edition. Chicago, IL: Benjamin Cummings.

A discussion on some of the issues of niche theory: Chesson, P. (1991) A need for niches? *Trends in Ecology and Evolution*, Volume 6, 26–28.

Linking niche theory to competition and biodiversity in plants: Silvertown, J. (2004) Plant coexistence and the niche. *Trends in Ecology and Evolution*, Volume 11, 605–611.

2.5 Competition.

Whether it is bacteria growing on a Petri dish or humans towing nets through fisheries, resources are not infinite. Because of this, the requirement for those resources will sometimes exceed their availability, and at this point organisms must compete with each other for those resources. Bacteria on the Petri dish must compete for space and therefore nutrients on agar in the laboratory, plants in woodland compete for light, and prides of lions compete with each other for prey.

Using the concept of niche can be useful when considering competition. If the niches of all **individuals** within a **community** (a group of interacting **populations** of individuals from the same species) were plotted, some niches would not overlap at all, many would overlap in some aspects, and some would overlap completely. Those that overlap completely would be individuals from the same species, as these usually have identical niches. However, once again the complexity that is inherent in ecology rears its head. It is not always the case that members of the same species share a niche. Males may have different ecologies to females; juvenile life-stages often differ from adults. The latter difference is particularly obvious in species that have larval stages, where the juvenile stages may have a radically different mode of life to the adult—for example, adult dragonflies have a short terrestrial life stage (albeit generally near to water), whereas larvae have a long larval aquatic life stage, which can extend to years in some species. The closer two organisms' niches are, the more similar will be their resource requirements.

Whilst some dimensions of niche will be largely unaffected by resource availability (such as temperature range or salinity) other components will be directly or indirectly affected by the availability of resources (such as space, light, oxygen, and food/nutrients).

2.5.1 Intraspecific competition.

Regardless of the individual properties of the niche, individuals from the same species will likely have the most overlap and this leads to intraspecific competition ('within species' competition). Intraspecific competition is important in determining some of the patterns within populations. In particular, it gives rise to the sigmoidal ('S'-shaped) growth curves that are typical of many populations. Population size is initially small, but starts to increase as individuals reproduce. Growth is initially only slow, but reaches a stage of rapid growth as more individuals produce yet more offspring. This stage of rapid growth, if extrapolated over time, would give the impression that the population could continue to grow indefinitely. However, at some point the population growth will start to slow as resource limitation becomes important, eventually reaching a value known as the **carrying capacity**. As demand grows the finite nature of resources becomes apparent and individuals begin to compete for them.

Competition can, on occasion, be direct. Vultures on a carcass, for example, may fight to displace each other and this type of competition is termed **contest** or **interference competition**. However, competition more typically operates indirectly. Maggots developing on a carcass may not physically compete with each other, but each maggot's feeding activity reduces the food supply available to other maggots. This type of competition is termed **scramble competition**. Regardless of the nature of the competition, the effects are similar, with competition generally leading to decreased rate of resource uptake, decrease in growth and development, decrease in physical condition, and an increase in predation and parasite load. Together, these negative effects result in a decrease in reproductive rate, which leads to reduced population growth and potentially population decrease.

Intraspecific competition has **density-dependent effects**, which is to say that the magnitude of the effects increases with greater density or crowding of the population. For example, at higher densities individuals are more likely to make contact with each other and so are more likely to transmit parasites. The patterns of population growth, the exact effects of density, and the mechanisms by which these feed back through individuals to population level changes, can be complex, but developing an understanding of the powerful but sometimes subtle nature of intraspecific competition is essential for understanding the ecology of populations.

2.5.2 Interspecific competition.

Intraspecific competition is crucial for understanding populations, but it is the concept of interspecific competition ('between species' competition) that is the key when considering the ecology of communities. The level of competition between individuals of different species is likely to be less intense than competition between individuals of the same species because their niches tend to overlap less. However, the separation of those niches might be the consequence of natural selection acting to push organisms with similar niches apart precisely to reduce competition. Thus, interspecific competition can have similar population level effects to intraspecific competition, while also having a profound effect on the evolution of species, their distribution and their success and dominance within a given habitat.

Demonstrating competition can be difficult (Case Study 2.2). If a pair of species is thought to compete then a removal experiment, whereby one of the pair is removed from an experimental plot, and the responses of the remaining species observed and measured, is one possibility. In practice, however, such experiments are practically or ethically difficult to carry out. Despite this, experimental and theoretical work on competition and its relationship with niche has led to the concept of the **competitive exclusion principal**, which states that:

- If two species compete but co-exist in a stable environment then they must have niches that are different in some way.
- If there is no such differentiation between niches then competition will result in one species eliminating or excluding the other.

The concept can lead to difficulties in its interpretation. For example, two similar birds feed on insects on similar trees, but one specializes on caterpillars, while the other specializes on small flies. It is possible that this demonstrates niche differentiation as a consequence of previous competition, the 'ghost of competition past', but this is not proof that past competition has led to the current situation. Another issue is the use of the word 'stable'. The environment is not stable and is **heterogeneous** (variable) in both space and time. Thus, the conditions that favour one competitor might change and favour another in the future. Such changes might take place locally and globally, and across different timescales. This means that there might not be one clear 'victor' in a suite of competing species. This, in itself, can help increase species richness by preventing a few species from gaining competitive advantage and dominating as a result of outcompeting others.

FURTHER READING FOR SECTION

An introduction to competition: Beeby, A. & Brennan, A.M. (2008) *First Ecology: Ecological Principles and Environmental Issues*, Third Edition. Oxford: Oxford University Press.

Competition in plants: Keddy, P.A. (2012) Competition in plant communities. In D. Gibson (Ed.) *Oxford bibliographies online: ecology*. New York: Oxford University Press. Available at: http://www.oxfordbibliographies.com/view/document/obo-9780199830060/obo-9780199830060-0009.xml?rskey=LNVAHJ&result=1&q=keddy#firstMatch

Competition between microbes and animals: Burkepile, D.E., Parker, J.D., Woodson, C.B., Mills, H.J., Kubanek, J., Sobecky, P.A., & Hay, M.E. (2006) Chemically mediated competition between microbes and animals: microbes as consumers in food webs. *Ecology*, Volume 87, 2821–2831.

2.6 The importance of evolution in ecology.

'Nothing in biology makes sense, except in the light of evolution.'

(Theodosius Dobzhansky (1973), Russian–American evolutionary biologist, emphasizing the importance of evolution in the title of an essay.)

To understand the ultimate cause of biotic patterns and relationships, it is necessary to appreciate that such patterns and relationships have usually evolved and that evolution occurs through **selection**. Ecologists can interpret many relationships

CASE STUDY 2.2

Investigating competition in action

Competition as a descriptor of a process has the ring of action, confrontation, and even violence about it. As with other ecological terms that conjure up such notions, like predation and parasitism, it is tempting to think of competition as being something intrinsic to larger animals that are capable of rapid action. Such a large animal-centric view of the world is increasingly common, as subjects like entomology and botany take a back seat in modern science education, but in ecology plants and insects (and other invertebrates like mites) are valuable experimental systems. They are vital components of ecosystems, numerous and diverse, but they also have some important advantages over larger vertebrate animals as experimental organisms—they can be relatively cheap and straightforward to culture, it is easier to control conditions around them, and studies rarely have a legal or ethical dimension. For Applied Ecologists interested in plants and insects as 'problems to be solved' (perhaps because such organisms are invasive or causing economic harm) the insights gained from studies of these organisms can be invaluable.

An interesting example of interspecific competition that involves both insects and plants was carried out by Inbar et al. (1995), who studied two closely-related species of aphids. These aphid species both feed on (graze) the phloem of *Pistacia palaestina* (a shrub common in Israel where the study was carried out) and release chemicals that cause the plant to produce a 'growth' of plant tissue, called a gall, in which the aphids live. The aphids differ in where the gall forms. *Geoica* forms a spherical gall on the leaflet midrib, whereas *Forda formicaria* forms a crescent-shaped gall on the leaf margin. Galls of both species can co-occur on the same leaflet.

Using ^{14}C labelling it was possible for the experimenters to follow the fate of carbohydrates within the plant. They found that *Geoica* galls on the midrib of leaflets were diverting resources away from *F. formicaria* and, by the end of the season, 84% of *F. formicaria* galls and their aphids were dead when both species were present on the same leaflet. In surviving galls, reproductive output of *F. formicaria* was reduced by 20%.

Not only does *Geoica* divert resources, but it also causes early leaf drop, which reduces the opportunity for competing species to reproduce. The two species show niche differentiation in terms of gall placement (although the study does not show this was the result of competition) and do not have interference competition for galling sites, but they nonetheless suffer exploitative interspecific competition over resources with a clear pattern of winner (*Geoica*) and loser (*F. formicaria*).

REFERENCE

Inbar, M., Eshel, A., & Wool, D. (1995) Interspecific competition among phloem-feeding insects mediated by induced host-plant sinks. *Ecology*, Volume 76, 1506–1515.

and patterns using evolutionary arguments and insight, explaining how such patterns and relationships evolved, and how and why they change through time and space. An example of how the relationships between evolution and ecology can be understood is discussed in the Online Case Study for Chapter 2.

You can find the online case study at www.oxfordtextbooks.co.uk/orc/goodenough

The tremendous variation in the natural world exists not just between species, but also within a species. This point is easily appreciated by looking at the large variation that exists in just the physical appearance of the human species (other variations exist in the biochemistry, physiology, and behaviour of humans that are less easy to observe). What Charles Darwin, and his contemporary Alfred Russell Wallace, came to realize was that variation within a species is an important factor in understanding how variation between species can evolve.

In modern biology, evolution is defined as a change in gene frequency over time. That change can be relatively rapid (such as the evolution of antibiotic resistant strains of bacteria) or relatively slow (for example, in the evolution of new species), but regardless of the timescale involved, both these types of change are evolution in action and both are the consequence of ecological interactions. Darwin was able to formulate the mechanism by which these changes occur, which he termed **natural selection.**

Natural selection is a logical argument developed from careful observation. Crucially, it is a hypothesis that generates testable predictions, and the testing of those predictions has preoccupied many evolutionary biologists from the mid to late 20th century to the present day. The argument proceeds as follows:

- Individuals within a species are not identical and sometimes exhibit great variation.
- Offspring generally resemble their parents and, therefore, some variation is heritable.
- Individuals typically produce far more offspring than survive to reproduce themselves.
- If the heritable variation observed between individuals in any way affects their ability to survive and reproduce, then some individuals within a population will be better able to survive and reproduce, and will have offspring better able to survive assuming their environment stays the same.
- Consequently, some individuals and their progeny (and their progeny's progeny, and so on) are favoured and will come to dominate within that population—in other words, evolution will have taken place.

Natural selection acts to weed out (or 'selects against') those individuals less able to survive and reproduce in the prevailing environment and favours ('selects for') individuals better able to survive. Assuming that there is sufficient variation on which it can operate, and that the variation is **heritable** (i.e. genetically based) then the process of natural selection produces populations with traits for survival and reproduction within the selection environment. **Artificial selection** is the term used when human-imposed criteria are the selective pressure, and artificial selection for domestic pets, farm animals, and crops reveals the tremendous variation that exists for selection to work on. A Chihuahua dog has not evolved through natural selection, but an environment of strong artificial selection on wild canids has nonetheless created this animal within a blink of an eye in evolutionary time.

Changes can accrue in subpopulations exposed to different conditions to such an extent that individuals within the subpopulation can no longer successfully reproduce with other members of the wider population; these types of evolutionary changes can lead to the creation of new species through the process of **speciation.**

Other forms of selection are also important in evolution. Darwin recognized that not all animal traits seem especially useful in terms of survival. These traits are generally only observable in one sex (usually the male) and are associated with better reproductive output. Males bearing these exaggerated traits are either better able to fight other males for access to females (for example, the large antlers of a male red deer) or better able to attract females who find the male's trait attractive. Assuming that the trait is heritable then successful males will pass on the genes for this success to their male offspring; such selection was termed **sexual selection** by Darwin.

Genes can increase in frequency in a population if they cause bearers of those genes to act altruistically (i.e. in such a way that they provide benefit to the recipient at a cost to themselves) towards other bearers of those genes; in other words, close relatives that are likely to possess copies of those genes. This led to the formulation of **kin selection** theory. Kin selection has proved to be a powerful explanatory tool in understanding social behaviour, especially that of the animals that Darwin recognized as a 'special problem' for his theory—the social insects, such as ants and some bees. These insects have more-or-less sterile workers and highly fecund queens. Darwin realized that since such workers do not reproduce, they could not be subject to natural selection in a way that would lead to their evolution. What actually happens is that such traits can evolve through kin selection, with social insect nests being highly related family groups, where sterile workers assist in rearing their close kin.

2.6.1 Life history theory and r/K strategists.

Intraspecific and interspecific interactions generate a range of selection pressures on organisms that result in the evolution of adaptations to those pressures. Some of these adaptations involve development, growth, reproductive rate, and lifespan. These are called life history traits. Such traits, like age to first reproduction, the existence of larval stages, juvenile growth rate, and longevity, can vary greatly between organisms as they evolve in response to their biotic and abiotic environment. Some organisms, for example, mature rapidly (perhaps because of strong predation pressure on juvenile stages), trade-off longevity for large-scale reproductive output, and invest little in parental care.

In some extreme cases, **semelparous** organisms may only breed once at a single 'big bang' breeding event (e.g. mayflies, Ephemeroptera). The adults are highly vulnerable to predation and consequently have evolved to mate rapidly and lay large numbers of eggs in a short time. Typically, adults live for less than 24 hours and in some cases for only a few minutes (e.g. *Dolania americana*). Organisms that adopt this 'live fast, die young' strategy are known as **r-strategists** because their life histories focus on high growth rates (r). They are often found in unstable or unpredictable environments, or as early colonizers (Section 2.7.2) in an **ecological succession.**

In contrast, other organisms are selected for a different 'live slow, live long' approach that emphasizes slower growth, larger body size, multiple breeding cycles, fewer and larger offspring at each breeding attempt, parental care, and longer lifespan. Such species are termed **K-strategists** and they tend to predominate in more stable environments and communities, where the ability to compete for limited resources is critical. 'K' comes from the German word *Kapazitätsgrenze*, which means capacity limit or carrying capacity. K-strategists tend to have more-or-less constant populations that are close to the maximum number the environment can carry, whereas r-strategists can have far more dynamic populations.

FURTHER READING FOR SECTION

Introduction to evolution: Stearns, S.C. & Hoekstra, R.F. (2005) *Evolution: an Introduction*. Oxford: Oxford University Press.

Natural selection as outlined by Darwin: Darwin, C.R. (1859) *On the Origin of Species by Means of Natural Selection, or the Preservation of Favoured Races in the Struggle for Life*. London: John Murray.

The centrality of evolution to biology: Dobzhansky, T. (1973) Nothing in biology makes sense, except in the light of evolution. *American Biology Teacher*, Volume 35, 125–129.

Selection, evolution and ecology: Westneat, D. & Fox, C. (Eds) (2010) *Evolutionary Behavioral Ecology*. Cary, NC: Oxford University Press.

2.7 Patterns in ecology.

Interactions are the central theme of ecology. However, these interactions result in patterns of species composition and abundance, and such patterns are of great interest to ecologists. Like the ecological interactions that produce them, ecological patterns can vary across space and time, and be apparent at different scales.

2.7.1 Biogeography.

Different physical environments, climates, and seasonality across the planet combine to produce a remarkable range of habitats and, within those habitats, an array of potential niches for organisms to exploit. A consequence of this variation and diversity of opportunity is that communities of organisms vary greatly between habitats (**spatial variation**), and within habitats through time (**temporal variation**). The study of why organisms are found where they are is biogeography.

Although primarily spatial in its approach, biogeography must also consider temporal patterns and changes. For example, grassland in East Africa has a very different species composition to grassland in Central Russia (Figure 2.10), but there might also be differences between blocks of grassland within each

Figure 2.10 Grassland in East Africa (left) is much more species-rich than grassland in Central Russia (right); the species composition is also different.

Source: Photographs by Anne Goodenough (left) and Zabara Alexander/CC BY 2.0 (right).

location, perhaps as a consequence of recent burning history. Both grasslands will vary through the year as seasonal patterns of rainfall and sunlight affect growth of the grass, and thus, the biomass available to grazing animals, but their different latitudes and geographical location mean that the effects could be very different. Medium-term climatic patterns, such as droughts in El Niño years and long-term climate change patterns, also play a role in determining the number, diversity, and identity of the species present at any given time.

2.7.2 Ecological succession.

Whether it is new land created by volcanic activity or spaces opening up in a crowded forest when trees blow down in a storm, new opportunities for organisms able to exploit them are created constantly. Some organisms (usually r-strategists) are exceptionally good at initially exploiting such gaps. Their early and high reproductive rate means they and their offspring are quickly able to take advantage. If the environment remains unstable, then the community present may be dominated by these types of species. However, given some stability, early colonizers are eventually out-competed by K-strategists and the community composition shifts towards an equilibrium community.

The change in community structure over time is called succession. Given sufficient stability, a community may change in a more-or-less predictable way, producing a succession of communities that finishes with a **climax community.** Intermediate communities on the way to the climax are termed **seral communities** or **seres.** This process is illustrated in Figure 3.2. Successions that begin on land not previously occupied by any ecological community (such as land newly created by volcanic activity) are termed **primary successions,** whereas successions that follow disturbances in established communities are termed **secondary successions.**

FURTHER READING FOR SECTION

Introduction to biogeography: Cox, C.B. & Moore, P.D. (2010) *Biogeography: an Ecological and Evolutionary Approach*. Oxford: Wiley.

A classic text on biogeography: Simberloff, D.S. (1974). Equilibrium theory of island biogeography and ecology. *Annual Review of Ecology and Systematics*, Volume 5, 161–182.

Linking ecological succession with management: Walker, L.R., Walker, J., & Hobbs, R.J. (2007) *Linking Restoration and Ecological Succession* (Springer Series on Environmental Management). New York, NY: Springer.

A discussion of r and K selection: Reznick, D., Bryant, M.J., and Bashey, F. (2002) *r*- and *K*-selection revisited: the role of population regulation in life-history evolution. *Ecology*, Volume 83, 1509–1520.

2.8 Conclusions.

Organisms interact with each other and with their environment in a variety of ways. Feeding relationships can be complex, including herbivory, predation, grazing, and decomposition, and blurred boundaries can exist between categories. Parasitic relationships are asymmetrical, but not all ecological interactions have winners and losers. In mutualistic relationships both parties gain, while in commensal relationships one party gains with no loss to the other. Competition is a strong ecological force, and competition within and between species serves to limit populations and to constrain niche (the ecological 'space' in which a species can exist). Interactions also produce patterns at different scales of space and time. Such patterns include the location of organisms in different habitats and the composition of communities in different places or through time, and are ultimately understood by appreciating the underlying ecological interactions that cause them.

CHAPTER 2 AT A GLANCE: THE BIG PICTURE

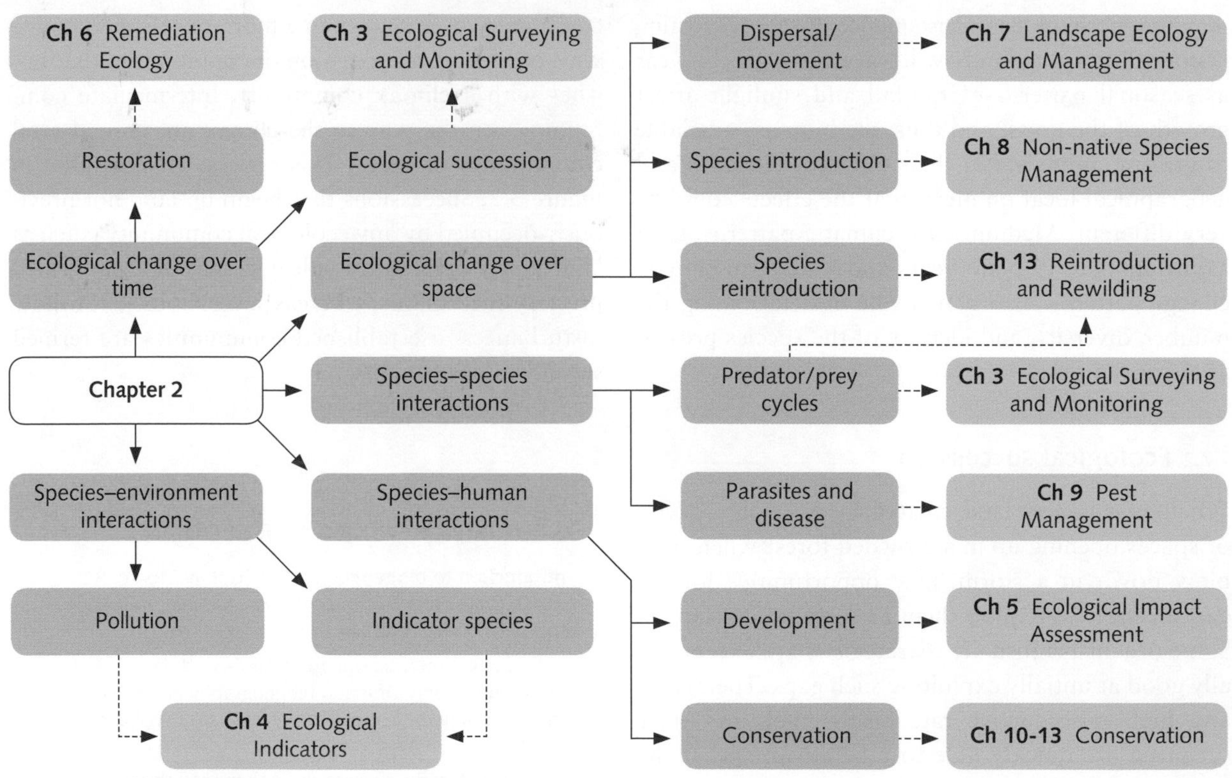

PART 2

Monitoring

Monitoring species, habitats, and wider environments is a fundamental part of Applied Ecology. Information can be important in understanding ecological processes, species–habitat interactions, population dynamics, ecosystem function, and ecological change. It also has a vital role in informing management and setting conservation goals and priorities: it is generally not possible to manage or conserve effectively until Applied Ecologists understand what species and habitats are present, and their abundance or extent. Moreover, monitoring is also the primary mechanism through which the effectiveness of ecological management can be evaluated and optimized.

Chapter 3 considers surveying and monitoring aims, approaches, and guiding principles, drawing on a range of examples from different environments across the globe. **Chapter 4** leads on from this by considering what the presence and abundance of certain species can tell ecologists about environmental conditions through the concept of ecological indicators.

3 **Ecological Surveying and Monitoring**

4 **Ecological Indicators**

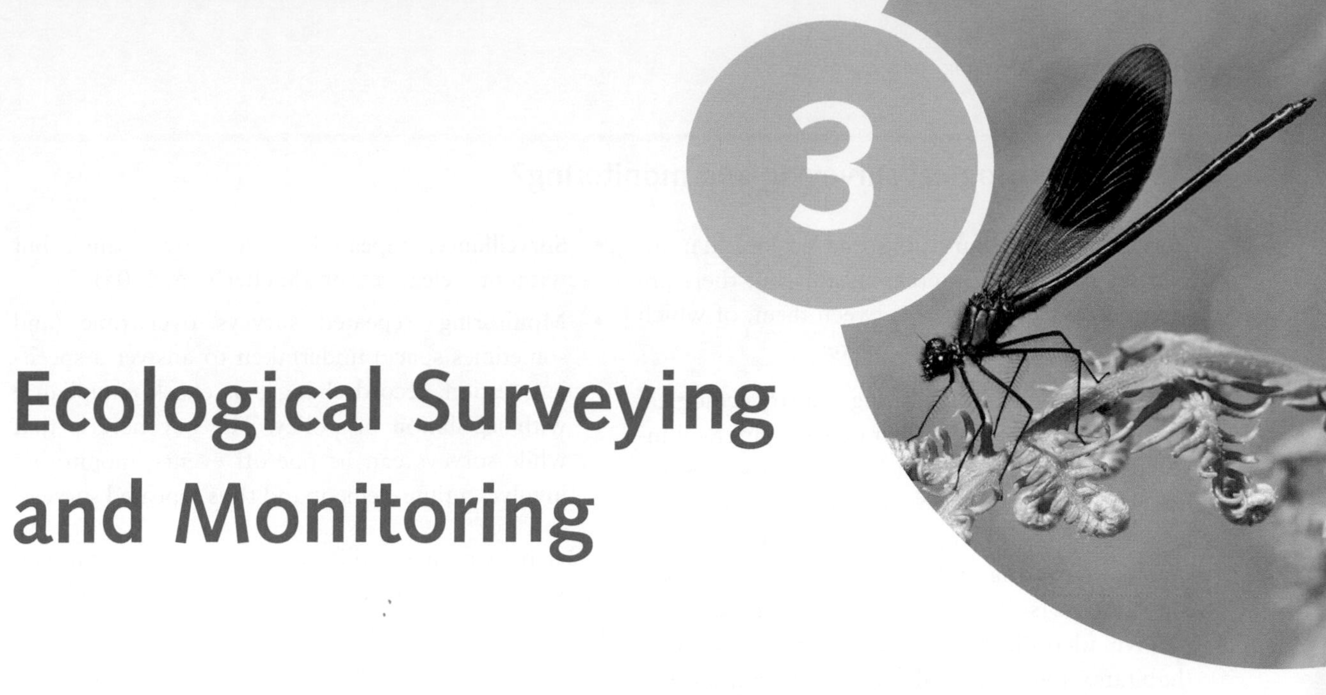

3 Ecological Surveying and Monitoring

3.1 Introduction.

It is hard to overstate the importance of surveying and monitoring in Applied Ecology. Virtually all topics or questions have a surveying and monitoring component. It links to assessing impacts of development through **Ecological Impact Assessment** (Chapter 5), **remediation** of previously contaminated land (Chapter 6), understanding spatial aspects of ecology and landscape-scale **management** (Chapter 7), prioritizing **conservation** needs (Chapter 10), and then devising, refining and evaluating conservation strategies (Chapters 11–13), and management of both **non-native species** (Chapter 8) and **pests** (Chapter 9). The understanding of the effects of environmental parameters on species through monitoring also allows those species to be used as indicators of features of that environment, such as soil fertility, air quality, or water pollution (Chapter 4).

Given the importance and broad scope of surveying and monitoring, it is not surprising that most Applied Ecologists spend a considerable amount of their time either undertaking primary surveying or considering information gleaned from monitoring programmes. This might involve, for example, establishing whether a species is present at a specific site, assessing change in the population of a species over time, mapping the distribution of a habitat at national level, alerting the relevant bodies to a pollution incident that has ecological implications, checking the compliance of the management of a nature reserve with environmental legislation and codes of practice, and monitoring the effects of climate change on ecosystem services such as pollination.

This chapter considers aspects of ecological surveying and monitoring that are crucial parts of many Applied Ecologists' jobs (see 'Interview with an Applied Ecologist' for this chapter). We will outline the purposes of the various types of monitoring frameworks, and the different approaches to monitoring habitats, species, and environmental parameters, including automated and semi-automated data collection. The aim is not to provide an exhaustive manual of methods, but rather to draw out key principles and practical considerations, and to consider how these are related to underlying theory. To achieve this, links to theoretical concepts such as distribution patterns and population dynamics will be made, and different monitoring schemes from around the world will be used to exemplify key points.

3.2 What is ecological surveying and monitoring?

The terms 'ecological surveying' and 'ecological monitoring' are often used interchangeably, but there are some important differences between them, of which Applied Ecologists should be aware:

- **Surveying:** the act of collecting primary ecological data. For example, invertebrate surveying will involve using one or more survey methods to establish what species are present and, in many cases, how many there are. Some surveying methods involve capturing individuals (animals) or taking samples (soil, water, etc.). Most surveys will involve identification (species) or classification (habitats, environmental features). Common subtypes of surveying include:
  - **Censusing:** generally refers to assessing species' population size.
  - **Mapping:** establishing and displaying spatial arrangement of species/habitats.
  - **Measuring:** typically refers to quantifying an aspect of the physical environment using an established scale, often an SI unit or derivative of this, for instance the depth of soil (cm), flow rate of a stream (m/s), water concentration of nitrate (kg/l). Can also be used for habitat measurements (area or length for linear habitats).
- **Surveillance:** repeated surveys over time, but without a clear reason (Spellerberg, 2005).
- **Monitoring:** repeated surveys over time (and sometimes space) undertaken to answer a specific question, record change, or check compliance with legislation or policy. The key here is that while surveys can be one-off events, monitoring involves a time element and thus repeated surveys.

There are many different parameters that can be surveyed/monitored, but they can generally be grouped into one of four categories:

- Habitats.
- Species.
- Interactions and ecosystem services.
- Environmental parameters that influence ecology.

See Table 3.1 for more details.

FURTHER READING FOR THIS SECTION

Introduction to monitoring frameworks: Spellerberg, I. (2005) *Monitoring Ecological Change.* Cambridge: Cambridge University Press.

Designing monitoring schemes: Lindenmayer, D.B. and Likens, G.E. (2010) *Effective Ecological Monitoring.* Clayton: CSIRO Publishing.

Table 3.1 What can Applied Ecologists survey and monitor?

Habitats	Species
Landscape-level • Habitat heterogeneity—how varied the habitats are at a site (high heterogeneity = highly variable; the opposite is homogeneity, which means the habitat is uniform). • Fragmentation—whether a landscape has habitats covering a contiguous area or is broken up into discontinuous patches.	**Community-level** • Species richness—number of different species. • Species diversity—a mathematical expression of species richness related to abundance of each species.
Habitat-specific • Presence. • Absence. • Relative abundance—whether there is more of Habitat A or Habitat B. • Size—area (or length for linear habitats such as streams). • Connectivity—how well habitat patches are linked together by corridors (e.g. hedgerows). • Condition—how 'healthy' a habitat is.	**Species-specific** • Presence. • Likely absence (rarely possible to be completely certain). • Relative abundance—whether there is more of Species A or Species B. • Population size—how many individuals there are. • Population density—how many individuals there are per unit area (e.g. 50 individuals per km^2). • Population change—growth rates or decline.

Table 3.1 Continued.

Interactions and processes	Environmental variables
Interactions and functions	**Abiotic parameters**
• Phenology—the timing of seasonal events such as leaf burst, bird breeding, or amphibian spawning. • Ecological succession—change in habitat structure and species over time in an ecosystem. • Predator–prey interactions. • Herbivore–grazer interactions.	• Climatic parameters—temperature, precipitation, wind, humidity, barometric pressure, sunlight. • Topographical (land shape) parameters—altitude, latitude, aspect, slope angle. • Edaphic (soil) parameters—depth, type, hardness, structure, texture, moisture content, organic content.
Ecosystem processes and services	**Chemical parameters**
• Pollination. • Seed dispersal. • Water filtration or hydrological regulation. • Nutrient cycling. • Photosynthesis and decomposition rates.	• pH—acidity or alkalinity. • Nutrient levels (especially nitrogen, phosphorous, and potassium; needed for plant growth, but in excess cause eutrophication). • Multi-element profiling to shown natural and artificially elevated heavy metal and trace elements. • Radioactive and stable isotopes. • Oxygen—amount of or demand for.

3.3 Purposes of surveying and monitoring.

Before delving any further into what aspects of Applied Ecology can be surveyed or monitored, and how this can be done, it is first necessary to consider *why* ecological surveying and monitoring might be undertaken.

The **purpose** of surveying and monitoring affects both the overall approach (how many samples are taken and over what period) and the actual methods used. Knowledge of the main reasons for ecological monitoring is, therefore, a vital prerequisite for designing an effective monitoring strategy. A brief description of the main motivations for surveying/monitoring is given below and then explored in more detail in subsequent sections:

1. **Baseline surveying:** an initial investigation into the species, habitats, or environmental conditions of an area. It is a 'snapshot' quantification of an area at one moment in time, which literally provides a baseline for further work.
2. **Spatial surveying:** undertaken to assess patterns in ecology over space and how geographical patterns change over time.
3. **Temporal monitoring:** baseline surveying followed by repeated surveys is used to assess change longitudinally over time; for example, species population change, change in environmental parameters, and the interaction between these (i.e. how species or habitats are affected by, or respond to, environmental change).
4. **Alert monitoring:** provides a 'first warning' of potential problems. Such problems might be inherently biotic (e.g. non-native species invasion, disease outbreaks) or might involve problems in the physical environment that could have an effect on ecology (e.g. pollutants).
5. **Compliance monitoring:** undertaken to ensure conformity with legislation, and/or agreed codes of practice. Compliance monitoring often focuses upon abiotic environmental parameters, particularly pollutants, but can cover biota (e.g. wildlife crime, management of protected areas, etc.).
6. **Impact, mitigation, and compensation monitoring:** Impact monitoring is used following an Ecological Impact Assessment (Chapter 5) to establish the accuracy of predictions made regarding the impact of a new development on local ecology. Mitigation and compensation monitoring is undertaken to establish the success or otherwise of measures put in place to mitigate (eliminate or reduce) or compensate (make up for) negative effects of a development or land use change. Again,

this is most often used within an Ecological Impact Assessment framework (Chapter 5).

3.3.1 Baseline surveying.

One of the most common reasons for surveys being undertaken is to quantify **baseline conditions.** This might include classifying habitats, quantifying a species' population size, or formulating a basic understanding of the environmental parameters that affect habitats and species, for example, soil or water conditions. Baseline surveying can be undertaken at any spatial scale, from a small individual site, to national or even international levels. There are several reasons why baseline surveys might be undertaken:

1. To create a **biodiversity audit** to identify what species/habitats are present in an area. This can be carried out at a specific site (e.g. a school) or at a regional scale to document biodiversity within a county or province. At its simplest, it is a basic inventory of what species/habitats are present. More detailed audits also identify areas where more detailed surveying would be useful, areas where management might be appropriate to increase biodiversity value, and areas that have high biodiversity value. An example is the biodiversity audit that was done in Cheltenham, a medium-sized town in Gloucestershire, UK, in 2006 (Case Study 3.1).

   In many instances, a biodiversity audit will be able to draw on **past records** and, if necessary, supplement these with **new survey data** to get a detailed understanding of the biodiversity of the target site (Sections 3.4.1 and 3.4.2). In other cases, a biodiversity audit might be the first time a site's biodiversity has been formally surveyed. Even then, though, the species and habitats are likely to be familiar. Indeed, it is often possible to predict likely species by looking at aerial habitat photographs and national species distribution maps. In rare cases, a biodiversity audit can be the first time ecological work has been done in an area. This is most likely in remote regions. Biology expeditions to such areas often involve expert taxonomists who catalogue as many species as possible. In these circumstances, there is the very real possibility that some species new to science will be recorded (Hot Topic 3.1).

2. To form part of an **Ecological Impact Assessment** (EcIA). The first step of any EcIA is to identify the site-specific ecological conditions. It is not possible to predict the impacts of a new development until the current conditions at that site are known. This will normally involve undertaking a habitat survey and a range of species surveys to identify species present and, sometimes, provide abundance estimates. The site and proposal itself will dictate exactly what surveys are appropriate. The EcIA, and the guidelines from the Chartered Institute of Ecology and Environmental Management (CIEEM) that guide the process, are considered in more detail in Chapter 5.

3. To gain an understanding of current conditions to inform **management**. Surveys are a vital part of determining the need for management in the first place, and then to formulate an appropriate management protocol. For example, at a local scale, a baseline butterfly survey might show that there is a rare species of butterfly on a grassland site that has just been taken over by a new farmer. This might suggest the need for management at the site. If it was noted in the baseline survey that the butterfly only occurred in areas that supported a specific larval food plant, and that the larval food plant only grew in areas with a particular grazing regime, it might be appropriate to examine the possibility that modifying grazing in other areas might encourage the larval food plant to colonize more areas and allow the range of the butterfly population to expand. The links between surveying/monitoring and management are explored more fully in Section 3.7.

4. To improve **knowledge**. This rather vague category encompasses all baseline surveying undertaken without a specific applied reason at that time. The importance of this is highlighted by Spellerberg (2005) in reference to the example of radioactivity. In April 1986, reactor 4 at the nuclear power plant at Chernobyl, in present-day Ukraine, exploded. Over the next few days, a radioactive cloud moved over Europe and, partly due to heavy rainfall, radioactive fallout occurred over a wide area, including in the River Rhine. Radioactivity levels in the river by mid-May 1986 were estimated to be thousands or millions of

CASE STUDY 3.1

Cheltenham biodiversity audit

In 2006, a biodiversity audit was undertaken by ecological consultants for Cheltenham, a medium-sized town in the county of Gloucestershire, UK, by Miller (2008). The audit is typical of the hundreds (possibly thousands) of biodiversity audits carried out globally and is used here as an example.

The Cheltenham audit covered 132 sites including allotments, churchyards, green corridors, open spaces, parks, and playing fields. Several different methods were used as shown in Table A.

Data were summarized to give an overall indication of the town's biodiversity. For example, it was noted that 52 habitats were present, of which 10 were listed as UK **priority habitats** and 12 were on the countywide priority habitats list. Of the hundreds of species listed, 28 had national protection and 7 were on the UK's priority list. Data were then used to assess:

- Biodiversity 'hot spots' (i.e. areas with particularly high species richness).
- Wildlife corridors (i.e. areas used for species movement, or that could potentially serve this function)—for example, streams, hedgerows, and old railway lines.
- Areas with scope for biodiversity enhancement.

Each individual site was also graded as to its biodiversity quality on a scale from A (large site with moderate/high existing biodiversity value or small site with high existing biodiversity value) to C (small site with low biodiversity value and minimal potential). **Value** was determined considering:

- Frequency of the habitat at local and national levels.
- Conservation designations of the habitats at local and national level.
- Spatial extent of the habitat at local and national level.
- Species diversity within the site.
- Presence of rare or protected species within the site.

REFERENCES

Miller, H. (2008) Biodiversity Audit Cheltenham Borough Council Report Number: RT-MME-3879-rev01. https://www.cheltenham.gov.uk/downloads/file/976/biodiversity_audit

Table A Different data sources used in the audit.

Data type	Aim	Resources
Secondary	To investigate land designations and identify areas with statutory protection	Used MAGIC website (www.magic.gov.uk), which provides map data of all conservation designations used in the UK (e.g. National Parks)
Secondary	To locate past records of species held within the county	Local Biodiversity Record Centre; local wildlife organizations such as the Gloucestershire Wildlife Trust; council records
Secondary	To locate past records of species from data held within the county	National records networks such as the National Biodiversity Network and national wildlife organizations such as the British Trust for Ornithology
Primary	To map habitats across the town	Phase One Habitat Surveying undertaken by ecological consultants

times higher than the natural background level. The key word here is 'estimated'. There were no records of the natural level because there was, until that time, no perceived need for baseline data. (The effect of this event on wildlife, including species reintroduced to the area, is considered in Hot Topic 13.3.)

3.3.2 Spatial surveying.

At its simplest, spatial surveying involves **mapping** where a particular species or habitat occurs or where a specific ecological phenomenon occurs. From an applied perspective, this can be used to focus further survey work or management/conservation effects.

HOT TOPIC 3.1

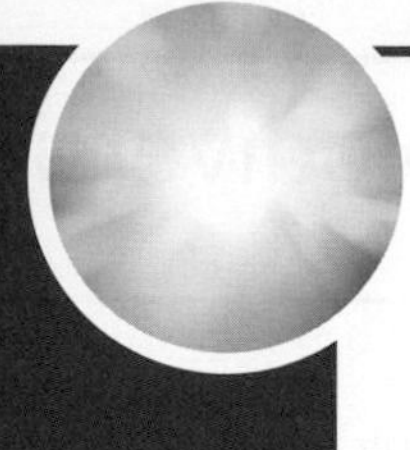

Modern discovery and classification of new species

The terms 'species discovery' or 'biological expeditions' tend to conjure up images of the past: Charles Darwin on his five-year voyage on HMS Beagle in the 1830s discovering new species in Patagonia and the Galapagos islands; Alfred Russell Wallace and the thousands of new species he collected in the Malay Archipelago; and Alexander von Humboldt's species discoveries in Latin America in the early 1780s.

All species found by those early great biological explorers were classified taxonomically and given scientific names. This uses the concept of binomial nomenclature; a hierarchical classification system much like a family tree (Figure A). The idea was first proposed by Linnaeus in *Systema Naturae* (*The System of Nature*), which he published in 1735, but has been amended and updated many times since.

Domain	• New "super-level" of 3 overarching groups: Archaea (non-bacterial single-celled organisms), Bacteria, Eukaryota.
Kingdom	• Top rank used by Linnaeus. Originally 2 biological kingdoms (flora and fauna), plus minerals. Now 6 generally recognized (Bacteria, Protozoa, Chromista (algae and algal-like species), Plants, Fungi, Animals).
Phylum	• Based on presence/absence of major traits such as backbone. Sometimes called divisions in botany. Typically end in the letter "a" as in Bryophyta (mosses and liverworts).
Class	• Often thought of as the highest level in taxonomy as this is where the first familiar groupings tend to be found – for example Mammalia (mammals), Aves (birds), and Insecta (insects). Typically end in the letter "a".
Order	• Subgroups based on morphology. Examples are Diptera (di = two; ptera = winged – species in Insecta with two flight wings) and Passeriformes (species in the Aves with one back toe and three front toes on each foot).
Family	• Usually plant families can be recognized in nomenclature by the ending "aceae" – as in Magnoliaceae (magnolia) but there are exceptions. Usually animal families end "idae" as in Gammaridae (freshwater shrimp).
Genus	• Groups of closely-related species that share obvious morphological similarities (e.g. *Betula* or birch tree species). Separates individuals similar but not close enough genetically to routinely produce viable, fertile and fit offspring.
Species	• Groups of individuals similar enough genetically to interbreed routinely to produce viable, fertile, and fit offspring. 2-part names (e.g. *Parus major* for great tit – *Parus* = genus; *major* = species). Always italicised.
Subspecies	• Groups within species that interbreed but look or behave differently. Often occur in different geographical locations of the overall range. Denoted by an italicised 3-part name (e.g. *Parus major newtoni* for the UK-occurring great tit).

Figure A Taxonomy classification hierarchy.

Source: Reproduced from Deryabina, T. G. et al. (2015) 'Long-term census data reveal abundant wildlife populations at Chernobyl'. *Current Biology*, 25(19), R824–R826/ CC BY 4.0

There are two approaches to classifying new species. The first is the **Biological Species Concept**, which is based on morphology and, importantly, breeding data. Breeding data are essential, since it is the ability of a group of individuals to breed together to produce viable, fertile, and fit offspring that defines a species. Indeed, the term 'species' should not be applied to an organism whose breeding behaviour is unknown. There are problems with this approach though:

- Can miss **cryptic species**—species that look very similar to one another in such a way that their precise relationships to one another cannot be easily determined through consideration of morphology or behaviour. This often occurs with species that mimic others. Cryptic species are usually closely related to one another, but this is not always the case.
- Can miscount species that undergo **metamorphosis,** including many invertebrates and amphibians, as different life-history stages can be misclassified as different species. The same is true for individuals that look different in different situations, such as lianas (climbing plant species), where individuals can have very different leaf shapes in the understory compared with in the canopy.
- Can miscount **polymorphic** species—this is when different individuals of the same species look very different from one another. Differences are usually based on colour or patterning rather than body shape—a good example is individual Harlequin ladybirds *Harmonia axyridis* which can be red, orange, yellow, or black, and have a variable number of spots, but are one species genetically.
- Can be hard to apply where breeding data are unknown or where species undertake **asexual reproduction**, such as the plant Japanese knotweed *Fallopia japonica*, which can reproduce vegetatively.

It is also possible to classify species using molecular techniques. This is the **Phylogenetic Species Concept**, which is based on DNA similarities and is much more accurate. Until recently, this was beyond the reach of most taxonomists, but, with costs falling rapidly, it is ever more feasible. Molecular analysis focuses on parts of the DNA structure known to vary between species, rather than conservative regions known to share great similarity with other species (50% of human DNA is shared with bananas). The variable *Cytochrome-b*, which is part of the mitochondrial DNA, is often used. This is DNA passed down the maternal line such that every offspring should—barring mutations—have an exact copy of its mother's mitochondrial DNA. Molecular analysis should theoretically yield the same results as other methods for sexually reproducing life forms, but this rarely holds in reality.

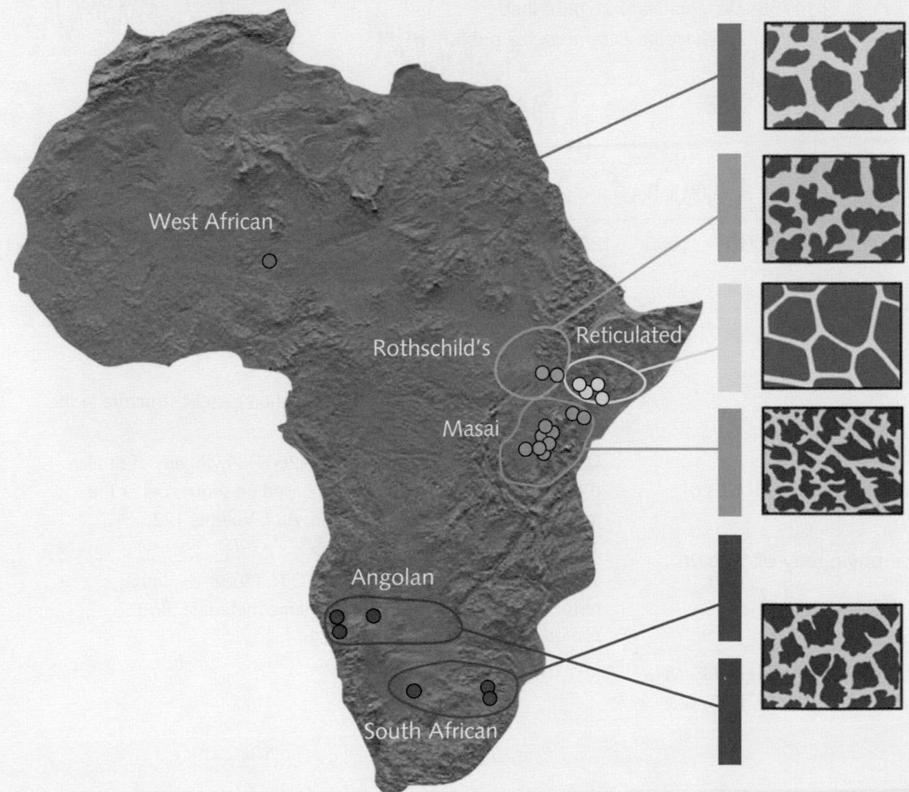

Figure B Different subspecies of giraffe that possibly should be elevated to species level.

Source: Reproduced from Brown, D. M., et al.(2007) 'Extensive population genetic structure in the giraffe'. *BMC Biology*, 5(1), 57/ CC BY 2.0.

There are two main ways that new species are 'discovered' today. The first is using molecular approaches to re-analyse species groups that have been previously classified biologically. This often causes re-classification, as happened with birds in the Paridae (tit) family in 2005 when multiple new genera were created (Gill et al., 2005), and giraffes in 2008 (Brown et al., 2008; Figure B). This latter study found that, whereas previous analyses had suggested a single species with six subspecies, there was good evidence to elevate those subspecies to species level.

The other method by which new species are added to the global species list is true discovery—that is the finding of a life form previously unknown to science that is then classified as a new species. A good example of this is the po'o-uli *Melamprosops phaeosoma* discovered in 1973 on the island of Maui. It was the first species of Hawaiian honeycreeper to be discovered since 1923, and was so different from all other described species that it was not only classified as a new species, but a new genus was created.

The number of species discovered each year is about 18,000, which has been relatively consistent for the last 120 years except for dips during the two world wars (Figure C). Many of these are discovered through baseline monitoring—specifically biodiversity audits—in poorly studied or remote locations. For example, a new species of rat, the Bosavi woolly rat, was discovered in a remote volcanic crater in Papua New Guinea as part of a biodiversity expedition by the Smithsonian Institution and the British Broadcasting Corporation. This has provisionally been included in the genus *Mallomys*,with the precise scientific name still to be confirmed as of early 2017. It should also be noted that there are considerable opportunities to discover new species in less charismatic or more obscure taxa even in well-studied areas such as the UK and the USA.

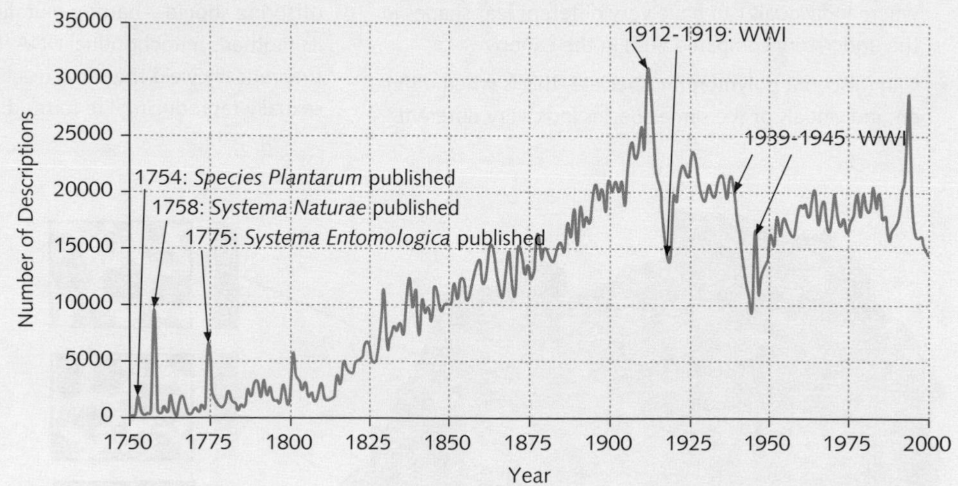

Figure C Discovery of new species over time.

Source: Reproduced from Sarkar, I. N., et al. (2008) 'Exploring historical trends using taxonomic name metadata'. *BMC Evolutionary Biology*, 8, 144/ CC BY 2.0.

QUESTIONS

How can species be defined and classified?

What are the problems with the different methods of defining species?

Why do biologists reclassify the phylogeny of 'known' species?

REFERENCES

Brown, D.M., Brenneman, R.A., Koepfli, K.P., Pollinger, J.P., Milá, B., Georgiadis, N.J., Louis, E.E, Grather, G.F., Jacobs, D.K., & Wayne, R.K. (2007) Extensive population genetic structure in the giraffe. *BMC Biology*, Volume 5, 57.

Gill, F.B., Slikas, B. & Sheldon, F.H. (2005) Phylogeny of titmice (Paridae): II. Species relationships based on sequences of the mitochondrial cytochrome-b gene. *Auk*, Volume 122, 121–143.

Sarkar, I.N., Schenk, R., & Norton, C.N. (2008) Exploring historical trends using taxonomic name metadata. BMC *Evolutionary Biology*, Volume 8, 144.

Spatial surveying can also help set the ecology of a specific area in broader **landscape context.** For example, when considering site-based conservation for the chequered skipper butterfly *Carterocephalus palaemon* in Scotland, it is vital that management of individual colonies is planned in a wider landscape. Even if the habitat is ideal, the chances of natural colonization (or recolonization) of this poorly dispersing species is very low if the next nearest colony is more than 5 km away (Forestry Commission Scotland, 2009). At an even more pragmatic level, it might not be sensible to spend considerable conservation resources managing a species at a marginal site on the edge of the species' range when the species has a healthy population throughout the rest of its distribution. (This is debated in more detail with examples in Hot Topic 10.1.)

More complex spatial surveying involves analysis of geographical patterns, rather than simply describing patterns. One approach is to use mapping to compare, for instance, the distribution of a species and of a habitat. This can be extremely informative, especially for species for which there is little information on **habitat requirements**. For example, when Adams (2008) examined the distribution of the crested cow-wheat *Melampyrum cristatum* in relation to habitat in Eastern England, the habitat preference for chalky boulder clay was clear (Figure 3.1). This kind of mapping is often undertaken using computer-based Geographical Information Systems, which are covered in more detail in Section 7.3.3.

An alternative approach to mapping is recording the occurrence of one feature (e.g. crayfish abundance) in relation to environmental and biotic parameters (e.g. substrate type and density of fish predators), and comparing graphically or **statistically.** This approach was used to analyse distribution

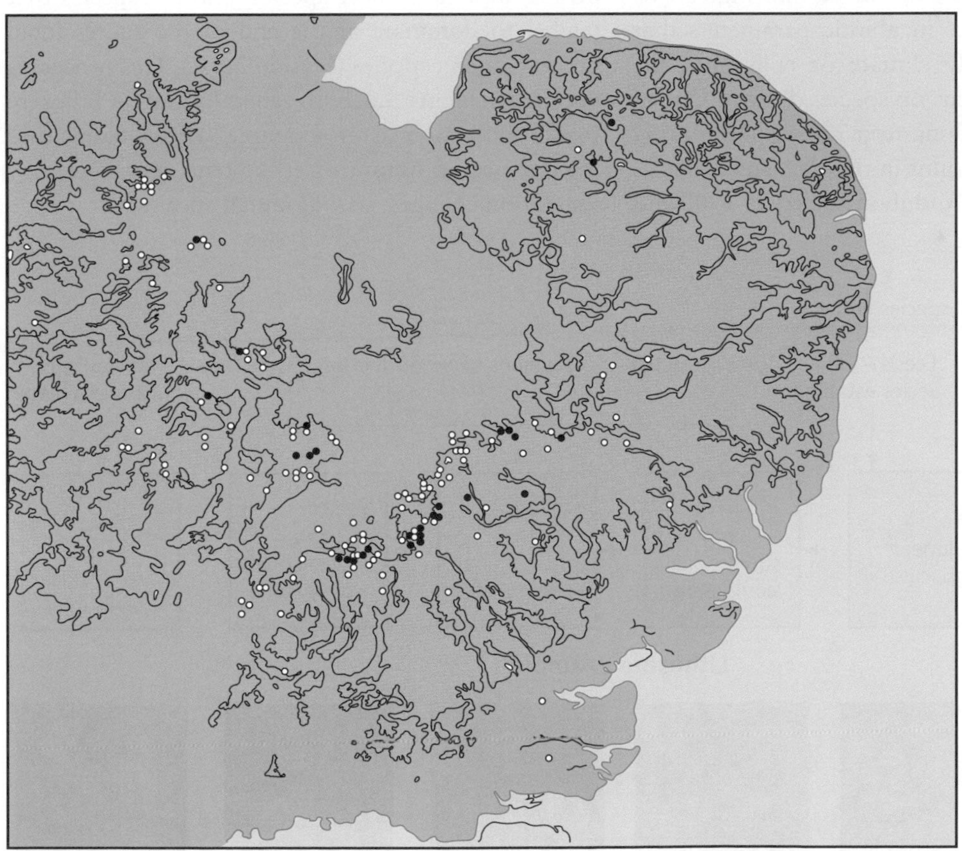

Figure 3.1 The recorded occurrences of the rare crested cow-wheat *Melampyum cristatum* in Eastern England, showing its preference for the chalky boulder clay (grey shading). Filled circles – post 1990 records; open circles – historical sites.

Source: Adams, K. (2008) 'The status and distribution of crested cow-wheat *Melampyrum cristatum* L. in Britain, now largely confined to Essex'. *Essex Naturalist* 25, 120–127; used with kind permission from Ken Adams.

of non-native crayfish in Wisconsin (USA), which showed crayfish were more likely to use habitat with cobble-based substrates, especially when predator density was high (Kershner & Lodge, 1995). This also demonstrates just how complex and interrelated ecology can be—in this case predation risk influenced habitat usage, thereby generating structured spatial patterns of crayfish distribution.

3.3.3 Temporal monitoring.

Once baseline surveys have been undertaken, temporal monitoring can be used to assess **change over time**. Ideally, temporal monitoring involves regular surveys so that gradual changes can be detected. However, any form of repeat surveying (including a survey before and after a specific event, such as a disease outbreak or a change in fire regime) can constitute temporal monitoring.

Although temporal ecological monitoring can be used to monitor change in abiotic parameters that affect species, especially climate or pollutants, the majority of schemes focus on species directly. One of the oldest examples of **long-term species monitoring** is the Christmas Bird Count in the USA and Canada. This was started by the Audubon Society in 1900 and now attracts thousands of participants each year. It provides comprehensive data for many species for well over 100 years. These data have been used in numerous studies such as change in waterbird populations (Wilson et al., 2013) and decline in marine species (Vilchis et al., 2015).

Monitoring natural change

Change in ecology over time can be a natural part of a dynamic ecosystem. For example, habitats and species communities are naturally in a state of flux until (theoretically) they reach a stable state. This is the process of **ecological succession**, which was first conceptualized by Clements (1916). Clements noticed that new land in any part of the world tended to go through a series of vegetative stages, culminating in a fairly predicable vegetation community based on the location of that area. He called the first plants to invade an area pioneer species, and the species that tended to dominate at the end of the successional process climax species (Section 2.8.2). This process is outlined in Figure 3.2, both generally and for lithosere (rock) succession as an example. The dynamic and interconnected nature of ecosystems means that as vegetation changes, so will faunal species.

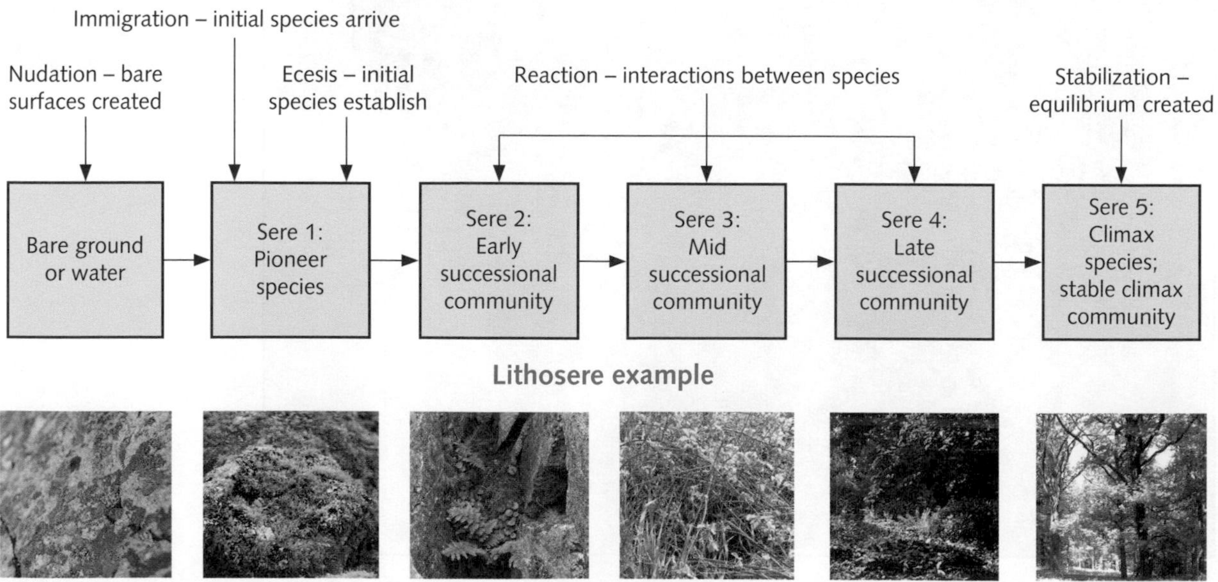

Figure 3.2 The general process of ecological succession through seral stages from pioneer to climax and an example for succession starting on bare rock.

Source: Photographs by Anne Goodenough.

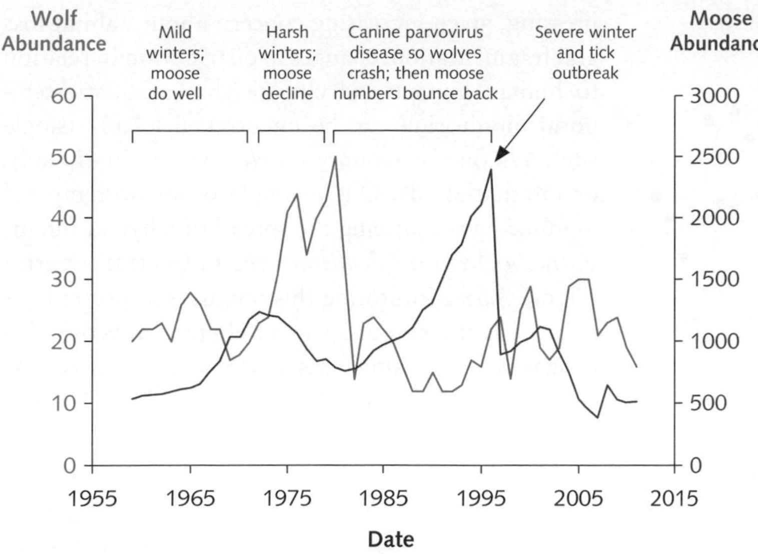

Figure 3.3 Predator–prey cycle for moose and wolves in Isle Royal, USA.

Source: Original data from Vucetich, J.A. and Peterson R.O. (2012) The population biology of Isle Royale wolves and moose: an overview. www.isleroyalewolf.org

Another form of natural change is **predator–prey population cycles**. The simplest model describing such change is the Lotka–Volterra model, originally proposed for ecology by Alfred Lotka (1920) and then independently following an analysis of fish catches in the Atlantic Ocean by Vito Volterra (1926). The model is a cyclical one, whereby a predator population increases in response to an increase in prey supply until the point at which the high predator population causes the prey population to decrease. Ultimately, demand for food outstrips supply and the predator population crashes. With this reduction in predation—a concept termed **predator release**—the prey population increases once again, which triggers an increase in prey, and the whole cycle starts again.

One of the best-known examples of the Lotka–Volterra model is that of moose *Alces alces* and wolf *Canis lupus* populations in Isle Royale National Park, USA. However, although the general cyclical nature of the populations reflects Lotka–Volterra, the most dramatic changes have been caused by external events. For example, the wolf population crash in 1980 occurred due to the anthropogenic introduction of canine parvovirus disease, while the moose population collapsed in 1996 due to a severe winter and a tick outbreak (Vucetich & Peterson 2012; Figure 3.3). These sorts of events are called **stochastic processes**—random events that can have a substantial and unpredictable effect on the overall population, and reconfirm the importance of long-term temporal monitoring.

Monitoring anthropogenic change

Temporal change can also be **anthropogenic** either intentionally through conservation or management (Chapters 11 and 13), or unintentionally, for example, as a result of human-induced climatic change or pollution. Indeed, the fact that pollutants cause change in some species is the founding principle of many monitoring systems that use indicator species (Chapter 4).

An important recent unintentional change has been shifts in bird breeding **phenology** (the timing of seasonal events) in many areas that have had a rise in spring temperature since the 1970s, partly as a result of human-accelerated climatic change. Birds should synchronize breeding so peak demand for food from their young coincides with peak supply. For many species this peak supply is caterpillars, which have been hatching earlier in later (warmer) years (Buse et al., 1999). Because individuals that synchronize with food supply tend to fledge more young, the change in caterpillar timing has caused directional selection pressure—sustained impetus to breed earlier (Goodenough et al., 2010). Many temperate bird species are breeding earlier now than a couple of decades ago, including the nuthatch *Sitta europea*. Figure 3.4 shows how dramatic such change can be over a comparatively short period and how important temporal monitoring can be over short timescales, as well as longer ones.

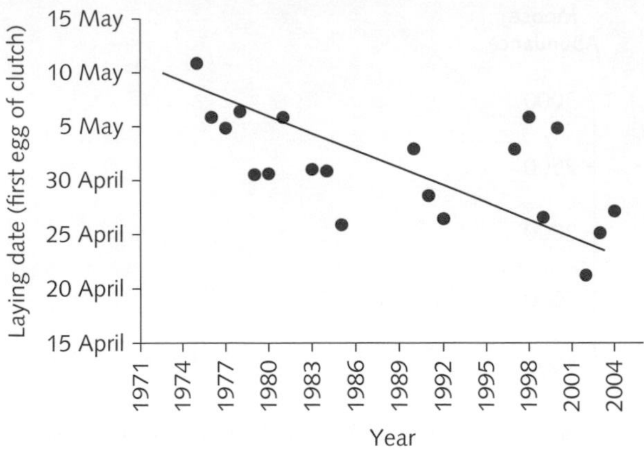

Figure 3.4 Phenological change in the annual mean timing of egg laying by nuthatch, *Sitta europea*, in a UK woodland.

Source: Data from Anne Goodenough (unpublished).

Spatio-temporal monitoring

Spatio-temporal monitoring involves studying how the distribution of species changes over time. The need for such monitoring is becoming ever more pressing, given increasing concern about **habitat loss** and **fragmentation**, changes in distribution in relation to human impact and climate change. Spatio-temporal monitoring can be undertaken locally (single site), regionally (county/state/province), nationally, or internationally. One example of spatio-temporal monitoring is mapping the spread of Chytrid fungus *Batrachochytrium dendrobatidis* in Central America (Figure 3.5). Monitoring this fungus is important because it is the causal agent of chytridiomycosis **disease**, which is causing mass mortality across Australia and the Americas, including, for example, in frogs in Panama (Lips et al., 2006). The high virulence and large number of potential hosts of this emerging infectious disease mean that it is now a threat to global amphibian diversity. This is discussed in more detail in Case Study 12.2.

Comparing distributions over time is only sensible if there is **consistent sample effort.** The spatial distribution of plants was studied in the UK by surveying 2 × 2 km survey areas in 2003–04 and comparing

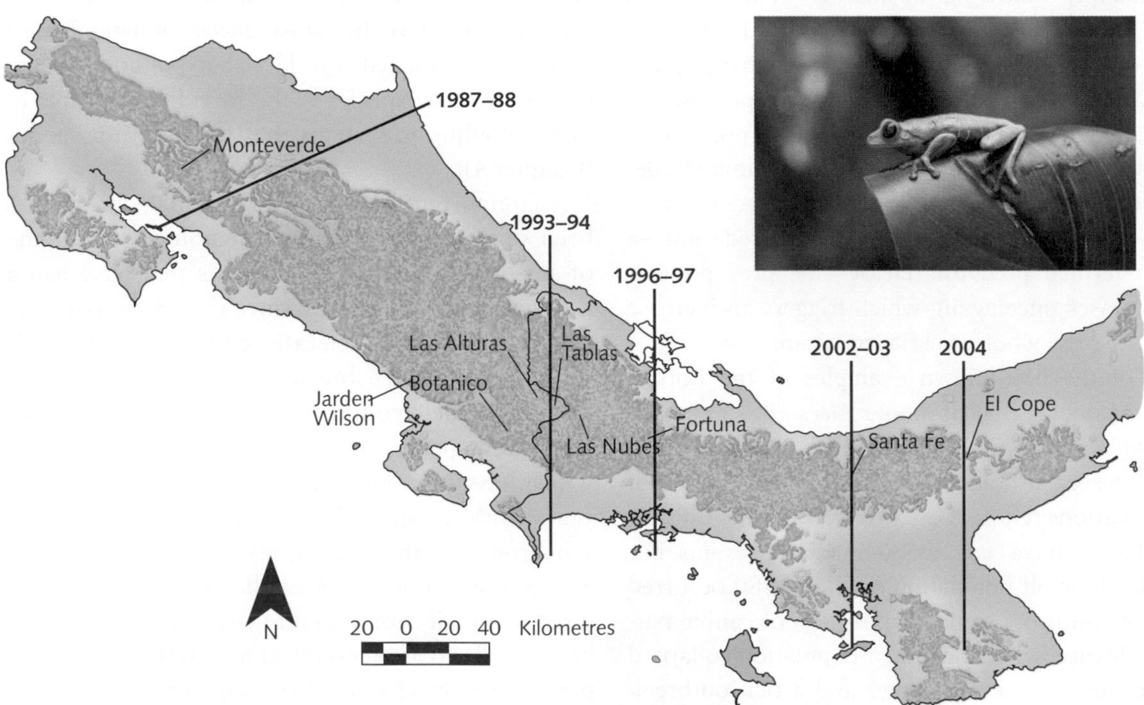

Figure 3.5 The spread of Chytrid fungus in central America is having a devastating effect on amphibian species, such as the red-eyed tree frog *Agalychnis callidryas* which lives throughout Central America.

Source: Map adapted from Lips, K.R., et al. (2006) 'Emerging infectious disease and the loss of biodiversity in a Neotropical amphibian community'. *PNAS of the United States of America*, 103(9), 3165–3170 © 2006 National Academy of Sciences, U.S.A. Photograph by Ayub Khan. Both used with kind permission.

Figure 3.6 Harebells *Campanula rotundifolia* grow in nutrient-poor grassland and heathland across large parts of the temperate zone of the northern hemisphere.

Source: Photograph by Anne Goodenough.

the results to data from 1987–1988 (Braithwaite et al., 2006). However, although the same areas were surveyed, the second survey involved greater survey effort. The problems this causes were apparent in two consecutive censuses of harebell *Campanula rotundifolia* (Figure 3.6) in the UK (Braithwaite & Walker, 2012). Because the second survey was more intensive, there was a higher probability of the species being found. In other words, the risk of false absence (i.e. a species not being recorded in an area despite actually being there) was higher for the first survey. The survey data indicated a decline in the prevalence of the species (mapped change = −5%), but this was actually an underestimate, with the real change likely to be closer to −11% once the unequal survey effort was taken into consideration. Recognition of the kinds of issues causes by unequal sample effort has led to establishment of a number of survey protocols that specifically aim to control this, such as the constant effort site (CES) scheme for bird ringing in the UK (Conway, 2010).

3.3.4 Alert monitoring.

Alert monitoring is the use of monitoring frameworks to give **'early warnings'** of potential ecological problems. This is often used in cases of introduction of non-native species (Chapter 8). For example, Australasian soybean rust disease first occurred in the USA in 2004 following the introduction of *Phakopsora pachyrhizi*—the microbe that causes the disease. Because of the effect of this pest on farmers (Chapter 9), an outbreak monitoring programme run by the US Department of Agriculture (USDA) was started in 2005. The framework's linchpin is a website providing real-time, district-level information on disease spread so that detection is more likely in at-risk areas (Roberts et al., 2006). Figure 3.7 shows an example map of rust disease outbreaks in southeast USA in 2005.

3.3.5 Compliance monitoring.

The main focus of compliance monitoring is on compliance with formal **legislation**, although it can also include **policy** initiatives and agreed codes of practice. The vast majority of countries have several major governmental Acts that cover species protection and protection of specially designated protected areas. The main US Act is the Endangered Species Act 1973, while the main Act in Australia is the Environment Protection and Biodiversity Conservation Act 1999, and the main UK Act is the Wildlife and Countryside Act 1981 (as amended). Although specific clauses differ, generally such Acts have two main parts:

1. **Species-focused:** covers species that have 'standard' and 'extra' legal protection, species that are exempt from the standard legal protection in that country (pest species; Chapter 9), and species that are non-native, which it can be an offence to spread (Chapter 8).

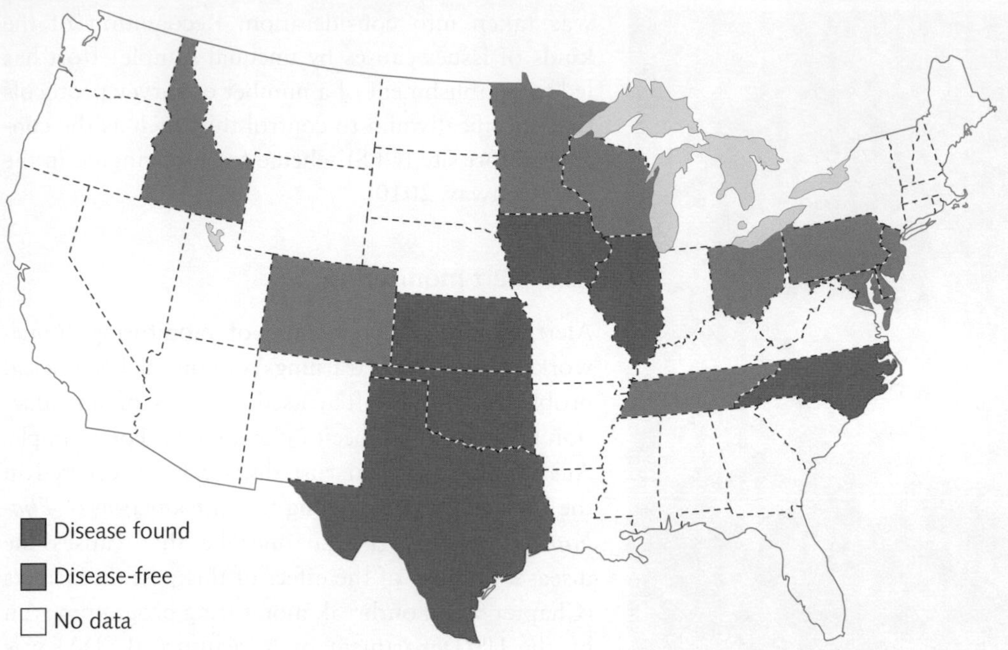

Figure 3.7 Spatial monitoring of Australasian soybean rust disease, caused by the fungus *Phakopsora pachyrhizi* in the USA in 2007. States where the disease was recorded are shown in red; states where recording has taken place and the disease was not found are shown in green. States where there is no survey information are uncoloured.

Source: Data from National Agricultural Pest Information System (NAPIS) collected via the Cooperative Agricultural Pest Survey and collated by the Center for Environmental and Regulatory Information Systems (CERIS).

2. **Site-focused:** covers geographical areas with special protection (Section 11.4.1), activities that can be carried out in protected areas, and activities that require special permission.

There are usually **legally enforceable punishments** for people found breaking these laws and most countries have wildlife crime units tasked with investigating suspected breaches of the law. Depending on the country and the crime, punishments can vary from a formal warning to a fine or even imprisonment. In many cases, punishments can often be bestowed on both perpetrators and those who ordered any illegal activities to take place. This reduces the risk that developers could, for example, order a tree supporting a roost of legally protected bats to be destroyed by an employee without being culpable themselves. There is more detail on wildlife crime, including international conventions, in Hot Topic 3.2.

In the case of European countries, there are also **European directives,** such as the Convention on the Conservation of European Wildlife and Natural Habitats (the EU Habitats Directive), the European Union Directives on the Conservation of Wild Birds (79/409/EEC), and the Natural Habitats and Wild Fauna and Flora Directive (92/43/EEC). Enforcement of these at a national level is undertaken through the appropriate national legislation (i.e. the Wildlife and Countryside Act in the UK, the Law on Conservation and Environmental Care in Germany, and the Law on the Protection of Nature and Landscape in the Czech Republic).

Major national legislative measures are often supplemented by other laws, particularly those covering waste, pollution, and development. There are also specific Acts on other topics; for example, in the UK, trees of particular biodiversity or cultural value can be subject to a Tree Protection Order (TPO) under the Town and Country Planning (Tree Preservation) (England) Regulations 2012. An example of national pollution compliance monitoring is regular testing of oxygen levels in Cardiff Bay, South Wales, to comply with specifications detailed in the Cardiff Bay Barrage Act 1993 (the construction of this barrage is discussed in detail in Case Study 5.1).

HOT TOPIC
3.2

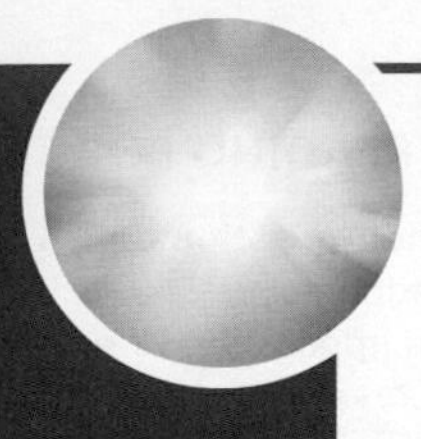

Fighting international wildlife crime

CITES (the Convention on International Trade in Endangered Species of Wild Fauna and Flora) is a **multilateral agreement** between national governments. Drafted in 1963 and adopted in 1973 by an initial 80 countries, its aim is to ensure that international trade of wild animals and plants does not threaten their survival.

Because the trade in wild animals and plants crosses borders between countries, the effort to regulate it requires international cooperation to safeguard certain species from over-exploitation. There are now 180 signatories to CITES, which agree to **regulate trade** of more than 35,000 species and subspecies of animals and plants. Species covered by CITES provision are listed on one of three Appendices:

- Appendix I lists species threatened with extinction. Trade in specimens of these species is permitted only in exceptional circumstances. Examples include pangolin *Manis* and manatees *Trichechus*.
- Appendix II includes species not necessarily threatened with extinction, but in which trade must be controlled in order to avoid utilization incompatible with their survival. Examples include black stork *Ciconia nigra* and king cobra *Ophiophagus hannah*.
- Appendix III contains species that are protected in at least one country, which has asked other CITES parties for assistance in controlling the trade. Examples are walrus *Odobenus rosmarus* in Canada and masked palm civet *Paguma larvata* in India.

In some cases, different subspecies, or even different populations, are included in different appendices. For example, the African elephant *Loxodonta africana* is included in Appendix I except the populations of Botswana, Namibia, South Africa, and Zimbabwe, which are included in Appendix II. There can also be differences for wild-caught individuals versus individuals bred in captivity from domesticated stock. For example, wild chinchillas are listed in Appendix I, but specimens of the domesticated form are not covered by CITES. Importantly, CITES covers listed species whether they are traded as live specimens, fur coats, or dried herbs.

CITES is, by definition, a form of **compliance monitoring**. It aims to control unsustainable use, rather than promote sustainable use—in other words it has a 'stick', rather than 'carrot' approach. The focus on trade also means that it does not address other forms of unsustainable exploitation, such as within-country hunting, habitat loss, or pollution. Other limitations include the fact that it is hard to police and also prone to **corruption**. Despite these limitations, however, CITES still has a vital part to play in tackling international wildlife crime.

QUESTIONS

Why is wildlife crime so hard to detect and punish?

How effective is CITES and how could its effectiveness be improved?

How fair is it to criminalize people 'on the ground' with very low or no income for making use of their environment to feed their family?

FURTHER READING

CITES: *What is CITES?* Available at: http://www.cites.org/eng/disc/what.php and *CITES World*. Available at: http://www.cites.org/eng/news/world/15.pdf

3.3.6 Impact, mitigation, and compensation monitoring.

These monitoring types relate to Ecological Impact Assessment (EcIA) and will be considered more fully in Chapter 5. Briefly, however, impact monitoring involves quantifying impacts of a development and comparing these to preconstruction **impact predictions.** This is essential to monitor how species respond to change and improve future assessments. Another aspect of EcIA is devising measures to reduce negative impacts (**mitigation**) or to make up for these (**compensation**). The only way to assess mitigation/compensation effectiveness is through post-development monitoring. This is vital to build-up knowledge for improving future assessments, and best practice for reducing or buffering any negative impacts.

FURTHER READING FOR THIS SECTION

Question and non-question driven approaches to monitoring: Lindenmayer, D.B. & Likens, G.E. (2010) The science and application of ecological monitoring. *Biological Conservation*, Volume 143, 1317–1328.

Long-term monitoring: Gitzen, R.A. (2012) *Design and Analysis of Long-term Ecological Monitoring Studies*. Cambridge: Cambridge University Press; and Magurran, A.E. (2010) Long-term datasets in biodiversity research and monitoring: assessing change in ecological communities through time. *Trends in Ecology & Evolution*, 25, 574–582.

Strategic use of biodiversity audits: Dolman, P.M., Panter, C.J., & Mossman, H.L. (2012) The biodiversity audit approach challenges regional priorities and identifies a mismatch in conservation. *Journal of Applied Ecology*, 49, 986–997.

Example schemes: Christmas Bird Count http://birds.audubon.org/christmas-bird-count/ and soybean rust mapping http://pest.ceris.purdue.edu/map.php?code=FJAAPNC

3.4 Approaches to monitoring.

In addition to there being many different reasons for monitoring, there are also many different approaches to how that monitoring is undertaken. Any monitoring programme should start by considering what, if any, previous (secondary) data exist at the site in question. If there are previous data, these can inform additional (primary) data collection to make it more targeted and useful, whereas a lack of data highlights the gaps that need to be filled. In some cases, it might also be appropriate for primary data to be collected at the same locations or in the same way as any previous data collection to allow comparability.

3.4.1 Use of secondary data for monitoring.

Secondary or previously collected data can be invaluable in ecological monitoring. Information gleaned can provide insight into a site's ecological history and conditions at different times of the year. Indeed, many major scientific advances are based on long-term data from monitoring undertaken out of interest, rather than for scientific study (e.g. research into phenology shifts with climatic change—Figure 3.4), or even old naturalist notebooks and letters. Such data can be immensely valuable, but there are often obstacles to their use, including standardizing the data for use in a scientific context and securing the appropriate permissions. There are four main types of secondary data:

1. **Anecdotal evidence:** this is previous records of site conditions or species present that are unconfirmed and uncorroborated. Such records are usually given by non-experts and typically relate to highly visible and easy-to-identify species, as well as species that are popular locally, such as flagship species (Section 10.5.2). Such data might not be especially robust, but can be very valuable in providing background information and informing what sort of surveys need to be completed.
2. **Data from local recorders or local biological records offices—casual records:** in many countries, there is a network of experienced recorders who

submit species records to local biological records offices. Data are usually fairly robust, since recorders generally have considerable knowledge both of specific taxonomic groups and their local area. Such data can be invaluable in conservation (as recently reviewed by Pocock et al., 2015).

3. **National volunteer data:** for some countries there are long-running and well established national surveys for specific taxa. Examples include the North American Amphibian Monitoring Programme in the USA (NAAMP, launched 1997) and the UK Butterfly Monitoring Scheme (UKBMS, run by Butterfly Conservation, launched 1976). These are large national surveys, with standardized methods, and are undertaken by relatively experienced and trained volunteers; habitat data are often collected too.
4. **Previous surveys:** sometimes there have been focused ecological surveys at a given site previously, by trained ecologists. This is particularly the case for sites with conservation designations (national parks, etc.) or sites with strong development potential (Chapter 5).

Despite its value, using secondary data is not without challenge. Such data have often been collected in a non-standardized (and non-scientific) way. If survey effort has changed over time this confounds results. In some cases the quality of the data is also unclear, and it is not obvious how much confidence can be placed in records, species identification skills, and so on.

An even bigger issue for secondary species datasets is that they are generally **presence-only**. In other words, people submit records of when they see a species, but do not submit records when they do not see that species. This contrasts with scientific surveys when absence (or, more correctly, failure to find a species) is also recorded. This is most likely to occur in *ad hoc* data from casual observations and local records centres, and can be a particular problem for rare species when **low sampling effort** could easily miss a species. If the amount of effort applied is largely unknown, the lack of presence of a species could relate to a real absence or simply a lack of effort. This is an important issue with volunteer-collected data, including those from citizen science initiatives (Section 3.4.4).

Online repositories for ecological data

Within a specific country there are usually numerous monitoring schemes. Some of these focus on particular taxonomic groups; others consider multiple taxa, but focus on a specific spatial area, such as a site (e.g. nature reserve) or a region (e.g. local environmental records offices). With so many different schemes, it can be hard to locate specific information in a desk-based study. 'One stop shops' for environmental records are extremely useful in allowing Applied Ecologists to readily access and synthesize data. A good example is the UK's National Biodiversity Network (NBN), which holds over 100 million species and site records.

NBN data are available electronically through the NBN Gateway online portal. This allows searches by species, so that distribution at any scale from local to national can be easily accessed. One of the most useful features here is an interactive map that can be customized in terms of location and scale. It is also possible to search by site—for example, to locate species recorded in a specific area, by dataset, or by site designation (e.g. National Nature Reserve).

The NBN has expanded considerably since it started in 2000, as shown in Figure 3.8, and now has many different contributors, from statutory regulators, such as Natural England and Scottish Natural Heritage, and national organizations like the British Trust for Ornithology and the National Trust, to local groups. The data are used in numerous studies, including on distribution of wildcats *Felis silvestris* (Silva et al., 2013), the habitat needs of ground beetles Coleoptera (Gillingham et al., 2012), and bryozoan diversity in coastal waters (Rouse et al., 2014).

3.4.2 Use of primary data for monitoring.

Although secondary data can be very useful, especially for baseline monitoring (Section 3.3.1) or retrospective analysis, the majority of ecological monitoring involves primary data.

The main method of collecting data is **manual surveying** using standardized methods. Specific examples for habitat and species monitoring are discussed in Sections 3.5 and 3.6. It is also possible to collect primary data for a specific survey through casual observations. For instance, an ecologist might be

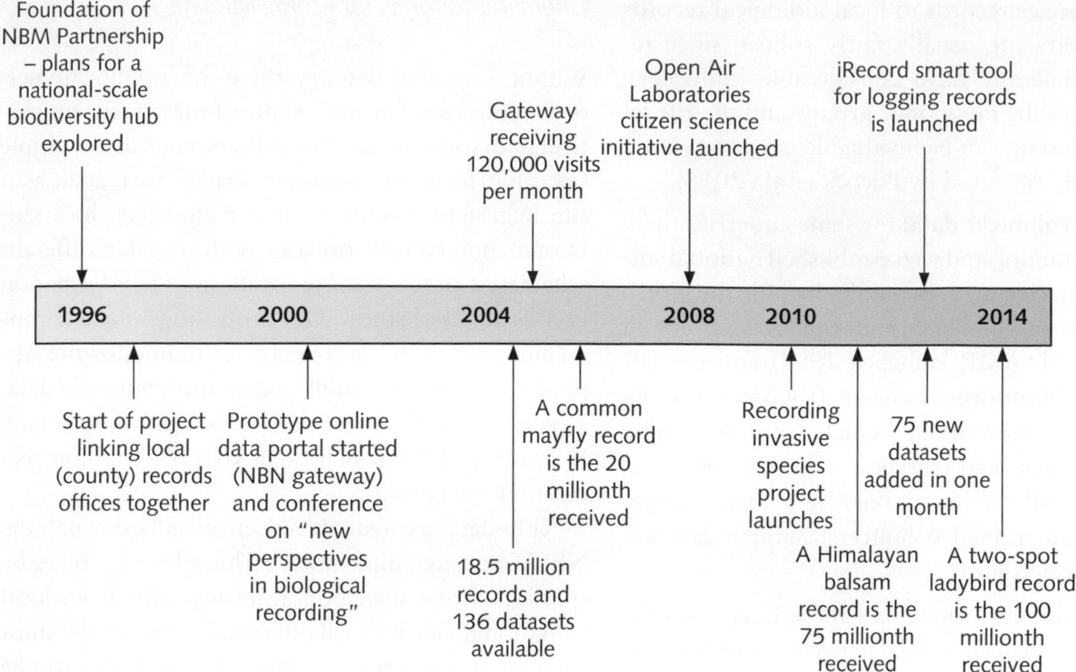

Figure 3.8 Development and growth of the National Biodiversity Network.
Source: Information synthesized from National Biodiversity Network.

undertaking a mammal survey, but encounters a rare bird, which is recorded. These are useful sightings as they are verified, although they are not as robust as data from specific surveys, since sample effort has not been controlled.

Automated approaches to monitoring

It is sometimes possible to undertake **automated monitoring** when there is no regular human involvement after initial set-up. This is mainly used for monitoring of environmental parameters and usually increases the amount of data recorded, because it is possible to survey more locations or with higher frequency. This allows a more comprehensive picture to be constructed compared with manual sampling, which tends to be more irregular.

In most automated monitoring systems, data are automatically collected and transferred via satellite to a receiver. An example is the Automated Water Quality Monitoring Network, which is used in North American lake ecosystems, especially at transboundary sites (i.e. those on the border of two or more countries—in this case Canada and the USA). The aims are to determine long-term trends in water quality (temporal monitoring), and allow rapid identification of pollution incidents so action can be taken (alert monitoring). Similar surveys can be undertaken for rivers such as the Danube and its tributary the Sava (Mijović and Palmar, 2012). Most automated aquatic monitoring stations collect data on dissolved oxygen, pH, and specific chemicals using a multiparameter probe. Manual sampling allows more parameters to be studied than is possible using a multiprobe, but samples are much more 'snapshot' in nature.

Automated systems are not restricted to water pollution. For example, on the coast of San Diego, USA, data loggers monitor landslides using horizontal time-domain reflectometry, which is similar to radar. If a movement threshold is reached, an automated telephone system is used to raise the alarm. This is alert monitoring (Section 3.3.4) as data are only transmitted if a specified threshold is reached (Campbell Scientific, 2014). Automated transmitting technology can also be mounted on animals so they become mobile biological sensors. For example, Banzi (2014) suggested that African elephants could be platforms for visual and infrared cameras and GPS, to monitor classic 'panic' movement to be fed back to rangers. This could especially help anti-poaching efforts for big game species in Africa (Hot Topic 11.2)

Semi-automated approaches to monitoring

While automated systems are often the 'gold standard', they are not always feasible or cost effective. In such circumstances, **semi-automated monitoring** is often used, which typically involves data being logged automatically, but human intervention is needed for download. Such monitoring is less immediate, but the equipment is much cheaper than transmitting devices.

Semi-automated monitoring is often undertaken using a multifunction data logger, which changes function according to the probes that are connected to it. This means that the logger can record, for example, temperature with a thermistor probe, air pressure with a barometric sensor probe, and carbon dioxide with a gas analysis probe. In some cases, data loggers have ports for several sensors so they can monitor several environmental parameters at the same time. Usually, units can be checked and programmed in the field without the need for a computer interface. This approach was used by Pérez et al. (2012) to study the environmental differences in different parts of forest ecosystems in Puerto Rico. Significant differences in temperature, relative humidity, and light intensity were observed between edge and forest interior, which affected abundance of mosses.

While automated data loggers tend to be bulky, expensive, and susceptible to human interference, microloggers are small enough to be concealed and cheap enough that loss of a few units, although never desirable, is not a major problem. Such systems use a computer chip to record and store data—all programming is done via a computer on units as small as a shirt button. The environmental parameters that can be recorded tend to be simple—temperature and humidity being the most common—but the small size and all-in-one nature of these devices mean they can be easily used in studies of species–habitat interactions. Insights gained have direct applicability for Applied Ecologists in terms of management. For example, it is standard practice when a bat roost is destroyed (under licence, for example, due to development) that a new, artificial, roost is built. To maximize the similarity between the natural and artificial roosts, data loggers are used in the former to understand the microclimate so that it can be replicated in the latter (Bat Conservation Trust, 2014).

In certain circumstances, semi-automated monitoring is done without data loggers. For example, diffusion tubes offer an inexpensive and easy way of monitoring air pollution. This involves using tubes positioned in the air with one end open to enable pollution uptake. Tubes are left in place for a known period of time—usually several weeks. An absorbent is packed into the other end that converts the nitrogen dioxide to nitrate, which is trapped in steel gauze that can ultimately be analysed in a laboratory after collection. This approach was used by Lee et al. (2012) to monitor the effects of car exhaust fumes on nearby plants.

3.4.3 Proxy monitoring.

Proxy monitoring involves monitoring an ecological parameter indirectly by measuring another aspect of the environment. Use of **proxy data** tends to occur either when it is difficult or time-consuming to measure the actual parameter(s) of interest or where it is not possible to measure the parameter(s) of interest because it no longer exists. In ecological monitoring, use of proxies tends to involve using species as ecological indicators. This might involve, for example, using plants to infer soil conditions, using lichens to monitor air quality, or using aquatic invertebrates to monitor oxygen levels (Section 4.3.2). It is also possible to use species richness or diversity in one taxonomic group to infer overall species richness in an ecosystem (Section 4.5.1).

3.4.4 Using citizen science for monitoring.

An alternative to Applied Ecologists collecting monitoring data is to work with the public to collect scientific data in a citizen science framework. **Volunteer participation** has always been an important component of ecological monitoring, especially of charismatic species, such as birds (e.g. Christmas Bird Count in the USA and the British Trust for Ornithology's Breeding Bird Survey in the UK). In the past two decades, however, use of volunteers has expanded greatly and is increasingly evolving to become 'citizen science', a term first coined by

Irwin (1995). There is a blurred distinction between volunteer-collected monitoring data and citizen science, but volunteer-collected monitoring data frequently involves volunteers with considerable knowledge and devoting a considerable amount of time to a project, often over many years. In contrast, citizen science, involves untrained or 'non-expert' individual members of the public submitting one-off records, often using technology such as smartphones or tablets (Catlin-Groves, 2012). This is sometimes known as **crowd sourcing**.

Citizen scientists respond to specific calls for participation in surveys, which may themselves be short-term and designed to answer a specific ecological question. The approach is becoming increasingly common in recording where species occur (spatial surveys; Section 3.3.2) and, where issues of standardizing sample effort can be overcome, temporal monitoring of species trends too (Section 3.3.3). Examples are given in Table 3.2.

In addition to running programmes of research that encourage users to engage in a data submission process, there is also the underexplored option of **mining data** from social networks and taking a more opportunistic approach. Many images, especially those taken on mobile phones, contain GPS information and can readily be searched and mapped, which can be useful in seeing species distribution. This is a passive approach as it does not require direct engagement with citizens, rather it piggybacks on their previous use of social networks.

Citizen science has the potential to allow ecological monitoring programmes to be undertaken successfully where constraints such as money and time would otherwise make studies unfeasible or impossible for an individual organization. Advantages of citizen science data collection for Applied Ecology include:

1. Ease and speed with which data can be gathered.
2. High sample sizes that are possible, often with wide geographic reach.
3. Low cost.

However, there can be concerns about the quality, reliability, and overall value of these data. There can be uncertainty about the skill level of participants submitting data, and there is rarely any opportunity to verify or validate records. The more information that can be determined automatically, the better. For example, it is often possible to obtain date and time from the EXIF data in a photograph and, if the photograph was taken with a GPS-enabled camera or a smart phone, often the location can be determined automatically too. An alternative strategy is asking people to submit photographs of a target species, rather than a text-based record, so identification can be corroborated by a professional ecologist.

Ecological monitoring using citizen scientists is most successful when the surveys are simple. Surveys that ask people to report sightings of popular and charismatic species often work well, for example, while surveys on species that are obscure or hard to identify will usually get a low take-up and results might be more prone to error. Ecological monitoring using citizen scientists is still reasonably new and constantly evolving, but there have already been some notable successes. For example:

Table 3.2 Examples of citizen science projects.

Monitoring	Survey	Country	Aim	Survey type	Years/records
Spatial	BeeID (bumblebees)	UK	To establish the distribution of bumblebee species	Flickr photos uploaded with a specified tag and related to location	2010; 203 tagged images uploaded
Temporal	Project budburst (plants)	USA	To track the timing of plant species' bud burst in spring to relate to climate change	Mobile phone application	
Spatio-temporal	Flying Ant Survey	UK	To examine the synchronicity of ant flying behaviour in different spatial areas and identify possible stimuli	Web-based survey	2012–2014; 13,000 records in total

AN INTERVIEW WITH AN APPLIED ECOLOGIST 3

Name: Dr Elizabeth Pimley
Organization: Worcestershire Wildlife Consultancy (part of Worcestershire Wildlife Trust)
Role: Senior Ecologist
Nationality and main countries worked: British; worked in UK and West Africa

What is your day-to-day job?
I work as Senior Ecologist for Worcestershire Wildlife Consultancy, part of the Worcestershire Wildlife Trust. About 70% of my job involves undertaking a variety of ecological surveys (for both habitats and protected species) and preparing a range of ecological reports, including preliminary ecological appraisals, specific protected species survey reports, Ecological Impact Assessments, and BREEAM/Code for Sustainable Homes reports. The other part of my job involves project management, production of quotations for future work, and client liaison. This last involves ensuring that clients are aware of any ecological constraints and helping them find a way that a given project can proceed in an ecologically sound and legally acceptable manner.

What is your most interesting recent project and why?
I have undertaken a variety of projects that were particularly interesting. One such project was a water vole monitoring survey of the Battlefield Brook in Bromsgrove, Worcestershire, for the Environment Agency (EA). This involved the need to assess the impact of EA work to maintain flows within the catchment on water voles, comparing survey information collected over several years. It was enjoyable to walk the whole section of brook, looking for signs of water voles and noting how the brook habitat changed along its course. It was also rewarding to have the opportunity to use the survey information to advise the EA on how to manage the brook in the future for water voles.

Another interesting project involved setting up a dormouse monitoring programme for a woodland managed by the Herefordshire Wildlife Trust. The survey information was used to provide recommendations for future management of the woodland to benefit the local dormouse population and other woodland fauna. The data collected were submitted to the National Dormouse Monitoring Programme and will also provide a baseline for a longer-term dormouse monitoring programme of the woodland.

What are the most satisfying parts of your job?
I enjoy being able to spend a reasonable amount of my time getting out and about undertaking a variety of ecological surveys, as well as having the opportunity to learn more about our native flora and fauna as part of my daily work. I also like working for a Wildlife Trust and thereby being part of 'the bigger picture' for conservation nationally.

What do you see as the main challenges in your field and how can they be overcome?
One of the main challenges in the area of ecological consultancy concerns the fact that during times of economic downturn, ecology tends to be one of the first things to fall by the wayside, as developers, local governments, and so on, look to maximize profits and minimize 'unnecessary expenditure'. While all ecological consultants in the UK should follow a code of professional conduct set out by the Chartered Institute of Ecology and Environmental Management (CIEEM), this does not always happen, with some consultants being more 'developer friendly' than others, to the detriment of our wildlife and natural habitats. To avoid this happening, the industry needs to be more effectively monitored by local planning authorities and CIEEM.

What's next for you, and why?
I have been involved in a few mammal projects both in the UK (on Bechstein's bat *Myotis bechsteinii*) and abroad (on nocturnal primates), and am hoping to find some time to analyse the data and get some of these projects written up in peer-reviewed journals. It is so important that such information gets out there into the public domain, rather than just remaining on a dusty shelf in a naturalist/ecologist's notebook.

Finally, how did you get into species monitoring and what advice would you give to others?
I did a degree in Zoology at the University of Bristol and really wanted to do my own research following this, which led to a PhD in Zoology at the University of Cambridge, where I studied the behaviour and ecology of nocturnal primates in Cameroon. I then worked for the Animal and Plant Health Agency, studying the ecology and ranging behaviour of badgers *Meles meles* to assess the impacts of culling badgers on TB spread in badgers and cattle. I gained considerable experience during this period in ecological survey techniques and analytical methods, which gave me an appropriate background to move into ecological consultancy, first with Cresswell Associates/Hyder Consulting and lately with Worcestershire Wildlife Consultancy.

Figure 3.9 The rare nine-spotted ladybird *Coccinella novemnotata* was rediscovered using citizen science monitoring in America.

Source: Image courtesy of Rod Haley/CC BY-SA 2.0.

- **New records of rare species:** the rare nine-spotted ladybird *Coccinella novemnotata* was rediscovered through citizen science in eastern North America in 2006, after 14 years of absence (Losey et al., 2007). The charismatic nature of this species (Figure 3.9) was crucial to the success of this survey.
- **Identification of migration pathways:** monarch butterfly *Danaus plexippus* in the USA recorded on the citizen science project Journey North (Howard and Davies, 2009).
- **Changes in policy:** alterations in marine policy in Jamaica following a citizen science project that revealed the extent and severity of bleaching on coral reefs (Crabbe, 2012).

FURTHER READING FOR THIS SECTION

Automated aquatic monitoring: Macekova, L. & Ziga, M. (2014) The wireless sensor network concept for measurement of water quality in water streams. *Acta Electrotechnica et Informatica*, Volume 14, 60–67.

Automated species monitoring: Wall, J., Wittemyer, G., Klinkenberg, B., & Douglas-Hamilton, I. (2014) Novel opportunities for wildlife conservation and research with real-time monitoring. *Ecological Applications*, Volume 24, 593–601.

Automated acoustic monitoring: Blumstein, D.T., Mennill, D.J., Clemins, P., Girod, L., Yao, K., Patricelli, G., Clark, C., Cortopassi, K.A., Hanser, S.F., McCowan, B., Ali, A.M., & Kirschel, A.N.G. (2011) Acoustic monitoring in terrestrial environments using microphone arrays: applications, technological considerations and prospectus. *Journal of Applied Ecology*, Volume 48, 758–767.

Proxy monitoring: Siddig, A.A., Ellison, A.M., Ochs, A., Villar-Leeman, C., & Lau, M.K. (2016) How do ecologists select and use indicator species to monitor ecological change? Insights from 14 years of publication in *Ecological Indicators*. *Ecological Indicators*, Volume, 60, 223–230.

Citizen science monitoring (general): Sutherland, W.J., Roy, D.B., & Amano, T. (2015) An agenda for the future of biological recording for ecological monitoring and citizen science. *Biological Journal of the Linnaean Society*, Volume 115, 779–784.

Citizen science monitoring—marine-specific: Pecl G., Brodribb, F., Brown, R., Walsh, P., Frusher, S., Edgar, G., Lyle, J. Poloczanska, E., & Stuart-Smith, R. (2014) Developing citizen science as a communication and research tool for monitoring ecological change in the marine environment. Available at: https://www.nccarf.edu.au/conference2010/wp-content/uploads/Stewart-Frusher.pdf

National Biodiversity Network: NBN Gateway. Available at: https://data.nbn.org.uk

3.5 Monitoring habitats.

There are two main aspects to monitoring habitats. The first is to establish where specific habitats occur and whether there is any change in their **distribution** over time (spatio-temporal monitoring; Section 3.3.3). This requires methods to classify habitat types so that their distribution and any change can be quantified. The second aspect is monitoring habitats is to assess their ecological **condition.** This might be undertaken to track change in condition over time (temporal monitoring; Section 3.3.3), to provide an early warning of change (alert monitoring; Section 3.3.4), or to ensure conformity with legislation or directives (compliance monitoring; Section 3.3.5). The following section considers both habitat classification and condition assessment.

3.5.1 Classifying and quantifying habitats.

There are many ways that habitats can be classified, but methods are usually based on vegetation structure. For example, woodland might be distinguished from grassland based on its higher structural complexity and multilayer structure. Thus, the height of the vegetation and its density/openness is important, as is its age. Origins—natural or anthropogenic—and management are also important.

Most systems rely, at least to some extent, on vegetation type. At an international scale, one of the best-known classification systems is that of **biomes**. This splits the terrestrial areas of the world into broad categories based on the climax vegetation community that is the endpoint of **ecological succession** (Sections 2.8.2 and 3.3.3; Figure 11.6). It is based on Clements' idea that each climatic area would have one vegetation community. In reality, things are not this simple. As originally pointed out by Tansley (1939), any broad climatic type is likely to have one of several possible 'end points'—different climax communities based on non-climate factors, such as topography and soil. However, although Tansley's polyclimax (multiple-end) model reflects reality more accurately than Clements' monoclimax (single-end) model, most maps rely, for simplicity, on monoclimax communities.

At a finer scale, there is considerable **habitat heterogeneity**. The climax community of much of Africa, for example, is savannah grassland, while most of North America has temperate woodland as its climax community. In both cases, not only are there numerous other habitats, but the climax community is often virtually non-existent. Succession can be interrupted before the climax is reached (a subclimax) or the climax can be disturbed (a plagioclimax; Figure 11.6).

Sometimes, when only broad habitat information is needed, a simple classification system can be employed. This is especially useful if habitat data are being collected by non-expert volunteers or, increasingly, citizen scientists (Section 3.4.4). For example, in the UK, volunteers for the non-governmental organization Butterfly Conservation survey butterflies according to a **standardized method**. This involves them walking two 500 m transects within a 2 × 2 km map grid square, while also collecting habitat data to allow species–habitat interactions to be analysed. Rather than asking untrained volunteers to collect complex habitat information, which might be unreliable, there is a simple coding form with 22 different habitats, including four types of woodland (deciduous, coniferous, mixed, or orchard/planation/commercial) and four types of grassland (calcareous, acidic, unimproved meadow, and fertilized/improved/reseeded).

It is obviously possible to come up with a **bespoke habitat classification** method for a specific survey, as Butterfly Conservation has done. The advantage is that the habitat classification system can be tailored to the specific species or aims of the study, and the likely skills of surveyors. Often, though, more complex systems are used by professional ecologists, and these tend to be industry standard.

Phase One Habitat Surveying

Phase One Habitat Surveying (P1HS) is the industry standard for habitat classification in the UK. It allows a rapid and 'broad-brush' record of land use and habitats to be made, and gives initial insight into how a particular site looks and functions. The key point is that a P1HS is standardized so data can be collected and interpreted by any Applied Ecologist. The system is based primarily on vegetation plus key landscape and linear features, such as rock escarpments, fences, ditches, and hedges. It does not require advanced plant identification skills.

The P1HS system has a **hierarchical approach**. In total there are 155 habitat types, but these are grouped into 10 broad categories, each denoted by a letter. Once the broad habitat type is identified, the fieldworker 'drills down' through the options within that category to get to more detailed and accurate description for that particular habitat. Two examples, one terrestrial habitat and one aquatic, are given in Figure 3.10. The system also allows fieldworkers to record extra detail on key features, and species and management techniques, using a system of target notes.

3.5.2 Assessing habitat condition.

Monitoring the quality of habitat patches is often important, especially for protected areas. Such monitoring can serve various purposes. It serves as a baseline for assessment of future projects (baseline monitoring; Section 3.3.1) and can be used to monitor change (temporal monitoring; Section 3.3.3), which is particularly useful for site managers as a means by which to assess the effectiveness of a change in management initiatives. Also, by combining the results of multiple **condition assessments** at a national level, it is a means of fulfilling that country's obligations

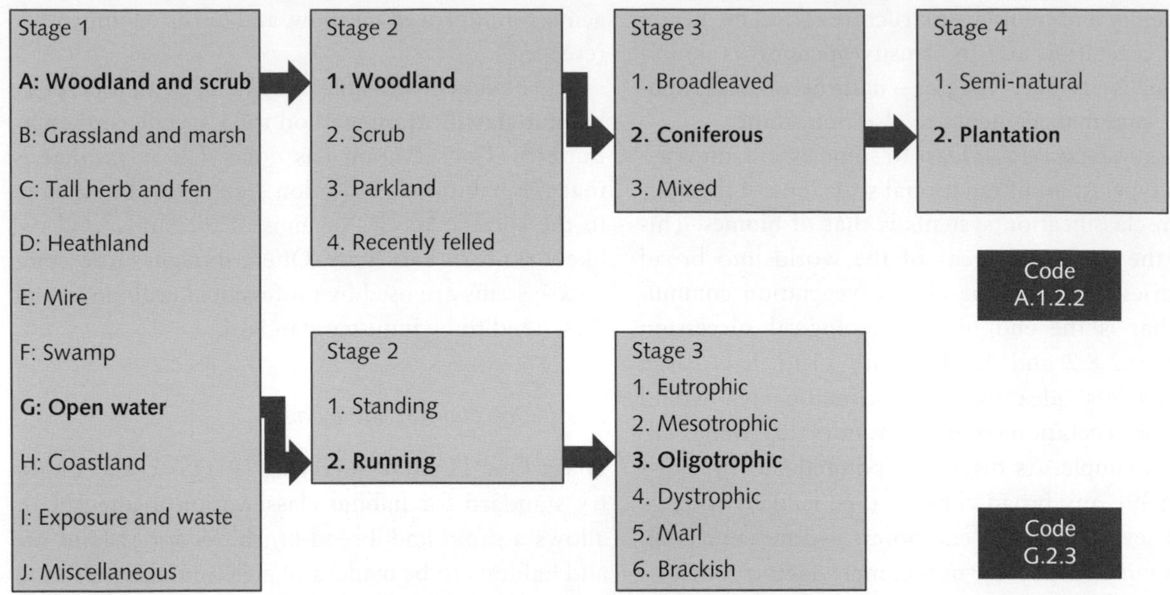

Figure 3.10 Phase One Habitat Surveys. The hierarchical, alpha-numeric, approach to habitat classification used in Phase One Habitat Classification is shown on the left for examples of coniferous plantation woodland and oligotrophic (low nutrient) running water.

under European statute (compliance monitoring; Section 3.3.5).

Article 17 of the EU Habitats Directive requires member states to report progress made in implementing the Directive every six years. This **Common Standards Monitoring (CSM)** framework involves quantifying the effectiveness of legal protection, conservation policy, and management for protected areas. In each member state, the statutory nature conservation agency is responsible for reporting condition of sites protected under European statutes (e.g. NATURA 2000 sites, including Special Areas of Conservation (SACs) and Special Protection Areas (SPAs)) or national legislation. Although only national-level data are needed, most member states assess condition at site (or even sub-site) level. This means data aggregated for Common Standards Monitoring reports are also useful for informing management at each site.

Individual member states have taken different approaches to condition assessment. Most involve consideration of any change in spatial area and then assess site condition using a series of standardized but **habitat-specific criteria.** Typically, three indicator types are considered—habitat structure, vegetation species, and disturbance. The actual indicators themselves and their threshold values can be adapted to the specific habitat. For instance, in Belgium, a key condition indicator is variation in vegetation (e.g. variation in tree stem diameters in woodland or sward height in grassland). Belgium's vegetation indicators usually focus on the presence of key species (e.g. the grass *Corynephorus canescens* for dry heath), while disturbance indicators typically include shrub encroachment or overgrazing in open habitats, and invasive alien species (European Environment Agency, 2014).

The UK system is similar: standard questions on change in range and presence of non-native species are followed by habitat-specific questions on vegetation structure and species composition. As an example, the criteria for lowland heathland are given in Figure 3.11. For Sites of Special Scientific Interest (SSSIs), reference will be made to the **notifiable features** (i.e. the reasons why that particular site was originally designated a SSSI). So, for example, if a site was designated because of a high diversity of bankside invertebrates, the diversity of that taxonomic group would be assessed.

Ultimately, there are five categories in Common Standards Monitoring:

1. Favourable.
2. Unfavourable recovering.
3. Unfavourable no change.
4. Unfavourable declining.
5. Destroyed (or part destroyed).

Figure 3.11 Characteristics of good quality lowland dry heath in England—all criteria have to be met for a lowland dry heath site to be classified as being in favourable condition in the Common Standards Monitoring framework.

Source: Photographs by Anne Goodenough.

Some national-level assessments use different outcomes for site-level assessments (e.g. Belgium uses A = Excellent, B = Good/Moderate, C = Bad). Generally, though, national systems use the favourable/unfavourable system to avoid the need to translate site data to the rubric used in the Common Standard Monitoring report.

For site-level assessments in the UK, 'favourable' is the optimal condition category. In order for this to be achieved, every mandatory habitat-specific criterion must be met (i.e. all those listed in Figure 3.11 for lowland heathland). A status of 'unfavourable recovering' condition is given where some criteria are not met, but management is in place to address issues. Together, favourable and recovering constitute target conditions, while the other condition categories constitute adverse conditions. There are many reasons why a site might attain unfavourable status, including undergrazing, overgrazing, presence of or failure to control non-native species, pollution, development encroachment, poor agricultural practice, coastal squeeze, and human disturbance.

The attributes used to make a site assessment are chosen because they allow rapid assessment of condition without the need for specialist equipment or expert (as distinct from experienced) surveyors. There are criticisms of this standardized approach though, particularly that it does not allow for the variation that makes ecology so diverse. Habitats are influenced by multiple variables, including microclimate, topography, and soils, and even historical land use. Having a series of rigid (albeit habitat-specific) criteria for the 'ideal' version of a specific habitat type could undermine the inherent—and valuable—complexity of natural ecological systems.

FURTHER READING FOR THIS SECTION

Overview of Phase One Habitat Surveys: Joint Nature Conservation Committee (2010) *Handbook for Phase 1 Habitat Survey—a Technique for Environmental Audit.* Available at: http://jncc.defra.gov.uk/PDF/pub10_handbookforphasc1habitatsurvey.pdf

Example of condition assessment reporting: Newton, A., Cantarello, E., Myers, G., Douglas, S.J., & Tejedor, N. (2010) The condition and dynamics of New Forest woodlands. In: A.C. Newton (Ed.) *Biodiversity in the New Forest*, pp. 132–147. Newbury: Pisces Publications.

Common Standard Monitoring: *Common Standard Monitoring: Introduction to the Guidance Manual.* Available at: http://jncc.defra.gov.uk/pdf/CSM_introduction.pdf

3.6 Monitoring species.

Monitoring species is probably the single largest aspect of ecological monitoring. Baseline surveys might be necessary in order to establish whether a species is present (Section 3.3.1), spatial surveys might focus on how species abundance changes geographically (Section 3.3.2), while temporal monitoring might quantify changes in species' abundance over time (Section 3.3.3). To meet such aims there are several parameters that can be used, including, in order of complexity:

1. **Species presence:** this is the easiest parameter to measure—simply whether a species occurs in a given area. It is frequently recorded in volunteer and citizen science datasets and forms an important part of baseline surveying, especially for biodiversity audits (Case Study 3.1) and EcIAs (Chapter 5).
2. **Likely absence:** it can be important to know where a species does not occur. It is never possible to prove absence categorically, as not finding a species at a given site could be because it isn't there or it could simply mean it has not been detected. Not finding a species that is actually present is a **false absence.** False absences are most common with species that are hard to survey, possibly because they are disturbance-sensitive, cryptic or nocturnal (Section 3.6.2). Generally, the risk of false absence decreases as survey effort increases. In other words, the more time that is spent surveying for a species, the more likely it is that an apparent absence is a real absence rather than a false absence. The ultimate examples of false absences are so-called **'Lazarus species'**—species declared extinct because repeated surveying had failed to find them, before ultimately being rediscovered. Two examples are shown in Figure 3.12, one of which, the Lord Howe Island stick insect *Dryococelus australis*, is discussed in more detail in Chapter 12.
3. **Prevalence:** prevalence is similar to presence/absence data, but whereas that can only ever equal 1 (present) or 0 (absent), prevalence data is the proportion of occasions a species was found and is thus a measure of frequency. For exam-

Figure 3.12 Two Lazarus species believed extinct and then rediscovered. (Left) The Chacoan peccary *Catagonus wagneri* was first described in 1930, based on fossil records, and believed extinct, before live individuals were found in Paraguay in 1975. (Right) The Lord Howe Island stick insect *Dryococelus australis* was believed extinct in 1930 after being lost from its only known habitat on Lord Howe Island, but rediscovered in 2001 on Ball's Pyramid, the world's tallest and most isolated sea stack; see Chapter 12 for more details on the conservation of this species.

Source: Photograph of Chacoan peccary courtesy of 'Jean'/CC BY 2.0. Photograph of Lord Howe Island stick insect courtesy of Rohan Cleave, Melbourne Zoo, used with kind permission.

ple, if 20 ponds were surveyed and smooth newt *Lissotriton vulgaris* were present in 18 ponds but palmate newt *Lissotriton helveticus* were present in only two ponds, the prevalence would be 90 and 10%, respectively, regardless of the numbers found in each pond.

4. **Relative abundance:** this is the abundance of species relative to one another based on count data, percentage coverage (for vegetation and species such as lichens and barnacles that colonize habitat surfaces), or abundance ranks, such as DAFOR or DOMIN (Section 4.3.3).
5. **Absolute abundance:** this is the actual number of individuals present in an area. Absolute abundance might be expressed as population (the number of individuals) or density (the number of individuals per unit of area), often using capture–mark–recapture methods.

Together, surveys of presence/absence provide **nominal data** (counts of sites where species are present and absent), whereas relative abundance surveys will give **ordinal data** (data in ranks, but not proper numbers), and prevalence and absolute abundance surveys will yield **ratio data** (proper numbers that can be added together, divided, etc.). Frequency statistics such as chi square and Fisher's exact test are needed to analyse nominal data, non-parametric statistics such as Spearman's rank test and Mann–Whitney can be used for ordinal data, and a full range of parametric statistics such as *t*-tests, ANOVA, and regression, can usually be used on ratio data. Generally, the power of the analysis is highest for parametric statistics.

3.6.1 Collection and use of direct data.

The majority of surveys are based on direct data, when species are seen or heard. **Sight** is important for the vast majority of target species and survey types. However, sound pattern identification can be important when surveying vocal species; for example, song recognition is extremely important when surveying birds. This is especially true in dense habitat, such as reed beds, when it is difficult to see species but comparatively easy to hear vocalizations.

For other species, such as bats, **sound** is even more important. Many bats hunt using echolocation: producing high-frequency sounds and listening to the echo to locate prey. Bat detectors are small devices that transform these high frequency noises to a lower frequency that is audible to humans. Species can be identified in the field using the sound pattern (by ear) and call frequency. Alternatively, sound can be visualized using computer software such as BatScan or AnaLookW (Russ, 2012).

Occasionally, other senses are used, for example, **smell**. This can be particularly useful for identifying plants, especially those that produce essential oils such as rosemary *Rosmarinus officinalis*, lavender *Lavandula spica*, and eucalyptus *Eucalyptus obliqua*, but can also be used for other species. For example, the fungus *Tricholoma virgatum* smells like mouldy cucumber, while the presence of foxes *Vulpes vulpes* can often be deduced from the musky smell in urination areas. An interesting extension of the successful use of smell in ecological surveying is the use of sniffer dogs to detect the scats of small Indian mongoose *Herpestes palustris* on Okinawa Island, Japan, where they are invasive (Fukuhara et al., 2010).

Basic species surveying methods

There are four basic types of survey, as outlined in Table 3.3. They have different functions, advantages, disadvantages, and biases, and part of the skill of an Applied Ecologist is to choose the right survey type to answer a specific question that is appropriate to the taxa and spatial scale concerned.

The methods involved in each survey technique listed in Table 3.3 are covered comprehensively in Sutherland (2006). It should be noted that there are many contexts where survey methods have to be adapted. For example, kick sampling for aquatic species can be impossible in deep water; bird territory mapping is only appropriate during the breeding season, so outside of this another technique would be needed; Longworth traps might need to be mounted on tethered rafts when surveying streamside small mammals, to prevent flooding if the water level rises.

Observational or invasive? Destructive or non-destructive?

The vast majority of species survey methods are **observational** and **non-destructive**. This means that the surveyor uses sight, sound, or smell without trapping,

Table 3.3 The four main sampling approaches used in ecological surveying with some examples of the taxa for which they are appropriate.

Sampling approach	Specific examples	
Point sampling Species at or near a specific point are recorded. This encompasses situations where individuals are trapped. Points can be based on actual occurrence (e.g. nests) or chosen randomly, systematically, or strategically.	**Vegetation** Point frame **Insects** Pitfall traps (ground-dwelling species) Sweep netting (field layer species) Vacuum sampling (field layer species) Bait attractant traps (butterflies, flies) Water traps (flies, bees/wasps) Flight interception trap (flying insects) Malaise trap (flying insects) Light traps (moths) Kick sampling (aquatic species) Tullgren funnel (leaf letter/soil species)	**Birds** Point counts Nest records **Mammals** Longworth/Sherman traps (small mammals) Nest records (dormice) Roost emergence (bats) **Reptiles/Amphibians** Terrestrial—use of refugia (reptiles) Terrestrial—pitfall buckets (reptiles/amphibians) Aquatic—bottle traps (newts) Aquatic—lamping (amphibians) Aquatic—netting (amphibians)
Quadrat sampling Species abundance in a specific defined area is recorded. Usually focuses on percentage coverage; sometimes prevalence (number of sub-squares in which a species is present).	**Vegetation** Quadrat (size depends on vegetation type from 25 cm*25 cm for mosses to 50 m*50 m for trees; can be nested where vegetation is layered) **Lichen** Coverage of trees or rocks (see Section 4.3.2 for more details)	**Coral** Underwater quadrat **Algae** Seaweed in intertidal zone (usually 2 m*2 m) **Rocky shore molluscs** Limpets, chitons, etc.
Transect sampling Species are recorded as the surveyor moves through the site on a predetermined route. Transects are usually walked, but can be undertaken by car, aeroplane, or boat.	**Line transects** Recording is undertaken continuously. Useful for linear habitats, such as rivers or hedgerows, or sampling across a habitat or environmental gradient. Also used by some national monitoring schemes (e.g. Butterfly Monitoring Scheme in UK).	**Point transects** Recording is undertaken at specific points along the route, sometimes using quadrats: Bats (activity surveys) Vegetation change along environmental gradient
Mapping Recording the spatial location of species.	**Vegetation** Phase One habitat mapping (Section 3.5.1) National vegetation classification maps Field-based mapping (walk-over survey) Maps from remote sensing data (Section 7.3.3)	**Birds** Territory mapping

restraining, or killing survey species. Such methods are the most straightforward in terms of ethics and legal considerations.

Some direct methods do involve trapping, for example using Longworth traps to catch small mammals, mist nets to catch birds, harp nets to catch bats, or vacuums to catch field layer invertebrates. These techniques are all **invasive surveying methods**, since individuals are captured and manipulated. However, assuming that the fieldwork is done following best-practice protocols, there should be virtually no injury or mortality risk. Despite this, licenses might still be required for some surveys, depending on species and country. These permit activities that would otherwise be illegal under the environmental legislation of that country. For example, in the UK, a license is needed to survey legally protected species, such as dormice

Muscardinus avellanarius bats when handling or roost filming is involved, or great crested newts *Triturus cristatus* all of which are protected under the Wildlife and Countryside Act 1981.

There are some survey methods that are inherently **destructive.** An example is wet pitfall trapping, where invertebrates are killed and preserved, usually using ethanol. Other methods are not themselves destructive, but destructive methods are necessary to identify the species. For example, many invertebrates are only identifiable following dissection under the microscope. Obviously, destructive sampling, or killing of samples, is always something that needs careful consideration in terms of necessity, implications, ethics, and legality.

Capture–mark–recapture

One technique that is invasive but not destructive is **capture–mark–recapture (CMR)** (also called mark–release–recapture), which is used to estimate population size in numerous taxa, including mammals, fish, reptiles, and insects. CMR is often used to for temporal monitoring (Section 3.3.3) or to generate knowledge of an initial population to inform offtake or fish quotas in compliance monitoring (Section 3.3.5; Hot Topic 11.3).

The basic premise is to quantify population size using the ratio of marked to unmarked individuals. To do this, multiple individuals of the same species are captured and marked before being released back into the original, wild population. The marks do not necessarily have to identify one individual from another (individual marking), but they must indicate individuals that have been caught previously (batch marking). After release, the population is resampled by retrapping and the ratio of previously caught individuals to new individuals is recorded. Because this ratio in the sample should be the same as in the population as a whole, it can be used to estimate overall population size.

The simplest CMR method is the Lincoln–Peterson index, which only requires one capture stage, one mark stage, and one recapture stage. The equation is:

$$N = (((M + 1)*(C + 1))/(R + 1)) - 1$$

where N is the number of individuals in the population; M is the number of individuals captured initially and *marked*; C is the total number of individuals *captured* in the second sample; and R is the number of individuals *recaptured* (i.e. the number of marked individuals found in the second sample). So, if 20 individuals were captured and marked initially, and in the second capture 18 were caught of which 10 were marked, the equation would be:

$$N = (((20 + 1)*(18 + 1))/(10 + 1)) - 1$$

$$N = 35.27$$

There are several assumptions with this approach. The major assumption is that there is a closed population—in other words, there is no immigration (individuals entering the population) or emigration (individuals leaving the population). It also assumes that N is stable so either there is no reproduction or mortality, or the birth and death rates are equal and so cancel one another out.

Some populations can conform to this (see the example of red-eared terrapins *Trechemys scripta elegans* in Case Study 3.2), but this is very unusual. Accordingly, more complex methods have been developed with multiple recapture periods. These usually have higher accuracy and can be used when different individuals have different 'trapability' (e.g. if juveniles have a different likelihood of entering traps relative to adults: Burnham and Overton method). There are specialist methods for open populations including the Jolly–Seber method (Sutherland, 2006).

Another key assumption is that the initial trapping and marking process does not **bias** the second sample. This means that individuals should not be more likely to be trapped (trap-happy), which could happen when traps are baited if individuals associate traps with food, or less likely to be trapped (trap-shy), which could happen if individuals associate traps with stress. Trap-shyness, which results in overestimates of population size, can be reduced by reducing trap stress. Reducing trap-happiness, which results in underestimates of population size, is harder, but it is possible to reduce its effect by prebaiting with traps unset so animals will not be caught for a while before the first catch is made, so

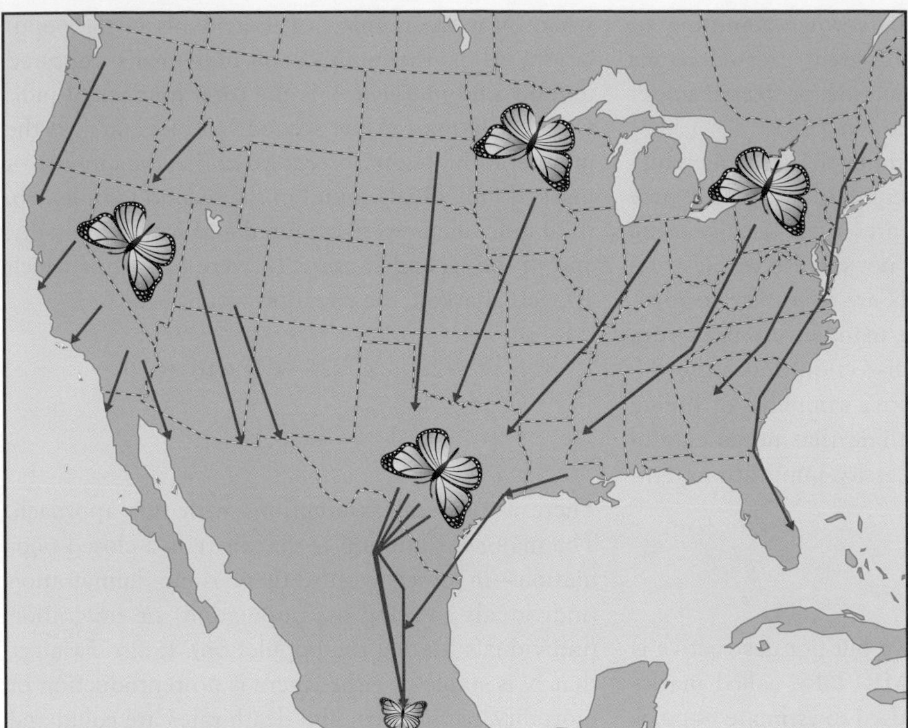

Figure 3.13 The autumn migration pathway of monarch butterflies in the US.
Source: Based on tag recoveries complied by MonarchWatch.org

all animals are equally trap-happy before catching starts.

The type of tagging mark depends on species and length of study, as well as ethical and legal considerations. Marks are either permanent/temporary and unique/batch. Examples of **permanent** unique marks are bird leg rings and mammal ear tags, as well as tags for insect wings as used to study movement patterns of monarch butterflies in America, shown in Figure 3.13 (spatial surveys—Section 3.3.2). Examples of **temporary** tags include using non-toxic dye or removing a small area of fur. It is important to ensure that the mark does not increase predation risk or affect mate choice.

3.6.2 Collection and use of indirect data.

Gathering direct data on species is not always as straightforward as it might sound as some species are particularly hard to survey. For example:

- Some species are **nocturnal**: many bats, land mammals such as the aye-aye *Daubentonia madagascariensis* and caracal *Caracal caracal*, and insects such as oil beetles *Meloe*.
- Other species are **crepuscular** (from the Latin for twilight) and are therefore most active at dawn (matutinal species) and/or dusk (vespertine species). Examples are mammals such as the European rabbit *Oryctolagus cuniculus*, spotted hyena *Crocuta crocuta,* and African wild dog *Lycaon pictus*, insects such as some species of mosquito Culicidae, and birds such as woodcock *Scolopax rusticola.*
- Some species occur in dense habitat, often being secretive and prone to disturbance. Many such species also use **camouflage** very effectively.

Indirect data can be extremely helpful in confirming species presence of the hard-to-survey species. Indirect evidence can take many forms, as outlined in Table 3.4.

Table 3.4 Types of indirect evidence used in species surveying with examples.

Type of evidence	Examples
Physical Physical signs of species being in a location	• Body remains (including road kill) and skeletal remains • Eggs • Hair (mammals)/feathers (birds) • Skins (snake)
Physiological Signs left as a result of physiological processes, such as eating and digestion	• Feeding remains—characteristically eaten seeds or nuts; depredated remains • Feeding evidence—damage to living plants (e.g. tree damage by deer; leaf damage by insects) • Ground disturbance when foraging, for example, uprooting turf by wild boar • Faeces—droppings, spraints, dung • Pellets (owls + birds of prey—can provide information on presence of prey species too)
Behavioural Indirect evidence that comes as the result of a species' behaviour	• Breeding areas—nests • Roosting or sleeping areas • Footprints • Trails/pathways through vegetation • Earthworks—setts, dens, holts, etc.

Indirect evidence can be collected through specific surveys. **Hair tubes** (basically a piece of pipe with some bait inside and some sticky tape along the top) can be used to collect hair, while **footprint traps** (tubes containing bait in the middle, an ink pad near the entrance, and paper that any animal must cross, with inky paws, to reach the bait) can also be useful. However, most indirect evidence is collected in an *ad hoc* or casual way by simply recording, for example, hair caught under fences, the presence of an animal den, or footprints in a muddy area.

Indirect evidence is an excellent way of confirming species presence at a site, but is of limited use for providing abundance estimates. As no live individuals are seen, it is not usually possible to say whether multiple footprints were made by multiple individuals, or just one individual crossing and recrossing an area. One recent way of getting around this issue is demonstrated by Proctor et al. (2010), who extracted **DNA** from hair caught on barbed wire hair traps. This allowed them to uniquely identify 1412 individual grizzly bears *Ursus arctos* occurring in different area of British Columbia and in Alberta, and thus attain abundance and density estimates.

The power of DNA is also being harnessed by a cutting-edge technique of sampling environments to find traces of DNA of species that have been in the area. This is called environmental DNA or eDNA and is discussed in more detail in Hot Topic 3.3.

3.6.3 Collection and use of remote data.

Remote collection of data can be useful to confirm species presence or examine patterns in abundance over time of space. The main method uses remote cameras. Depending on the specification, these might record continuously or be set so that a photograph or a short video sequence is recorded at the approach of an animal, using motion sensitivity technology. The advantage of **camera trapping** is that animals are not captured or disturbed in any way, making this a perfect example of non-destructive and non-invasive sampling (Section 3.6.3). Camera traps operate continually and silently, and many have a night mode using an infrared light source that is not usually detectable by mammals or birds.

Camera traps are excellent for confirming species presence (see examples for Asian tapir *Tapirus indicus* in Case Study 3.2). When used for species where individuals have unique features or have previously been captured and marked in some way, camera trap data can also be used to allow abundance to be estimated (see example for red-eared terrapins *Trachemys scripta elegans* in Case Study 3.2). Camera traps can also be used to verify provisional identification of footprints so that indirect evidence (Section 3.6.2 and Section 3.6.3) can be interpreted correctly. Sites can be baited, for example, using meat for carnivores, or olfactory (smell) cues such as ammonia or overripe fruit.

HOT TOPIC 3.3

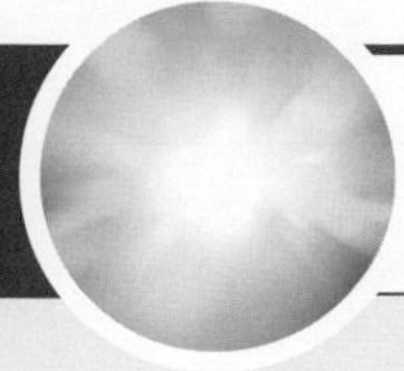

eDNA

A recent addition to the survey and monitoring tools for Applied Ecologists is use of environmental DNA (eDNA). As species interact with their environment, they will continuously expel DNA to their surroundings. This can come from biological material, such as urine, faeces, hairs, skin, feathers, or sweat, and through the process of decomposition. This means that it can be possible, through **molecular analysis**, to establish which species have been in an area through these trace fragments.

The eDNA approach involves finding traces of the DNA of target species in the wider environment as a form of indirect evidence. The technique has recently been used successfully to detect and monitor not only common species, but also those that are endangered, invasive, or elusive (Bohmann et al., 2014).

Samples are collected through **environmental sampling** (water samples, vegetation samples, soil samples, etc.). The DNA present in the sample is then extracted using a process that is suitable for the type of sample. The DNA extracted will be total DNA (including genomic DNA from chromosomes and mitochondrial DNA passed directly down the maternal line) from all species that have interacted with that environment to the point that their DNA is transferred to that environment. This is effectively a DNA 'soup' and, on its own, is not much use.

The next stage involves detecting the target DNA from the rest of the DNA 'soup' to determine whether it is there or not. This is done through a process called **polymerase chain reaction (PCR).** PCR works by amplifying one part of the DNA so that it dominates the sample and can thus be extracted from the sample in isolation. This is done by creating multiple copies of a specific gene (or DNA region) of the target species (Figure A).

A PCR reaction contains several ingredients, the main ones being:

1. The DNA soup—if the target species has interacted with the environment, that gene should be part of the DNA soup. This is referred to as the target sequence.
2. A solution containing free base nucleotides—in other words, lots of free A, G, C, and T nucleotides in solution, not attached to one another or anything else. These can be thought of as building blocks that can be used in different ways and in different sequences to create different patterns.
3. Two **specific primers** (one forward and one reverse)—that bind each of the two strands of the DNA and act as a starting point for the synthesis of new strands. Following the building block analogy, these can be thought of as the plan for the foundations of whatever structure is being built.

When heat is applied to the PCR reaction, the DNA of the target sequence denatures. This means that the two intertwined strands of DNA split apart to form two incomplete sides of the ladder. It is then possible, as the reaction cools, for one specially designed primer to attach to one strand of the target sequence and the other specially designed primer to attach to the other strand. The whole target sequence might be, for example, 600 nucleotides long. The primers might only be, say, 30 nucleotides long, but they are designed to fit to the first part of strand one (the forward primer) and the last part of strand two (the reverse primer). For this to work, the target sequence needs to be known in advance so that primers can be designed. A primer is basically the reverse of the actual sequence, so a primer for the sequence CCGATCGAT would be GGCTAGCTA (because nucleotide C binds to G and nucleotide A binds to T). This means that the DNA of the target species—or a portion of it at least—needs to have been sequenced in advance.

Once the primers have attached, the free nucleotides use this foundation to fill in the gaps so that the target sequence is reassembled. The key thing here is that there are now two copies of this target sequence—the original one was split into two incomplete parts and the other ingredients in the PCR reaction have acted together to replicate both of these. If this cycle is repeated, the two copies of the target sequence will split apart into four incomplete strands and the PCR ingredients will be used again to create four complete target sequences. This can then be repeated again and again, each time doubling the number of strands of target sequence. If the process is completed 30 times, as is typical, there will be over one billion copies of that target sequence (1,073,741,824 to be precise).

If the technique works, if the target section is amplified, this is good evidence that the species was in the environment from where the sample was taken (and can be confirmed if the amplified section of the DNA is extracted and sequenced). If the technique is known to work and no amplification happens (a **null result**), this suggests that the species was not present (or at least not for a long period of time or in high enough numbers to leave DNA). To have confidence in a null result, the technique must be known to work for that specific target sequence in the type of sample

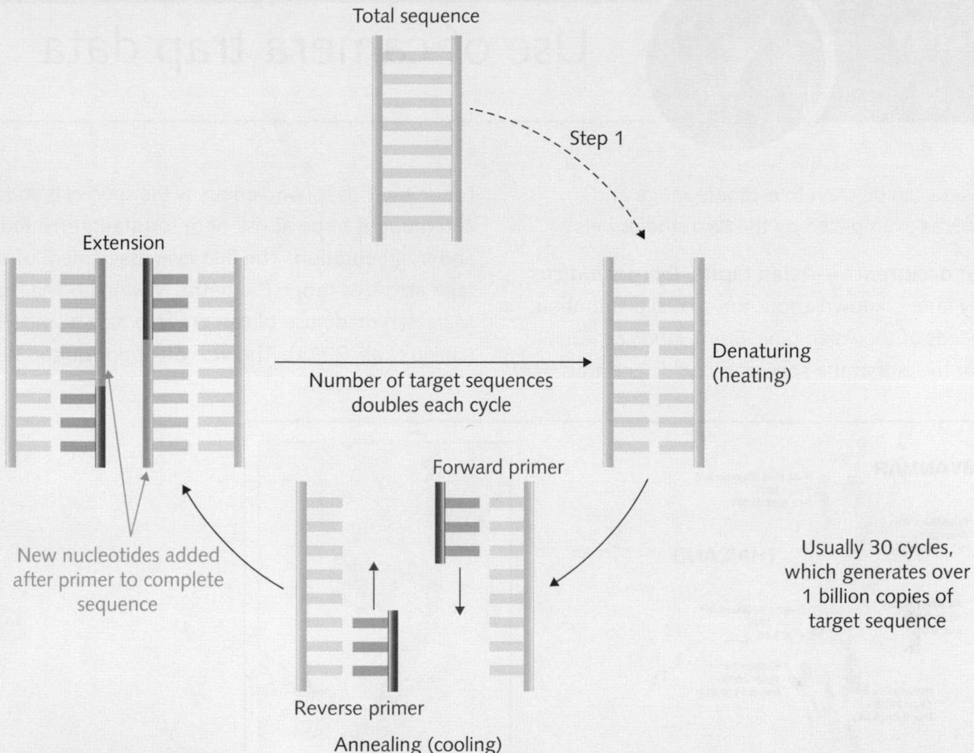

Figure A The process of polymerase chain reaction (PCR).

used, otherwise it is possible that the species is not being detected because the PCR is not working properly.

The eDNA technique is particularly useful in baseline studies to detect species, as well as in spatio-temporal monitoring to detect changes in where species occur over time. The technique has been used in a range of different environments and sample types, for example:

Seawater: detection of harbour porpoise *Phocoena phocoena* and long-finned pilot whale *Globicephala melas* detected in the western Baltic (Foote et al., 2012)

Freshwater: detection of non-native American bullfrog *Lithobates catesbeianus* as a form of alert monitoring (Dejean et al., 2012)

Browsed or grazed plants: detection of herbivores such as moose *Alces alces*, red deer *Cervus elaphus*, and roe deer *Capreolus capreolus*, from saliva on plant remains (Nichols et al., 2012)

Honey: plants in an area used by bees as nectar sources (Schnell et al., 2010)

QUESTION

How might Applied Ecologists make more use or better use of eDNA to inform practice?

REFERENCES

Bohmann, K., Evans, A., Gilbert, M.T., Carvalho, G.R., Creer, S., Knapp, M., Yu, D.W., & de Bruyn, M. (2014) Environmental DNA for wildlife biology and biodiversity monitoring. *Trends in Ecology & Evolution*, Volume 29, 358–367.

Dejean, T., Valentini, A., Miquel, C., Taberlet, P., Bellemain, E., & Miaud, C. (2012) Improved detection of an alien invasive species through environmental DNA barcoding: the example of the American bullfrog *Lithobates catesbeianus*. *Journal of Applied Ecology*, Volume 49, 953–959.

Foote, A.D., Thomsen, P.F., Sveegaard, S., Wahlberg, M., Kielgast, J., Kyhn, L.A., Salling, A.B., Galatius, A., Orlando, L., Thomas, M., & Gilbert, T.P. (2012) Investigating the potential use of environmental DNA (eDNA) for genetic monitoring of marine mammals. *PLoS One*, Volume 7, e41781.

Nichols, R.V., Koenigsson, H, Danell, K., & Spong, G. (2012) Browsed twig environmental DNA: diagnostic PCR to identify ungulate species. *Molecular Ecology Resources*, Volume 12, 983–989.

Schnell, I.B., Fraser, M., Willerslev, E., & Gilbert, M.T.P. (2010) Characterisation of insect and plant origins using DNA extracted from small volumes of bee honey. *Arthropod-Plant Interactions*, Volume 4, 107–116.

CASE STUDY 3.2

Use of camera trap data

Camera trap data can be used to estimate range and population size, as exemplified by the two studies below.

Probability of occurrence—Asian tapirs *Tapirus indicus*
Comparatively little is known about the spatial distribution and habitat needs of the Asian tapir *Tapirus indicus*. One of the reasons for this is that the species is cryptic and thus hard to survey. This is concerning as the species is endangered, and thought to be at risk of deforestation-related habitat loss and fragmentation. The first ever assessment of the Asian tapir across its range (Sumatra, Myanmar, Thailand, and Malaysia) made use of camera trap data from 1997 to 2011 (Linkie et al., 2013). This was an opportunistic study, as the

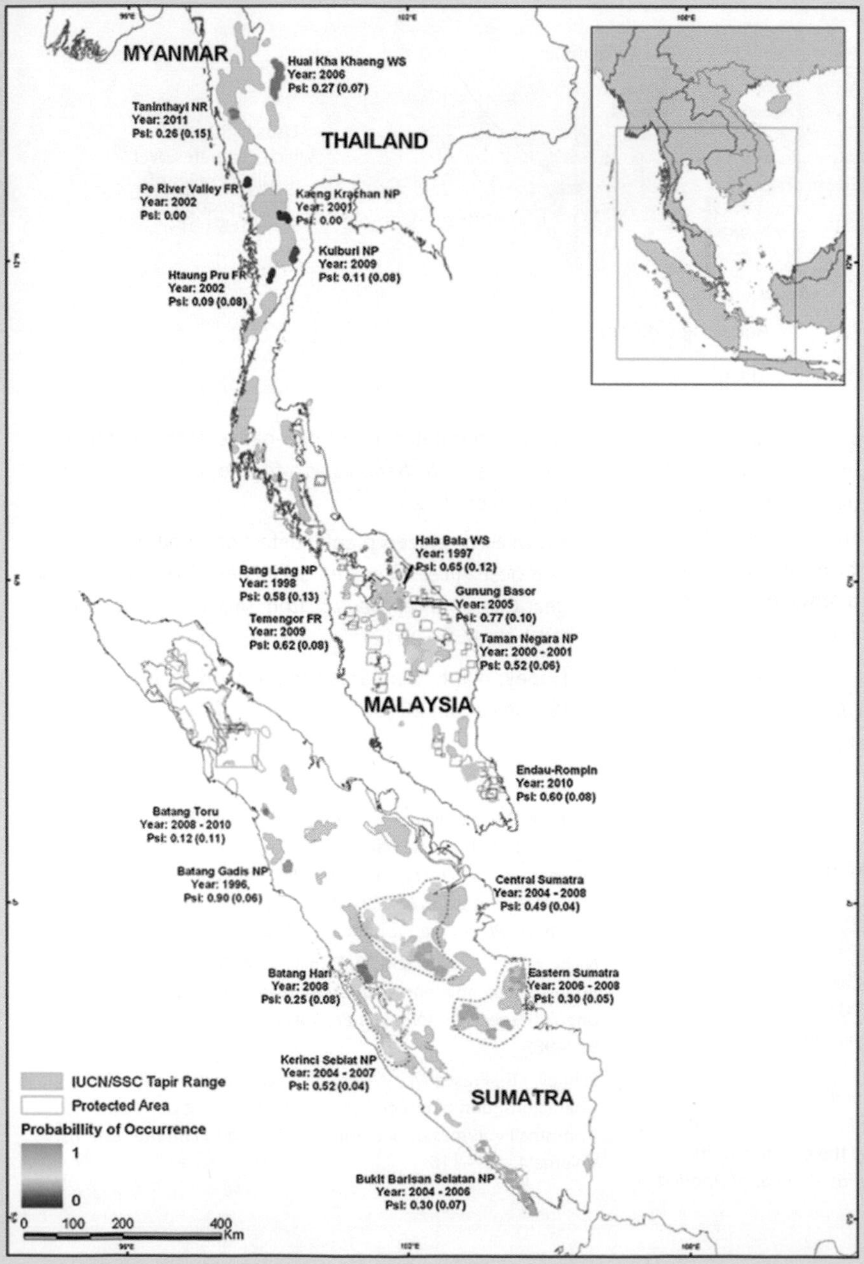

Figure A Probability of occurrence of Asian tapirs *Tapirus indicus* based on camera trap data.

Source: Reproduced from Linkie, M., et al. (2013) 'Cryptic mammals caught on camera: Assessing the utility of range wide camera trap data for conserving the endangered Asian tapir'. *Biological Conservation*, 162, pp. 107–115, with permission from Elsevier.

Figure B Red-eared terrapin, *Trachemys scripta elegans*.
Source: Photograph by Anne Goodenough.

camera trap surveys had generally been for other purposes, including surveying tigers *Panthera tigris*.

There were 1128 camera trap sites in total, and tapirs were recorded at 295, including 180 sites outside the known range as per Medici et al. (2003). The probability of tapir occurrence was mapped from this camera trap data (Figure A). This showed that occurrence probability was lower in Myanmar, where tapirs had been subject to poaching within the preceding years, and highest in Malaysia and Sumatra. There were also no detections at two key areas that were thought to have tapir populations—Pe River Valley in Myanmar and Kaeng Krachan National Park in Thailand. It should be remembered that absence of detection does not equal absence (it could be that the species was present and not detected, in which case this is a false absence, rather than a true absence), but this does highlight the need for additional surveys to be undertaken in these locations.

Population size—terrapins *Trachemys scripta elegans*

Camera trap data were used to determine population size of red-eared terrapins *Trachemys scripta elegans* in Illinois, USA (Bluett & Schauber, 2014; Figure B). Camera traps were placed on man-made rafts that the terrapins used for basking. The population of 25 individuals all had unique markings. In a 20-day period, 114 detections were made, of which 110 yielded photos where the terrapin could be identified. In total, 23 terrapins from the 25 known to be at the site were detected. This shows how a capture–mark–resight protocol (Section 7.3.2) can be used in a semi-automated way using camera trap data, rather than direct observations.

REFERENCES

Bluett, R.D. & Schauber, E.M. (2014) Estimating abundance of adult *Trachemys scripta* with camera traps: accuracy, precision and probabilities of capture for a closed population. *Transactions of the Illinois State Academy of Science*, Volume 107, 19–24.

Linkie, M., Guillera-Arroita, G., Smith, J., Ario, A., Bertagnolio, G., Cheong, F., Clements, G.R., Dinata, Y., Duangchantrasiri, S., Fredriksson, G., Gumal, M.T., Horng, L.S., Kawanishi, K., Khakim, F.R., Kinnaird, M.F., Kiswayadi, D., Lubis, A.H., Lynam, A.J., Maryati, Maung, M, Ngoprasert, D, Novarino, W, O'Brien, T.G., Parakkasi, K, Peters, H, Priatna, D, Rayan, D.M., Seuaturien, N. Shwe, N.M., Steinmetz, R., Sugesti, A.M., Sunarto, Sunquist, M.E., Umponjan, M., Wibisono, H.T., Wong, C.C.T., & Zulfahmi, (2013) Cryptic mammals caught on camera: assessing the utility of range wide camera trap data for conserving the endangered Asian tapir. *Biological Conservation*, Volume 162, 107–115.

Medici, E.P., Lynam, A., Boonratana, R., Kawanishi, K., Hawa Yatim, S., Traeholt, C., Holst, B., & Miller, P.S. (2003) *Malay Tapir Conservation Workshop Final Report*. IUCN/SSC Conservation Breeding Specialist Group, Apple Valley, USA. Available at: http://www.tapirs.org/Downloads/action-plan/malay-tapir-workshop-report.pdf

Camera traps should be deployed in situations where they are more likely to capture species activity. Depending on the species and the location, this might be:

- Near a known den or feeding area, or at a water hole.
- Along a pathway or wildlife corridor.
- Along a boundary, such as a hedge or ditch.
- At a site where there are unconfirmed sightings or indirect evidence of species' presence.

In a similar approach to using cameras, sound loggers can be used in **acoustic surveying.** These can be as simple as a microphone in the case of amphibians, such as frogs and geckos. More sophisticated examples include surveying bats with automated sound detectors, such as Anabats™. These units record sound, but only when ultrasonic sound is detected, in much the same way that motion-sensitive camera traps only record images when there is movement. The resultant sound files can be analysed digitally to discount non-bat records such as wind noise, and identify the bat species.

Techniques used to monitor individuals, such as satellite tracking (automated) and radio tracking (semi-automated) will be considered in Chapter 7.

3.6.4 Key surveying principles and links back to theory.

Whatever the method of surveying, there are several key principles:

- Ensure surveying takes place at an **appropriate time of year.** Many species are migratory, others hibernate. For many plant species, even if there is still above-ground growth in winter, species identification can be challenging without leaves or flowers/fruit.
- Ensure surveying is undertaken at the **best time of day.** All species have a circadian rhythm (circa = about; dian = corruption of diem = day). This determines when an individual undertakes specific activities, such as vocalizing and feeding, which in turn affects how conspicuous and detectable it is. Some methods can only be used at particular times (e.g. 'lamping' a pond for amphibians has to occur at night). Be aware that even within a taxonomic group there can be differences in the time that specific species are around. For example, bat species often have different emergence times relative to sunset to reduce inter specific competition (competition between species).
- **Weather conditions** can influence behaviour, so it is vital that weather does not prevent any behaviour necessary for the survey to work. In many butterfly survey protocols, for example, a minimum temperature is stipulated to ensure butterflies are on the wing (and thus visible). There can be practical considerations too, such as it being hard to hear bird calls in strong wind.
- Increase accuracy, precision, and representativeness by ensuring that there is **adequate replication** in both time and space. That typically means having multiple sample units (i.e. multiple points, transects, or quadrats) and surveying each one more than once. Indeed, some techniques, such as bird territory mapping, cannot be applied without replication over time. Attaining a large enough sample size is crucial. It is also necessary to ensure that surveyors have appropriate skills and training to ensure data accuracy.
- Maximize the chances of recording all individuals present, while **minimizing double-recording** (i.e. inadvertently recording the same individual more than once). Generally, the number of individuals recorded increases with survey effort. Based on this, it might be thought that it is best to maximize survey time. However, this is not necessarily true because the risk of double recording also increases; this is especially true for species that are highly mobile, such as birds (unless they are uniquely marked).
- Use of a **consistent survey effort** over time and space. Standardizing methods minimizes inter-observer variability (variation in data collected by different people) and means results are directly comparable.

FURTHER READING FOR THIS SECTION

Examples of species monitoring worldwide: Pereira, H.M., Belnap, J., Brummitt, N., Collen, B., Ding, H., Gonzalez-Espinosa, M., Gregory, R.D., Honrado, J., Jongman, R.H.G., Julliard, R., McRae, L., Proença, V., Rodrigues, P., Opige, M., Rodriguez, J.P., Schmeller, D.S., van Swaay, C., & Vieira, C. (2010) Global biodiversity monitoring. *Frontiers in Ecology and the Environment*, Volume 8, 459–460.

Species surveying techniques: Sutherland, W. (2006) *Ecological Census Techniques*. Cambridge: Cambridge University Press.

Capture-mark-recapture methods: Amstrup, S.C., McDonald, T.L., & Manly, B.F.J. (2010) *Handbook of Capture-Recapture Analysis.* Princeton: Princeton University Press.

Using indirect evidence in ecological surveying (tigers in India): Karanth, K.U., Gopalaswamy, A.M., Kumar, N.S., Vaidyanathan, S., Nichols, J.D., & MacKenzie, D.I. (2011) Monitoring carnivore populations at the landscape scale: occupancy modelling of tigers from sign surveys. *Journal of Applied Ecology*, Volume 48, 1048–1056.

Overview of camera trapping: Rovero, F., Zimmermann, F., Berzi, D., & Meek, P. (2013) 'Which camera trap type and how many do I need?' A review of camera features and study designs for a range of wildlife research applications. *Hystrix, the Italian Journal of Mammalogy*, Volume 24, 148–156.

Introduction to statistical analysis: Dytham, K. (2010) *Choosing and Using Statistics: a Biologist's Guide.* Oxford: Wiley-Blackwell.

3.7 From monitoring to management.

Monitoring and management should go hand-in-hand. The initial need for management can be highlighted by baseline surveying, while the need to change management can be shown by temporal monitoring. It is simply not possible to manage environments, ecosystems, habitats or species without first understanding what is there, how it functions, and how it is affected by the surrounding landscape.

The relationship between monitoring and management should be two-way. Management interventions should be **evidenced-based** and monitoring-informed. Moreover, when any management measure is put in place, there needs to be ongoing monitoring to assess the effectiveness of that management intervention, and inform future management strategies to make management more effective (Section 11.7). The ideal process is shown diagrammatically in Figure 3.14. When management involves establishing legislation or codes of practice, compliance monitoring is often vital to ensure the effectiveness of such measures.

A management strategy employed to address one ecological challenge can have unintended impacts on other species. This might mean that it is appropriate to monitor change in other geographical areas or other taxonomic groups to determine whether any alteration to the original management would be beneficial. The Online Case Study for Chapter 3 gives an example of this in relation to excluding deer from certain areas, the effect on moss–lichen competition, and ultimately its effect on community structure.

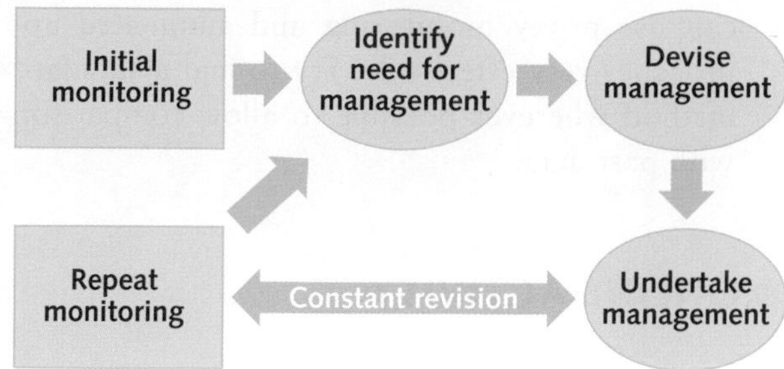

Figure 3.14 The relationship between monitoring and management.

You can find the online case study at www.oxfordtextbooks.co.uk/orc/goodenough

● FURTHER READING FOR THIS SECTION

Cost effectiveness of monitoring-informed management: Possingham, H.P., Wintle, B.A., Fuller, R.A. & Joseph, L.N. (2012) The conservation return on investment from ecological monitoring. In: D. B Lindenmayer & P. Gibbons (Eds) *Biodiversity Monitoring in Australia*, pp. 49–58. Melbourne: CSIRO Publishing.

3.8 Conclusions.

There are many different types of surveying and many different aspects of ecosystems that can be monitored. When designing a surveying or monitoring scheme, it is important that the objectives are well understood so that suitable data can be collected. This will involve balancing the work involved (number of samples, number of sites, sampling frequency, study duration) with time and budget constraints.

The main types of monitoring, and the main considerations when using them, are:

- **Baseline surveying:** must be as comprehensive as possible; can draw on lessons from other sites. Needs to be managed in such a way that maximum insight can be gained, which often involves combining both primary and secondary data, and both direct and indirect evidence.
- **Spatial surveying:** again needs to be standardized, in this case both over space and time. Patterns need to be examined for underlying bias, such as a greater number of records of a species being submitted in areas with higher human population.
- **Temporal monitoring:** needs to be standardized and consistent over time so any changes detected

are 'real', rather than being due to methodological bias. It should ideally be as continuous as possible rather than a series of snapshots, and can use proxy monitoring and automated approaches very effectively. Try to find a standard method wherever possible to allow comparison with past data.

- **Compliance monitoring:** a vital way of ensuring that ecological legislative measures are translated from paper to practice.
- **Impact/mitigation monitoring:** a key aspect of Ecological Impact Assessment frameworks (Sections 5.5 and 5.6) and link through into management of ecosystems, habitats, and species.

ONLINE ACTIVITY

Go to www.oxfordtextbooks.co.uk/orc/goodenough/ to download the activity that accompanies this chapter.

CHAPTER 3 AT A GLANCE: THE BIG PICTURE

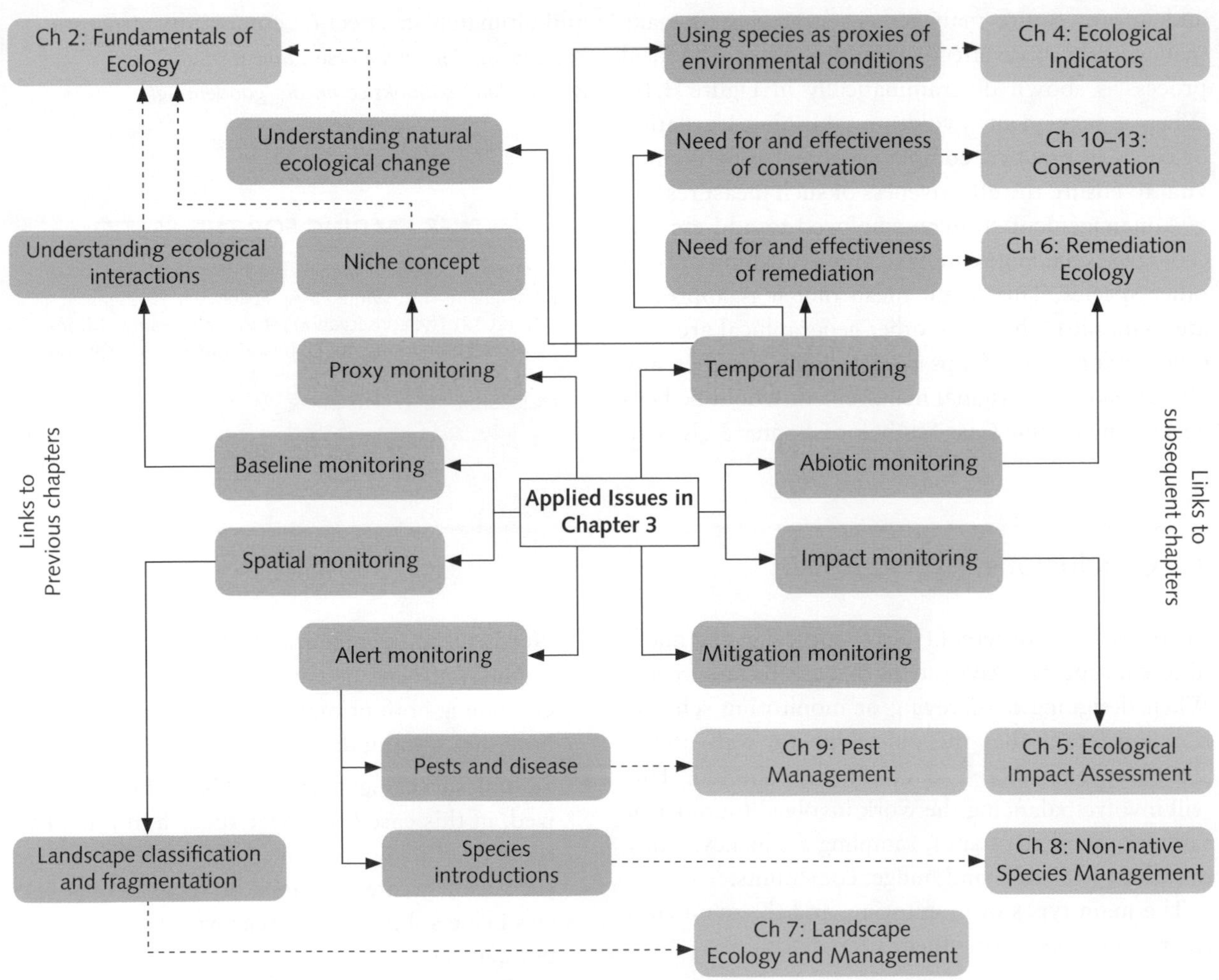

REFERENCES

Adams, K (2008) The status and distribution of crested cow-wheat *Melampyrum cristatum* L. in Britain, now largely confined to Essex. *Essex Naturalist*, Volume 25, 120–127.

Banzi, J.F. (2014) A sensor based anti-poaching system in Tanzania National Parks. *International Journal of Scientific and Research Publications*, Volume 4, 1–7.

Bat Conservation Trust (2014) Requirements for roost retention and creation. Available at: http://roost.bats.org.uk/principles/requirements-roost-retention-and-creation

Braithwaite M.E., Ellis R.W., & Preston C.D. (2006) *Change in the British Flora 1987–2004*. London: Botanical Society of the British Isles.

Braithwaite, M. & Walker, K. (2012) *50 Years of Mapping the British and Irish Flora 1962–2012*. London: Botanical Society of the British Isles.

Buse, A., Dury, S.J., Woodburn, R.J.W., Perrins, C.M. and Good, J.E.G. (1999) Effects of elevated temperature on multi-species interactions: the case of pedunculate oak, winter moth and tits. *Functional Ecology,* Volume 13, 74–82.

Campbell Scientific (2014) *San Diego: Slope Monitoring*. Available at: https://www.campbellsci.co.uk/slope-monitor

Catlin-Groves, C.L. (2012) The citizen science landscape: from volunteers to citizen sensors and beyond. *International Journal of Zoology*. Article ID 349630. Available at: http://www.hindawi.com/journals/ijz/2012/349630/

Clements, F.E. (1916) *Plant Succession: an Analysis of the Development of Vegetation*. Washington: Carnegie Institution of Washington.

Conway, G. (2010) *Constant Effort Sites Scheme: Instructions*. Available at: http://www.bto.org/sites/default/files/u17/downloads/ringing-surveys/CES/ces_instructions.pdf

Crabbe, M.J.C. (2012) From citizen science to policy development on the coral reefs of Jamaica. *International Journal of Zoology*. ID 102350.Available at: http://www.hindawi.com/journals/ijz/2012/102350/

European Environment Agency. (2014) *Terrestrial Habitat Mapping in Europe: an Overview*. Available at: http://spn.mnhn.fr/spn_rapports/archivage_rapports/2014/SPN%202014%20-%2010%20-%20Terrestrial_habitat_mapping_in_Europe_-_an_overview.pdf

Forestry Commission Scotland (2009) *Action for Chequered Skipper*. Available at: http://www.forestry.gov.uk/pdf/fcs-species-chequered-skipper.pdf/$FILE/fcs-species-chequered-skipper.pdf

Fukuhara, R., Yamaguchi, T., Ukuta, H., Roy, S., Tanaka, J., & Ogura, G. (2010) Development and introduction of detection dogs in surveying for scats of small Indian mongoose as invasive alien species. *Journal of Veterinary Behavior*, Volume 5, 101–111.

Gillingham, P.K., Palmer, S.C., Huntley, B., Kunin, W.E., Chipperfield, J.D., & Thomas, C.D. (2012) The relative importance of climate and habitat in determining the distributions of species at different spatial scales: a case study with ground beetles in Great Britain. *Ecography*, Volume 35, 831–838.

Goodenough, A.E., Hart, A.G., & Stafford, R. (2010) Is adjustment of breeding phenology keeping pace with the need for change? Linking observed response in woodland birds to changes in temperature and selection pressure. *Climatic Change*, Volume 102, 687–697.

Howard, E. & Davis A.K. (2009) The fall migration flyways of monarch butterflies in eastern North America revealed by citizen scientists. *Journal of Insect Conservation*, Volume 13, 279–286.

Irwin, A. (1995) *Citizen Science: a Study of People, Expertise and Sustainable Development*. London: Routledge.

Kershner, M.W. & Lodge, D.M. (1995) Effects of littoral habitat and fish predation on the distribution of an exotic crayfish, *Orconectes rusticus*. *Journal of the North American Benthological Society*, Volume 14, 414–422.

Lee, M.A., Davies, L., & Power, S.A. (2012) Effects of roads on adjacent plant community composition and ecosystem function: an example from three calcareous ecosystems. *Environmental Pollution*, Volume 163, 273–280.

Lips, K.R., Brem, F., Brenes, R., Reeve, J.D., Alford, R.A., Voyles, J., & Collins, J.P. (2006) Emerging infectious disease and the loss of biodiversity in a Neotropical amphibian community. *Proceedings of the National Academy of Sciences of the United States of America*, Volume 103, 3165–3170.

Losey, J.E., Perlman, J.E., & Hoebeke, E.R. (2007) Citizen scientist rediscovers rare nine-spotted lady beetle, *Coccinella novemnotata*, in eastern North America. *Journal of Insect Conservation*, Volume 11, 415–417.

Lotka, A.J. (1920) Analytical note on certain rhythmic relations in organic systems. *Proceedings of the National Academy of Sciences of the United States of America*, 6, 410–415.

Mijovic´, S. & Palmar, B. (2012) Water quality monitoring automation of rivers in Serbia. *Facta Universitatis Series: Working and Living Environmental Protection* Volume 9, 1–10.

Pérez, M.E., Jesús, S. D., Lugo, A.E., & Abelleira Martínez, O.J.(2012) Bryophyte species diversity in secondary forests dominated by the introduced species *Spathodea campanulata* in Puerto Rico. *Biotropica*, Volume 44, 763–770.

Pocock, M.J., Roy, H. E., Preston, C.D., & Roy, D.B. (2015) The Biological Records Centre: a pioneer of citizen science. *Biological Journal of the Linnaean Society*, Volume 115, 475–493.

Proctor, M., McLellan, B., Boulanger, J., Apps, C., Stenhouse, G., Paetkau, D., & Mowat, G. (2010) Ecological investigations of grizzly bears in Canada using DNA from hair, 1995–2005: a review of methods and progress. *Ursus*, Volume 21, 169–188.

Roberts, M.J., Schimmelpfennig, D., Ashley, E., & Livingston, M. (2006) *The Value of Plant Disease Early-warning Systems*. Available at: http://www.ers.usda.gov/media/330972/err18_1_.pdf

Rouse, S., Spencer Jones, M.E., & Porter, J.S. (2014) Spatial and temporal patterns of bryozoan distribution and diversity in the Scottish sea regions. *Marine Ecology*, Volume 35, 85–102.

Russ, J. (2012) *British Bat Calls: a Guide to Species Identification*. Exeter: Pelagic Publishing.

Silva, A.P., Kilshaw, K., Johnson, P.J., Macdonald, D.W., & Rosalino, L.M. (2013) Wildcat occurrence in Scotland: food really matters. *Diversity and Distributions*, Volume 19, 232–243.

Spellerberg, I. (2005) *Monitoring Ecological Change*. Cambridge: Cambridge University Press.

Sutherland, W. (2006) *Ecological Census Techniques*. Cambridge: Cambridge University Press.

Tansley A.G. (1939) *The British Islands and their Vegetation*. Cambridge: Cambridge University Press.

Vilchis, L.I., Johnson, C.K., Evenson, J.R., Pearson, S.F., Barry, K.L., Davidson, P., Raphael, M.G., & Gaydos, J.K. (2015) Assessing ecological correlates of marine bird declines to inform marine conservation. *Conservation Biology*, Volume 29, 154–163.

Volterra, V. (1926) Variazioni e fluttuazioni del numero d'individui in specie animali conviventi [Variation and fluctuation in the number of individuals of animal species cohabitants] *Memorie del Academia dei Lincei Roma*, **2**, 31–113

Vucetich, J.A. & Peterson R.O. (2012) *The Population Biology of Isle Royale Wolves and Moose: an Overview*. Available at: www.isleroyalewolf.org

Wilson, S., Anderson, E.M., Wilson, A.S., Bertram, D.F., & Arcese, P. (2013) Citizen science reveals an extensive shift in the winter distribution of migratory western grebes. *PLoS One*, Volume 8, e65408.

Ecological Indicators

4.1 Introduction.

Central to ecology is the concept that species interact, both with each other (their biotic environment) and their physical abiotic environment. With increasing knowledge and understanding of these interactions comes the potential for Applied Ecologists to use the presence or abundance of species to indicate ecosystem characteristics or biological processes. This is the concept of **ecological indicators**—the use of specific indicator species, multiple species within a taxonomic group, or even a whole community of species, to indicate something about the chemical, physical, and biological parameters of the environment in which they are found.

The first recorded use of ecological indicators involved **animal sentinels**—the use of species to detect conditions dangerous to humans. For example, in natural environments, indigenous people have long used the warning calls of birds to alert them to potential danger. It is also possible to introduce animal sentinels to artificial environments; as long ago as the 1890s, canaries *Serinus* were taken into mines to warn miners of dangerously high levels of carbon monoxide through cessation of singing or death. Figure 4.1 shows this once-common practice, which possibly gave rise to the expression 'falling off the perch' as a euphemism for death.

Today, most of the monitoring that relates directly to human health uses technological approaches. However, the concept of using species as indicators of their environment, or ecological processes in that environment, is both widespread and extremely relevant. For example, aquatic invertebrates can be used to indicate dissolved oxygen levels in water bodies, plants can be used to estimate the pH and nutrient content of the soils in which they grow, and lichens can be used to infer levels of airborne pollutants. It is even possible to use pollen to indicate previous environmental conditions and infer past habitats and climates.

This chapter will explore the concepts behind ecological indicators, debate the attributes that make a species a good indicator, and consider the advantages and challenges of the approach. Because of the sheer number and diversity of species used as indicators, and the great many environmental variables they can indicate, this chapter will not be able to consider all indicator systems used globally. Instead, we will discuss a cross-section of indicator systems useful to Applied Ecologists to highlight key concepts, and demonstrate the theory that underpins their use.

Figure 4.1 It was common practice for miners to take canaries down mines as indicators of poor air.
Source: Jeff Morgan 03/Alamy Stock Photo.

4.2 Types of ecological indicator.

Species can be used as indicators in two main ways:

1. **Biosurveys**: these rely on the presence of specific species or communities in the real world to draw inferences about local conditions. For example, the presence of salmon *Salmo* in a river is indicative of good oxygenation, while the presence of stinging nettles *Urtica dioica* in a field is indicative of high soil fertility. Generally, biosurveys are based on species being observed, and deductions being made from those observations, rather than using a standardized method (see the 'Interview with an Applied Ecologist' for this chapter). Such an approach can be invaluable for providing rapid insights into environmental parameters without the need for highly specialized knowledge or equipment. However, because of the lack of standardized methods and protocols and the focus being on very simple, largely qualitative, data (presence or a general impression of abundance), the data generated are not particularly robust and can vary substantially between fieldworkers.

2. **Biotic indices**: these are designed specifically to use species communities (nearly always communities, rather than individual species) to answer key questions about an ecosystem. Biotic indices usually involve the use of a **standardized** or semi-standardized method to collect data. They are also usually **numerical** in some way. Normally, a formal protocol is used to calculate a score, which is then compared with a predetermined standard scale to aid interpretation. An example is the Estuarine Fish Biotic Index that was developed for Belgium in 2007 to define the quality status of an estuarine area (Breine et al., 2007). The final index was derived from an equation based on the species found within different age classes and dietary groupings—omnivorous (consuming both plant and animal food sources) and piscivorous (fish-eating).

AN INTERVIEW WITH AN APPLIED ECOLOGIST 4

Name: Alice Trevail

Organisation: University of Liverpool and South West Fulmar Project

Role: PhD student and Instigator, respectively

Nationality and main countries worked: British; worked in UK and Antarctica

What is your day-to-day job?

I am studying for my PhD in seabird ecology and oceanography at the University of Liverpool. My PhD aims to investigate the interactions between top marine predators and their environment. I am also interested in how seabirds can be used to monitor plastic pollution. I work together with researchers around Europe to quantify plastic ingestion by northern fulmars *Fulmarus glacialis* both in the North Atlantic and the Antarctic to inform international policy.

What is your most interesting recent project and why?

I set up the South West Fulmar Project as a geographic expansion of an existing North Sea collaboration to use seabirds as monitors of marine litter. We study northern fulmars because they forage solely at sea and feed ubiquitously on anything floating at the ocean surface, including plastic. Marine litter is broken down inside the stomach after approximately 1 month, and therefore quantifying ingested plastic can provide an interesting and current snapshot of oceanic pollution levels. Furthermore, fulmars are widely distributed and extensively monitored across the northern hemisphere, and consequently any new data can instantly form part of spatial and temporal comparisons.

What are the most satisfying parts of this project?

There are two elements of the project that I find most satisfying. First, the South West Fulmar Project has direct policy implications. The number of fulmars with plastic in their stomachs is written into EU environmental targets that our government is committed to achieving. Therefore, every bird found and dissected from our beaches can influence government action on marine litter mitigation.

Secondly, volunteers are inherent to the project as they collect the beached birds. I love being able to talk to passionate local people and conservation groups about the project, as well as their ongoing marine conservation work.

What's next for you, and why?

I've got three years left of my PhD, during which I am hoping to maximize my field seasons to collect as much valuable data as possible. I hope that my project will reveal insights into the origins of individual animal behaviour, as well as having conservation implications in the context of our changing climate and requirement for additional protected marine areas. Alongside this, I continue to run the South West Fulmar Project because long-term studies are most valuable for marine monitoring.

What do you see as the main challenges in your field and how can they be overcome?

Studies of indicator species are often directed towards achieving targets that signify 'good environmental status'. The challenge is thus defining what we actually mean by 'good' environmental status. It is also imperative that once reached, targets are re-evaluated, and the actual negative effects of anthropogenic activity on the environment are defined.

Finally, how did you get into species monitoring and what advice would you give to others?

For me, the application of my research towards an aim that I feel passionate about is important. This is reflected in my research projects from undergraduate onwards, which have incorporated species monitoring for policy and conservation. My advice to others would be to choose projects that interest you as you will be much more motivated to pursue them to a high level.

4.2.1 What can ecological indicators be used to indicate?

Species can be used to indicate many different abiotic and biotic parameters. Table 4.1 sets out some examples of the types of parameter that ecological indicators can be used to indicate, many of which are considered in greater depth later in the chapter.

4.2.2 Single- and multispecies approaches.

Some ecological indicator systems are based on the presence or abundance of a **single species**. For example, at a very simple level, if sea thrift *Armeria maritima* occurs in a given area, that area is usually close the sea (although it can occur on mountains). Other examples are less obvious; for example, the

Table 4.1 Examples of environmental, biological, and biodiversity indicators. Headings revised and extended from McGeoch (1998).

Indicators used to indicate environmental conditions (environmental indicators) (Section 4.3)	
Example 1: soil properties	Presence or abundance of plant species can be used to indicate soil parameters such as soil fertility, moisture, and pH.
Example 2: pollution	Presence of species that can tolerate high levels of specific pollutants (often coupled with absence of pollutant sensitive species). Lichens are often used.
Example 3: oxygen levels	Presence of species that can tolerate low levels of oxygen coupled with absence of species that require high oxygen levels. Invertebrates are often used.
Indicators used to indicate biological processes or ecosystem processes (biological indicators) (Section 4.4)	
Example 1: grazing pressure	The relative abundance of grasses in different functional groups to establish level of grazing pressure in savannah environments.
Example 2: habitat age	The use of the presence of slow-dispersing and disturbance-sensitive species to indicate ancient habitats. Used for ancient woodland in the UK.
Indicators used to indicate biodiversity parameters (biodiversity indicators) (Section 4.5)	
Example 1: species richness	The number of species in one taxonomic group can be used to indicate richness in other taxonomic groups at the same site. Beetles are a common proxy.
Example 2: sustainability	The number of species with increasing, decreasing, or stable population trends (or overall population trend from a group of species) can indicate ecological sustainability. Birds and bats are used in the UK.

plant cream milkvetch *Astragelus racemosus* which is native to the USA and Canada, has been used to indicate selenium concentration, while the abundance of the moss *Hylocomium splendens* has been used to indicate heavy metal pollution in Alaska, USA (Trelease & Trelease, 1938; Hasselbach et al., 2005).

As the name implies, single-species indicator systems focus on one species. For example, the lichen species *Usnea cornuta*, which is shown in Figure 4.2, is only present in clean-air areas. Thus, if it is present, the air must be clean enough to support it. This is the simplest type of indicator system, as it relies on **presence-only data** for one species. It should, however, be noted that absence of the species from an area does not necessarily mean that the air is heavily polluted—the species might be absent for other reasons.

Figure 4.2 *Usnea* is a genus of lichen that is indicative of clean air. Source: Photograph by Anne Goodenough.

A slightly more complex single-species framework involves focusing on one species relative to the overall community. For example, another lichen, *Hypogymnia physodes*, is used to identify areas of high atmospheric sulphur. The species can actually be present in any area, so its presence does not automatically indicate sulphur pollution. However, it tends to occur in only low numbers in clean-air areas where it is has to compete with many other species. Thus, if *H. physodes* is found in abundance and few other species are present, that is good evidence of high sulphur levels. This is because *H. physodes* can tolerate such conditions, whereas other species cannot. Despite still focusing on a single species, this environmental indicator framework is slightly more complex because the **relative abundance** of the focal species is used to infer environmental conditions.

Single-species frameworks are generally biosurveys, rather than full biotic indices, and species are usually being used as environmental indicators, rather than biological or biodiversity indicators (Table 4.1). Such frameworks also tend to use specialist species (species that only occur in certain habitats and under certain specific conditions: Section 2.6) because generalist species that can occur in a range of conditions are unlikely to be good proxies for specific environmental conditions (as shown in Figure 4.3).

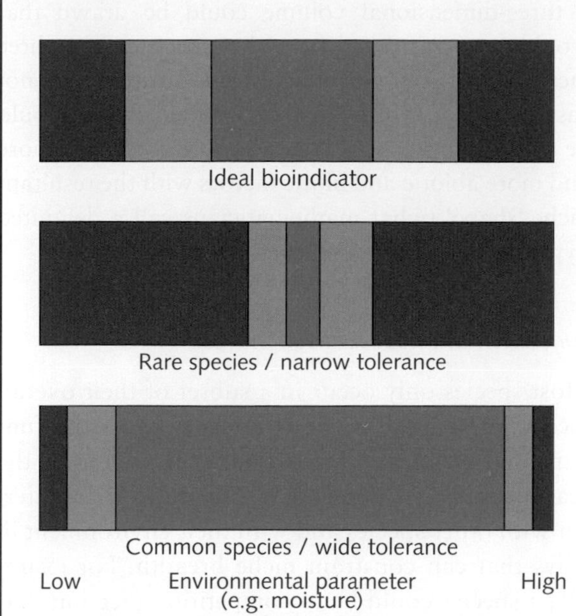

Figure 4.3 All species have tolerance range for numerous environmental parameters. Red areas represent portions of an environmental gradient (e.g. temperature) where a species can survive. The green zone includes the optimum range, the amber zone is a physiological stress zone where the species occurs in lower numbers, and the red zone is outside the species' survival range. Ideal bioindicators are not too specialized, as the tolerance range will be very small, and not too generalized, as the tolerance ranges will be very broad and thus not specific.

Although single-species indicators can sometimes be effective, full **community-based approaches**, which use the presence and/or abundance of multiple species, are often more robust. In most situations, results that are based on several lines of evidence (multiple species in this case) tend to be more robust than those based on just one line of evidence (a single species). This is referred to as a **multiproxy approach.** In indicator systems, using multiple species also means that the system is less prone to outliers or atypical results, such as when species occur outside their normal range. The use of multispecies approaches also makes it possible to ascertain the abundance of different species, which is central to the use of many biotic indices. Examples of community-based systems include the use of lichens to monitor air quality and aquatic invertebrates to indicate oxygen levels (Section 4.3.2).

FURTHER READING FOR THIS SECTION

General overview of bioindicators: Holt, E.A. & Miller, S.W (2010) Bioindicators: using organisms to measure environmental impacts. *Nature Education Knowledge*, Volume 3, 8.

General approaches to indicators using protists as an example: Payne, R.J. (2013) Seven reasons why protists make useful bioindicators. *Acta Protozoologica*, Volume 52, 105–113.

Selection of indicator species using terrestrial invertebrates as an example: McGeoch, M.A. (1998) The selection, testing and application of terrestrial insects as bioindicators. *Biological Reviews of the Cambridge Philosophical Society*, Volume 73, 181–201.

4.3 Environmental indicators.

Worldwide, hundreds of indicator systems have been devised, which are used to indicate all manner of physical and chemical environmental parameters. These span terrestrial, freshwater, and marine environments. This section provides a brief overview of the foundations upon which environmental indicators are founded, before outlining some examples of their use. Areas of current debate are also discussed.

4.3.1 Theory underpinning environmental indicators: niches and tolerance ranges.

Environmental indicator species can be used as indicators of the physical environment in which they occur (Holt & Miller, 2010). This concept is founded on the principle of **ecological niches** (Hutchinson, 1957), and the fact that all species have a **tolerance range** for a whole host of environmental parameters, as outlined in Section 2.5. For example, a plant species might only be able to survive in environments in which the temperature is between –5° and 30°C, which offer reasonably fertile soil that has a high level of available water for most of the year, and where light levels are sufficient to allow photosynthesis to occur for at least 6 hours per day during the growing season. The plant thus has a tolerance range for each individual parameter—temperature, moisture, soil fertility, light, and so on—within which it can survive and reproduce, but outside of which it cannot.

As shown in Figure 4.3, the tolerance range itself is split into two parts. Using temperature as an example, the central part of the overall tolerance range will be the **optimal range** where the species is able to thrive. Near to the extremes of the tolerance range (i.e. close to the minimum and maximum temperatures in this example), there will be physiological stress. This means that normal physiological processes, such as photosynthesis and respiration, are occurring under stress and either demand more energy, or are less effective, than under ideal conditions. A species can survive under these conditions, but it does so at the edge of its physiological capabilities. Population size is likely to be lower in this physiological stress zone than under optimal conditions because it is likely to be outcompeted by species better suited to such conditions.

If a species' niche is controlled by just two factors, say temperature and water availability, it is conceptually straightforward to define that niche. This can be visualized on a graph with two axes, where the x-axis is one factor (temperature) and the y-axis the other (water). It is possible, as shown in Figure 4.4, to draw a shape on the graph that encompasses the theoretical space with respect to temperature and water availability in which the hypothetical species can live. In other words, the shape would depict a two-dimensional niche.

It would then be possible to add in a third factor/dimension—for example, the amount of sunlight needed to survive. Returning to the graph, this third dimension would be represented by the z-axis, and a three-dimensional volume could be drawn that would represent the niche with respect to these three factors. This is shown in Figure 4.4. Although it is not easy to plot a fourth or higher dimension, it would be possible to keep adding axes to incorporate more and more abiotic and biotic factors with the resultant niche 'shape' (what mathematicians call a delimited hyper-volume) becoming ever more complex.

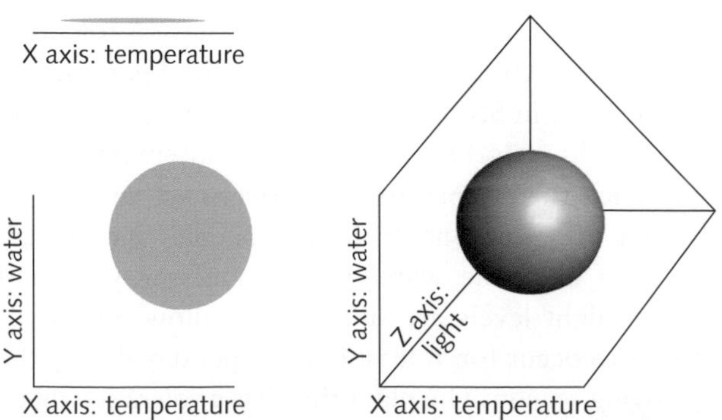

Figure 4.4 Hypothetical diagram of a one-dimensional niche (top left), a two-dimensional niche (bottom left), and a three-dimensional niche (right).

Fundamental versus realized niches

Most species only occur in a subset of their overall niche. The overall niche is referred to as the **fundamental niche**, and the subset is referred to as the **realized niche**. As noted in Section 2.5, species interact with other species and with their environment in ways that can constrain niche breadth. For example, a species could be in competition (Section 2.4) for a particular resource, which might mean its full fundamental niche cannot be exploited. Thus, while the fundamental niche is based on the physiological capabilities of the species concerned and species adaptability, the realized niche is based on physiology, adaptability and interactions with other species (and thus which other species are present and their relative competitive ability). Figure 4.5 explains this visually.

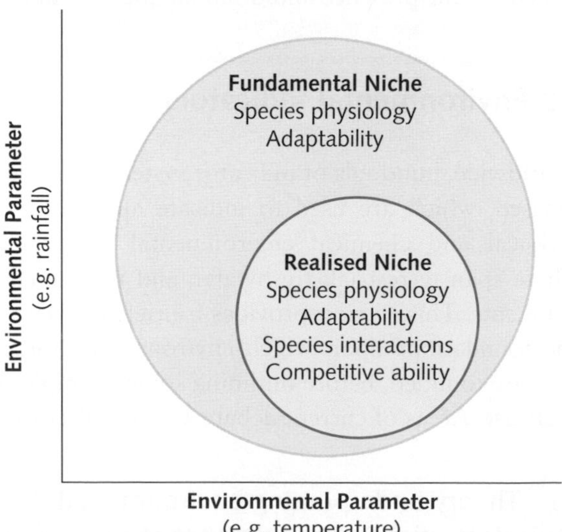

Figure 4.5 A species' realized niche (where it actually occurs) is almost always a subset of its fundamental niche (where it could occur physiologically).

4.3.2 Exemplar environmental indicators.

Once tolerance ranges are known, and fundamental and realized ecological niches are quantified, it is possible to use species presence or abundance as a proxy indicator for environmental conditions. The following subsections detail three different indicator systems. These have been chosen to highlight the diversity of taxa used as indicators (plants, lichens, and aquatic invertebrates) and the range of environmental parameters of which they can be indicative (soil properties, airborne chemical pollutants, and organic pollution of water).

All of the examples below are non-destructive. In other words, they involve observing and recording what species are present, sometimes taking account of abundance as well, without damaging those individuals in any way. However, there can be situations where environmental indicators involve tissue analysis or bioassays, both of which can be **destructive**. Whether such approaches can be justified is debated in Hot Topic 4.1.

Using plants to indicate soil properties

Many terrestrial plant species are tightly constrained by edaphic (soil) factors. Plant communities therefore have the potential to be proxies for several parameters including soil nutrient content, moisture, and pH.

At its simplest, the use of plant species to indicate soil properties such as soil fertility and moisture can involve noting the presence of single species indicators in a biosurvey. Sometimes this requires species-specific identification. For example, the stinging nettle *Urtica dioica*, a plant that occurs throughout much of Europe and Asia, is indicative of fertile soil that is typically fairly damp. Correct species identification is vital as other species in the same genus, such as the annual or dwarf nettle *Urtica urens* need different environmental conditions.

An example of an **indicator genus**, rather than an indicator species, is the rush, *Juncus*. The genus has a cosmopolitan distribution (i.e. it occurs over a very large spatial scale, in this case occurring on every continent except Antarctica) and all species in the genus are confined to areas where soils are moist or wet most of the year. This means that the presence of any *Juncus* species can be used to indicate areas that are often damp, even if conditions at the time of the survey suggest otherwise.

Occasionally, it is possible to use even broader taxonomic ranks than genus within environmental indicator systems. For example, Payne et al. (2012) found that total moss bryophyte cover decreased as nitrogen levels rose on wet heathland. Use of mosses in total is a **whole-phylum approach** and, because no species- or genus-specific information is needed, the approach is very rapid.

As noted in Section 4.2.2, many indicator systems use multiple species to draw conclusions about environmental parameters. A common plant-based community indicator system is the Ellenberg system (Ellenberg, 1974). This uses the presence of plants to describe nutrient levels, moisture, and pH. The system is a full biotic index since values are assigned for each species and an overall value is calculated for a site, which is then compared with predetermined reference values. Although this has been around for some time, having first been devised in the 1970s, the approach is still widely used today throughout mainland Europe (including Germany, France, Italy, and Estonia), the UK where it was rescaled by Hill et al. (1999), and parts of Asia. The Ellenberg system, and some of the Applied Ecology questions it has been used to help answer, is outlined in more detail in Case Study 4.1.

Using lichens to indicate air quality

Plants are not the only environmental indicators. Lichens make ideal indicators of air pollution as they are highly sensitive to subtle changes in environmental conditions. Lichens are complex species: they are not single organisms, but rather are stable symbiotic associations between a fungus and algae and/or cyanobacteria that exist in mutualistic relationships (Section 2.3.1). Some species only occur in very clean air environments, while other species are nitrogen-loving (so indicate eutrophication—the process by which natural systems become more nutrient-rich) or are sulphur-tolerant (so indicate acidification).

The potential use of lichens as environmental indicators was first noticed when mass emission of sulphur dioxide (SO_2) from coal burning and industry led to a major loss of lichens. These 'lichen deserts' were most common near industrial areas, while rural areas (especially upwind of industry) had more lichen cover. However, further work established that most so-called lichen deserts actually still contained some

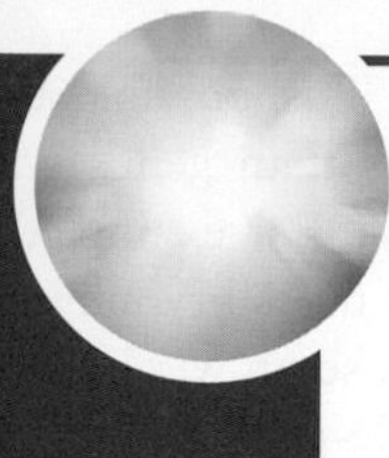

HOT TOPIC 4.1

Are destructive environmental indicator systems justified?

Although most indicator frameworks involve sampling that does not injure or kill indicator species, this is not always the case. Some indicator systems involve **tissue analysis** to determine levels of heavy metals and persistent organic compounds. Such analysis is based on the ecological concept of **bioaccumulation**, whereby non-degradable chemicals build up in biological tissues.

A classic case of bioaccumulation involved the insecticide dichlorodiphenyltrichloroethane (DDT). DDT occurred in low concentrations in the aquatic environment to which it was applied to control insect vectors of diseases, such as malaria (see also Chapter 9). Concentrations remained low in the first trophic level (plants), but built up through successive trophic levels as herbivores consumed plants and carnivores consumed herbivores (Figure A). This happened because some chemicals, such as DDT, cannot be metabolized or broken down by the body and so accumulate in body tissues through repeated consumption. Bioaccumulation can cause problems for any species, particularly those at the top of the food chain where concentrations are usually greatest. In the case of DDT, top carnivores experienced neuromuscular problems and, famously for birds, eggshell thinning due to a change in the way calcium carbonate moved from blood into the eggshell gland.

Species that bioaccumulate readily, either because of their position at or near the top of the food chain or because they are long-lived, can be used to indicate levels of non-biodegradable chemicals and heavy metals. This usually requires destructive sampling (Section 3.6.1) as individuals are dissected for analysis of tissues. An example is quantification of polybrominated diphenyl ethers in muscle samples of white-tailed eagles *Haliaeetus albicilla* from West Greenland (Jaspers et al., 2013). The exact methods adopted depend on the species, tissue type, and chemical(s) compound(s) under consideration. Broadly speaking, however, tissue is homogenized and analysed, often using atomic absorption spectroscopy (AAS) for metals or gas chromatography with either mass spectrometry (GC-MS) or a halogen-sensitive detector (GC-HSD) for organic compounds.

Can destructive environmental indicator species even ever be justified? There are obvious **ethical issues** here as there is no getting around the fact that individuals are killed to answer a scientific question. Depending on the species and the country in which the work is being undertaken, there can be the need to get a licence for such work to be **legally permissible**. The view of many scientists is that destructive sampling should not be undertaken lightly, but can be justified if the answer has important implications for applied management. In this way, the individuals on whom analysis is being performed might be regarded as being sacrificed to inform policy or practice that has benefits for that species, or the wider ecosystem. If this argument is accepted, it is still vital to ensure that the sample size is as low as possible and that individuals are euthanized quickly and without pain between capture and analysis.

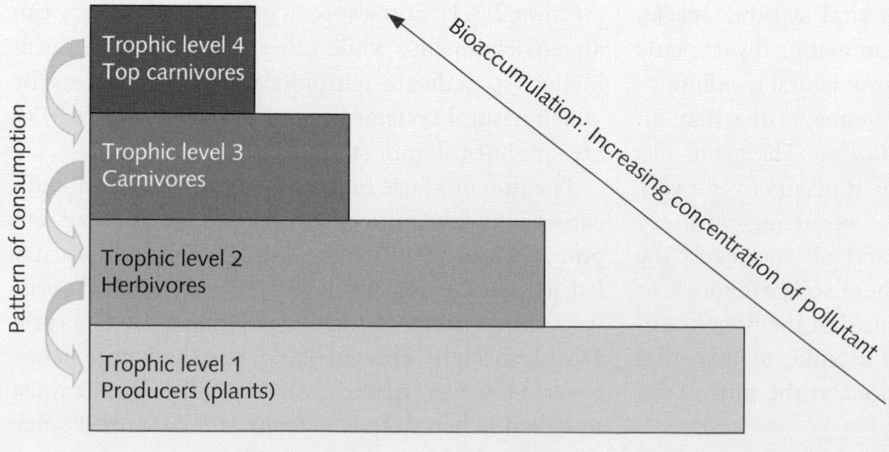

Figure A The process of bioaccumulation of a pollutant through the food chain.

It is also important to note that sometimes tissue analysis can be non-destructive. For example, Jaspers et al. (2006) analysed feathers of common buzzards, *Buteo buteo*, to assess levels of polychlorinated biphenyls (PCBs) in Belgium. As only a single tail feather was needed from each bird, which can be easily regrown, this technique is not only non-destructive, but was also relatively non-invasive as it required a single capture event. Similarly Flache et al. (2015) analysed hair samples from bat species, such as noctule *Nyctalus noctula* and common pipistrelle *Pipistrellus pipistrellus*, to quantify heavy metals, such as lead, zinc, and cadmium.

QUESTIONS

How can the three Rs approach of 'Replace, Reduce, Refine' applied to the use of animals in animal testing be translated to ecology?

Is there a difference in the acceptability of destructive sampling between common and rare species?

REFERENCES

Flache, L., Czarnecki, S., Düring, R.A., Kierdorf, U., & Encarnação, J.A. (2015) Trace metal concentrations in hairs of three bat species from an urbanized area in Germany. *Journal of Environmental Sciences*, Volume 31, 184–193.

Jaspers, V.L.B., Sonne, C., Soler-Rodriguez, F., Boertmann, D., Dietz, R., Eens, M. Rasmussend, L.V., & Covaci, A. (2013) Persistent organic pollutants and methoxylated polybrominated diphenyl ethers in different tissues of white-tailed eagles (*Haliaeetus albicilla*) from West Greenland. *Environmental Pollution*, **175**, 137–146.

Jaspers, V.L., Voorspoels, S., Covaci, A., & Eens, M. (2006) Can predatory bird feathers be used as a non-destructive biomonitoring tool of organic pollutants? *Biology Letters*, Volume 2, 283–285.

lichen species, so it was not possible to use overall lichen presence/absence as an indicator of of air pollution. Moreover, some heavily polluted areas still had reasonable species richness, albeit with very different species compared with unpolluted sites. Although this continued presence prevented species richness being used as an air quality indicator, it suggested that examining the actual species in a given community might lead to a useful indicator system.

As reviewed by Conti & Cecchetti (2001), several biotic indices now use lichens to indicate air quality. For example, lichens have recently been used as air quality bioindicators in Argentina (Estrabou et al., 2011), urban parts of the Mediterranean (Llop et al., 2012), and Alaska (Schirokauer et al., 2014). Much of this work builds on a seminal system developed in the UK: the lichen zonation system (Hawksworth & Rose, 1970). This gives values to many lichen species found in the UK on a scale of 1 (poorest air quality) to 10 (purest air), with values being cross-calibrated with actual sulphur dioxide levels (Figure 4.6). This cross-calibration is very unusual for a biotic index. Most other indices are based on a ranking scale (dry–wet, poor–excellent, cool–hot, etc.).

While **cross-calibration** is a key strength of the index, paradoxically, one of its other main strengths is also its major disadvantage: it is based on numerous species and thus requires accurate identification of those species. This is a particular problem for lichen, which is a very difficult group to identify. Moreover, species in the same genus are not only easy to confuse, but can be indicative of very different sulphur levels. For example, the genus *Parmelia* has numerous species, some of which are present in 'clean' areas (zone 8; mean winter SO_2 levels 35 μg/m^3), while others occur in relatively polluted areas (zone 4; mean winter SO_2 levels of 70 μg/m^3) (Hawksworth & Rose, 1970). The misidentification of species is a distinct possibility for non-experts and could thus result in very different conclusions about the air pollution levels at a particular site.

A different index takes a middle ground between a single-species approach and the **full community approach** used by Hawksworth and Rose. The trunk and twig method requires information on two sub-communities of key indicator species—those that indicate high atmospheric nitrogen and those that indicate high airborne sulphur (Larsen Vilsholm et al., 2009). The system uses the number of sample plots in which the indicator species are present (prevalence) relative to overall lichen abundance. The technique itself, and example applications, are outlined in Case Study 4.2.

CASE STUDY 4.1

Ellenberg biotic index

The technique

The Ellenberg system comprises values for 2726 central European vascular plants, which represent their tolerance for soil moisture (*F*), soil fertility (*N*), and reaction pH (*R*). All parameters have an **ordinal ranking scale** describing conditions along that environmental gradient, as shown in Table A. Each plant species has a value indicating the position along the specific environmental gradient at which it reaches peak probability of occurrence (the optimum).

To establish the Ellenberg profile of a site, a **plant species inventory** is undertaken. This involves identifying all plants in a plot of a predetermined size—10 × 10 m is often used. The bigger the sample plot, the better in terms of the data collected. However, larger plots take longer to survey to the point where the sample effort becomes prohibitive. Once a species is identified, it is allocated the score from the Ellenberg database for the parameter(s) of interest. The site profile is then determined by calculating the average score for each parameter.

Plants often lag behind a change in environmental parameters, such that it may take years before abiotic change is visible in plant species assemblages, and thus reflected in the Ellenberg score. For example, a change in hydrology that makes a site wetter will not result in an instant change in species. Therefore, the Ellenberg score might indicate that conditions are drier than they actually are until the vegetation profile has 'caught up' with the change in conditions. It is for this reason that Ellenberg indicator values were found to be useful environmental predictors in ancient woods, but much poorer in recent woods (Dzwonko, 2001). This is partly due to ecological succession, coupled with wind-dispersed species being over-represented in recent woodland.

Example applications

The Ellenberg system was used by Bennie et al. (2006) to examine temporal change in eutrophication and habitat fragmentation on unimproved calcareous grassland (i.e. grassland on calcium carbonate bedrock, such as limestone or chalk, which does not have direct application of artificial fertilizers). They found that the naturally oligotrophic grasslands were tending towards mesotrophic grassland communities, potentially resulting in rare stress-tolerant species that specialize on oligotrophic grasslands being outcompeted by grasses that dominate in fertile conditions.

A similar study on woodland understory vegetation (Verheyen et al., 2012) focused less on the effect of change and more on the factors driving change. Contrary to expectations, although sites showed increasingly fertile soils as indicated by the vegetation, change did not correlate with increases in atmospheric nitrogen. Instead, one of the main driving factors was increases in leaf litter and increased inputs of nutrients from that source. This seemed to link to strategic decreases in forest management (clearing, thinning crown management, deadwood removal) caused by an increasing focus on conservation.

Table A Ellenberg values for soil moisture, pH, and fertility.

	Minimum	Median	Maximum
Moisture	Score = 1 Very dry soil Example: Somerset hair grass *Koeleria vallesiana*	Score = 6/7 Moist (not wet) soil year-round Example: sessile oak tree *Quercus petraea*	Score = 12 Floating/submerged aquatic plant Example: duckweed *Lemna*
pH	Score = 1 Highly acidic conditions Example: small cranberry *Vaccinium microcarpum*	Score = 5 Mildly acidic/neutral conditions Example: beech tree *Fagus sylvatica*	Score = 9 Highly alkaline conditions Example: waterweed *Hydrilla verticillata*
Fertility	Score = 1 Oligotrophic (very low nutrients) Example: limestone bedstraw *Galium sterneri*	Score = 5 Mesotrophic (medium nutrients) Example: hard rush *Juncus inflexus*	Score = 9 Hyper-eutrophic (very high nutrients) Example: broad-leaved dock *Rumex obtusifolius*

REFERENCES

Bennie, J., Hill, M.O., Baxter, R., & Huntley, B. (2006) Influence of slope and aspect on long-term vegetation change in British chalk grasslands. *Journal of Ecology*, Volume 94, 355–368.

Dzwonko, Z. (2001) Assessment of light and soil conditions in ancient and recent woodlands by Ellenberg indicator values. *Journal of Applied Ecology*, Volume 38, 942–951.

Verheyen, K., Baeten, L., De Frenne, P., Bernhardt-Römermann, M., Brunet, J., Cornelis, J., Decocq, G., Dierschke, H., Eriksson, O., Hédl, R., Heinken, T., Hermy, M., Hommel, P., Kirby, K., Naaf, T., Peterken, G., Petřík, P., Pfadenhauer, J., Van Calster, H., Walther, G.-R., Wulf, M., & Verstraeten, G. (2012), Driving factors behind the eutrophication signal in understorey plant communities of deciduous temperate forests. *Journal of Ecology*, Volume 100, 352–365.

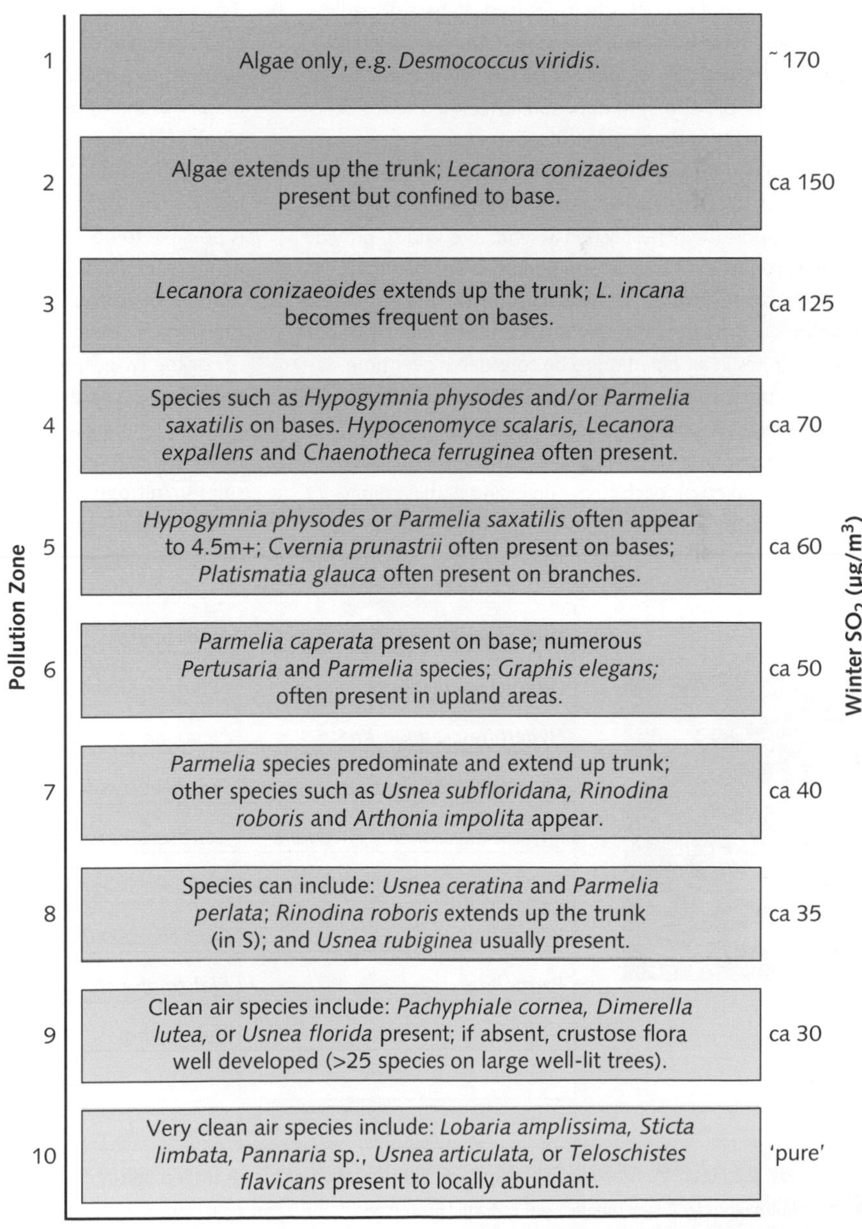

Figure 4.6 Hawksworth and Rose system—a ranking scale of 1–10 of the lichen species expected to be present at high (zone 1) to very low (zone 9) or absent (zone 10) levels of sulphur dioxide.

CASE STUDY
4.2

Monitoring air quality with lichens using the trunk and twig method

The technique

The trunk and twig method is a relatively quick method of surveying lichens on trees that focuses on a small group of indicator species for sulphur, which are called **acidophytes**, and nitrogen, which are called **nitrophytes** (Figure A). Five trees are sampled at each site of interest. Where possible, trees should have acid bark (e.g. oak, *Quercus*, and birch, *Betula*) and the trunk should not be heavily shaded. This is to avoid bark pH and light exposure affecting the results of the survey in complex or unpredictable ways.

Both trunk and twigs are surveyed because they are important in different ways (Seed et al., 2013). Tree trunks usually support the richest lichen assemblage, and so provide the most robust data over a long period. Twigs, being a much younger substrate, provide a useful indication of recent conditions. Comparing results from both trunk and twig surveys allow changes in air pollution to be considered over time.

For the trunk survey, a **ladder quadrat** is used with several squares, each 10 × 10 cm, arranged vertically. Because orientation can have a substantial effect on lichens, four samples are taken on each tree, one facing each cardinal direction (north, east, south and west). The circumference of the tree must thus be >40 cm to avoid counting the same lichen twice. For each sample point, the presence or absence of each indicator species is recorded. The species used are based on research in the Netherlands by van Herk (1999, 2001).

Prevalence is calculated as a proportion. For example, if one species was in half the sample squares on all trees, mean prevalence would be 0.5; if a second was found in every square on one of the five sample trees, but not at all on any of the others, mean prevalence would be 0.2. The trunk acidophyte index is the sum of the prevalence of all acidophyte indicator species, while the trunk nitrophyte index is the sum of all the nitrophyte indicator species. It is possible to decouple the indices from the lichen species richness by dividing the acidophyte index value by the number of acidophyte species found (or total lichen richness); the same is possible for nitrophytes.

For the twig survey, 10 twigs are examined for the same indicator species and the twig acidophyte index and twig nitrophyte index are calculated based on the prevalence of the indicator species. A recent increase in atmospheric sulphur/nitrogen is indicated by acidophyte/nitrophyte indices that are higher on twigs than trunks. A recent

Nitrophytes	Acidophytes
Diploicia canescens	*Evernia prunastri*
Hyperphyscia adglutinata	*Flavoparmelia caperata*
Phaeophyscia orbicularis	*Pseudevernia furfuracea*
Physcia adscendens; P. tenella	*Cetraria chlorophylla*
Xanthoria parietina; X. polycarpa	*Parmeliosis ambigua*
Physconia	*Parmelia saxatilis*
Parmelina	*Platismatia glauca*
	Hypogymnia
	Cladonia
	Usnea

Case Study 4.2 Figure A Example lichen species that indicative of high nitrogen (left) or high sulphur (right).

Source: Photographs by Oliver Moore, used with kind permission.

decrease in atmospheric sulphur/nitrogen is indicated by acidophyte/nitrophyte indices that are lower on twigs than trunks. If values are approximately equal, pollutant levels are likely to have been fairly consistent over time.

Example application

A slightly simplified version of this index was used in a UK-wide citizen science initiative as part of the Open Air Laboratories scheme, which was run through the Natural History Museum in 2009–2011 and reported in Seed et al. (2013). Volunteers recorded the presence/abundance of nine lichen species/genera located on tree trunks between 50 and 200 cm above ground. The target lichens were three N-sensitive indicators, three N-tolerant indicators, and three intermediate lichens. If twigs of the same tree could be reached, participants were also asked to spend no more than five minutes to record the presence or absence of each species on twigs under 2 cm diameter; no abundance data were collected for twigs.

The results demonstrated a general negative relationship between the N-sensitive lichen variables and nitrogen deposition, and a significant positive relationship between most N-tolerant species and nitrogen deposition. Lichen indices which incorporate data for both N-sensitive and N-tolerant lichens showed a significant negative relationship with the dry deposition of both NOx and NHx.

FURTHER READING

Seed, L., Wolseley, P., Gosling, L., Davies, L., & Power, S.A. (2013) Modelling relationships between lichen bioindicators, air quality and climate on a national scale: Results from the UK OPAL air survey. *Environmental Pollution*, Volume 182, 437–447.

van Herk, C.M. (1999) Mapping of ammonia pollution with epiphytic lichens in the Netherlands. *Lichenologist*, Volume 31, 9–20.

van Herk, C.M. (2001) Bark pH and susceptibility to toxic air pollutants as independent causes of changes in epiphytic lichen composition in space and time. *Lichenologist*, Volume 33, 419–441.

Using aquatic species to indicate organic pollution

The use of biota as a means of monitoring water has a long history, having been first proposed by Kolkwitz & Marsson (1902) to identify areas of sewage pollution. Indeed, the natural parameter now most frequently assessed through environmental indicators is water quality, especially in relation to organic pollution, with over 30 specific indices having been developed.

Well-oxygenated water is needed for a healthy ecosystem. Under natural conditions, most water bodies sustain good oxygen levels year round. However, an increase in the amount of dead organic matter in a water body can cause problems. This is because the microbes involved in decomposition are aerobic and thus use up oxygen. This demand for oxygen is referred to as **biological oxygen demand** (BOD). High BOD can cause dissolved oxygen levels to drop, especially in still water bodies, such as lakes and ponds, where there is little mechanical mixing of atmospheric oxygen into the water. It can also happen in water bodies that have comparatively little vegetation, so the addition of oxygen as a by-product of plant photosynthesis is limited.

High BOD can occur for a number of reasons, including the accumulation of dead leaves (which explains why BOD is usually higher, and dissolved oxygen is lower, in autumn) and input of sewage, which is of longer-term concern. The eutrophication status of the water can also influence the effect of dead organic matter on BOD because decomposition microbes reproduce more quickly in nutrient-rich water. This means that BOD can become extremely high (and thus oxygen levels extremely low) within a few days of organic matter being released.

Most biotic indices used to determine BOD use aquatic invertebrates, which are quick and easy to sample. However, the huge number of different species and difficulties surrounding identification at the species level means that most systems are family-level biotic indices (FBIs). Consequently, species only need to be identified to family level, rather than genus or species level.

Several common indices exist. Some of these are based only on species presence and do not factor in abundance—for example, the British Monitoring Working Party (BMWP) system developed in the UK. This index has been modified for use in many different countries, including Spain, Czech Republic, Poland, Hungary, Australia, India, Thailand, Brazil, Ghana, and Rwanda (e.g. Junqueira & Campos, 1998; Wronski et al., 2015). It is also used in Egypt, where it has been renamed the Nile Biotic Pollution Index (Fishar & Williams, 2008). The UK BMWP index is explained in more detail in Case Study 4.3.

CASE STUDY
4.3

Monitoring water oxygen with invertebrates using the BMWP method

The technique

The British Monitoring Working Party (BMWP) first met in 1976 and started work on a new monitoring system for biological condition of aquatic environments. Aquatic invertebrates were chosen because they responded predictably to differences in oxygen level, which mirrored levels of pollution. After one unsuccessful pilot and considerable revision, the BMWP system was launched in 1981 (Table A). This gave scores to 85 families between 1 (highly tolerant of pollution, very low demand for oxygen, found in any conditions) and 10 (highly intolerant of pollution, very high demand for oxygen, found in very clear water only) (Hawkes, 1998).

The basic method involved collecting aquatic invertebrates either using a metal-framed net to dislodge the substrate and any benthic (bottom-dwelling) invertebrates, or using a Surber net or stream barrier net to collect the sample when sediment is dislodged by kicking. Once collected, individuals were identified to family level to allocate the appropriate score. This was a **presence-only system**: if one freshwater shrimp (Gammaridae) was found, a score of 6 was given; if 100 individuals were found, the score was still 6. Once all scores were allocated, they were summed to give the BMWP score (higher = less polluted).

There were some issues with this original iteration of the method:

- Scores were summed so the score was linked to species richness (and possibly sample effort).
- Scores for individual families were taken as being consistent in all habitats, whereas most species can actually survive at higher pollution levels in flowing environments as some oxygen is mixed into the water mechanically, especially if there is some surface disturbance.
- All species within a given family were given the same family-level score.

The first issue was addressed by the original working party by introducing a companion index—ASPT or Average Score Per Taxon. This is simply the BMWP score divided by the number of families represented in that score. This is usually equal to the number of families in the sample, unless any unusual families that are not on the BMWP list have been

Table A Original BMWP values for invertebrates.

Invertebrates	Score
Siphlonuridae, Heptageniidae, Leptophlebiidae, Ephemerellidae, Potamanthidae, Ephemeridae, Taeniopterygidae, Leuctridae, Capniidae, Perlodidae, Perlidae, Chloroperlidae, Aphelocheiridae, Phryganeidae, Molannidae, Beraeidae, Odontoceridae, Leptoceridae, Goeridae, Lepidostomatidae, Brachycentridae, Sericostomatidae	10
Astacidae, Lestidae, Agriidae, Gomphidae, Cordulegasteridae, Aeshnidae, Corduliidae, Libellulidae, Psychomyiidae	8
Caenidae, Nemouridae, Rhyacophilidae, Polycentropodidae, Limnephilidae	7
Neritidae, Viviparidae, Ancylidae, Hydroptilidae, Unionidae, Corophiidae, Gammaridae, Platycnemididae, Coenagriidae	6
Mesovelidae, Hydrometridae, Gerridae, Nepidae, Naucoridae, Notonectidae, Pleidae, Corixidae, Haliplidae, Hygrobiidae, Dytiscidae, Gyrinidae, Hydrophilidae, Clambidae, Helodidae, Dryopidae, Elminthidae, Chrysomelidae, Curculionidae, Hydropsychidae, Tipulidae, Simuliidae, Planariidae, Dendrocoelidae	5
Baetidae, Sialidae, Piscicolidae	4
Valvatidae, Hydrobiidae, Lymnaeidae, Physidae, Planorbidae, Sphaeriidae, Glossiphoniidae, Hirudidae, Asellidae	3
Chironomidae	2
Oligochaeta	1

found. If this does happen, such families are excluded from BMWP (and thus ASTP) calculation.

The second issue was addressed in 1996, when the system was re-evaluated by Walley & Hawkes (1996) to differentiate between riffles (areas of moving water), pools (areas of still water), and riffles/pools (areas with mixed flow dynamics). A computer model was used to adjust values and this resulted in some scores being >10.

The third issue has not been addressed and it would be counter-productive to do so. Although it is a simplification to say that all species in a family respond to pollution in the same way, and should thus have the same score, taking the classification framework down to genus or species level would result in a lot of extra work, which is a problem for an index designed to be relatively quick and simple.

The issues that remain with the system are:

- Its assumption that all species are equally catchable (and detectable in a muddy sample).
- It takes no account of abundance.

It should be noted that the overall system is based not on objective testing of the different oxygen levels needed by different species (which could have been calculated using bioassays or correlating field data with laboratory measurements of dissolved oxygen), but by subjective assessment of the types of environments in which different species were found.

An example application

An interesting example of how BMWP can be used to answer topical questions is provided by Hatami et al. (2011) in Iran. This examined the possibility that flow-through aquaculture systems, which are increasingly used to meet growing global demands for fish, result in effluent discharge that has a negative effect on oxygen in the wider aquatic environment. They tested this using BMWP and found decreased water quality immediately after effluent outfall, where the invertebrate community was dominated by midge larvae (Chironomidae) relative to upstream, where the invertebrate community was dominated by freshwater shrimps (Gammaridae). Water had resumed its upstream quality within 1 km (Figure A).

REFERENCES

Hawkes, H.A. (1998) Origin and development of the Biological Monitoring Working Party score system. *Water Research*, Volume 32, 964–968.

Walley, W.J. & Hawkes, H.A. (1996) A computer-based reappraisal of the Biological Monitoring Working Party scores using data from the 1990 river quality survey of England and Wales. *Water Research*, Volume 30, 2086–2094.

Hatami, R., Mahboobi Soofiani, N., Ebrahimi, E., & Hemami, M. (2011) Evaluating the aquaculture effluent impact on macroinvertebrate community and water quality using BMWP index. *Journal of Environmental Studies*, Volume 37, 13–15.

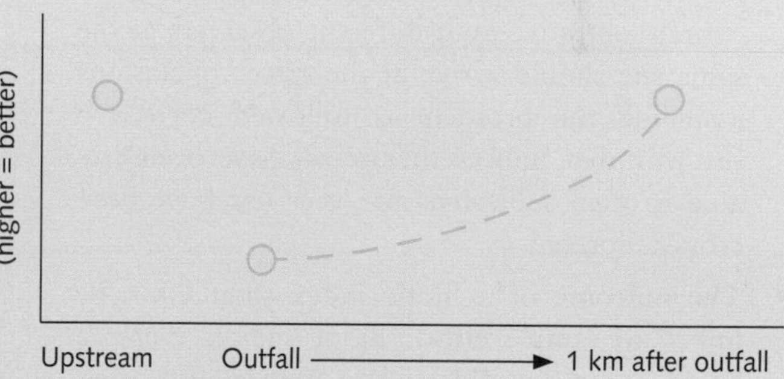

Case Study 4.3 Figure A Water quality upstream of a through-flow fish farm in Iran (medium water quality) decreases substantially at the outfall (low water quality). Water quality remains low 50 m after the outfall, but returns to upstream levels after 1 km.

Source: Findings from Hatami et al. (2011).

Other aquatic indices take account of abundance. Some are **semi-quantitative**, which means that an abundance category is given to each species (e.g. Chandler's score, which has five abundance categories: Chandler, 1970). The most complex indices are **fully quantitative**, where it is necessary to count the exact number of individuals in each family (e.g. Hilsenhoff's index in the USA: Hilsenhoff, 1977). Such indices are more accurate and detailed, but involve a considerable increase in the sample time and effort needed.

4.3.3 General lessons from environmental indicator schemes.

Use of environmental indicators can be an extremely powerful method within Applied Ecology, as exemplified earlier, but it is not without challenge. Several general lessons can be taken from the indicator systems discussed previously:

- **Indicator systems should use species that are easy to identify** to the required level (family-level for BMWP and most other aquatic invertebrate biotic indices; genera-level for the majority of the lichen systems; species-level for the Ellenberg system using vascular plants). This consideration is especially important when misidentification of species is likely to result in very different conclusions being drawn (e.g. in the case of the Hawksworth and Rose lichen zonation system).
- **Biotic indices tend to be more robust and objective than biosurveys, even though the same basic principles are used.** Biosurveys are useful for basic observations, but are not standardized. It is often not, therefore, possible to compare results between areas or over time reliably; indeed, even the comparison of results between different observers at the same site can be problematic. Biotic indices, with their standardized methods and quantitative outputs, tend to be more objective and comparable over space and time, but generally take longer to complete. It should also be noted that although the quantitative output suggests that the index is entirely objective (and different observers at the same site should arrive at the same conclusions assuming the protocol is followed correctly), the way that indices themselves have been created is often subjective and very few have been cross-calibrated.
- **The outcome of a biotic index should not be linked to sample effort.** Biotic indices that are based on scores allocated to individual species, and thus based on the presence of those species, need to allow for variations in the amount of time people spend sampling and over what area (sample effort). This is to avoid the risk that a greater sample effort will likely result in a longer species list at a given site and, *ergo*, an inflated score. The most common ways to de-couple sample effort and index result are:
  - to use an index whereby even if scores are given to each individual species/genus/family found on site, the ultimate score is the average of these, rather than the total. Some indicator systems do this as standard (e.g. Ellenberg: detailed in Case Study 4.1), while others use a companion index to the main score (e.g. Average Score Per Taxon (ASPT) developed for BMWP as explained in Case Study 4.3);
  - to use a standardized method to ensure sample effort is consistent (e.g. the standard assessment protocol used in the trunk and twig lichen index).
- **The outcome of a biotic index should not be linked to species richness.** This is a similar problem to that outlined previously. If an index is based on scores allocated to each species found at a site, the score will not be independent of species richness (i.e. more species will give a higher index value if the index is a sum of all the species found). This can be avoided by using averages as detailed above.
- **Biotic indices should be used for the exact purpose for which they were developed.** There can sometimes be 'creep' in terms of what specific indices are used to indicate. For example, aquatic indices, developed to assess organic pollution are sometimes used to infer overall water quality including chemical and heavy metal pollution, and eutrophication. Such usage extrapolates findings substantially: it would be possible, for example, for an aquatic invertebrate index to suggest 'good' water quality in terms of dissolved oxygen levels in situations with high levels of some heavy metals. If indices are applied to monitor new variables, their suitability should be tested first and the indices modified as necessary. This was done to good effect for Spanish BMWP, which was modified to detect increases in water sulphate that occur with coal mine effluent (García-Criado et al., 1999).
- **Lag time can determine the usefulness of indicator species and how they are used.** Lag time is the time period between an environmental parameter changing and that change being indicated by (and detectable using) indicator species. In some cases, such as with Ellenberg, indicator species only give reliable results in stable ecosystems. In other cases, lag time is exploited to give additional information—for example, in the case of lichens

and air quality, systems can use a difference or similarity in score between old substrates (the main trunk) and new substrates (relatively newly-grown twigs) to give data on air quality change (Case Study 4.2).

- **It is most effective to use r-strategist species to detect environmental change.** Species that are r-strategists are quick growing, reach sexual maturity quickly, have multiple offspring and are relatively short-lived. Floral examples include most annual plants and tree species, such as silver birch *Betula pendula* and willow *Salix*. Faunal species include many insects, mice, rabbits, and many small songbirds. (The opposite of r-strategists are K-strategists, which are slow growing, reach sexual maturity slowly, have few offspring, and are generally fairly long-lived. These include oak *Quercus* and beech *Fagus* trees, elephants, blue whales, and albatrosses.) An example of an r-strategist indicator is algae as an indicator of eutrophication. Algae respond very quickly to eutrophication as they grow fast and multiply quickly. This causes 'algal blooms'—sudden mass increases in algae—a useful early warning of eutrophication events.

Presence/prevalence versus abundance

Many biotic indices require the collection of **abundance** data, rather than presence data. There is a trade-off between index accuracy and robustness on the one hand, and ease of use and speed of data collection on the other. Where species abundance is not considered and a biotic index is calculated simply on presence, a species occurring in very low numbers at a site is rated the same as a species that occurs in very high numbers. However, counting every individual is often prohibitively time consuming. It is thus common for abundance categories to be used to make abundance semi-quantitative. In some cases—for example, Chandler's index for aquatic biological pollution—these categories are specific to that index. In other cases, standard abundances are used. Common measures include:

- **DAFOR scale:** a five-point ordinal scale where each species is designated as Dominant, Abundant, Frequent, Occasional or Rare. The DAFOR scale has no objective basis so personal interpretation is needed when using it. This is a relative scale rather than a scale of absolute abundance as species that occur at reasonably low overall abundance might still be 'dominant' in the specific system under consideration. A similar system is ACFOR—Abundant, Common, Frequent, Occasional, or Rare.
- **Braun-Blanquet scale:** a five-point scale based on percentage cover (1 = 1–5% cover; 2 = 6–25% cover, 3 = 26–50% cover; 4 = 51–75% cover; 5 = 76–100% cover). Technically this is a cover scale rather than an abundance scale, as it looks at the percentage of an area covered by a species, rather than the number of individuals present, and so is generally only used for vegetation.
- **DOMIN scale:** a 10-point scale based on percentage cover and, again, only suited to vegetation (1 = 1–2 individuals with no measurable cover; 2 = several individuals, but no measurable cover; 3 = 1–5% cover; 4 = 5–10% cover; 5 = 11–25% cover; 6 = 26–33% cover; 7 = 34–50% cover; 8 = 51–75% cover; 9 = 76 = 90% cover; 10 = 91–100% cover).

In cases when abundance data are collected in some way a decision must be taken as to how these data are used. This is a contentious topic and is discussed further in Hot Topic 4.2.

4.3.4 Species that make good indicators.

Several attributes make a species or taxonomic group a good environmental indicator:

1. **Be closely linked to the environmental parameter for which it is a proxy:** if only a weak relationship exists between the indicator and the environmental variable(s) under consideration, the index will simply not be robust, regardless of how carefully data are collected. This sounds extremely basic, but a number of indicator systems, including Ellenberg and BMWP, have been devised using expert advice, rather than objectively quantified relationships between species and environmental parameters.
2. **Be predictably linked to the environmental parameter for which it is a proxy:** the relationship between an indicator species and the

HOT TOPIC 4.2

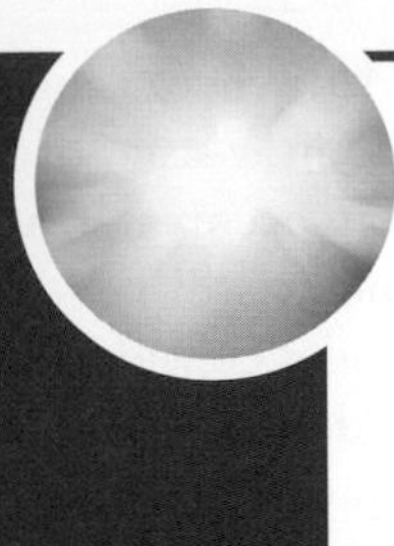

What is the best way of calculating 'the average' in environmental indicator systems? Thoughts on an ongoing controversy

Many environmental indicator biotic indices involve calculating an average, which raises the question of which measure of the average is best.

The **arithmetic mean** is used most often. For example, the Average Score Per Taxon companion index to BMWP involves summing the scores for each invertebrate family in a sample and dividing by the number of families found (see Case Study 4.3 for more details).

In some cases, though, the mean can be affected by one or two species with particularly high or low scores relative to the rest of the community. These are referred to as **outliers**. Let's imagine a hypothetical situation where five plant species have been found at a specific site, all of which have a species-specific Ellenberg score for moisture—see Case Study 4.1 for more details on the Ellenberg system. In this case, four of the five species have values that are very similar (2, 2, 2, and 3), but one species is an outlier with an atypically high score (10). The arithmetic mean for the community is 3.8, which suggests that there are some damp soil areas. However, this value—although mathematically correct—is skewed by the outlier (Figure A).

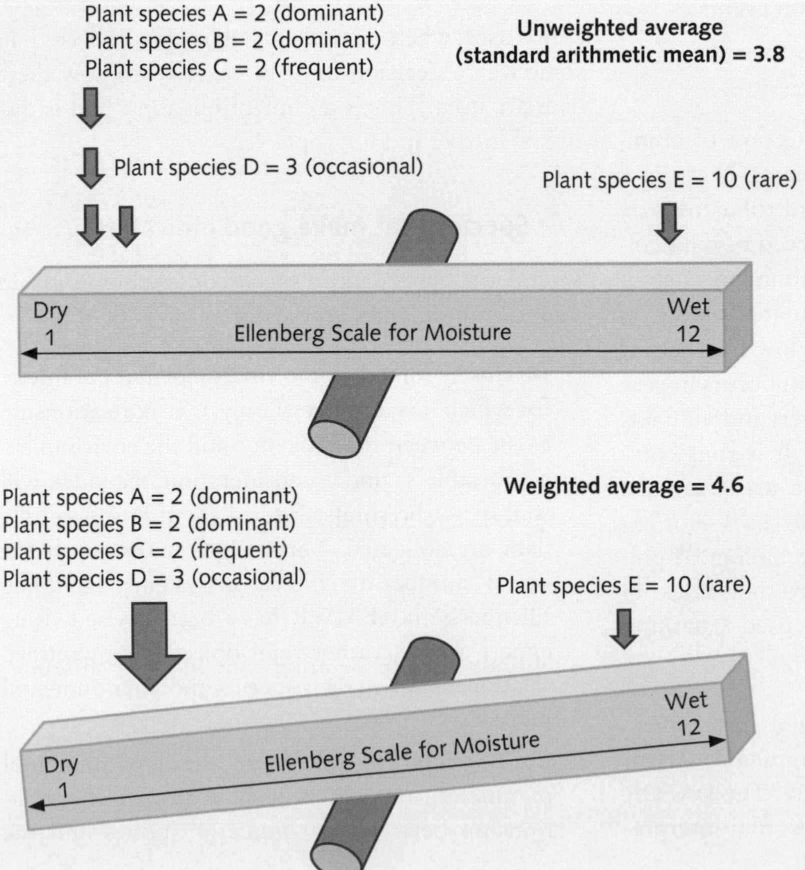

Figure A Differences between unweighted average using the arithmetic mean (top) and weighted average (bottom).

If, in this example, the middle value (the **median**) was used as a measure of the average, the result is a score of 2. Another possibility is to use the most frequently occurring value (the **mode**). Both these possibilities are more representative of the overall community and suggest that the conditions are fairly dry. Because of outliers skewing the mean like this, using the median or mode can often be better, although this is still not common practice.

In cases where species abundance has been recorded, either exactly or in abundance categories, it is possible to use a **weighted average**, whereby the contribution of each species to the average is weighted according to its abundance, rather than the contribution of each species to the average being equal, as in a standard (unweighted) average. In this way, species that occur in high numbers influence the final biotic index figure more than species that occur in low numbers. Returning to the example, two of the three species with an Ellenberg score of 2 were dominant on the DAFOR scale, the other species with a score of 2 was abundant, and the species with a score of 3 was frequent; in contrast, the species with an Ellenberg score of 10 was only rare (Figure A). If these DAFOR scores are translated so Dominant = 5, Abundant = 4, Frequent = 3, Occasional = 2, Rare = 1, a weighted average can be calculated thus. The equation is:

The equation is: $\dfrac{\sum \textit{Index score for each species} \times \textit{Abundance of each species}}{\sum \textit{Abundance of all species found}}$

$$\frac{(\text{Species A score} \times \text{abundance}) + (\text{Species B score} \times \text{abundance}) + (\text{Species C score} \times \text{abundance}) + (\text{Species D score} \times \text{abundance}) + (\text{Species E score} \times \text{abundance})}{\text{Species A abundance} + \text{Species B abundance} + \text{Species C abundance} + \text{Species D abundance} + \text{Species E abundance}}.$$

$$= \frac{(2 \times 5) + (2 \times 5) + (2 \times 4) + (3 \times 3) + (10 \times 1)}{5 + 5 + 4 + 3 + 1}.$$

$$= \frac{47}{18} = 2.6$$

QUESTION

To what extent should Applied Ecologists concern themselves with mathematical detail when using biotic indices?

FURTHER READING

Seminal paper on weighting averages in ecological indicators:
Ter Braak, C.J. & Barendregt, L.G. (1986) Weighted averaging of species indicator values: its efficiency in environmental calibration. *Mathematical Biosciences*, Volume 78, 57–72.

environmental parameter for which it is used as a proxy should not be altered by other environmental variables. Again, this sounds basic, but in many systems this is not tested. For example, there has been no testing of how nutrient levels or pH could affect the relationship between many of the aquatic invertebrate indices (including BMWP and Chandler's) and oxygen load. Before indices are used, they should ideally be tested in the specific environment in which they are to be used.

3. **Respond in proportion to magnitude of change in the environmental parameter for which it is a proxy:** the species needs not only to respond to environmental parameters and change in those parameters, but needs to do so at a fairly content rate. Having a species that is, for example, able to thrive in an environment until a certain level of sulphur concentration, at which point it is completely lost from the ecosystem, is much less informative than having a species that gradually declines as sulphur levels increase, until such time that it becomes absent completely.
4. **Be well studied:** the ecology and life history must be well understood, and the species needs to be well documented taxonomically.
5. **Be widespread, with a geographically consistent relationship to the environment:** indicator systems need to be applicable in many different geographical areas without the need for (extensive) modification. In the case of the Ellenberg system, values were designed for use in central Europe. Their use elsewhere usually requires location-specific recalibration (undertaken for

Estonia, the Faroe Islands, France, Italy, Russia, and the UK, for example). Similarly, the BMWP system was devised for the UK, and has since been adapted for other countries, including Spain, Czech Republic, Poland, Hungary, and Turkey. Although in both cases the basic concept was not geographically based, the reliability with which values could indicate environmental parameters declined with increasing distance from initial location. This occurred because species adapt to local conditions, and the relationship between any one species and its environment can be influenced by other species in the same community.

FURTHER READING FOR THIS SECTION

Monitoring environmental change using biotic indices: Spellerberg, I. (2005) *Monitoring ecological change*. Cambridge: Cambridge University Press.

Different types of biomonitoring and comparability between systems: Cao, Y. & Hawkins, C.P. (2011) The comparability of bioassessments: a review of conceptual and methodological issues. *Journal of the North American Benthological Society*, Volume 30, 680–701.

Aquatic biomonitoring: Friedrich, G., Chapman, D., & Beim, A (1996) The use of biological material. In: D. Chapman (Ed.) *Water quality assessments—a guide to use of biota, sediments and water in environmental monitoring*, Chapter 5. Cambridge: UNESCO/WHO/UNEP.

4.4 Biological indicators.

In addition to species being used to indicate the nature of the environment in which they occur (Section 4.3), they can also be used to indicate biological processes, including species–species interactions, species–habitat interactions, and species–human interactions.

4.4.1 Exemplar biological indicators.

This section considers two examples of biological indicators. Both systems use plants as the indicator taxa, but whereas the first examines species-specific interactions by using plant community composition as an indicator of grazing levels, the second extrapolates from plant data to allow us to infer habitat age.

Using plants to infer disturbance and animal grazing levels

The savannah grasslands of southern and eastern Africa are grazed by many different herbivores, including numerous antelope. In landscapes where grass is the dominant habitat, species move freely to find areas of good grazing. However, such landscapes are now often highly fragmented (broken up into small sections by urbanization and other land uses) such movement is much harder (Chapter 7). Fragmentation is also causing issues for ecosystem dynamics by affecting natural fire regimes. Regular burning is a key part of the maintenance of ecosystems as fire-induced disturbance prevents species that are extremely competitive in stable ecosystems dominating and allows less competitive species to survive.

The importance of regular burning in savannah grasslands is a real-world example of the **intermediate disturbance hypothesis**. This hypothesis—often attributed to Connell (1978) but actually originally discussed back in the 1940s by the father of the ecosystem concept, Arthur Tansley (1949)—states that local species diversity is maximized when ecological disturbance is neither too rare nor too frequent. At an intermediate level, species that typically invade quickly after disturbance (pioneer species) and species that need more stable environments (climax species) can co-exist. This is in contrast with the situation if disturbance occurs very regularly when only pioneer species will be able to tolerate fairly harsh abiotic conditions and high exposure, and the situation that occurs if disturbance is rare when climax species will out-compete pioneers and dominate the ecosystem. This is shown diagrammatically in Figure 4.7.

Because of the fragmented nature of savannah, a lot of species conservation relies on national parks and wildlife reserves, many of which are fenced (Section 11.5.1). In such reserves, it is essential that

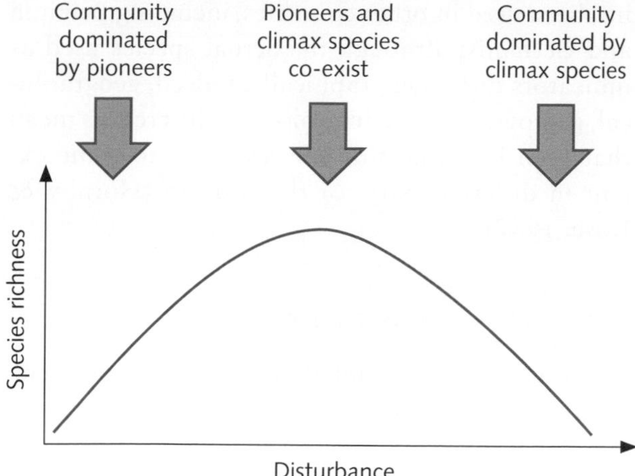

Figure 4.7 Relationship between disturbance levels and species richness. Richness is higher at intermediate level of disturbance.

grassland is managed, often using an artificial fire regime, to ensure herbivores (whose movement is also artificially restricted) are able to graze satisfactorily. The community composition of grasses in an area can indicate both the optimum fire disturbance level and grazing pressure.

One method of assessing fire disturbance level and grazing pressure is a biotic index called the Veldt Condition Index as described in Tainton (1999), which involves comparing the relative abundance of grasses in four ecological groups:

- **Decreaser grasses:** dominate when grassland is in good condition and decline in abundance when grassland is in poor condition. Poor condition can result from over-utilization and/or over-disturbance (too many grazers or too many fires) or can result from under-utilization and/or under-disturbance (too few grazers or too few fires).
- **Increaser I grasses:** increase in abundance where grassland is under-grazed and/or fire is excluded.
- **Increaser II grasses:** should be occasional or rare when grassland is in good condition, but replace Decreaser species where grassland is overgrazed. Increaser II species include pioneer and invader species that increase in abundance with severe overgrazing or very frequent disturbance.
- **Increaser III grasses:** should be occasional or rare when grassland is in good condition, but increase in abundance when selectively overgrazed. Here, selective grazing implies that palatable species are grazed, so unpalatable Increaser III species gain a competitive advantage and increase in abundance.

By comparing the relative proportion of Decreaser, Increaser I, Increaser II, and Increaser III species, it is possible to assess whether:

- The relationship between grassland and grazers is optimal (~60% Decreasers to 40% Increasers), and current management is working well.
- Habitat management needs to be undertaken to allow Decreaser species to reappear in the ecosystem. This is usually done by burning areas and is indicated when Increaser I prevalence is >25%.
- Animal management (translocation, population control, etc.) is needed as there are too many herbivores in the area. This is indicated when Increaser II and III prevalence together is >60%.

Using plants to infer habitat age

It is sometimes possible to use species as indicators of habitat age. A simple approach is to establish species richness on the basis that older habitats usually contain more species than younger ones. However, while this is a useful rule of thumb (and, indeed, is one of the hypotheses often advanced to explain why tropical rainforests are so species rich), many factors other than age affect species richness. For example, the species–area relationship (Preston, 1960) means that the number of species in an area increases as the size of that area increases, as shown in Figure 4.8. Habitat complexity can also be important, with habitats such

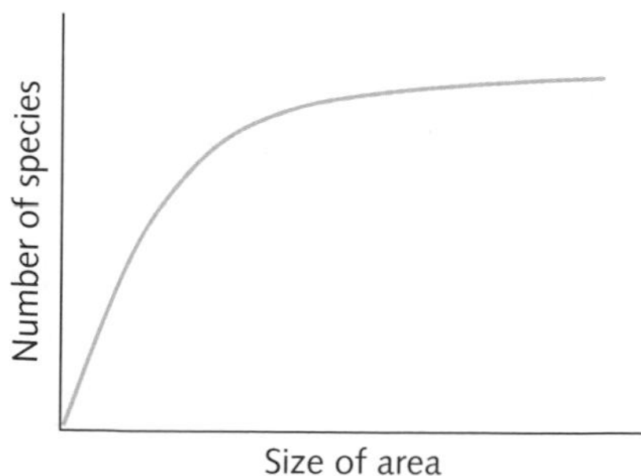

Figure 4.8 Species–area relationship. The number of species increases with the size of area considered, but this relationship gets less pronounced as the area itself increases.

as woodlands with high structural complexity, and thus more niches, likely to have a higher species richness than simpler.

To obtain an estimate of habitat age that is less influenced by factors such as area and complexity, and thus more reliable, it can be useful to examine the presence or abundance of specific indicator species. More than a century ago, naturalist F. Buchanan White recognized that the Caledonian pine forests of Scotland have a suite of plants, such as twinflower *Linnaea borealis*, creeping lady's-tresses *Goodyera repens*, and one-flowered wintergreen *Moneses uniflora*, which are much less common in disturbed woods. However, as Rose (1999) pointed out, such observations seemed to say more about the ecological limitations of these species to grow in recent woodlands, rather than allowing people to make general inferences about habitat age.

It took work by Peterken (1974) in Lincolnshire on the flora of known old and recent woodlands to discover that around one third of woodland plant species are strongly, and consistently, found in ancient woodland sites. This paved the way for a list of **ancient woodland vascular plants** (AWVPs), around 100 species that indicate ancient woodland sites, and that are common and widespread enough to be acceptable indicators. Ecologically, such species tend to be:

- Woodland specialist species.
- Poor dispersers: they cannot easily disperse to new habitat areas, even if the habitat itself is within the species' ecological niche.
- Tending towards K-strategists (slow growing and long-lived: Section 2.7.1).
- Climax species that need a stable environment (unlike pioneer species that occur near the start of succession and tolerate disturbance well).

AWVP systems are now commonly used for identifying ancient woodlands (in the UK, this means woods that pre-date 1600). Because ancient woodlands often have special characteristics and high ecological value, this is useful both in the context of Ecological Impact Assessment (Chapter 5) and *in situ* conservation (Chapter 11). A similar approach has been used in other countries, including Belgium and Germany, although the actual species used as indicators differ geographically. Indeed, geographical, geological, and climatological differences mean that even UK indicator species differ to some extent in different parts of the country (Hornby & Rose, 1987).

Using marine animals to infer fishing levels

Overfishing is a substantial and increasing problem, whereby off-take of fish occurs above the level at which fish stocks can replenish naturally. Where this occurs, populations of some fish species can be well below the replacement level. This can have substantial effects on community structure and, in extreme cases, can even leave some species vulnerable to extinction. It is possible to use indicator marine species to provide an early warning that fishing levels are too high within an alert monitoring programme (Section 3.3.4). This is discussed in more detail in the Online Case Study for Chapter 4.

You can find the online case study at www.oxfordtextbooks.co.uk/orc/goodenough

FURTHER READING FOR THIS SECTION

Review of the Intermediate Disturbance Hypothesis: Wilkinson, D.M. (1999) The Disturbing History of Intermediate Disturbance. *Oikos* Volume 84, 145–147.

Recent refinements to VCI: Wesuls, D., Pellowski, M., Suchrow, S., Oldeland, J., Jansen F., & Dengler, J. (2013) The grazing fingerprint: modelling species responses and trait patterns along grazing gradients in semi-arid Namibian rangelands. *Ecological Indicators*, Volume 27, 61–70.

Review of species–area relationships: McGuinness, K.A. (1984) Species–area curves. *Biological Reviews*, Volume 59, 423–440.

More detail on traits of ancient woodland indicators: Kimberley, A., Blackburn, G.A., Whyatt, J.D., Kirby, K., and Smart, S.M. (2013) Identifying the trait syndromes of conservation indicator species: how distinct are British ancient woodland indicator plants from other woodland species? *Applied Vegetation Science*, Volume 16, 667–675.

4.5 Biodiversity indicators.

Environmental indicators and biological indicators are aimed at allowing Applied Ecologists to infer how a particular environment looks and functions, based on the species within it. In contrast, biodiversity indicators utilize the abundance of, or diversity in, a specific taxonomic group as an indicator of the overall biodiversity of a specific ecosystem or change in that biodiversity over time. Thus, they are much more focused on describing the species components of an ecosystem, rather than the processes and environmental parameters within it.

4.5.1 Exemplar biodiversity indicators.

The basic idea behind biodiversity indicators is that it can be possible to quantify or monitor biodiversity generally though studying a restricted aspect of that biodiversity. For example, Kati et al. (2004) found that woody plant species in Greece were a good proxy for other species, including birds and amphibians, because sites that had target plant species usually also supported target bird and amphibian species. Different biodiversity indicators use different biodiversity measures as the proxy variable. Most common is the presence or abundance of specific target species in a taxonomic group (as in Kati et al., 2004) or the richness of a focal taxonomic group, but sometimes **temporal trends** in these measures provide a much more powerful indicator system than snapshot records (Section 3.3.3).

Despite being conceptually simple, biodiversity indicators are probably the most recent, and still the most contentious, type of ecological indicator. The two examples in this section outline the opportunities and challenges of using biodiversity indicators.

Using terrestrial invertebrates as a biodiversity proxy

Undertaking a full biodiversity audit to establish total species richness can take substantial time and effort, especially if the area is in a **biodiversity hotspot** (like a tropical forest), or when little is known about the species that make up the overall community. A lack of knowledge is especially problematic since knowledge of species richness can be extremely important for Ecological Impact Assessment (Chapter 5) and prioritizing distribution of conservation resources (Chapter 10).

One approach to this issue is to undertake a detailed analysis of species richness in one taxonomic group and consider this to be indicative of richness in other taxonomic groups (and thus the whole ecosystem). For this approach to be sensible, the chosen taxonomic group has to be widespread, abundant, and diverse. It also needs to be reasonably easy to sample and there has to be a biological reason to assume that species richness in the chosen taxonomic group will correlate with overall species richness. This means the chosen taxonomic group has to contain species that use a wide range of different niches, so that the number of niches is linked to species richness.

Invertebrates are widely regarded as powerful indicators, as they are abundant and sensitive to the local ecosystem. Moreover, other species rely on them either because of ecosystem services, such as pollination and decomposition, or as direct or indirect food sources. However, invertebrates are themselves a large taxonomic group and there is no one sampling method that covers all groups and niches. Because of these problems, specific insect taxa tend to be used, rather than the entire taxonomic group.

In the northern hemisphere, the most widely used invertebrate bioindicators are beetles, especially the Carabidae (ground beetles). For example, GLOBENET (Global network for monitoring landscape change) is a global initiative for assessing landscape change that uses ground beetles as its indicator taxon (Niemelä et al., 2000). Other invertebrate taxa can also be used, for example:

- **Ants in Australia**: an ant monitoring system was initially developed for assessing restoration success following mining, with the ants effectively being used as environmental indicators. However, research showed that ant richness and abundance also reflected changes in other invertebrate groups, such that they can be a useful biodiversity indicator (Andersen et al., 2004). This is an example of a **cross-taxonomic biodiversity indicator**, whereby one taxon is assumed to be indicative of other taxa.

- **Butterflies in Switzerland**: Pearman & Weber (2007) found that a number of specific butterfly species in Switzerland was closely related to overall butterfly species richness. This is useful because it is comparatively quick and easy to survey overall butterfly species richness, whereas it takes much longer to survey rare species because they are encountered much less often. With the knowledge that overall butterfly species richness correlates well with that of rare species, it is possible to use overall richness as a proxy for monitoring rare species, even if these are not encountered. This is an example of a **within-taxa biodiversity indicator**, whereby a subset of one taxon is used as indicative of another subset of the same taxonomic group.

Using species' population trends as a measure of sustainability

Population trends of indicator species can be used as measures of sustainability. For example, a **composite index** of bat populations in the UK is calculated by summing the national population of eight different species of bat. This composite index is then monitored over time as a Headline Biodiversity Indicator as part of the UK's commitment to the global **Convention on Biological Diversity**. Similarly, bird populations are monitored over time as another Headline Biodiversity Indicator. In this case, four separate guilds (i.e. functional species groups, often related to habitat type or food source) are monitored: farmland birds, woodland birds, waterbirds, and seabirds. The population size was converted to a standardized index, whereby the first year of monitoring has been translated to an arbitrary value of 100, such that increases over time are shown as values >100 and decreases are shown by values <100 (Figure 4.9).

Both birds and bats have been chosen because they are indicative of conditions in the wider countryside. For example, bats are sensitive to changes in land use practices, such as agricultural intensification, development, and habitat fragmentation, which affect roost sites and populations of the nocturnal insects on which they feed. Many bird species occur exclusively or largely in certain habitats, such as in woodland, on farmland, or at sea. Examining the trends of avian guilds separately gives information on topics such as the effects of agricultural and climatic change, woodland management, and fish stocks. Also, on a more pragmatic level, the data on both taxa are widely available, mainly through volunteer and citizen science data collection schemes.

4.5.2 Caveats when using biodiversity indicators.

It is important not to overstate conclusions based on using one taxonomic group, or a selection of species from one taxonomic group, as a proxy for overall

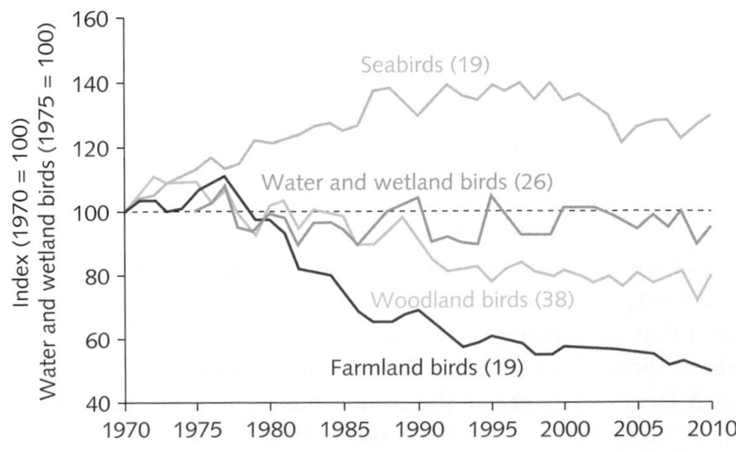

Figure 4.9 Sustainability indicators for UK – change in combined population size of birds in different guilds relative to an index of 100 at the start year (1970).

Source: DEFRA, RSPB, BTO under Crown Copyright 2013. Contains public sector information licensed under the Open Government Licence v3.0.

species richness or population change. Ecology is complex and dynamic, and it must be recognized that simplifying ecosystems so they can be characterized by one parameter will never give a fully comprehensive picture. As with environmental indicators, there is a considerable trade-off between index accuracy, complexity, and feasibility.

It should also be noted that composite metrics made up of information on multiple species, such as the sustainability indicators that set an arbitrary start date, are prone to misuse. This is because small increases in some species (e.g. Eurasian skylark *Alauda arvensis*) populations are lauded as a conservation success, despite the fact that populations are still tiny compared with the past. Moreover, increases in one, relatively common, species can offset declines in several species of conservation concern. This has been seen in the woodland bird example, where decline in rare migrants, such as pied flycatchers, *Ficedula hypoleuca*, and redstarts, *Phoenicurus phoenicurus*, are being offset by increases in resident birds, such as great tits, *Parus major*.

FURTHER READING FOR THIS SECTION

Comparison of different types of biodiversity indicator: Kati, V., Devillers, P., Dufrêne, M., Legakis, A., Vokou, D., & Lebrun, P. (2004) Testing the value of six taxonomic groups as biodiversity indicators at a local scale. *Conservation biology*, Volume 18, 667–675.

Using biodiversity indicators on a global scale: Pearson, D.L. & Cassola, F. (1992) World-wide species richness patterns of tiger beetles (Coleoptera: Cicindelidae): indicator taxon for biodiversity and conservation studies. *Conservation Biology*, Volume 6, 376–391.

UK sustainability indicators: Department for Environmental, Food, and Rural Affairs (2013) Sustainable Development Indicators. Available at: https://www.gov.uk/government/uploads/system/uploads/attachment_data/file/223992/0_SDIs_final__2_.pdf (accessed 19 July 2016).

4.6 Reconstructing historic landscapes.

Most ecological indicator systems are used because they speed up the process of collecting environmental data, reduce the costs, or because they confer some other practical or pragmatic advantage (see Section 4.3). However, it is also possible to use ecological indicators to study environments that no longer exist. Historical records often provide information on habitat and ecological changes at a particular site, but where these records are not available, changes in ecological and environmental conditions can be reconstructed using **palaeo-biomonitoring**.

Monitoring fluctuations and variations in fossilized plants can be a powerful tool for inferring changing ecological or environmental conditions (for example, moisture, temperature, nutrient availability). The most commonly used technique is **palynology**, which is the study of pollen grains. Pollen grains are physically robust, and can be retrieved from peat and lake sediments. These soft sediments accumulate slowly over thousands of years, preserving evidence of the local and surrounding vegetation in discrete layers throughout the sequence. Coring techniques can be employed to retrieve a column of these sediments, which preserves the record without disturbing the evidence.

The pollen present at different depths of a core can be used to indicate the vegetation, and therefore the environmental conditions, at the time the sediment was deposited. For example, an abundance of tree pollen in one layer and a lot of sedge *Carex* pollen in a lower (older) layer provides good evidence that the site was once wooded and, before that, marshy. **Dating methods** such as radiocarbon dating can be used to form a chronology of the changing vegetation. This means that it is possible to go beyond relative dating (deeper = older) to get actual dates that are fairly accurate.

Analysis of pollen from a core at Nelson Lake, Illinois, USA, that had been deposited over 17,000 years shows the environment changing from a cool and wet climate (indicated by abundant sedge and

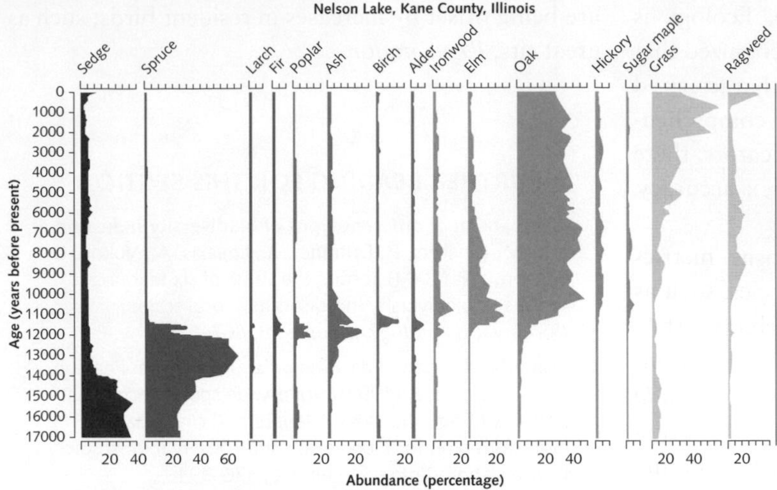

Figure 4.10 Pollen diagrams for Nelson Lake, Illinois, USA. For interpretation, see main text.

Source: Data from Illinois State Museum.

spruce *Picea*), through a warmer but still moist climate (indicated by ash *Fraxinus*, aspen *Populus*, and birch *Betula*) to a drier woodland environment with grassy clearings (indicated by oak *Quercus* and grass Poaceae pollen), but probably still with wetter areas and/or high rainfall (indicated by ash and elm *Ulmus* still being present). This is shown most clearly on a pollen profile with time/depth on the *y*-axis and the different species shown along the *x*-axis. The profile for this site is shown in Figure 4.10.

It is also sometimes possible to infer human activity patterns, such as farming, land drainage, and forestry practices, as well as the arrival of non-native species. For example, in New Zealand, analysis of peat cores has shown marked change from forest pollen to reed *Phragmites* pollen and bracken *Pteridium* spores in the late 1200s and early 1300s (McGlone, 2009). Along with evidence of charcoal burning, this change in the vegetation pattern suggests that forests were deliberately burned by human settlers to make way for crop production. The arrival with European settlers of non-native species such as dandelion *Taraxacium* is also evident.

● FURTHER READING FOR THIS SECTION

Introduction to using palynology to infer past environmental conditions: Traverse, A. (2008) *Paleopalynology*. New York, NY: Springer.

4.7 Conclusions.

Many different ecological indicator frameworks allow species presence and abundance to be used as a proxy for environmental conditions, biological processes, and biodiversity patterns. All indicator systems are founded on key ecological concepts, such as niche, tolerance range, adaptation, bioaccumulation, and ecological succession, and it is vital to understand these concepts in order to understand how ecological indicators work and how new, or improved indicators might be developed. Ecological indicators have many advantages, and are an important and expanding part of the Applied Ecologist's toolkit. However, it is also important to be aware of the potential pitfalls of using ecological indicators so that they are used appropriately and results are interpreted correctly. A brief summary is given in Table 4.2.

Table 4.2 Advantages and disadvantages of using ecological indicator systems.

Advantages	Disadvantages
Data can often be gathered quickly and without specialist equipment.	Ecological complexity is simplified considerably in many ecological indicator frameworks.
Results provide a long-term insight into environmental parameters, rather than the 'snapshot' often provided by chemical/physical testing. For example, if a 3-year-old dragonfly is found in a pond, the conditions in that pond must have been within the tolerance range for that species throughout that whole time period.	Data might not be as accurate or robust as chemical/physical measurements (environmental indicators), or consider ecological processes and biodiversity trends directly (biological and biodiversity indicators).
Data can be used to demonstrate effects of an environmental parameter, or change therein, on biota directly (rather than this needing to be quantified separately).	Ecological indicator frameworks are often developed based on expert opinion, rather than actual data, and are often not fully tested.
Ecological indicators can be used in situations when it is not possible to quantify a specific environment directly (because it no longer exists or its location is unknown).	The effectiveness of ecological indicator frameworks can be affected by other variables or time of year—this is often unknown and untested.
	Lag time can be problematic if it is not allowed for, especially in systems used to assess abiotic change.

ONLINE ACTIVITY

Go to www.oxfordtextbooks.co.uk/orc/goodenough/ to download the activity that accompanies this chapter.

CHAPTER 4 AT A GLANCE: THE BIG PICTURE

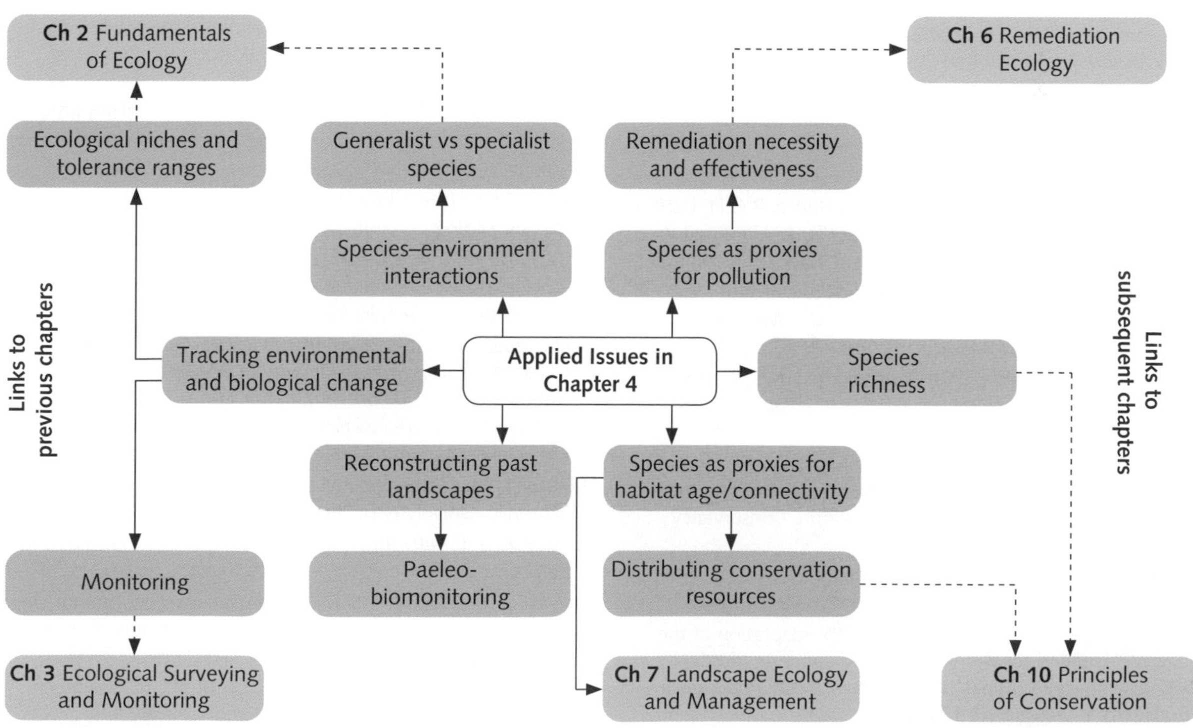

REFERENCES

Andersen, A.N., Fisher, A., Hoffmann, B.D., Read, J.L., & Richards, R. (2004) Use of terrestrial invertebrates for biodiversity monitoring in Australian rangelands, with particular reference to ants. *Austral Ecology*, Volume 29, 87–92.

Breine, J.J., Maes, J., Quataert, P., Van den Bergh, E., Simoens, I., Van Thuyne, G., & Belpaire, C. (2007) A fish-based assessment tool for the ecological quality of the brackish Schelde estuary in Flanders (Belgium). *Hydrobiologia*, Volume 575, 141–159.

Chandler J.R. (1970) A biological approach to water quality management. *Water Pollution Control*, Volume 69, 415–421.

Connell, J.H. (1978) Diversity in tropical rain forests and coral reefs. *Science*, Volume 199, 1302–1310.

Conti, M.E. & Cecchetti, G. (2001) Biological monitoring: lichens as bioindicators of air pollution assessment—a review. *Environmental Pollution*, Volume 114, 471–492.

Ellenberg, H. (1974) Indicator values of plants in Central Europe. *Scripta Geobotanica*, Volume 9, 3–122.

Estrabou, C., Filippini, E., Soria, J.P., Schelotto, G., & Rodriguez, J.M. (2011) Air quality monitoring system using lichens as bioindicators in Central Argentina. *Environmental Monitoring and Assessment*, Volume 182, 375–383.

Fishar, M.R. & Williams, W.P. (2008) The development of a biotic pollution index for the River Nile in Egypt. *Hydrobiologia*, Volume 598, 17–34.

García-Criado, F., Tomé, A., Vega, F.J., & Antolin, C. (1999) Performance of some diversity and biotic indices in rivers affected by coal mining in northwestern Spain. *Hydrobiologia*, Volume 394, 209–217.

Hasselbach, L., Ver Hoef, J.M., Ford, J., Neitlich, P., Crecelius, E., Berryman, S., Wolk, B., & Bohle, T. (2005) Spatial patterns of cadmium and lead deposition on and adjacent to National Park Service lands near Red Dog Mine, Alaska. *Science of the Total Environment*, Volume 348, 211–238.

Hawksworth, D.L. & Rose, F.R. (1970) Qualitative scale for estimating sulphur dioxide air pollution in England and Wales using epiphytic lichens. *Nature*, Volume 227, 145–148.

Hill, M.O., Mountford, J.O., Roy, D.B., & Bunce, R.G.H. (1999) *Ellenberg's Indicator Values for British Plants*. Huntingdon: ECOFACT Technical Annex, Institute of Terrestrial Ecology.

Hilsenhoff, W.L. (1977) Use of arthropods to evaluate water quality of streams. *Technical bulletin of the Wisconsin Department of Natural Resources*, Volume 100, 1–15.

Holt, E.A. & Miller, S.W. (2010) Bioindicators: using organisms to measure environmental impacts. *Nature Education Knowledge*, Volume 3, 8.

Hornby R.J. & Rose F. (1987) *The Use of Vascular Plants in Evaluation of Ancient Woodlands for Nature Conservation in Southern England*. Peterborough: Nature Conservancy Council Report, UK.

Hutchinson, G.E. (1957) Concluding remarks. *Cold Spring Harbor Symposia on Quantitative Biology*, Volume 22, 415–427

Junqueira, V.M. & Campos, S.C.M. (1998) Adaptation of the 'BMWP' method for water quality evaluation to Rio das Velhas watershed (Minas Gerais, Brazil). *Acta Limnologica Brasiliensia*, Volume 10, 125–135.

Kati, V., Devillers, P., Dufrêne, M., Legakis, A., Vokou, D., & Lebrun, P. (2004) Testing the value of six taxonomic groups as biodiversity indicators at a local scale. *Conservation Biology*, Volume 18, 667–675.

Kolkwitz, R. & Marsson, M. (1902) Grundsätze für die biologische Beurteiling des Wassers nach seiner Flora und Fauna [Principles for the biological assessment of water bodies according to their flora and fauna]. *Mitt Königlischen Prüfungsanstalt für Wasserversorgung und Abwässerbeseitigung zu Berlin*, Volume 1, 33–72.

Larsen Vilsholm, R., Wolseley, P.A., Søchting, U., & Chimonides, P.J. (2009) Biomonitoring with lichens on twigs. *The Lichenologist*, Volume 41, 189–202.

Llop E., Pinho, P., Matos, P., Pereira, M.J., & Branquinho, C. (2012) The use of lichen functional groups as indicators of air quality in a Mediterranean urban environment. *Ecological Indicators*, Volume 13, 215–221.

McGeoch, M.A. (1998) The selection, testing and application of terrestrial insects as bioindicators. *Biological Reviews of the Cambridge Philosophical Society*, 73, 181–201.

McGlone, M.S. (2009) Postglacial history of New Zealand wetlands and implications for their conservation. *New Zealand Journal of Ecology*, 33, 1–23.

Niemelä J., Kotze J., Ashworth A., Brandmayr, P., Desender, K., New, T., Penev, L., Samways, M., & Spence, J. (2000) The search for common anthropogenic impacts on biodiversity: a global network. *Journal of Insect Conservation*, Volume 4, 3–9.

Payne, R.J., Thompson, A.M., Standen, V., Field, C.D., & Caporn, S.J. (2012) Impact of simulated nitrogen pollution on heathland microfauna, mesofauna and plants. *European Journal of Soil Biology*, Volume 49, 73–79.

Pearman, P.B. & Weber, D. (2007) Common species determine richness patterns in biodiversity indicator taxa. *Biological Conservation*, Volume 138, 109–119.

Peterken, G.F. (1974) A method of assessing woodland flora for conservation using Indicator Species. *Biological Conservation*, Volume 6, 239–245.

Preston, F.W. (1960) Time and space and the variation of species. *Ecology*, Volume 41, 611–627.

Rose, F. (1999) Indicators of ancient woodland: the use of vascular plants in evaluating ancient woods for nature conservation. *British Wildlife*, Volume 10, 241–251.

Schirokauer, D., Geiser, L., Bytnerowicz, A., Fenn, M., & Dillman, K. (2014) *Monitoring Air Quality in Southeast Alaska's National Parks and Forests: Linking Atmospheric Pollutants with Ecological Effects*. Natural Resource Technical Report. Fort Collins: National Park Service.

Tainton, N.M. (1999) *Veld Management in South Africa*. Scotsville: University of Natal Press.

Tansley. A.G. (1949) *Britain's Green Mantle: Past, Present and Future*. London: George Allen and Unwin.

Trelease, S.F. & Trelease, H.M. (1938) Selenium as a stimulating and possibly essential element for indicator plants. *American Journal of Botany*, Volume 25, 372–380.

Wronski, T., Dusabe, M.C., Apio, A., Hausdorf, B., & Albrecht, C. (2015) Biological assessment of water quality and biodiversity in Rwandan rivers draining into Lake Kivu. *Aquatic Ecology*, Volume 49, 309–320.

PART 3

Managing

Ecological management can take many forms and be undertaken for a plethora of different reasons. Sometimes management can be undertaken to control or reduce threats, for instance, those caused by human development. This is discussed in **Chapter 5**, which explores the role of Ecological Impact Assessment in buffering development impacts on ecological value.

In other situations, management can be undertaken to 'put things right'—to restore good ecological conditions to an area that, for example, has become polluted. This is discussed in **Chapter 6**, which focuses on the value of remediation and the role of many different types of species, particularly plants and microbes, in achieving this aim.

While Chapters 5 and 6 generally consider site-level management, the importance of managing species at a landscape scale becomes the focus of **Chapter 7**. This examines spatial patterns and how these can be influenced by humans negatively through habitat loss, degradation or habitat fragmentation, and positively though habitat creation and initiatives to improve ecological connectivity to allow species to move through the landscape more easily.

In many cases, improving connectivity is ecologically beneficial, but **Chapter 8** explores management of non-native species that have moved into new areas as a result of human activity, travel or trade. **Chapter 9** follows on from this to explore management of any species—native or non-native—that is detrimental to human activity, often economically, and is thus regarded as a pest. Such management can involve chemical, physical and biological techniques, all of which are considered in this chapter.

5 **Ecological Impact Assessment**

6 **Remediation Ecology**

7 **Landscape Ecology and Management**

8 **Non-native Species Management**

9 **Pest Management**

5 Ecological Impact Assessment

5.1 Introduction.

Assessing the ecological impacts of proposed changes in land use, such as development of new houses or industrial units, is one of the fastest-growing areas of Applied Ecology and a key aspect of ecological consultancy. This appraisal is typically undertaken through a formalized Ecological Impact Assessment (EcIA) procedure. Through this process, **current ecological conditions** can be assessed, likely **impacts** predicated, and recommendations formulated to avoid, reduce, or buffer negative effects through the process of **mitigation**. If implemented properly, EcIA can be a powerful tool in sustainable development planning, while also providing a scientifically defensible approach to ecology and ecosystem management (Treweek, 1999), and it is much better to take a preventative approach than for remediation (Chapter 6) to be necessary.

The assessment of development impacts on the environment, including ecology, is becoming ever more important in a global context. There is a formal framework for such assessments in over 140 countries worldwide (Donnelly et al., 1998), and the number of assessments undertaken per year is substantial and growing. In 2011, an estimated 16,000 assessments were undertaken in the EU, with around 600–700 in the UK alone in the early 2000s (up from an estimated 300–400 a decade earlier; Institute for Environmental Management and Assessment, 2011, and references therein). This growth is due to two interlinked factors:

1. increasing urbanization, industrialization and infrastructure development, all of which are underpinned by human population expansion;
2. growing concern for the environment, partly due to greater understanding of ecosystem services and the rise of the sustainable development agenda.

This chapter provides an overview of Ecological Impact Assessment and considers how baseline conditions are assessed, impact predictions are formulated, and mitigation measures are applied. The focus will be on the practical application of EcIA with examples to highlight the realities of using the process in the real world, and links to **legislation** and **policy** as appropriate. The chapter draws on two case studies—one to highlight the entire EcIA process from initial plans to ongoing monitoring (Case Study 5.1 on Cardiff Bay Barrage, UK) and a second to focus on more unusual mitigation strategies (Case Study 5.2 on green bridges in Banff, Canada).

5.2 What is Environmental Impact Assessment?

An Environmental Impact Assessment (EIA) is a formal process through which the impact of a proposed development is considered. Because an Ecological Impact Assessment (EcIA) is usually undertaken as a discrete part of this wider EIA, and is set within the EIA legislative framework, it is worth considering the EIA process to get some broader context (this section) before moving on to EcIA (Section 5.3 and onwards).

5.2.1 Aim and purpose of EIA.

The overall purpose of an EIA is to objectively assess the impacts of a proposed development on many different aspects of the environment. It involves desk-based research and synthesis of any relevant secondary data, followed by collection of primary data at the proposed site, sometimes over a substantial time period of months or even years. There are three overarching aims of an EIA:

1. to provide information on whether a proposal conforms to legislation;
2. to assess whether a proposal is environmentally acceptable;
3. to ensure that adverse environmental impact is minimized.

It is important to note that 'development' is used in EIA (and EcIA) terminology very broadly. Focal developments can include building homes or factories, or creating new infrastructure such as road/rail networks, energy generation plants, waste disposal sites, and quarrying/mining areas.

5.2.2 Legislative history and current context of EIA.

The first environmental legislation that established a requirement for EIA anywhere in the world was the US National Environmental Policy Act 1969. The adoption of this act, and subsequent publicity that it received, led to other countries adopting similar legislation. By the early 1990s, over 40 countries had EIA legislation (Robinson, 1992), with key legislative acts and processes at that time, including the Resource Management Act 1991 in New Zealand and the Environmental Assessment and Review Process (EARP) in Canada (later replaced with the Environmental Assessment Act 2012).

In most countries, EIA legislation is solely a national responsibility, both in terms of creation and application. Within the EU, however, the situation is a little more complex as the legislative framework is **two-tiered:**

- The **first (higher) tier** was originally the Environmental Impact Assessments Directive 85/337/EEC as amended in 1997, 2003, and 2009. Because the amendments to the original directive created a somewhat unwieldy legislative structure, the original directive and all amendments were brought together and codified in 2011 as Directive 2011/92/EU. As with other EU directives, member states become signatories of the directive, and are required to meet its targets and thresholds. Importantly though, the legislative mechanism through which each county achieves this is entirely a national decision.
- The need for one or more national acts is the **second (lower) tier** in the legislative framework. For instance, in the UK, the administration of the EU directive, plus national-level thinking on EIA, is undertaken through the Town and Country Planning (Environmental Impact Assessment) Regulations 2011, while in Austria the relevant legislation is the Environmental Impact Assessment Act 2000.

The EU two-tier legislative framework has the advantage of allowing individual countries to customize their position on EIA in relation to national priorities. It also means that individual countries can embed EIA policy in existing legislation, where it exists. However, the downside is that, despite the Europe-level requirements being consistent, there is a considerable difference in the way that different member states address the EIA process, especially in terms of deciding what projects need an EIA (Section 5.2.3) and exactly how the process is undertaken.

In addition to direct EIA legislation, environmental assessment is usually subject to a whole host of other national Acts, and, for EU member states, other European Directives too. In most cases, such legislation has not been put in place specifically for the purposes of EIA, but is still relevant to the EIA process. For example, whereas country-specific EIA legislation often highlights the need to consider legally protected species, the species thus covered are normally listed in national legislation (in the UK the primary legislation is the Wildlife and Countryside Act 1981 (as amended), while in the USA it is the Endangered Species Act 1973, and in Australia it is the Threatened Species Conservation Act 1995). National Acts can also be important in surveying and mitigation processes that involve handling species that are specially protected as these can only legally be undertaken by licensed personnel.

5.2.3 What developments need an EIA?

EIAs are not needed for every development—they only need be undertaken in situations where there could be substantial impacts on the environment. Typically, countries have a list of developments that always need an EIA (referred to as a mandatory EIA) or, conversely, that are always exempt from an EIA. In the latter group will be minor developments, especially in circumstances where there is no new building involved (for example, converting a corner shop into a residential property). There is also a 'grey area' between these two extremes where developments need to be **screened** to determine if an EIA is necessary or not. This will depend on:

- **Characteristics of the development proposal itself:**
  - size;
  - specific use of resources;
  - amount and type of waste produced (and pollution potential).
- **Location of the proposed development:**
  - existing land use;
  - **environmental sensitivity**—presence of rare or valued ecological/environmental components;
  - **presence of statutory designations**—green belt land, conservation areas, etc.;
  - **surrounding land use**—how the site fits into surrounding landscape and how it relates to nearby human populations, especially densely populated areas.
- **Characteristics of the potential impact:**
  - probability;
  - importance (severity);
  - geographical extent (magnitude);
  - complexity.

In the UK, developments always needing an EIA are referred to as **Schedule 1** projects, so-called because they are listed under Schedule 1 of the Town and Country Planning (Environmental Impact Assessment) Regulations 2011. The underpinning criteria that determine whether a project is Schedule 1 within the national legislation are exactly the same as listed in Annex 1 of the European EIA Directive 2011/92/EU. This means that those developments subject to a mandatory EIA in the UK is dictated by directly what is stipulated at European level, as is standard in all member states. Examples of Annex 1 projects are given in Table 5.1.

If a development does not fall under the auspices of Schedule 1, it does not automatically require an EIA. In such cases, a decision still needs to be made about whether the project requires a discretionary EIA, through a process called screening.

Screening of proposals

Most countries that have an EIA framework have screening procedures. In the EU, the types of projects that need to be screened are listed in Annex 2 of the European EIA Directive 2011/92/EU, but there are two main ways such projects can be screened, depending on a country's individual preference:

1. **Development-based:** some countries require a screening procedure for all projects of a certain type (e.g. Romania for shopping centres; Denmark for all non-hazardous waste landfill sites; Bulgaria for road construction) (IMPEL, 2012). These are sometimes called catch-all approaches.
2. **Threshold-based:** some countries compare proposals to pre-decided thresholds to determine whether an EIA is needed. Crucially, the

Table 5.1: Example developments within the EU that would fall under the remit Annex 1 (mandatory EIA) and Annex 2 (discretionary EIAs based on nationally determined thresholds or catch-all approaches) of Directive 2011/92/EU.

Annex	Type	Examples
1	Energy	Crude oil refineries, nuclear power stations, dams for hydroelectric power
	Industry	Iron/steel smelting, chemical production/processing
	Infrastructure	Long-distance train lines, motorways, and other major roads
	Waste	Landfill of hazardous waste, chemical treatment, or incineration of any waste
	Resources	Large-scale abstraction of groundwater, petroleum, gas or aggregates through open-cast removal (i.e. not mines); also major pipelines to move resources
2	Agriculture	Use of uncultivated land, large-scale irrigation, intensive livestock installations
	Energy	All medium-scale energy projection, including wind farms
	Industry	Large-scale light industry; medium-scale heavy industry (e.g. motor vehicle assembly, paper mills)
	Infrastructure	Industrial estates; small railway/road/tram developments, shipping yards and marinas, medium-scale motorway service stations
	Leisure	Medium-scale permanent camp/caravan sites, golf courses, theme parks, ski runs
	Resources	Medium-scale abstraction of groundwater, petroleum, gas, or aggregates through open-cast removal or mining; also medium-scale pipelines to move resources

thresholds for discretionary EIAs are not set at European level; they are all set nationally and so differ between member states. In the UK, screening thresholds are listed in Schedule 2 of the Town and Country Planning Regulations for all the types of projects listed in Annex 2 of the European EIA Directive 2011/92/EU.

Examples of Annex 2 project are given in Table 5.1. In the UK, guidance for screening Annex 2 projects is provided in Schedule 3 of the Town and Country Planning Regulations. It is important to note that while thresholds in Schedule 2 are fixed and objective, Schedule 3 falls more into the realms of subjective guidelines. Projects going through the screening process are sometimes referred to as **Schedule 3** projects, while those that are ultimately assessed as needing an EIA are referred to as **Schedule 2** projects.

For countries with threshold-linked screening for Annex 2 projects, developments with the potential to affect conservation areas might also examined to determine whether a formal EIA is necessary. In the case of the UK, this includes areas designated at European or global levels, including Special Protection Areas, Special Areas of Conservation, and World Heritage Sites, as well as nationally designated conservation areas, such as National Parks and Areas of Outstanding Natural Beauty (Section 11.4.1). This is another example of where the guideline criteria listed in Schedule 3 of the Town and Country Planning Regulations can be invoked.

This additional screening level means that the need for EIA can be considered for projects in sensitive areas, rather than the decision being based on scale thresholds alone. One of the main criticisms of this approach is that the success of this approach depends on sensitive areas having already recognized biodiversity or heritage value prior to the start of the EIA process.

Who should pay for EIA?

One key question in environmental assessment, which has implications for the type and number of developments that need EIAs, is who foots the bill. In some countries, the EIA process is both initiated and financed by the **statutory regulator** (sometimes referred to as the Competent Authority (CA)). The advantage of this is that it reduces the likelihood of bias compared with the alternative situation where the developer pays.

The '**developer pays principle**' can mean that, in theory, there is potential for environmental surveyors who make discoveries unwelcome to developers, such as the presence of rare or legally protected species, to be pressured into downplaying these (Treweek, 1999). There has been very little research on this, but in rare cases there can even be allegations of corruption. However, the problem with regulators paying is that it requires government, and ultimately tax-payer, funding. This might reduce the number of assessments that it is deemed appropriate to conduct, such that only developments with extremely high potential for causing environmental damage are formally assessed.

5.2.4 Key components and processes of EIA.

Because 'environment' comprises numerous dynamic and interlinked components, an EIA has a very broad scope and is often divided into subdisciplines, which are termed **receptors**. Some receptors relate to the physical environment, some to the human environment, and some to both as outlined in Figure 5.1. Given the broad nature of receptors, experts are often used to cover specific topics such as flora and fauna (ecology experts), water resources (hydrology experts) and landscape and visual impacts (landscape architects). Ideally, an EIA will not only consider these receptors individually, but will also examine interactions between them.

The EIA process starts with **scoping**. This is an initial overview of the proposed development and proposed site to establish what a full EIA needs to cover (as distinct from screening, which decides whether an EIA is needed) (Glasson et al., 2013). For example, if the proposal is an industrial development on marshland, a hydrological survey will be needed; if the proposal is a wind farm on top of a hill, a landscape and visual assessment will be needed. Once scoping is undertaken, the relevant surveys are completed. From this information it is possible to consider potential **impacts** and relevant **mitigation** measures, before providing **recommendations** as to whether or not the project should go ahead and, if so, whether any conditions should be attached to this. This process is explained in more detail in Figure 5.2.

FURTHER READING FOR THIS SECTION

Overview of EIA and EIA legislation: Glasson, J., Therivel, R., & Chadwick, A. (2013) *Introduction to Environmental Impact Assessment*. London: Routledge.

EIA in less economically developed countries: Appiah-Opoku, S. (2001) Environmental impact assessment in developing countries: the case of Ghana. *Environmental Impact Assessment Review*, Volume 21, 59–71.

EU overview: Available at: http://ec.europa.eu/environment/eia/eia-legalcontext.htm

Directive 2011/92/EU: Available at: http://eur-lex.europa.eu/LexUriServ/LexUriServ.do?uri = OJ:L:2012:026:0001:0021:EN:PDF

Differences in EIA between EU member states: IMPEL (2012) The implementation of the Environmental Impact Assessment on the basis of precise examples. Available at: http://ec.europa.eu/environment/eia/pdf/IMPEL-EIA-Report-final.pdf

UK Town and Countryside Planning Act: Available at: http://www.legislation.gov.uk/uksi/2011/1824/pdfs/uksi_20111824_en.pdf

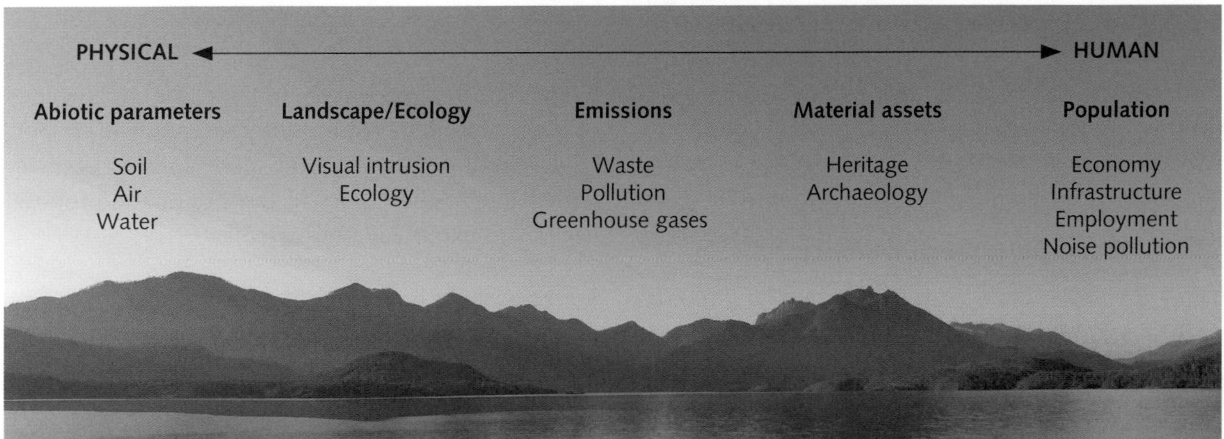

Figure 5.1 Topics (known as receptors) typically covered in an EIA.
Source: Photograph by Anne Goodenough.

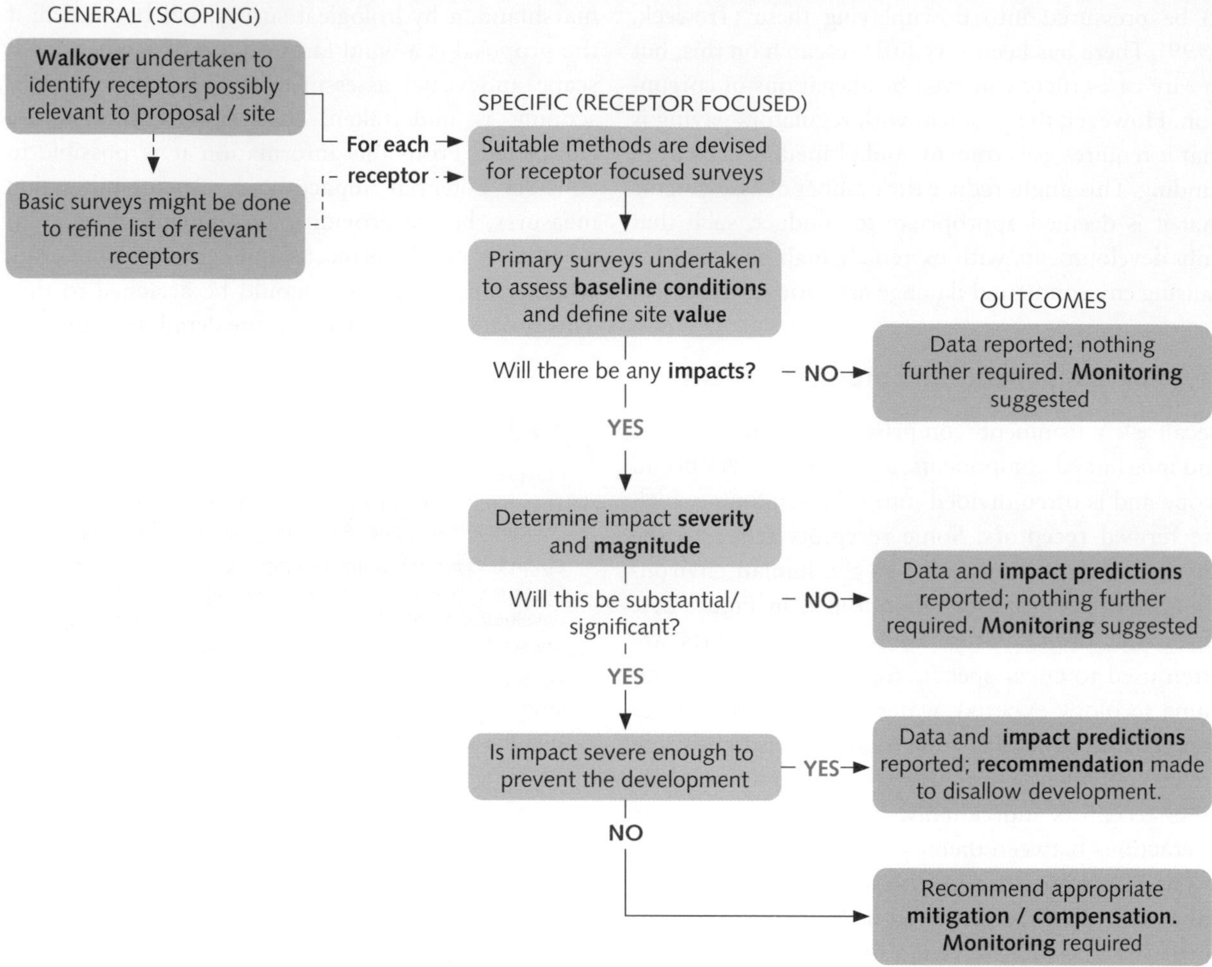

Figure 5.2 Basic process for undertaking an EIA – the standard receptors are listed in Figure 5.1. EcIA follows exactly the same structure but only for ecologically focused receptors (see Table 5.3 for a list of these).

5.3 What is Ecological Impact Assessment?

Just as EIAs exist to assess impacts of development on the environment in general, so EcIA seeks to consider the impact of new developments on ecology in particular.

5.3.1 Aim and purpose of EcIA.

The EcIA process is usually undertaken as a discrete part of a larger assessment into the impacts on the wider environment and landscape within an EIA framework. Ecology is one of many EIA receptors (Figure 5.1), but unlike some other EIA receptors it needs to be considered for virtually all proposals that are subject to an EIA. In other words, while it is comparatively common for, say, the archaeology or air receptors to require little attention in scoping studies and for no follow-on surveys to be necessary (either due to the nature of the site or the proposal itself), it is extremely unusual:

- for a site to have no ecological interest;
- for a proposal to have no potential to have an important impact on ecology.

Because of this, and the complexity of the receptor itself, EcIA is a well-defined parallel process to EIA, which exists as a discrete part of that broader

Table 5.2 Similarities and differences between EcIA and EIA.

Similarities	Differences
Exactly the same overall process: scoping to identify key receptor, quantifying baseline conditions, defining value, predicting impacts, suggesting mitigation, making recommendations, and suggesting post-development monitoring (as outlined in Figure 5.2).	Not *always* required as part of EIA if there is no ecological interest or threat (although this is unusual).
Almost always subject to the same legislative framework: the concepts of screening and thresholds are exactly the same.	EcIA *occasionally* carried out to support ecological management, particularly bridging the gap between **monitoring** (Chapter 3) and **conservation** when a change of management at a protected area (Chapter 11) or a reintroduction (Chapter 13) is considered. Also occasionally used to assess impacts of introducing a non-native species for biological pest control (Chapters 8 and 9). Thus, can (rarely) exist independently of EIA and its legislative framework.
Uses the same concept of receptors (topics to be considered).	Actual receptors will be different (EIA = broad and varied; EcIA = focused and specific).

process. The similarities and differences between the two processes are discussed further in Table 5.2.

The overarching aim for an EcIA is to minimize any negative effects of development or land use change on ecology and ecological processes. To achieve this, a successful EcIA should:

- understand the ecology of the site;
- identify highly valued ecological components;
- state impact predictions explicitly and justify them;
- have clear temporal and spatial contexts;
- contain a post-development monitoring plan.

(After Beanlands & Duinker (1984).)

In order to achieve this, Applied Ecologists involved in an EcIA need to:

- ensure data are accurate and of sufficient extent, volume, and quality to be used as the basis of decisions;
- provide an objective and transparent assessment of the ecological effects of the project;
- facilitate a clear, logical, transparent, and comprehensive assessment of the impacts of the development in local, regional, national, and (where appropriate) international contexts;
- set out what steps should be taken to adhere to legal requirements (as a minimum) and best practice (as a target) relating to designated sites, and legally protected, or controlled species;
- ensure all interested parties are involved and informed, including the general public.

(After The Institute for Ecology and Environmental Management (2006).)

The practicalities of achieving an effective EcIA are often more complex than commonly appreciated (see the Interview with an Applied Ecologist for this chapter). However, it is vital that all the stages of the process are completed successfully. An example of EcIA, which highlights these stages, is given in Case Study 5.1 on Cardiff Bay Barrage, a development with profound ecological impacts that necessitated considerable mitigation and compensation. The case study also highlights the way the legislative background can become even more complex than highlighted in Sections 5.2.2 and 5.2.3, with public enquiries and specific legislative Acts sometimes being required.

5.3.2 Ecological receptors used in EcIA.

The overall topic area of 'ecology' is one of the broad EIA receptors (see Figure 5.1 for the full list). Within EcIA there are also receptors, which are much more specific and specialized. These are listed in Table 5.3.

One very powerful EcIA approach is comparison of two or more **competing proposals** or alternative site options. This allows selection of the proposal that causes the least ecological damage and/or the site that has the lowest ecological value. (The concept of ecological value will be discussed further in Section

AN INTERVIEW WITH AN APPLIED ECOLOGIST 5

Name: Lorna Roberts

Organization: LUC (UK-based Environmental and Ecological Consultancy; http://landuse.co.uk/)

Role: Consultant Ecologist

Nationality and main countries worked: British; worked in UK

What is your day-to-day job?

I am a consultant ecologist with a medium-sized consultancy firm, LUC, in the UK. My work is necessarily very varied. A typical week could include carrying out a general Ecological Appraisal one day, undertaking surveys of protected species such as great crested newts *Triturus cristatus*, the next day, supervising ground investigation works the day after that, and ending the week writing technical reports. Project management is also an important part of the EcIA process. Sites often support a complex assemblage of habitats and species and careful planning and co-ordination of surveys is required. Attendance at regular planning meetings is often necessary, particularly with the larger and more complex sites.

What is your most interesting recent project and why?

For the past year I have been involved in the large scale redevelopment of an historic country estate. The site supports a range of protected species, and part of the EcIA focused on impacts on dormouse *Muscardinus avellanarius*, badger *Meles meles*, and great crested newts *Triturus cristatus*. Full protected species surveys were carried out for all these species. We also recorded bat roosts within 35 buildings on the site, and two rare species were recorded during activity surveys. The large number of roosts resulted in complex mitigation being proposed. This included retention of the larger roosts, provision of alternative roosting sites, timing and phasing of works to avoid works to roosts during the most sensitive times of year when bats are breeding/hibernating, part-demolition by hand of some buildings, and exclusion of bats from roosts that could not be retained.

The assessment process was complicated by the client's decision to submit three separate planning applications for the site, based on different change of use applications for the mansion. This essentially resulted in a separate assessment being carried out for each planning application, which is highly unusual, and made the assessment technically challenging for me. In addition, the site lies adjacent to a Special Protection Area (SPA), a site designated under European legislation for its bird interest. This meant that a Habitat Regulations Assessment was required in addition to the EcIA. A Habitat Regulations Assessment is an assessment to determine whether development would result in impacts on the designation features of a European Designated Site—the SPA in this case—and is required by law where development could affect the integrity of such a site.

What are the most satisfying parts of your job?

The knowledge that as a result of the surveys and assessments I have conducted, sensitive ecological features have been protected. Local planning authorities now place an emphasis on enhancement of a site's ecological features and input from ecologists such as me. The maximum benefit for ecology can be achieved through measures such as habitat creation, management of existing habitats to enhance their diversity or suitability for protected species, and design of buildings to include features such as bird boxes, bat boxes, or biodiverse living roofs.

What do you see as the main challenges in EcIA and how can they be overcome?

One of the main challenges in EcIA is engaging the client so that they are receptive to the mitigation and enhancement measures that you propose. All too often, clients can see ecology as merely a box they need to check. Thankfully, new survey methods are helping to provide visual evidence that can be supplied to the client. For example, LUC have recently begun to use infrared technology when carrying out bat surveys. This technology has allowed us to pinpoint roost entrances in internal spaces, and has resulted in some detailed footage of bat behaviour that the client wouldn't normally see. It has also helped to save the client money, which is always a good thing! Together, this helps engage clients in the EcIA process, and they have been happy to put in more comprehensive enhancement measures as a result.

You obviously spend a lot of time out in the field. What about the office-based side of things?

For large and potentially contentious projects, such as the redevelopment of the historic country estate that I mentioned earlier, it is likely that a public inquiry will be required. This is held by the Planning Inspectorate, which will scrutinize the evidence for and against the decision to grant planning permission. The ecologist's role in the public inquiry varies, but they can be required to either attend the public inquiry as an expert witness to give evidence on behalf of their client (this could either be on behalf of the developer, or sometimes on behalf of another interested party, such as the local Wildlife

Trust who may be opposing the development). If the ecologist is called as an expert witness they will be cross-examined by a barrister on potentially any part of the EcIA, including the survey methods, the reasoning behind valuations, or the assessment of impacts and mitigation proposals. In some cases, it is not necessary to attend the inquiry, but a written statement is prepared to be heard as part of the evidence.

What's next for you, and why?

I have now worked with the company for 3 years, and have taken up an exciting opportunity to move to the Bristol office. This will help expand our company's geographic range to the South-West of England and Wales.

Finally, how did you get into EcIA and what advice would you give to others?

I was unsure of exactly what I wanted to do after I finished my undergraduate degree. I went to my university careers officer, who asked me to describe what I would like in a job. It turns out I described ecological consultancy! The careers officer advised me to apply for a Master's degree that fitted with my career aspirations. I increased my volunteering activities during my MSc and met a consultant through the local bat group who offered me some seasonal amphibian and bat survey work. I got a job with LUC in 2013.

Ecological consultancies usually require a Master's degree. When choosing your Master's degree it is important to pick a course that is relevant. There are courses in Applied Ecology specifically targeted towards consultancy, but courses in biological recording or biodiversity are also relevant. Some companies will take on graduates straight from BSc courses, but they usually expect you to also have experience of the types of work involved in consultancy, such as botanical or protected species surveys. You may be able to find work as a seasonal surveyor, but volunteering with local wildlife groups is a good way to build up experience.

5.4; for now this can be thought of as ranking of sites in terms of their ecological importance based on their species, habitats, or ecological function.) However, with the exception of very large projects or projects that invoke considerable public interest, this **comparative approach** is generally underused. Much more typical is the consideration of a single proposal for a single site, which means that the focus has to be on mitigation to address impacts, rather than selecting from a series of potential candidates to minimize impact in the first place. This is contrary to the ecological (and common sense) concept that prevention is better than cure.

5.3.3 Collecting ecological data: scoping and follow-up.

The first stage of any EcIA after screening has been completed is scoping to start to assess **baseline ecological conditions.** It is not possible to predict any change

Table 5.3 EcIA receptors (after Canter, 1996). Processed-based receptors (final column) are those involving ecosystem services performed by species, changes made by species to the environment, or interactions between species that have an impact on wider ecology.

Habitat-based receptors	Flora-based receptors	Fauna-based receptors	Process-based receptors
Type	Trees/shrubs	Land mammals	Ecosystem engineering*
Condition (Section 3.5.2)	Grasses/sedges/rushes	Aquatic/marine mammals	Pollination*
Fragmentation	Flowering plants	Birds	Seed dispersal*
Corridors and connectivity	Aquatic plants	Insects/arachnids	Disease/insect vectors*
	Ferns	Amphibians	Symbiotic relationships*
	Mosses*	Reptiles	
	Lichens*	Fish	
	Fungi*	Molluscs*	
		Crustaceans*	
		Microbes*	

*Often under-considered.

CASE STUDY 5.1

Cardiff Bay barrage, South Wales

In 1913, Cardiff Bay in South Wales was the busiest coal exporting port in the world. It was home to five working ports and thousands of dock workers. However, with the increasing popularity and practicality of other fuel sources, many of the Welsh mines closed. The effect of this was compounded first by the Great Depression in the 1930s; then the Second World War. By the early 1980s the area around the former docks was derelict, the environment was polluted and unsafe, housing was in disrepair, the economy was in tatters, unemployment was at rates of over 50% and crime was extremely high.

In 1986 the Conservative government set up the Cardiff Bay Development Corporation (CBDC) with the aim 'To put Cardiff on the map as a superlative maritime city which will stand comparison with any such city in the world, thereby enhancing the image and wellbeing ... of Wales as a whole'. The task facing the CBDC was a Herculean one, especially as, although there was initial 'pump priming' funding, the area had to become self-sufficient economically very quickly. This required private investment in light industry, retail and leisure, as well as creating demand for, and supply of, residential property.

The main 'problem' as far as the CBDC was concerned was that the centre of the area—the Bay itself—was considered unsightly. The Bay, which was fed by the River Taff and the River Ely, opened into the Severn Estuary with the second biggest tidal range anywhere in the world. This meant the Bay comprised exposed mud flats for 12–14 hours a day, which was felt to be unattractive to investors.

In theory, a solution was easy; build a 1.1-km barrage spanning the entrance to the Bay and impounding the two rivers to create a 200 ha freshwater lake with 12.8 km of permanent waterfront to act as a key redevelopment stimulus. What looked simple on paper, however, was anything but due to the substantial impacts on the Bay's environment. Ecologically, the biggest issue was the loss of the important mudflat habitat. Impounding the rivers changed a saline environment to a freshwater one, altering the entire invertebrate community and, most importantly in terms of legal issues, destruction of high tide feeding ground for birds. Cardiff Bay was designated nationally (Site of Special Scientific Interest), at a European level for its bird assemblage (Special Protection Area) and internationally (Ramsar Convention as an internationally important wetland).

Figure A Cardiff Bay (top) pre- and (bottom) post-barrage.
Source: Images courtesy of Ben Salter/CC BY 2.0.

It took six years for the CBDC to win the case to build the barrage. This was done ultimately in the Cardiff Bay Barrage Act (1993), which was passed following public enquiry as the barrage was considered to be of overriding public interest. Following this legal process, the barrage was constructed (Figure A). The Act did, however, contain numerous conditions regarding environmental standards, monitoring, mitigation, and compensation, as detailed in Table A.

Was the scheme a success? Well, that depends on individual perspective. Economically, the area is now

Table A Ecological issues created by the construction of the Cardiff Bay Barrage, the mitigation/compensation strategies employed and any residual problems, and monitoring undertaken to ensure compliance.

Issue	Details	Mitigation/compensation	Problems and monitoring
Loss of tidal mudflats	Loss of 1% of the protected area, but one of the most important areas for bird feeding.	**Mitigation**: small freshwater wetland created on site (Figure B). **Compensation**: creation of what is now RSPB Newport Wetlands 20 km upstream. This included saline lagoons and saltmarsh.	**Mitigation problems**: did not replace the habitat that was lost; species assemblage different. **Compensation problems**: compensation site is now a fantastic area for wildlife, including rare, vulnerable, and priority species. However, the site does not replace the habitat lost (tide-washed mudflats being replaced with other wetland habitats) and was too far from the original site for the majority of displaced individuals to relocate to. The site took a long time to establish (only really establishing a decade after Cardiff Bay was lost). Cost of long-term management was not factored in so the site was created, but upkeep was not agreed. Site now managed by a charity.
Disrupted fish migration	The upper reaches of the Taff are used by ocean-living fish (500 salmon/800 sea trout) to spawn.	Cardiff Bay Barrage Act stipulated that a fish pass be installed and available at all times (one for high tide and another for low tide).	**Monitoring**: fish movement is tracked using a computerized system.
Risk of hypoxia in the lake	There were 14 sewage outfalls into the Bay. This was not really an issue under natural conditions as any increase in BOD was mitigated by the volume of water and rapid turnover in an open system, but could have caused mass deoxygenation in an impounded lake, killing species very quickly.	Sewage outfalls were rerouted to bypass the lake. Minimum dissolved oxygen level of 5 mg/l stipulated in the Cardiff Bay Barrage Act. Network of aeration pipes placed on Bay bed through which oxygen can be pumped (£3.5 million cost).	**Problem**: acute hypoxia could still occur on occasion despite mitigation, so in 2004 a 'bubbler barge' was purchased (£4.5 million) that could be towed to any part of the lake to supplement oxygen in acute situations. **Monitoring**: nine automated multi-parameter 'buoys' used by Cardiff Bay Harbour Authority to monitor water parameters, including dissolved oxygen concentration, pH level, and turbidity, every 15 minutes.
Eutrophication	High levels of nutrients causing algal blooms. The nutrient input has not gone up, but whereas the eutrophic water used to flow out at low tide, it is now retained.	Skimmers used to remove any algae—4000 m^3 removed in 2001.	
Sedimentation	Whereas sediment transported in the rivers used to flow out to sea as a high energy environment was maintained, now sediment-laden water enters the still lake, all energy is lost, and the sediment is deposited.	Regular dredging is undertaken—£5.5 million in the first 18 months.	Dredging has a detrimental effect on benthic (bottom-dwelling) invertebrates and crustacea.

Figure B Aerial photograph of Cardiff Bay today.

Figure C Water quality is monitored via nine buoys placed around Cardiff Bay; here also providing a place for a cormorant *Phalacrocorax carbo* to dry its wings. Data are in the public domain (https://stormcentral.waterlog.com/public/CHABarrage), ensuring that the monitoring process, and the result of this, are transparent.

thriving. Land prices are higher, unemployment is lower, the area is popular with locals and tourists, there are lots of business and retail developments, and crime rates have dropped. What was a derelict (and often dangerous) place has been revitalized. However, there have been many ecological and environmental issues, the majority of which require long-term management and monitoring (and financial input) to continue to mitigate (Figure C). A once naturally regulated ecosystem requires constant human input to remain healthy. Moreover, there was undoubted ecological and environmental damage that could not be mitigated and for which compensation was inadequate.

Was this the first government-sponsored destruction of a protected wetland in the UK as RSPB Wales claimed at the time? Or were the ecological and environmental issues of little consequence given the economic benefits as the chief executive of CBDC claimed in response? You decide …

FURTHER READING

Cardiff Bay Barrage Act: Available at: http://www.legislation.gov.uk/ukpga/1993/42/pdfs/ukpga_19930042_en.pdf

Cardiff Bay Harbour Authority: Available at: http://www.cardiffharbour.com/

in ecology arising from a development without first understanding that ecology. While more detailed surveys might follow later to deepen knowledge of specific species, habitats, interactions, or processes, initial baseline studies are fairly 'broad brush' in their approach:

- **Aim:** usually quantification of species/taxon presence, rather than more specific parameters, such as exact location, abundance/extent, quality/condition, etc.
- **Focus:** usually focus largely on communities or taxonomic groups, rather than individual species.
- **Method:** usually based on rapid walkover surveys—literally walking a site and noting features of interest—or through quick, simple methods, such as point counts for birds.

Ecological scoping should always start with desk-based research to collate **secondary data** before **primary data** collection is undertaken. This often starts with a basic habitat mapping exercise, which is often extended by classifying the vegetation community (or communities) within each habitat. In the UK this is a two-part process, with habitat mapping undertaken through a standardized method such as Phase One Habitat Surveying (Section 3.5.1), and a follow-on National Vegetation Classification (NVC) being used to classify vegetation communities. The same idea is used in many other European countries, although the names for the NVC stage differ—for instance, National Phytosociological Relevés is used in the Czech Republic. Outside Europe, these two processes are often combined—for example, both the USA and Canada have integrated habitat/vegetation protocols called NVC and CNVC (Canadian NVC), respectively.

National Vegetation Classification systems

Although the terminology and the specifics of NVC methodology differ to some extent between countries, the basic approach is very similar. The process involves making a species inventory of each habitat using a suitably sized plot or series of sub-plots per habitat type or vegetation unit. Abundance is also recorded **semi-quantitatively**, either using DOMIN scores (Section 4.3.3) or percentage cover.

Once field data have been collected, the vegetation community is keyed out using habitat-specific keys (Rodwell, 1991–2000 in the UK) in the same way as an unknown species would be keyed out. There is a series of steps and, at each step, the option that best summarizes the vegetation community is selected until a final classification is reached. In this way, the keying-out process is hierarchical, with each additional stage refining the vegetation classification. In the UK, for example, the final step is a three-part code that translates to the overall vegetation community and subcommunity; for example:

Overall Code = W8b

Habitat Type = W (Woodland)

Community (number) = Number 8 (*Fraxinus excelsior—Acer campestre—Mercurialis perennis* (ash, maple, dog's mercury))

Sub-community (letter) = b (characterized by *Anemone nemorosa* (wood anemone)).

Follow-up studies in EcIA

Once initial scoping has been carried out, it is then often appropriate to identify individual receptors to study in greater depth in follow-up studies, using more detailed field techniques and collecting data over a longer period. This is a hierarchical approach—there is little point, for example, in commissioning a detailed (and expensive) bat survey if a walkover survey has indicated that there are no bats on site. In

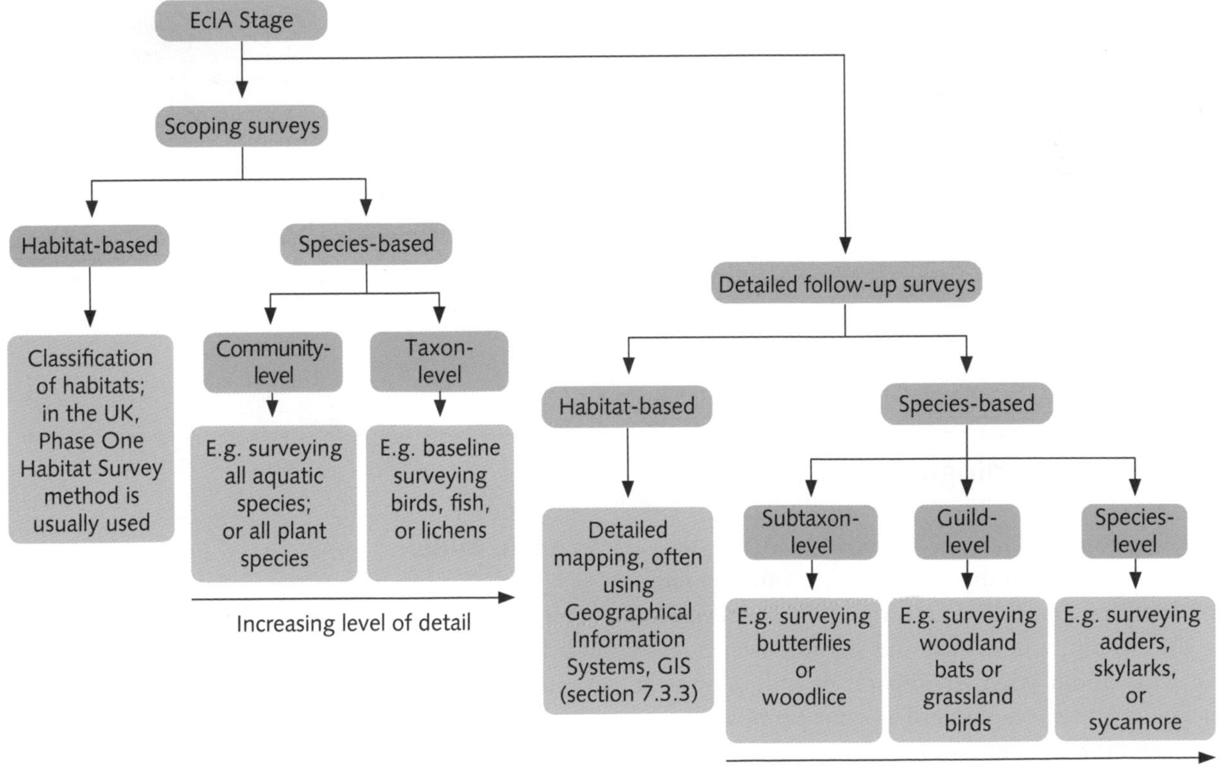

Figure 5.3 Focus of an EcIA at scoping and follow-up stages.

this way, the detailed follow-up surveys undertaken at a site will depend both on the site's ecology (ascertained through the initial baseline studies) and the proposal itself. The aims of such surveys can include:

- Mapping of rare or non-native species, and identification of key features (bat roosts, badger setts, etc.).
- Identification of wildlife corridors (Sections 7.2.3 and 7.6.2).
- Quantification of the abundance or population size of target species.
- Examination of interactions between species.
- Understanding ecology-based ecosystem services, such as pollination.

The focus of surveys that might be carried out within EcIA at both scoping and follow-up levels is detailed in Figure 5.3. The methods used span point-based surveys, quadrat-based surveys, transect-based surveys and full species mapping (Table 3.2), and can encompass both direct and indirect evidence (Sections 3.6.1 and 3.6.2). The survey pathway for one taxonomic group—bats—is shown in Figure 5.4

FURTHER READING FOR THIS SECTION

EcIA overview: Treweek, J. (1999) *Ecological Impact Assessment*. Oxford: Blackwell Science Ltd.

EcIA practitioner guidelines in the UK: Institute for Ecology and Environmental Management (2006) Guidelines for Ecological Impact Assessment in the United Kingdom. Available at: http://www.cieem.net/data/files/Resource_Library/Technical_Guidance_Series/EcIA_Guidelines/TGSEcIA-EcIA_Guidelines-Terestrial_Freshwater_Coastal.pdf

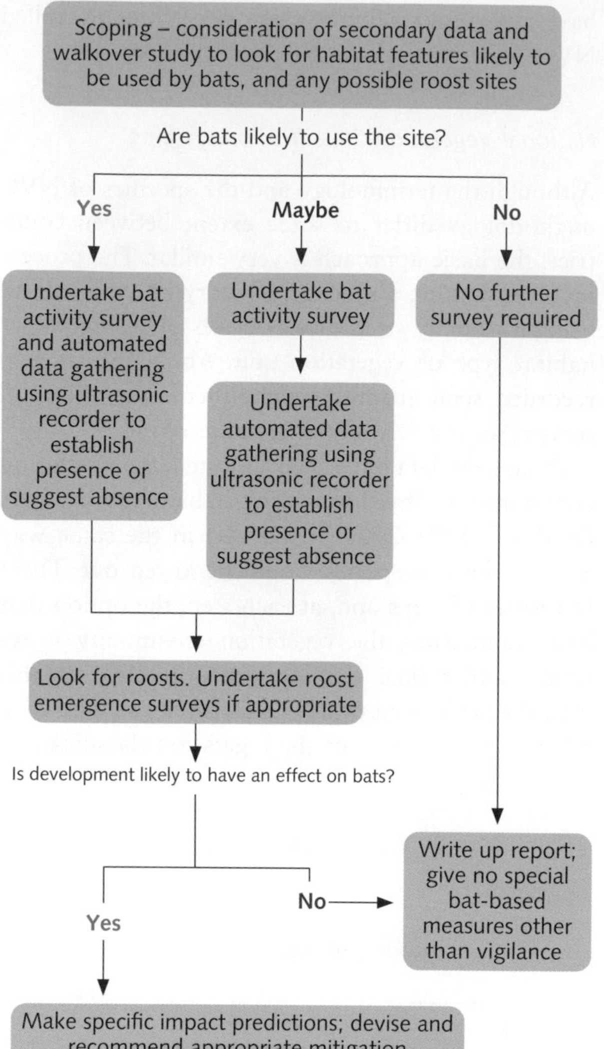

Figure 5.4 Progressing from a scoping study to a detailed follow-up survey – an example for bats.

5.4 Defining ecological value.

Once ecological data are collected, EcIA provides a framework for valuing sites together with the habitats, species, and individuals that they contain. This is vital to allow informed and sensible decisions to be made about whether impacts of a development on ecology are acceptable. The same proposal might be acceptable at some sites, with one specific habitat matrix and species assemblage, but not at a 'higher value' site.

5.4.1 Site value.

Some sites inherently have higher value than others. The single most important site-level criterion is whether any part of the site has a statutory designation limiting the activities that can be undertaken there. Designations can be ecology-driven (e.g. an area in the UK might be protected by **national designations,** such as Site of Special Scientific Interest, European designations such as Special Protection Areas, or international directives such as Ramsar for wetlands of international importance) (Section 11.4.1), or designations could be driven by **planning policy** (e.g. no-touch zones). The former clearly fall within the EcIA, rather than EIA, remit. The latter can fall within an EcIA remit where planning policies interrelate with landscape-scale conservation (Chapter 7).

Sites that have high habitat heterogeneity—a diverse range of different habitats—usually have higher value than sites with low heterogeneity. The reasons for this are linked to the concept of ecological niche (Section 2.5) as sites that have more habitats generally have more niches, and more niches usually means greater species richness. Sites with high **species richness**, whether or not caused by high habitat heterogeneity, are generally more valuable than sites with low species richness. Other reasons for high richness include a level of disturbance that allows pioneer and climax species to co-exist (Section 2.8.2), or careful management.

Sometimes, a site is considered especially important because it supports a substantial proportion of the national or international population of a species. This is a site-level value metric because it is based purely on population size at the site and does not depend on species characteristics such as rarity. Typically, a site supporting >1% of the international population of a species is regarded as **internationally important** and a site supporting >1% of the national population of a species is regarded as **nationally important.** It should be noted that in some cases a site might support a large proportion of a species' population at a national level (20%, say) but still have relatively low value internationally if the species is common in other countries. This type of debate between national and international priorities is considered further in Section 5.2.2 from an EIA perspective and in Chapter 10 from a conservation perspective.

It is worth pausing here to consider exactly what the concept of a site 'supporting' a species actually means, and thus under what conditions a site is viewed as especially valuable in this context. There are several key questions:

- Should the notion of a site supporting a species be restricted to specific 'types' of population, such as a breeding population?
- For how long does a population need to use a site before it is considered to support that population?
- What should happen for migratory species? For example, does a site that large numbers of individuals use as a stop-over site to refuel on a long migration count?
- Can any supporting role be considered? For example, could a hedgerow be considered to support a known (rare) bat colony roosting 5 km away?

In most contexts, 'support' refers to breeding populations or, on occasions, sites that are vital for supporting a species for a major part of its annual life cycle (e.g. wintering population for a migratory species). Stop-over sites would not normally be considered, but an argument could be made to do so if a specific site was essential, for example, by offering a unique and essential food resource. Generally, a site would only be considered as supporting a species if the species actually resides at that site (at least for part of its life cycle), but again, an argument could be made for this remit to be extended in situations where a site is essential to the viability of a species that uses the site but is not actually resident there.

Finally, sites might have a high value because of their **social value.** Although this might not be seen as an ecological consideration in its own right (an ecologist might view this as a social consideration to be considered through the population or material assets receptions of an EIA: Figure 5.1), public connections with a site, and thus the value that they place on it, can be driven by its level of biodiversity. For example, studies have shown that there is a correlation between a person's sense of well-being in a place and its species richness (e.g. Fuller et al., 2007; Dallimer et al., 2012).

5.4.2 Habitat value.

In the same way that sites can differ in ecological value, so too can habitats. Habitats with inherently high value include those that are:

- **Rare:** habitat rarity can be considered within an international, national, or local context. A terrestrial example of an internationally rare habitat is cork oak *Quercus suber* forests of the Mediterranean region. These woodlands have high plant biodiversity and are also extremely important not only for cork oak itself, but also faunal species such as the endangered Iberian lynx, *Lynx pardinus*, and rely on a delicate balance of management and sustainable use (Bugalho et al., 2011). In terms of aquatic ecosystems, a good example of a globally rare habitat is the chain of soda lakes along the East African rift valley, which have extremely high fish species richness. One stage down from international rarity is national rarity. This includes habitats that are uncommon within a countrywide context. In the UK, if an NVC survey has been undertaken (Section 5.3.3), NVC category can be compared with national data to give an objective classification of rarity. This allows habitats such as MG4, a flower-rich grassland community characterized by rare plants, such as snake's head fritillary *Fritillaria meleagris (*Figure 5.5), to be classified as rare in a national context (Jefferson et al., 2014).

- **Declining**: some habitats are declining, either in extent or condition, as a result of pressures such as urbanization, agricultural intensification, drainage, or climatic change. An example is lowland heath, which occurs on acidic, nutrient-poor soil and supports shrub species such as heather and gorse. It has declined in extent globally, and in many countries nationally (Joint Nature Conservation Committee, 2014).
- **Highly specialized/highly restricted location**: some habitats are very restricted spatially, such as Fynbos in the Western Cape of South Africa. This supports over 9000 plant species with 6200 endemic species that grow nowhere else in an area of under 50,000 km^2 (Goldblatt & Manning, 2002). Indeed, Table Mountain in Cape Town supports 2200 species, more than the entire United Kingdom.
- **Vulnerability**: some habitats are especially vulnerable to human or natural changes. For example, the Convention for the Protection of the Marine Environment of the North-East Atlantic has classified marine priority habitats that are vulnerable to pollution (Figure 5.6). The convention has 15 signatories, including countries with a Western European coast such as France, Iceland, Norway, and Portugal. Luxembourg and Switzerland are Contracting Parties as they are within the River Rhine catchment (OSPAR, 2015).
- **Difficult to recreate**: some habitats, such as ancient woodland, are difficult to recreate in sensible timeframes, so if they are damaged or destroyed it is hard to compensate. Other habitats, such as tide-washed mudflats, are difficult to recreate at all because they rely on ecosystem processes that are hard—if not impossible—to replicate effectively.
- **Useful to humans**: from a very anthropocentric viewpoint, some habitats are particularly useful to humans. For example, peat bogs, which are an important store of carbon, and wetland areas, which can be an important buffer against flooding.

Figure 5.5 Rare MG4 grassland classified through National Vegetation Classification showing typical species composition, including snake's head fritillary *Fritillaria meleagris*.

Source: Image courtesy of Dierk Haasis/CC BY-SA 2.0

Specific habitats may meet multiple importance criteria and may thus be designated as **priority habitats**. In the UK, this has historically been done through the Biodiversity Action Plan database; other counties often have similar frameworks, for example, the Priority Species and Habitats directive in Washington, USA, and the Priority Habitats and Corridors Strategy in Australia.

5.4.3 Species value.

In the same way that not all habitats have equal value, the value placed on different species can also vary. Species with high value include those that are:

- **Legally protected**: species that are legally protected are, in developers' eyes at least, simultaneously the species with the most value on the site and

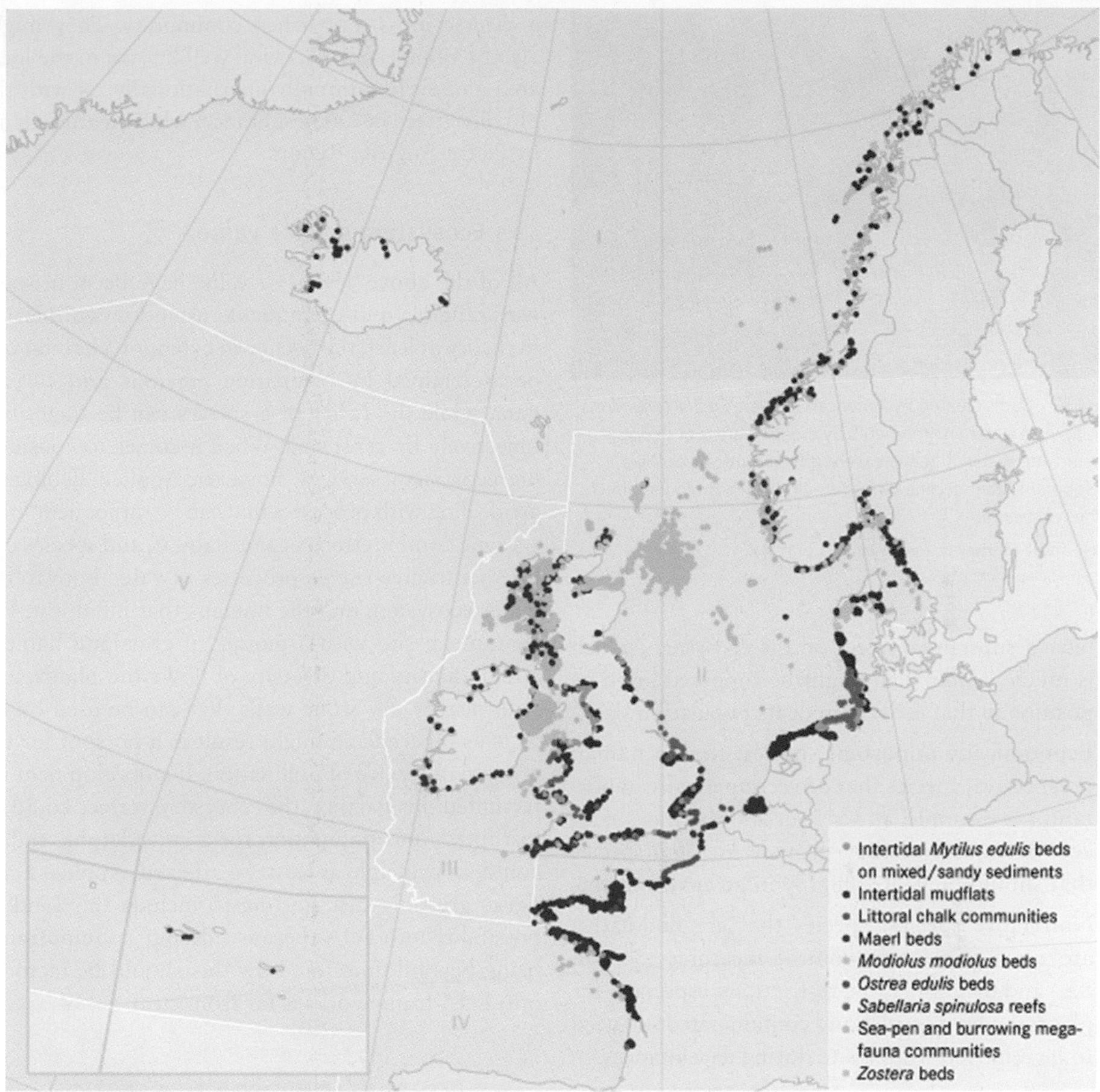

Figure 5.6 Vulnerable marine habitats as classified by the Convention for the Protection of the Marine Environment of the North-East Atlantic.

Source: Image used with permission via OSPAR commission and EMODnet Seabed Habitats.

the least desirable to find. Their presence on a site will, at the very least, require suitable mitigation measures to be initiated and, in some cases, can prevent development going ahead. Species can be protected under national legislation and, for European member states, European statutes that are administered through national legislation.

- **Rare:** like habitats, species can be are rare in international, national, or local contexts. Such species might be listed as endangered on the IUCN red list at an international scale or imperilled on the NatureServe global list (Section 10.3.2), or might be listed as rare at a narrower spatial scale (e.g. BAP-listed species or red/amber-listed birds in the UK, or species on the national or regional NatureServe lists: Section 10.3.3).
- **Ecosystem engineers:** some species modify the ecosystem in which they live and thus have a profound effect on that site. An example is the American Beaver *Castor canadensis* which alters aquatic environments by building dams (as shown in Figure 5.7). Loss of such a species from a site would

Figure 5.7 Dams created by American beavers *Castor canadensis* alter flowing water environments by creating areas of deeper, slow-moving water. If beavers were affected negatively by a development, their decline or loss would have a profound effect on the ecosystem.

Source: Image courtesy of Tony Beeman/CC BY 2.0.

have a substantial effect on the ecosystem, which is much greater than might be supposed given its position in that ecosystem or its population size.

- **Economically important species:** from a human perspective, species that are economically important, for example, as a resource or because of an ecosystem service they provide are also species that should be valued highly in an environment.
- **Non-native species:** species that are non-native are often subject to control measures (Section 8.6) and there might be restrictions, especially for plants, on how species and contaminated soil need to be removed from a site during development.

5.4.4 Individual organism value.

In some cases, specific individual organisms have a particularly high value. For example, veteran trees are often key habitats in their own right, supporting a diverse moss and lichen community. They might also be valued socially, being well known in the local area and even nationally/internationally, as with the old olive trees in Cagnes, France, which feature in art by Pierre-Auguste Renoir.

5.4.5 Ecosystem service value.

All of the above 'levels' of value have been, in some way, tangible and quantifiable using known metrics. In theory at least, the decline in extent of a habitat can be ascertained by comparing previous and current data, while the rarity of a species can be quantified objectively by censusing. When it comes to considering ecosystem services, however, Applied Ecologists are dealing with processes that can be rather nebulous.

Some combinations of site, habitat, and species can interact to give rise to processes of value both to the wider ecosystem and the humans that inhabit it. For example, a site with a mosaic of grassland habitat, a high density and diversity of flowering plants, and with nearby dry stone walls that can be used by insects as hibernacula might result in a hotspot for the ecosystem service of pollination. If a development interrupted this balance, the ecosystem service could be disrupted, and pollination for a considerable radius could, in principle at least, be affected. Applied Ecologists are only just starting to include this kind of possibility into EcIAs because, despite its importance being beyond question, how this should be factored into EcIA frameworks is far from clear.

FURTHER READING FOR THIS SECTION

Ecosystem services in impact assessment: Albert, C., Aronson, J., Fürst, C., & Opdam, P. (2014) Integrating ecosystem services in landscape planning: requirements, approaches, and impacts. *Landscape Ecology*, Volume 29, 1277–1285.

5.5 Assessing likely impacts of development.

After site-specific ecological components have been quantified and valued (Section 5.4), the EcIA process moves on to predicting impact. Impact predictions are usually split into different temporal stages of the development process. These almost always include a **construction** stage and an **operational** stage. In some cases, particularly for industrial developments and abstraction of raw materials, it is necessary to include **decommissioning** and **regeneration** stages.

5.5.1 Types of impact.

A development can have a direct, indirect, or cumulative impact on ecology. A **direct effect** is simple cause and effect, such as a factory discharging pollutants into an aquatic system. In contrast, an indirect effect involves additional processes occurring between cause and effect. For example, when trees are removed from an area by felling, soil erosion often increases because of the increased velocity of rainfall hitting the ground and, over time, a reduction in root structure binding soil together. Displaced soil can enter waterways, often some considerable distance away, leading to siltation that impacts fish spawning.

Indirect effects are much harder to predict than direct effects because of the additional processes involved and the potential for the effect to occur a considerable distance from the cause. An extreme example is aspergillosis disease in sea fans in coral reefs in the Caribbean and Florida Keys, caused by the soil fungus *Aspergillus sydowii* (Figure 5.8). The fungus seems to be transported via trade winds across the Atlantic in dust from the Sahel region of Africa as a result of desertification caused by urbanization in semi-desert regions (Shinn et al., 2000).

While indirect effects are hard to predict, predicting cumulative impacts is even harder. **Cumulative effects** occur when an ecological threshold, known as a **tipping point**, is reached. For example, power stations and industrial plants often remove water from rivers to use in cooling processes. Because the water is not chemically polluted, it is usually released back into the original water source. However, released water is usually warmer than under natural conditions. If an individual factory releases a small volume of warm water, the result is usually insignificant. However, if multiple factories release warm water, or if one factory releases warm water for extended periods, cumulative thermal pollution results, which can reduce oxygen levels and alter cell enzyme activity in fish causing physiological damage. Other examples of cumulative impacts are gradual habitat loss or fragmentation due to development over time, when each subsequent development only takes a small section of land or increases existing fragmentation by a small amount. The effects of habitat fragmentation are discussed further in Section 7.4. Cumulative impacts are so important that prediction of them, and the effect that they might have, has led to a subdiscipline of **Cumulative Impact Assessment**.

Cumulative impacts are hard to predict because tipping points are often unknown until they have been reached. Even for well-understood impacts, accurate prediction requires information about other

Figure 5.8 Sea fan showing a lesion caused by *Aspergillus* due to infection by wind-blown spores of a soil fungus from Africa. This type of ultra-long-distance indirect and cumulative impact of development or land-use change is almost impossible to predict or mitigate. In this case, the disease has caused the loss of around 60% of the living tissue of the same sea fan between the photograph on the left in July 2002 and the image on the right taken less than a year later.

Source: Image reproduced from Bruno, J. F., et al. (2011) 'Impacts of aspergillosis on sea fan coral demography: modelling a moving target'. *Ecological Monographs*, 81(1), 123–139, with permission from John Wiley & Sons, Inc.

similar pressures in the nearby area, which is often not within the knowledge base or remit of those undertaking an EcIA for one particular site or project. As well as considering whether impacts are direct, indirect, or cumulative, it might also be important to quantify whether they are:

- temporary or permanent;
- one-off, intermittent or continuous;
- immediate or delayed;
- avoidable or unavoidable;
- reversible or irreversible.

5.5.2 The potential for positive impacts.

Although one of the main objects of EcIA is to predict potential negative impacts so that they can be mitigated, it is important not to overlook the fact that a proposal, or one part of a proposal, might have ecological benefits. It is important that these are recognized and, where possible, maximized. For example, a plan to convert amenity grassland with low ecological value into allotments might have considerable **benefits** to wildlife in terms of providing habitat, food and shelter.

Positive impacts can result from an intentional part of the proposal, as in the allotments example above, or can occur because **remediation** of heavily polluted land is needed before development can proceed. The ecological benefits of remediation (Chapter 6) can sometimes outweigh any negative impacts of the development itself.

In some situations, positive impacts occur after the operational phase. For example, the upper Thames river catchment in England holds substantial quantities of glacial Jurassic limestone gravel that is valuable to the building industry as aggregate, primarily for use in concrete. Although the site was originally intensive farmland and had limited ecological value, quarrying areas that used to be fields still had a negative ecological impact, especially when hedgerows were removed. However, flooding of the gravel pits after extraction by groundwater flow has resulted in one of the largest marl lake systems (characterized by clear, alkaline, lime-rich water) in Europe, which support numerous species including rare water plants. The network of lakes, shown in Figure 5.9, contains water bodies that differ in size, depth, and degree of shelving. Between them, they support numerous priority and legally protected species, including great crested newts *Triturus cristatus*, and water vole *Arvicola amphibious*, and the site is protected by a series of protected areas (Harris & Pickering, 2008).

Figure 5.9 The marl lake habitat mosaic at Cotswold Water Park in the Upper Thames Valley, England, was created through gravel extraction.

Source: Photograph of Cleveland Lakes, Cotswold Water Park, view looking west in October 2012 by Robert Bewley, used with kind permission.

5.5.3 Impact magnitude and importance.

There are two fundamental concepts that must be considered when assessing impacts:

- the **magnitude** of the impact;
- the **importance** of the impact.

These two concepts are often confused, but are not the same. The magnitude refers to the spatial scale over which an impact occurs, from extremely localized to very widespread. The importance of an impact takes into account impact severity (how much damage it will cause) and the value of the ecological feature(s) suffering the damage. Impact severity can be considered for both species and habitats on a sliding scale (Figure 5.10) and can vary from negligible to extremely important.

The importance of an impact on a given species or habitat is determined as much by the value of that species/habitat as by the severity of the impact itself. For example, increased mortality of a rare or legally protected species is typically considered to be of greater importance than increased mortality of a very

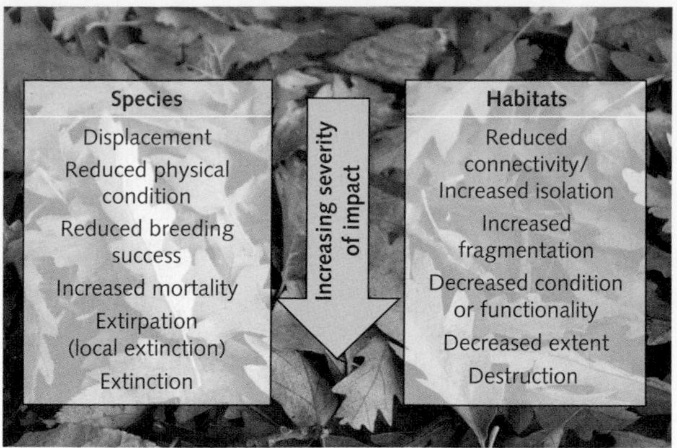

Figure 5.10 Generalized impact severity scale for impacts of species and habitats.

Source: Photograph by Anne Goodenough.

common species. Similarly, a decrease in the condition of a rare, declining, highly specialized, peatbog would normally be regarded as of greater importance than total destruction of a common, widespread amenity grassland with low species richness, even though 'destruction' is a more severe impact category than 'decreased condition' (Figure 5.10).

When determining importance, it is useful to consider local, national, or international contexts. For example, if a development involves destruction of a small section of native broadleaved woodland in Scotland that does not hold any particularly rare or legally protected species, this might be fairly negligible nationally/internationally, but important locally in a landscape where the majority of woodland is coniferous. The most concerning scenario is when there is a severe impact on a highly valued ecological feature, especially if impact is also high magnitude (i.e. occurring over a wide spatial scale).

5.5.4 Predicting impact.

Predicting the likely ecological impacts of a development is extremely challenging as ecosystems are dynamic and include many interacting factors. Impact prediction requires judgements to be made not only about how actions are likely to affect species and habitats, but also how and to what extent those species and habitats are likely to be able to buffer, or respond to, those actions. To make matters even more complex, the majority of the baseline data necessary to formulate predictions have to be collected though primary surveys within very tight timescales. **Uncertainty** is thus high in the majority of impact assessments (Treweek, 1999) due to:

- inherent complexity of ecological systems;
- incomplete understanding of ecosystem processes in general and the specific site involved;
- lack of long-term data, either on the site or on the actual impacts of similar developments on similar sites upon which to base impact predictions.

It is important to recognize that the same habitat or species can have a different level of **ecological resilience** in different landscapes. Ecological resilience can be thought of as the ability of a habitat or species to tolerate a change in conditions or recover from a major ecological disturbance. Variation in this resilience spatially means that the same disturbance can have more of an effect in some environments relative to others. For example, development-induced stress is likely to cause more negative impacts on fragmented landscapes, since fragmentation affects the ability of the ecosystem to buffer effects. Similarly, a population of a given species is more likely to decline as a result of development if it was already small, as phenotypic plasticity—the ability of individuals to change morphology, physiology, or behaviour in response to environmental change—is likely to already be compromised.

The fact that the resilience of any landscape or population is not easy to quantify adds further complexity into the impact assessment process. In addition, the impact prediction process tends to assume a starting position of no stress. This is often not the case, but it tends to be assumed because it is often almost impossible to know how close a (perhaps highly resilient) system already is to an ecological threshold, in order to factor this into the analysis. Because of this, such analysis tends to underestimate the potential for important negative effects because, in some situations, a comparatively minor change causes the tipping point to be reached.

Impact prediction methods

Impact predictions are usually based on reasonably low-tech approaches, such as straightforward description of changes and 'best guesses' of the impacts these will have, based on understanding of habitat functioning and species' ecology. For example, knowledge of the effects of organic pollution on the amount of dissolved oxygen in water systems, and understanding of species that need high oxygen levels (Section 4.3.2) coupled with a species inventory of a particular aquatic habitat, allows sensible predictions to be made within a straightforward descriptive framework.

The ecological knowledge that underpins prediction of impacts is usually based on published academic research within observational studies. Ideally, results of EcIAs undertaken at other sites where there has been follow-up impact monitoring (Section 3.3.6) feed into this process by allowing impact predictions to be based on actual evidence from similar sites. However, this is not always possible because **post-development monitoring** is not always undertaken. Even where post-development monitoring is undertaken, there is usually no national repository and no indexed way of drawing on previous reports, undertaken in different areas, by different consultancies or councils. Indeed, Zwart et al. (2015) have recently demonstrated that data from post-construction monitoring, if easily accessible, could be used to answer much bigger questions within Applied Ecology, such as how species are moving with climate change and which species adapt to change.

The descriptive/critical framework can be taken on a stage in Bayesian belief network modelling. Impact assessment usually generates lots of qualitative or semi-quantitative data on topics for which human judgements will be subjective and error prone. A Bayesian network is a modelling framework and analysis method that allows uncertainties to be factored in. For example, in their Evidence Reasoning Model, Wang et al. (2006) entered all assessment information - quantitative or qualitative, complete or incomplete, and precise or imprecise - to aggregate multiple factors and identify which impacts were likely to be greatest.

A very different framework to belief networks for analysis of impacts, especially of impacts that are hard to predict meaningfully in a traditional way, is that of scenario development, which uses **scenario analysis** techniques originally developed for strategic thinking (Schoemaker, 1995). Rather than trying to estimate what is most likely to occur, scenario analysis involves considering what might happen in a series of given scenarios. In this way, the analytical focus shifts to consideration of the consequences of different events and most appropriate responses under different circumstances, rather than 'just' putting probabilities on possibilities. It thus becomes a powerful tool for asking 'what if' questions to explore the consequences of uncertainty within EcIA (Duinker & Greig, 2007).

Sometimes, if EcIA analysis is focused on one specific type of impact, more complex prediction tools involving **numerical modelling** can be useful. An example is the Collision Risk Model (CRM) developed by Band et al. (2005) to estimate potential bird mortality at proposed wind farms. This first assesses the theoretical number of collisions likely to occur if birds take no avoiding action, which depends on flight activity, species, likely flight height and speed, and the dimensions and rotational speed of the turbines. An avoidance factor is then added to reflect that, in practice, birds often succeed in avoiding a turbine blade, either by changing their route or measuring the timing of their flight through the rotor. This avoidance factor is specific to the individual proposal, species, and likely flight paths. This model was found to be generally sound when applied to case studies by Chamberlain et al. (2006), but the estimates produced were not accurate enough to be meaningful, with avoidance rates being modified to become both species- and state-specific (i.e. breeding birds, recently fledged birds, moulting birds). This, again, underlines the complexity of ecological systems and the issues associated with trying to model them.

Impact prediction reporting

It is vital that impact predictions are as specific as possible in terms of habitats/species likely to be affected, the geographical location of the impact, and the likely timescale. This sounds obvious, but there is often over-reliance on vague statements about impacts, rather than specific predictions (Treweek, 1999).

One of the ways in which the importance and magnitude of potential impacts can be displayed is in a **Leopold matrix** (Leopold et al., 1971). This involves tabulating ecological receptors (Section 5.3.2) against key aspects of the proposal, which are often grouped into impacts that are likely during construction, operation, decommission, and regeneration phases. Scores are given to each combination on a 1–10 ranking for magnitude and a

Receptor	Details	Site clearance phase	Building phase	Operational phase
Dormice in hedgerow	Hedgerow removal affecting protected species	6 / −8		
Breeding birds in hedgerow	Hedgerow removal causing loss of nesting habitat	6 / −4		
Loss of species rich hedgerow	Loss of multiple rare plant species	6 / −10		
Removing Himalayan balsam *Impatiens glandulifera*	Regeneration of native plants following non-native removal	3 / +5		
Stream invertebrates	Runoff of pesticides etc. from gardens			3 / −2
Reduction in extent/quality of bat feeding grounds	Loss of parkland to create houses and gardens			6 / −4
Amphibians accessing foundation trenches	Inadvertent death of protected species		3 / −8	
Light disturbance to nocturnal species	Moths, bats, nocturnal land mammals			3 / −5
Noise disturbance	Could affect disturbance-sensitive species		1 / −1	

Figure 5.11 Leopold matrix showing impact of constructing a housing development on key ecological receptors for a parkland site with species-rich hedgerows and bordered by a stream. Magnitude is shown in the upper left of each cell in black on a 1 to 10 scale (1 = local, 10 = widespread). Importance in the bottom right in red for negative impacts (–10 to –1) and in green for benefit (+1 to +10).

–10 to +10 ranking for importance (the sign showing whether the impact is positive or negative). A less detailed but more visually intuitive version uses green, amber, and red colour codes. The matrix makes it easier to visualize what activities have a high risk of damaging which aspects of a site's ecology. Two separate tables can be drawn up, one for importance and one for magnitude, or both can be displayed in one table with magnitude shown in the upper left of each cell and importance in the bottom right. An example numeric Leopold matrix is shown in Figure 5.11.

FURTHER READING FOR THIS SECTION

Belief network modelling in impact assessment: Wang, Y-M., Yang, J-B., & Xu, D-L. (2006) Environmental impact assessment using the evidential reasoning approach. *European Journal of Operational Research*, Volume 174, 1885–1913.

Scenario modelling in impact assessment: Duinker, P.N. & Greig, L.A. (2007) Scenario analysis in environmental impact assessment: Improving explorations of the future. *Environmental Impact Assessment Review*, Volume 27, 206–219.

5.6 Mitigation and compensation strategies.

The actual impacts of a development on ecology depend not only on the development itself and the baseline conditions at the site, but also on the effectiveness with which any negative impacts are mitigated. Accordingly, once potential negative ecological impacts have been identified, the most important part of the EcIA process starts—finding ways of eliminating, reducing or buffering those impacts.

There are two main ways that negative impacts can be addressed:

1. **Mitigation:** measures taken to avoid or reduce negative impacts before they occur.

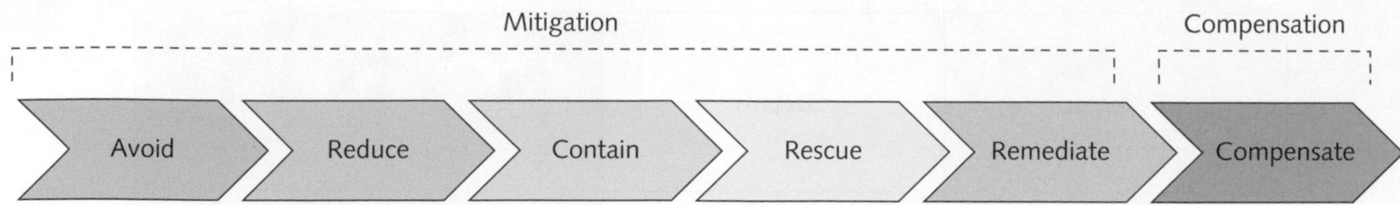

Figure 5.12 Mitigation/compensation hierarchy showing optimal strategy ecologically from the most 'green' solution onwards.

2. **Compensation**: measures taken to make up for the loss of, or permanent damage to, biological resources, often through the provision of replacement areas.

Since prevention of problems is always better than buffering the effects of development after they occur, mitigation is better than compensation and should always be the first option to be explored. Moreover, within mitigation, the aim should always be to avoid problems rather than simply minimize them. The overall mitigation/compensation hierarchy is shown in Figure 5.12. More detail about each type of mitigation/compensation is given in the following sections, together with examples.

Most countries that have an EIA or EcIA legal framework stipulate that mitigation is an essential part of the impact assessment process. Indeed, there is little point in assessing impacts if that information is not then used to formulate effective mitigation/compensation strategies. In theory, mitigation strategies could be developed for all negative impacts. However, in reality, mitigation is often restricted to impacts that have been judged to be high magnitude and/or of high importance (either because of the severity of the impact or the value of the species/habitat involved), while no action is taken for 'negligible' impacts. This further underlines the importance of the impact prediction stage of EcIA (Section 5.5.4).

5.6.1 Mitigation.

As shown in Figure 5.12, there are several different subcategories of mitigation, which differ both in their approach and desirability. These are explained in more detail in Table 5.4.

Table 5.4 Details of different subcategories of mitigation (colours align with colour scheme used in Figure 5.12).

Type	Details	Examples
Avoidance	Preventing all or part of a development going ahead	• Prevention of development • Choosing an alternative site for a factory; rerouting a road around a valuable habitat • Reducing the scale of a housing development so a sensitive area is not destroyed • Altering placement of specific parts of a development within a site—for example, siting chemical processing away from a vulnerable water source.
Reduction	Minimizing specific impacts, either for the whole site or specific (sensitive) areas	• Reducing emission of pollutants • Installation of a fish pass in a dam to reduce its impact on fish migration
Containment	Ensuring negative impacts only have a localized effect	• Installing chemical traps or a reedbed filtration system between an aquatic pollution source and a vulnerable wetland • Installing a noise barrier between a factory and a disturbance-prone species population
Remediation (Chapter 6)	Removal of, and recovery from, problems (e.g. clearing up contaminated water)	• Reseeding an area with wild flower mix after burying a new gas pipeline • Replanting trees on a site post-development to prevent soil erosion • Restocking of fish in a river affected by release of pollutants

Spatial approaches to mitigation

It is sometimes possible within an EcIA to consider **alternative sites** for a particular development; indeed, the main objective of an EcIA can be to decide which of a number of candidate sites is most appropriate for a development. Even in the more usual scenario when there is only one candidate site, it can be possible to alter the exact placement of development features to minimize negative impacts by locating development and access routes away from areas of high ecological interest, as exemplified in Table 5.4.

Impacts arising from a proposed development can have a spatial element and, in such cases, spatial approaches to mitigation are usually the most appropriate. For example, developments can result in habitat fragmentation. A classic example is a new road bisecting a site and fragmenting habitat into two patches, one on either side, with limited connectivity between them. The effects of this on the movement of animals can be mitigated by creating links between the two sides, either via green bridges ('ecoducts') or wildlife tunnels (Case Study 5.2). For large developments, it might be appropriate to ensure there are corridors or stepping stone habitats across the site to connect land on either side (these concepts are considered in more detail in Section 7.6.2).

While facilitating species' movement can be important, sometimes **preventing movement** can be equally vital. Fencing can be used to manage wildlife at development sites, during either construction or operation phases, or both. An example of exclusion fencing during construction is temporary drift fencing used to prevent amphibian species, especially rare or protected species such as the great crested newt *Triturus cristatus*, falling into building foundations or pipeline trenches. Drift fencing is a long ribbon of polythene, about 40 cm in height and dug down into the soil (Figure 5.13). It is held in place by stakes and can either be angled so it leans away from the

Figure 5.13 Temporary drift fencing erected to prevent amphibian and reptile species – in particular great crested newts – being trapped in development foundations and service trenches.

Source: Image courtesy of Jerry Kirkhart/CC BY 2.0.

CASE STUDY 5.2

Green bridges in Banff, Canada

Green bridges—or more correctly wildlife crossings—are a mitigation strategy designed to allow animals to cross a man-made barrier such as a road or railway line (Figure A). Banff National Park is bisected by the Trans-Canada Highway (TCH). As of 2014, there are six green bridges and 38 wildlife underpasses from Banff National Park's east entrance to the border of Yoho National Park, plus one underpass in Yoho National Park. These were constructed to increase habitat connectivity and, in conjunction with fencing, to reduce the number of possible vehicle versus large animal collisions.

In 1996, a long-term research project started to assess the effectiveness of the crossings. This research utilized several different methods including direct wildlife observations, indirect evidence such as animal tracks in the snow (Section 3.6.2), remote data collection using camera trapping (Section 7.3.2), and radiotelemetry monitoring. The crossings are used by a range of large mammals, but there are some species-specific preferences: grizzly bears *Ursus arctos*, wolves *Canis lupus*, elk *Cervus canadensis*, and moose *Alces alces*, prefer crossing structures that are high, wide, and short in length, while black bears *Ursus americanus* and cougars *Puma concolor* tend to prefer long, low, and narrow underpasses.

In the 16 years between 1996 and 2012, there were 120,000 animal crossings. The crossing rate per year increased over time, partly as a result of the habituation of individuals to the crossings and partly due to increases in

Figure A Green bridges can be used by individuals to move from one side of a barrier, such as a road, to the other side.

Source: Image courtesy of Dkrieger.

population sizes of some species such as wolves. It took up to five years for some wary species, like grizzly bears, to start using wildlife crossing structures.

The use of wildlife crossings and fencing reduced traffic-induced mortality on the TCH by more than 80% (for elk and deer alone this figure is >96%), but the relationship between vehicle–bear collisions was influenced only by the amount of traffic on the road, suggesting that while the crossings work for some species, they are less effective for others

FURTHER READING

Parks Canada Website: Available at: http://www.pc.gc.ca/eng/pn-np/ab/banff/plan/transport/tch-rtc/passages-crossings.aspx

Use of green bridges and underpasses in Canada: Clevenger, A.P. & Waltho, N. (2000) Factors influencing the effectiveness of wildlife underpasses in Banff National Park, Alberta, Canada. *Conservation Biology*, Volume 14, 47–56.

Use of green bridges and underpasses in Canada: Clevenger, A.P., Chruszcz, B., & Gunson, K.E. (2001) Highway mitigation fencing reduces wildlife-vehicle collisions. *Wildlife Society Bulletin*, Volume 29, 646–653.

Review of impacts of vehicles on wildlife: Gunson, K.E., Mountrakis, G., & Quackenbush, L.J. (2011) Spatial wildlife-vehicle collision models: A review of current work and its application to transportation mitigation projects. *Journal of Environmental Management*, Volume 92, 1074–1082.

exclusion area or can have a lip on the outer side to decrease the chances of newts climbing over. Alternatively, fencing can be permanent, for example, to prevent large mammal species straying onto a new road or railway (in which case having safe crossing points using green bridges or wildlife tunnels—Case Study 5.2—should be strongly considered to prevent populations becoming completely separated).

Temporal approaches to mitigation

Altering timing of construction can sometimes have a substantial mitigation effect. For example, although removing a hedgerow will always have some negative impacts, if work is timed for outside the bird breeding season, impacts will be reduced. Similarly, negative impacts associated with filling in a pond can be reduced by ensuring that this action does not take place during amphibian spawning.

The focus of mitigation actions

Mitigation actions are either focused on the development itself—its construction, operation, or decommissioning—or actions that influence what is in the receiving environment (development site) at the time the development goes ahead.

- **Actions that alter the proposed development:** Most mitigation relies on modification of the development itself.
  - **Design**—altering where aspects of the development are located, altering how new land uses fit together, or ensuring any important ecological features are replaced on-site (e.g. putting up bird nest boxes to mitigate loss of trees with nest cavities).
  - **Construction**—altering how or when construction occurs.
  - **Operation**—altering how, or to what level, a site operates. For industry, this might include reduction of operational impacts through emissions, waste or capacity controls. For residential housing, this might include reduction in street light illumination hours.
  - **Decommissioning**—altering post-use plans for the site to ensure that, long-term, there are no lasting negative impacts. That might include sealing a site to avoid pollution leakages, planning for how buildings could be dismantled or reconditioned for another use, or how a mineral extraction site could be reused.
- **Actions that alter the receiving environment:** if it is not possible to alter the development so that negative impacts are avoided, reduced, or contained, it can be possible to alter conditions at the development site. This is done through rescue mitigation—**translocating** valued ecological features to another site. The wisdom or otherwise of translocation within an EcIA context is considered

HOT TOPIC
5.1

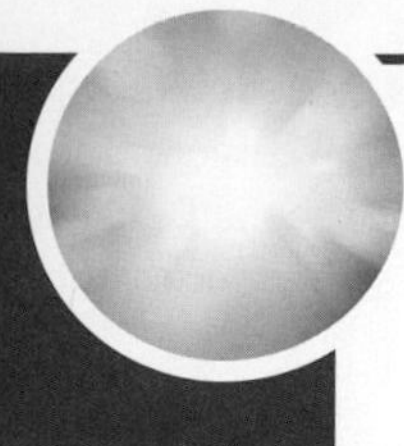

Is translocation a sensible mitigation strategy?

Translocation within an EcIA context is very different from species introduction (movement of a species outside of its native range: Chapter 8) and species reintroduction (movement of a species to part of its former range for conservation, aesthetic, or ethical reasons: Chapters 12 and 13). EcIA-motivated translocations are referred to as **rescue translocations**.

If it is not possible to avoid, reduce, or contain site-based impacts so that they do not have a detrimental effect on species of habitats at that site, the other option is to move the species or habitats themselves. Within the UK, it is the view of the statutory conservation agencies (Natural England, Countryside Council for Wales and Scottish Natural Heritage) that relocation of species within an EcIA context is not an acceptable alternative to impact-avoidance mitigation, but where a development has been given planning approval, relocation should be considered as a last resort (Joint Nature Conservation Committee, 2003). This view is due to the fact that translocation:

- Does not avoid the impact occurring; it simply reduces the effect of that impact.
- Is an invasive process with ethical considerations.
- Is time-consuming and expensive—in many countries requiring involvement of professional, licenced, and highly specialist ecologists.
- Relies on there being a suitable site to translocate species/habitats to.
- Often alters species composition when habitats are moved, potentially altering the very species composition that made the habitat valuable enough for translocation in the first place.

The success of a translocation scheme is far from guaranteed, even when sites are superficially similar (Hodder & Bullock, 1997). Indeed, according to Oldham et al. (1991), at least 33% of known failures of rescue translocations could have been predicted from current ecological knowledge. One of the main reasons that translocations are often unsuccessful (or at least, not known to be successful) is the lack of long-term monitoring, which reduces the knowledge base on which to build rigorous and robust translocation programmes.

Despite the logical issues, expense, and uncertain success, translocation can be the only way that habitat destruction is legal. This occurs in situations where it is an offence to destroy habitat used by legally protected species (as in the UK, for example, for dormice *Muscardinus avellanarius*) without suitable mitigation measures being in place. For legally protected species there are usually translocation guidelines—for example, for great crested newts *Triturus cristatus* (see http://www.froglife.org/wp-content/uploads/2013/06/GCN-Conservation-Handbook_compressed.pdf)—which maximize the chances of success.

QUESTIONS

Translocation is conceptually simple, but its application is far more complex. How might this process be undertaken in the field and how could each stage be optimized?

Why translocate? Why not just not develop?

REFERENCES

Hodder, K.H. & Bullock, J.M. (1997) Translocations of native species in the UK: implications for biodiversity. *Journal of Applied Ecology*, Volume 34, 547–565.

Joint Nature Conservation Committee (2003) Available at: http://jncc.defra.gov.uk/page-2920

Oldham, R.S., Musson, S., & Humphries, R.N. (1991) Translocation of crested newt populations in the UK. *Herpetofauna News*, Volume 2, 3–5.

in detail in Hot Topic 5.1. However, briefly, translocation might involve:

- **Movement of species** so that loss of habitat at a development site does not have a substantial negative impact. This strategy is usually reserved only for rare (usually legally protected) species when all other options have been exhausted.
- **Movement of whole habitats** to a new site can also sometimes be possible. This usually occurs for habitats such as grassland or heathland, which are the easiest to move logistically (moving a woodland would not be possible for obvious reasons).

5.6.2 Compensation.

Even when it is not possible to mitigate negative impacts, it is usually possible to compensate for them. This involves off-site actions to 'make up for' negative ecological impacts at a development site. The most common approach to off-site compensation is the purchase and donation of an existing site, which supports similar species and habitats as the development site. Alternatively, the relevant habitat is sometimes created at a new site, with the hope that, in time, it will support similar species.

The main issue is ensuring **like-for-like replacement.** Generally, if a site has a high enough ecological value for compensation to be seriously considered, it is because it supports rare or priority habitats or species. By definition, such sites are unusual, so finding a local example of a similar site is often far from straightforward. Similarly, many sites are considered high value because they contain habitats or systems that are vulnerable and hard to recreate. Such sites, again by definition, are hard to create through a compensation initiative.

Issues such as these mean that compensation sites often do not replace what was lost. This does not mean that compensation sites are worthless, or that compensation should not be undertaken, but it should be recognized that there are limitations to their effectiveness. Moreover, whereas acquisition of a compensation site, and any initial habitat creation or management work that is necessary, is often funded by the original developer, ongoing maintenance and management costs might not be covered. This is a particular issue for habitats, such as wetlands and grasslands that change naturally over time. One key form of ecological compensation is **biodiversity offsetting**, which is discussed further in Hot Topic 5.2.

HOT TOPIC 5.2

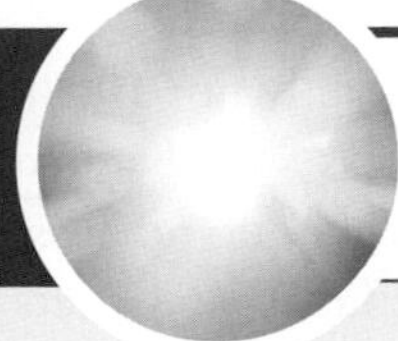

Biodiversity offsetting

Biodiversity offsetting is a type of indirect compensation that is designed to give biodiversity benefits to compensate for losses. The basic premise is to ensure that when a development has negative ecologic impacts that cannot be mitigated, new, bigger, or better nature sites will be created. This offsetting differs from other types of ecological compensation as they need to show measurable outcomes that are sustained over time. Moreover, there is an expectation that the offsetting should go over-and-above covering the loss to result in overall enhancement. There is also the possibility for the developer to pay someone else (called a third-party offset provider) to deliver the offset for them. This means that they have no ongoing responsibly for the offset.

One of the key features of biodiversity offsetting is that it is usually much more objective than other ecological compensation. In many countries, such as the UK, the developer uses a standardized formula to calculate the number of 'biodiversity units' to be lost as a result of their development, based on the habitat(s) affected, and its condition and extent. The pilot version of this need-for-offset equation is provided in DEFRA (2012a). Offset schemes are also subject to an equation, and if the result of that offset-possibility equation is equal to (or ideally greater than) the need-for-offset equation then that offset scheme is considered suitable and acceptable. The pilot version of this offset-possibility equation is provided in DEFRA (2012b).

This field is still new and evolving, but already there is considerable debate and uncertainty about the biodiversity offsetting process. Reviewing both the theory and practice Bull et al. (2013) biodiversity offset schemes have been inconsistent in meeting conservation objectives because of the challenge of ensuring full compliance and effective monitoring and because of conceptual flaws in the approach itself. Substantial concerns have also been raised by Curran et al. (2014), who believe that offset policies are not fit for purpose and actually cause net loss of biodiversity, and represent an inappropriate use of the otherwise valuable tool of ecosystem restoration. This is because time lags, uncertainty, and risk of restoration failure require offset ratios that far exceed what is currently applied in practice.

As Bull et al. (2013) highlight, the aims of the concept—increased objectivity and the aim for enhancement, rather than simply covering loss—are to be applauded. However, biodiversity offsetting is increasingly being used. This is despite concerns that the current way the concept is being employed is not fit for purpose. As hot topics go, this one is near the top of the temperature scale.

QUESTIONS

How realistic is the concept of 'no net loss'?

Biodiversity offsetting has been repeatedly criticized. Why do you think this is, and are such criticisms fair?

REFERENCES

Bull, J., Suttle, K., Gordon, A., Singh, N., & Milner-Gulland, E. (2013) Biodiversity offsets in theory and practice. *Oryx*, Volume 47, 369–380.

Curran, M., Hellweg, S., & Beck, J. (2014) Is there any empirical support for biodiversity offset policy? *Ecological Applications*, Volume 24, 617–632.

DEFRA (2012a) *Biodiversity Offsetting Pilots: Guidance for Developers*. Available at: https://www.gov.uk/government/uploads/system/uploads/attachment_data/file/69528/pb13743-bio-guide-developers.pdf

DEFRA (2012b) *Biodiversity Offsetting Pilots: Guidance for Offset Providers*. Available at: https://www.gov.uk/government/uploads/system/uploads/attachment_data/file/69530/pb13742-bio-guide-offset-providers.pdf

5.6.3 Enhancement.

There is a growing school of thought that the mitigation section of an EcIA should focus on ways in which a development could benefit the environment, rather than focusing solely on damage limitation. Ecological enhancement of a site could occur though initiatives within the development structures themselves (e.g. green roof spaces) or could involve habitat creation within the wider site—for example, digging ponds, planting native trees or hedges, or putting up nesting boxes/hibernacula, or incorporating these into construction.

Enhancement initiatives also reduce the risk of a development breaching the **'no net loss principle'**. This is the concept that, overall, a development should not cause ecological damage. Because mitigation is generally not 100% effective (and compensation even less so), there can be some slippage relative to the ideal of no net loss. This can also be exacerbated by the fact that aspects of the development that have negligible impacts are often not included in mitigation/compensation strategies. Having enhancement initiatives can thus be a way of 'banking' positive ecological management to offset any shortfall in mitigation.

FURTHER READING FOR THIS SECTION

Overview of mitigation types and success: Drayson, K. & Thompson, S. (2013) Ecological mitigation measures in English environmental impact assessment. *Journal of Environmental Management*, Volume 119, 103–110.

5.7 Recommendations and outcomes of EcIA.

Once an EcIA has been undertaken, and the all-important mitigation and compensation strategies devised, a recommendation must be given as to whether the development is ecologically acceptable. In principle at least, this recommendation feeds into the final outcome and informs the decision about whether a development should be allowed to go ahead or not. An example of how this process can unfold is given in the Online Case Study for Chapter 5, which also serves to highlight the value of the EIA process.

You can find the online case study at
www.oxfordtextbooks.co.uk/orc/goodenough

5.7.1 Forming recommendations following EcIA.

Following the EcIA process, and all the stages detailed above, the relevant planning authority must determine whether the *mitigated* project complies with legal requirements and meets national and local policy goals and objectives. All the relevant information must be contained within the EcIA formal report, and the developer will need to demonstrate commitment to undertaking the mitigation and/or compensation recommended. This can be done through submission of revised proposal plans or draft action plans, including details of when work is due to be undertaken, how, and by whom.

5.7.2 Outcomes of EcIA and what happens next.

Outcomes of the planning process that EcIA (and EIA) often sits within differ according to national legislation. However, there are four main possibilities regarding planning permission. It can be:

1. **Granted** (usually with conditions and legal agreements to enforce the implementation of mitigation, compensation and enhancement measures).
2. **Withheld** pending detail on impacts or mitigation measures.
3. **Withheld** pending consideration of alternative sites.
4. **Denied** because impact importance or magnitude is too high or mitigation is insufficient.

In most countries, it is possible that development is agreed because of imperative reasons of overriding public interest (IROPI), even if there will be substantial negative effects that cannot be mitigated effectively. In such cases a decision to go ahead with a development might conflict with ecological advice. For this to happen, generally, there needs to be:

- Evidence that alternative sites or alternative development proposals
  - do not exist; or
  - do not deliver the same intended results as the proposed site or proposed development; or
  - do not reduce ecological impact.
- Evidence that the proposal provides vital benefits to human health, public safety, or beneficial consequences of primary importance to the environment. The terms 'imperative', 'overriding', and 'public interest' must have been appropriately interpreted for use of this criterion to be robust:
  - **imperative**—essential (whether urgent or otherwise) that the plan or project proceeds to give the intended benefit;
  - **overriding**—the interest served by the plan or project outweighs the harm (or risk of harm) to the integrity of the site as identified in the appropriate assessment;
  - **public interest**—a public benefit, rather than a solely private interest.
- Evidence that all possible mitigation has been listed for implementation and that, where this is not possible or not effective, the best possible compensation has been proposed.

These points are not specific to any particular country or level of application; they are simply overarching decision-making principles.

5.7.3 Post-development monitoring.

As noted throughout this chapter and in Chapter 3, monitoring is a vital aspect of EcIA for several reasons:

1. If monitoring data are available for a site before an EcIA is conducted and afterwards, that improves the quality of that EcIA substantially, as it gives a larger dataset on which to base decisions.

2. Impact predictions need to be matched to reality after developments occur so insights can form part of the knowledge base that underpins subsequent EcIAs (**impact monitoring:** Section 3.3.6).
3. The effectiveness of mitigation and compensation measures is monitored (**mitigation monitoring:** Section 3.3.6) so knowledge can feed forward into future assessments, but also more immediately so that mitigation. can be adjusted if it is not having the desired effect.

FURTHER READING FOR THIS SECTION

The importance of post-developing monitoring within EIA: Dipper, B. (1998) Monitoring and post-auditing in environmental impact assessment: a review. *Journal of Environmental Planning and Management*, Volume 41, 731–747.

The role of post-development monitoring in improving EcIA: Hill, D. & Arnold, R. (2012) Building the evidence base for ecological impact assessment and mitigation. *Journal of Applied Ecology*, Volume 49, 6–9.

5.8 Limitations and challenges of the EcIA process.

There are several key challenges in current EcIA frameworks and the way assessments are undertaken and reported. For EcIA to grow as a discipline, these really need to be considered. The fundamental weakness is that although the EcIA/EIA process is in many cases a legal requirement (Sections 5.2.2 and 5.2.3), there is no guarantee that recommendations made will be reflected in the final decision about whether a development should be allowed to go ahead or not. This means that there can be a tendency for *some* developers to see EIA and EcIA as 'box-ticking' exercises, rather than really engaging with them to improve the environmental and ecological credentials of a development, and minimize adverse impact. This is especially true for mitigation and compensation. EcIA and EIA are essentially planning processes, and adherence to the EIA mitigation plan once the permission is granted can be patchy at best. Other key challenges are outlined in Table 5.5.

5.8.1 Strategic Environmental Assessment.

A broadly similar process to EIA that is worth mentioning briefly here is **Strategic Environmental Assessment** (SEA). Like EIA, SEA includes specific mention of ecology. The main difference between the EIA and SEA processes is that while EIA tends to focus on specific projects (as detailed in Sections 5.2.2 and 5.2.3), SEA looks at much broader situations, such as a series of new factories, developments of new houses that might increase the need to travel and thus travel-related impacts, etc. In some ways, this partly addresses concerns highlighted in Table 5.5 that EIAs and EcIAs are too project-focused and can ignore off-site and cumulative effects.

Within Europe, SEA legislation in set out in Directive 2001/42/EC on the Assessment of the Effects of Certain Plans and Programmes on the Environment (commonly called the 'SEA Directive'). SEA can either be conducted before a corresponding EIA is undertaken, or alongside a series of project-specific EIAs. The first scenario means that information on the environmental impact of a broad plan can cascade down into individual EIAs, while in the second scenario the SEA acts as a capstone to bring together all the project-specific impacts and mitigation measures so that the bigger picture becomes clearer.

FURTHER READING FOR SECTION

Overall quality of EcIAs and reasons for failures: Drayson, K., Wood, G., & Thompson, S. (2015) Assessing the quality of the ecological component of English Environmental Statements. *Journal of Environmental Management*, Volume 160, 241–253.

Issues and challenges in impact assessment: Fischer, T.B. & Noble, B. (2015) Impact assessment research: achievements, gaps and future directions. *Journal of Environmental Assessment Policy and Management*, Volume 17, 1501001.

SEA directive: http://ec.europa.eu/environment/eia/sea-legalcontext.htm

Table 5.5 Main challenges in existing EcIA frameworks.

Challenge	Details
To go beyond the site boundary in project-based assessments	Currently, most EcIAs are site-based. The focus is thus on local direct impacts rather than cumulative impacts. Even impacts on ecological conditions immediately next to a site can be overlooked (e.g. impact of street lighting on a bat roost immediately adjacent to the site).
Ecology can be traded off against other receptors	Ecology can sometimes be 'traded off' against other receptors. For example, if a proposal has benefits to human population and water quality, and no significant impacts on other receptors except ecology, it can be hard to argue that ecological problems should be weighted more highly in the decision-making process. This is particularly true when sites/habitats/species have no specific legal protection.
Planning/policy focus on brownfield sites	Target development sites are often brownfield sites (previously developed land), rather than greenfield sites (land that has not been developed in the past). In many ways this is a sensible policy, certainly in terms of infrastructure links, population, and visual impact. However, derelict brownfield sites are sometimes extremely ecologically rich, which can lead to a planning tension.
Lack of monitoring and an EcIA database	The lack of long-term monitoring, especially of impacts and mitigation, means EcIA experience is patchy, and the wheel is being reinvented multiple times, at multiple sites, by different ecologists.
Lack of formal independent review and funding	The amount of a developer's budget devoted to EIA is small. The amount devoted to EcIA is tiny. This, coupled with a lack of independent review, means that EcIAs are being undertaken with few resources and can lack the transparency and moderation of external review.
Some taxonomic groups are under-considered	EcIA can be biased towards species that are obvious and charismatic, or that particularly need to be surveyed because they are legally protected, when species in other taxonomic groups are just as important (sometimes more so!) to ecosystem functioning.
Mitigation and compensation not always implemented and not fool-proof	Mitigation/compensation is an essential part of EcIA, but whether and how mitigation suggestions translate into formal planning conditions is out of the hands of the ecologist. Moreover, if measures are not implemented, legal redress can be extremely complex. There is often a lack of mitigation for 'negligible' impacts, which does not recognize the potential for impacts to be cumulative.

5.9 Conclusions.

Ecological Impact Assessment is an area of Applied Ecology that is already important, and becoming ever more so in the context of environmental degradation as a result of human activity and the increasing emphasis placed on sustainable development. The decision whether a particular development should be permitted or not (and if so what mitigation is needed) is a product of multiple interlinked stages, such that EcIA really is more than the sum of its parts. All of those stages demand understanding of ecological concepts and how these translate to the real world. Mitigation can only be sensible if impact predictions are robust, and impact predictions can only be robust if the ecological data on which they are based are accurate and dependable. The importance of monitoring of the accuracy of impact predictions and mitigation effectiveness, to inform and improve future EcIAs, cannot be overstated.

ONLINE ACTIVITY

Go to www.oxfordtextbooks.co.uk/orc/goodenough/ to download the activity that accompanies this chapter.

CHAPTER 5 AT A GLANCE: THE BIG PICTURE

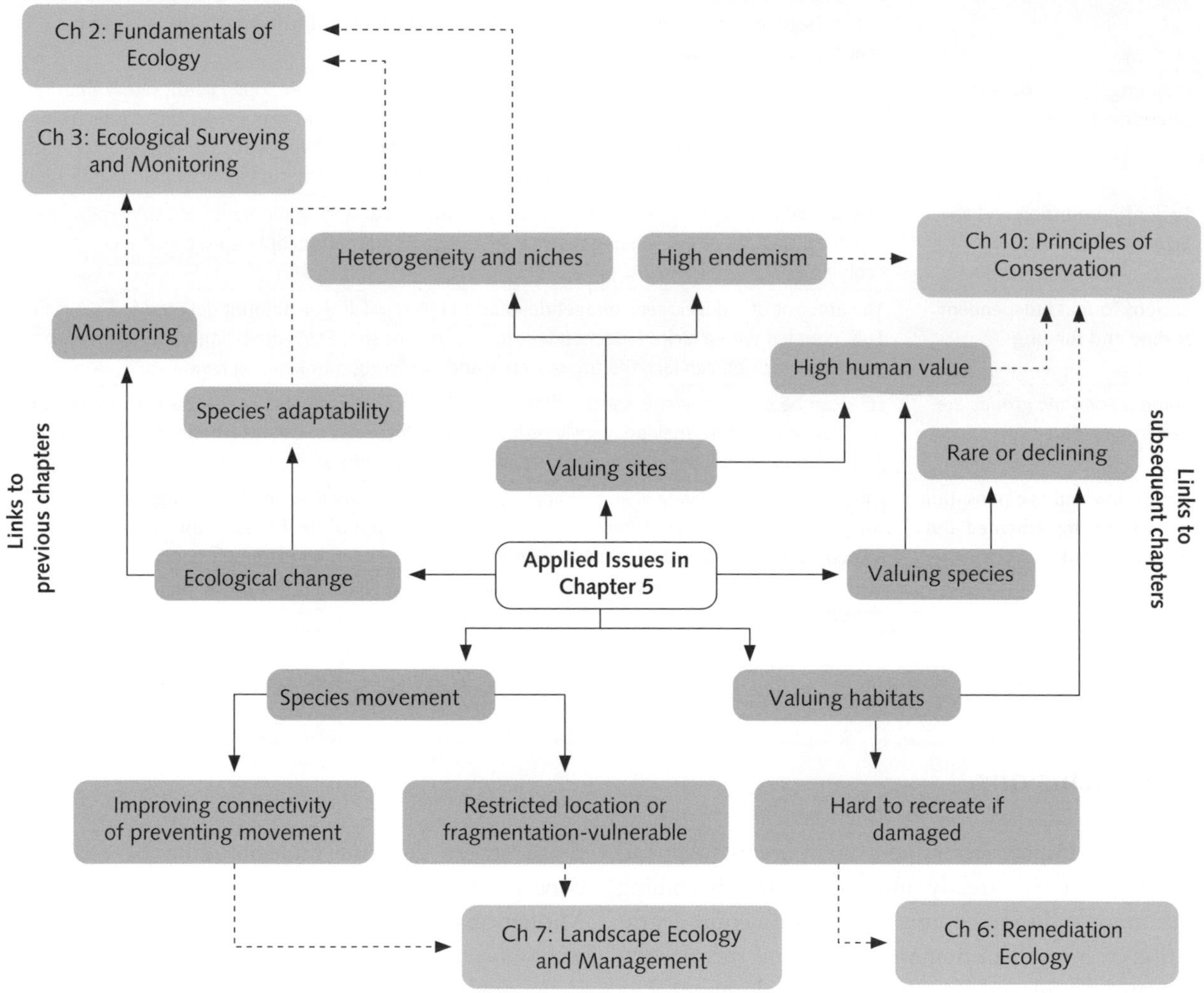

Ch 2: Fundamentals of Ecology

Ch 3: Ecological Surveying and Monitoring

Heterogeneity and niches

High endemism

Ch 10: Principles of Conservation

Monitoring

High human value

Species' adaptability

Links to previous chapters

Valuing sites

Rare or declining

Links to subsequent chapters

Ecological change

Applied Issues in Chapter 5

Valuing species

Species movement

Valuing habitats

Improving connectivity of preventing movement

Restricted location or fragmentation-vulnerable

Hard to recreate if damaged

Ch 7: Landscape Ecology and Management

Ch 6: Remediation Ecology

REFERENCES

Band, W., Madders, M., & Whitfield, D.P. (2005) Developing field and analytical methods to assess avian collision risk at wind farms. In: M. De Lucas, G. Janss, & M. Ferrer (Eds) *Birds and Wind Power*, pp. 259–275.. Barcelona: Lynx Editions.

Beanlands G.E. & Duinker P.N. (1984) An ecological framework for environmental impact assessment. *Journal of Environmental Management*, Volume 18, 267–277.

Bugalho, M.N., Caldeira, M.C., Pereira, J.S., Aronson, J., & Pausas, J.G. (2011) Mediterranean cork oak savannas require human use to sustain biodiversity and ecosystem services. *Frontiers in Ecology and the Environment*, Volume 9, 278–286.

Canter, L.W. (1996) *Environmental Impact Assessment*. New York, NY: McGraw-Hill.

Chamberlain, D.E., Rehfisch, M.R., Fox, A.D., Desholm, M., & Anthony, S.J. (2006) The effect of avoidance rates on bird mortality predictions made by wind turbine collision risk models. *Ibis*, Volume 148, 198–202.

Dallimer, M., Irvine, K.N., Skinner, A.M., Davies, Z.G., Rouquette, J.R., Maltby, L.L., Warren, P.H., Armsworth, P.R., & Gaston, K.J. (2012) Biodiversity and the feel-good factor: understanding associations between self-reported human well-being and species richness. *BioScience*, Volume 62, 47–55.

Donnelly, A., Dalal-Clayton, B., & Hughes R. (1998) *A Directory of Impact Assessment Guidelines*. International Institute for Environment and Development. Available at: http://pubs.iied.org/pdfs/7785IIED.pdf

Duinker, P.N. and Greig, L.A. (2007) Scenario analysis in environmental impact assessment: Improving explorations of the future. *Environmental Impact Assessment Review*, Volume 27, 206–219.

Fuller, R.A., Irvine, K.N., Devine-Wright, P., Warren, P.H., & Gaston, K.J. (2007) Psychological benefits of greenspace increase with biodiversity. *Biology Letters*, Volume 3, 390–394.

Glasson, J., Therivel, R., & Chadwick, A. (2013) *Introduction to Environmental Impact Assessment*. London: Routledge.

Goldblatt, P. & Manning, J. (2002) Plant diversity of the Cape region of South Africa. *Annals of the Missouri Botanical Garden*, Volume 89, 281–302.

Harris, G. & Pickering, S.J. (2008) *Cotswold Water Park Biodiversity Action Plan 2007–2016*. Cirencester: Cotswold Water Park Society.

IMPEL (2012) *The Implementation of the Environmental Impact Assessment on the Basis of Precise Examples*. Available at: http://ec.europa.eu/environment/eia/pdf/IMPEL-EIA-Report-final.pdf

Institute for Ecology and Environmental Management (2006) *Guidelines for Ecological Impact Assessment in the United Kingdom*. Available at: http://www.cieem.net/data/files/Resource_Library/Technical_Guidance_Series/EcIA_Guidelines/TGSEcIA-EcIA_Guidelines-Terestrial_Freshwater_Coastal.pdf

Institute for Environmental Management and Assessment (2011) *The State of Environmental Impact Assessment Practice in the UK*. Available at: http://www.environmentalistonline.com/article/state-eia-practice-uk

Jefferson, R.G., Smith, S.L.N. and MacKintosh, E.J. (2014) *Guidelines for the Selection of Biological SSSIs Part 2: Detailed Guidelines for Habitats and Species Groups—Lowland Grasslands*. Available at: http://jncc.defra.gov.uk/pdf/SSSI_Chptr03_revision_2014(v1.0).pdf

Joint Nature Conservation Committee (2014) *Priority Habitats: UK Lowland Heathland Habitats*. Available at: http://jncc.defra.gov.uk/page-1432

Leopold, L.B., Clarke, F.E., Hanshaw, B.B., & Balsley, J.R. (1971) *A Procedure for Evaluating Environmental Impact*. Washington: U.S. Geological Survey.

OSPAR (2015) Protecting and conserving the North-East Atlantic and its resources. Available at: http://www.ospar.org

Robinson, N. (1992) International trends in environmental impact assessment. *Boston College Environmental Affairs Law Review*, Volume 19, 591–621.

Rodwell, J.S. (Ed.) (1991–2000) *British Plant Communities*, Volumes 1–5. Cambridge: Cambridge University Press.

Schoemaker, P.J.H. (1995) Scenario planning: a tool for strategic thinking. *MIT Sloan Management Review*, Volume 36, 25–40.

Shinn, E.A., Smith, G.W., Prospero, J.M., Betzer, P., Hayes, M.L., Garrison, V., & Barber, R.T. (2000) African dust and the demise of Caribbean coral reefs. *Geophysical Research Letters*, Volume 27, 3029–3032.

Treweek, J. (1999) *Ecological Impact Assessment*. Oxford: Blackwell Science Ltd.

Wang, Y.M., Yang, J.B., & Xu, D-L. (2006) Environmental impact assessment using the evidential reasoning approach. *European Journal of Operational Research*, Volume 174, 1885–1913.

Zwart, M.C., Robson, P., Rankin, S., Whittingham, M.J., & McGowan, P.J. (2015) Using environmental impact assessment and post-construction monitoring data to inform wind energy developments. *Ecosphere*, Volume 6, 1–11.

6 Remediation Ecology

6.1 Introduction.

Pollution is anything that is introduced into the environment by humans that can, or does, cause harm. Many human activities cause pollution that can affect individual species, habitats, or entire ecological communities and ecosystems. Some pollution has only a local impact whereas some forms of pollution can have a global reach.

Reducing pollution is the subject of considerable effort, debate, legislation, and policy, and thus is an important area of environmental management. Applied Ecologists may be involved in pollution assessment and management through the use of bioindicators to assess polluted environments (Section 4.3.2) and through activities involved with efforts to remove pollution from contaminated sites through the process of **remediation**.

In this chapter we will briefly consider pollution and pollutants before examining the ways in which pollutants in the environment can be removed or remediated. We will focus on bioremediation, which is the use of organisms to assist in the removal of toxic compounds from the environment, or the breakdown of those compounds into less toxic (or non-toxic) compounds.

6.2 Pollution: an overview.

The development of modern society has gone hand-in-hand with technological developments that have greatly increased the emission of pollutants into the environment. The industrial revolution of the 18th century heralded the beginning of large-scale fossil fuel exploitation. With that exploitation came the development of heavy industry, and the beginning of concerns about local **environmental damage** and the impacts that pollution might be having on human health.

It was in the 20th century that our ability to pollute became globally significant. Concerted and varied industrial development, combined with unprecedented population growth, urbanization and globalization, caused a massive increase in the amount and diversity of chemical compounds released into the environment. Industrial manufacturing processes produce toxic products and by-products that find their way into water courses, farmers apply pesticides and fertilizers to soils, motor vehicles emit exhaust gases into the atmosphere: in fact there are few aspects of modern life that don't produce pollution either directly or indirectly, as shown in Figure 6.1.

Figure 6.1 Industry, agriculture, and our modern lifestyle produce atmospheric, soil, and water pollution.
Source: Images (L-R) courtesy of Peter Grima; Aqua Mechanical; Jason Woodhead/CC BY-SA 2.0.

6.2.1 Types and sources of pollution.

Broadly, pollutants can be classified as **inorganic** (including metals and compounds of metals) or **organic** (compounds based around carbon atoms, such as gasoline). Different compounds can occur in different states of matter: solid, liquid, or gas. Chemicals introduced to the environment are by far the most damaging form of pollution, and are the primary focus of this chapter. However, it is worth noting that our activities can also cause light, noise, or heat pollution that can all have serious ecosystem effects. Street lighting, for example, is thought to interfere with moths' ability to feed and mate and might be a driver in population decline (Macgregor et al., 2015).

Many industrial processes produce chemicals as a primary goal or as a by-product, but it is only when these products are released into the environment that they become pollutants. In many jurisdictions the release of pollutants into the environment is **regulated** and monitored in an attempt to reduce the harm they cause (compliance monitoring: Section 3.3.5). In many parts of the world, industry is under pressure to devise processes that reduce the production of pollutants, or that can render such pollutants less harmful before they are released.

In some cases, pollution emission can be **accidental** or the consequence of **negligence**. Small-scale accidental emissions may occur frequently and go largely unnoticed. However, large-scale emissions such as from tankers that have run aground, despite being relatively uncommon, can have far-reaching human and environmental consequences. They also tend to attract considerable media attention. Human error may play a role in pollution accidents, such as occurred in 1988 when aluminium sulphate was accidentally poured into the wrong tank at the Lowermoor Water Treatment Works, Cornwall, UK, contaminating the water supply to the town of Camelford (Committee on Toxicity of Chemicals in Foods, Consumer Products and the Environment, 2013).

An example of serious consequences of a single-event pollution emission is provided by the Bhopal disaster in India in 1984, where a series of failures in a number of safety systems, together with poor maintenance, led to the release of methyl isocyante gas from the Union Carbide India Ltd pesticide plant. The exposure of more than 500,000 people to the resultant toxic gas cloud led to thousands of deaths (with some estimates as high as 8000) and hundreds of thousands of injuries ranging from temporary and minor conditions to permanently disabling injuries (Eckerman, 2005).

Often, **legacy pollutants** exist in the environment as a consequence of former industrial processes that are no longer used or that have been superseded by cleaner, less polluting alternatives. Historical chemical manufacturing, mineral extraction, smelting, gasworks, landfill, and other sources produced pollution that remains in soil and can leach into aquifers. Collectively, the long history of industrialization in many economically developed countries, and a recent history of less regulated industry in economically developing countries, has resulted in a multitude of polluted sites. Often, historical pollution is only discovered when land used by former industry is to be reused. Historically polluted sites may require treatment to remove, or render less harmful, the chemical legacy of their previous occupiers.

6.2.2 Scale of pollution.

In common with many ecological interactions, the harm caused by pollutants can vary in **scale**. Scale in this context includes the severity of the effect on organisms, temporal persistence in the environment, and the geographical scale of the impact.

Harmful effects can be exacerbated if pollutants **persist** in the environment for a long time. For example, chlorofluorocarbons CFCs, used historically as refrigerant gases and as aerosol propellants, are atmospheric pollutants that can persist for 55–140 years and act to destroy the ozone layer of the upper atmosphere (Elkins, 1999). On the other hand, some pollutants might persist for a relatively short period. Black carbon particles (a major component of soot) caused by incomplete combustion are a pollutant that can influence cloud formation and cause atmospheric warming, but persist in the atmosphere for days rather than years.

Some pollutants might affect only the immediate locality of their production, whereas others (especially highly soluble waterborne pollutants and highly volatile airborne gases) have the capacity to affect ecosystems on a global scale. For example, the emission of sulphur dioxide and nitric oxides from industry into the atmosphere causes acidification of rainwater ('acid rain') that can fall over wide geographical areas, often considerable distances from emission sites. Some pollutants, such as mercury, can achieve a near-global distribution because of the ways that they are mobilized through ecosystems (Driscoll et al., 2013).

6.2.3 Tackling polluted sites: remediation and bioremediation.

In the long term, tackling environmental pollution involves developing processes that greatly reduce or eliminate the production of harmful polluting compounds, and seeking non-polluting alternatives. In the short to medium term, however, it is necessary to devise methods for removing pollutants from polluted sites. The treatment and removal of existing pollution is termed **remediation**.

In many cases it has been possible to clean up, or remediate, contaminated sites using physical or chemical methods, or a combination of both. Techniques include:

- **Immobilization**, also termed **stabilization:** this is when polluting compounds are converted in some way to a form that is less soluble or otherwise rendered 'immobile' and therefore incapable of doing harm to organisms. For example, hexavalent chromium is widely used in a range of industrial applications including chrome plating, paint manufacture, and glass making. In its natural state, hexavalent chromium is highly mobile in the environment and toxic. However, under the right conditions it can be reduced to solid, stable, and non-toxic trivalent chromium that becomes a harmless part of the soil matrix, effectively 'immobilizing' the harmful pollutant.
- **Removal of material or pollutants**: removal of polluted material, especially soil, sometimes termed 'dig and dump', may or may not include treatment away from the site and subsequent return to the site. If it does not involve returning the material to the initial site then steps need to be taken to ensure that the contaminated material is able to be contained in such a way to prevent further pollution problems at the dump site. Removal of pollutants can sometimes be achieved by the use of solvents to dissolve and extract some compounds, such as petroleum hydrocarbons, from contaminated soils. Since solvents are usually themselves potential pollutants, solvent extraction techniques need to be carried out carefully to prevent adding to the pollution problem rather than reducing it.

In many cases it is possible to make use of organisms to assist in the breakdown of environmentally toxic compounds into less toxic or non-toxic compounds, or to accumulate toxic compounds within 'harvestable biomass' (usually within plant material). The most commonly used organisms in bioremediation are bacteria, fungi, and plants. Taking advantage of organisms to remediate is termed **bioremediation.**

Bioremediation relies on living organisms and, with that, comes a requirement to understand the ecological interactions between the organisms being used for bioremediation, the polluted environment, and the wider environment. There is also a need to manage remediation sites (either on- or off-site, termed *in situ* or *ex situ*) for the benefit of bioremediators and other organisms that may be resident or passing through the area. Thus, bioremediation is a highly applied discipline that straddles the sometimes blurred boundary between ecology and environmental science. It also requires considerable input from the disciplines of microbial biology, botany, and, especially in the future, biotechnology.

It is not the intention of this chapter to provide the knowledge required to establish a bioremediation project. However, it is necessary for the Applied Ecologist to understand the scope and potential for bioremediation and the ways in which such projects might influence, and be influenced by, surrounding ecology. The details of bioremediation are ultimately the realms of the microbiologist or the plant physiologist, but understanding the wider implications of any given bioremediation project requires a knowledge and understanding of ecological interactions. An Applied Ecologist armed with some understanding of the strategies and methods of bioremediation is far better placed to give informed advice and will be better able to be involved in planning for and managing the wider impacts of bioremediation.

FURTHER READING FOR THIS SECTION

General overview of pollution and its effects: Hill, M.K. (2010). *Understanding Environmental Pollution*. Cambridge University Press, Cambridge, UK.

6.3 The general principles of bioremediation.

Bioremediation takes advantage of naturally occurring biological processes and the evolved adaptations of certain organisms. It is a fast-growing area and its development is supported by advances in **biotechnology** including the production of **genetically modified organisms** (Hot Topic 6.1).

6.3.1 Bioremediation through metabolic breakdown.

The most commonly used organisms for bioremediation are bacteria. Highly diverse and ecologically successful, bacteria are able to thrive in a wide range of physical and chemical environments, including extreme conditions such as the deep ocean, Antarctic ice, hot springs, salt lakes, and volcanic vents. Their ecological success comes in part because they have evolved a wealth of varied **biochemical pathways** that enable different species and strains to take advantage of, or to tolerate, the molecules and conditions found in different environments.

Bacteria that can **metabolize** (biochemically process) pollutant molecules can, if put into contact with those molecules, cause the natural biodegradation of harmful pollutants. In some cases, naturally present bacteria can be encouraged to grow, but in other cases it may be necessary to make use of specific organisms adapted to the pollutant in question.

Fungi can also be used to break down toxic compounds. Although their fruiting bodies (the familiar mushrooms and toadstools) are large and often highly conspicuous, the main 'body' or mycelium of fungi consists of microscopic filaments, called hyphae. These hyphae thoroughly penetrate soils, and it is this characteristic combined with a variety of evolved biochemical pathways that give fungi great potential for the bioremediation of pollutants in soil. Using fungi in bioremediation is termed **mycoremediation**. Example pollutants that fungi are able to remediate include aromatic hydrocarbons and aromatic amines (Harms et al., 2011).

6.3.2 Bioremediation through hyperaccumulation.

Another way to use organisms as bioremediation agents, especially plants, is to make use of their ability to **bioaccumulate** pollutants (building up the pollutant molecules within their bodies). Using plants to bioremediate is termed **phytoremediation.** Some plants can accumulate metals like cadmium, zinc, selenium, and arsenic from the soil in which they are growing, and once they have incorporated sufficient pollutants into their bodies they can then be removed from the site. It may be possible to make use of harvested material, for example, to make biofuel, and it may even be possible to extract the accumulated metals and reuse them—a bioharvesting process (largely still in development) known as phytomining (see Section 6.5; Sheoran et al., 2013).

Some plants, termed **hyperaccumulators**, can accumulate 100–1000 times more metal than other species (Rascio and Navari-Izzo, 2011). Growing hyperaccumulating plants on soils contaminated with metals can result in large amounts of pollutant being drawn out of the soil and into the plants, which can then be removed, taking the pollution with them. Phytoremediation is discussed in detail in Section 6.5.

● FURTHER READING FOR THIS SECTION

Information about hyperaccumulating plants: Rascio, N. & Navari-Izzo, F. (2011) Heavy metal hyperaccumulating plants: how and why do they do it? And what makes them so interesting? *Plant Science*, Volume 180, 168–181.

6.4 Bioremediation using microorganisms.

There are a great many polluted sites throughout the world and not all of these are suitable candidates for remediation. Remediation is not a 'cure-all' technique to remove pollution wholesale from the environment. Rather, it is a site-specific, complex, and often expensive process. Because of this it might not be cost-effective or technologically possible to clean up a specific site using bioremediation, even if this is deemed desirable.

If a site with contaminated soil or water has been identified as being suitable for bioremediation with microorganisms (rather than plants, discussed in Section 6.5), then there are four strategies that may be used. These strategies are not mutually exclusive and in cases where multiple pollutants are present it may be necessary to combine strategies to produce a tailored approach specific to the site:

1. Make use of **indigenous microorganisms**, i.e. those bacteria and fungi already present at the site (Section 6.4.1).
2. Encourage the growth of (**biostimulate**) indigenous microorganisms (Section 6.4.2).
3. Inoculate the site with naturally occurring microorganisms; a process termed **bio-augmentation** (Section 6.4.3).
4. Add genetically engineered microorganisms **GEMs** (Hot Topic 6.1).

Culturing bacteria or fungi on media containing a pollutant is a powerful technique for identifying species and strains of microorganism that are capable of metabolizing that pollutant. However, a great many species cannot, currently, be cultured in laboratories. Microorganisms can also be identified *in situ* using gene probe hybridization. These techniques involve labelling sequences of DNA and RNA in such a way that they will bind with specific sequences and genes in the target and enable researchers to detect when that binding has occurred. This enables the identification of species and strains, as well as confirming the presence of specific genes in a sample. From a bioremediation perspective, it is the genes responsible for breaking down a pollutant that are important, with the rest of the organism being merely a vehicle for those genes. Gene probe hybridization techniques can also be used to monitor bioremediation and to check the presence and abundance of species and genes.

6.4.1 Bioremediation using indigenous microorganisms.

Soils naturally contain a very large number of microorganisms. Bacteria can be present in numbers exceeding 40 million cells per gram, with considerable species diversity (Fierer and Jackson, 2006), while fungal hyphae also thoroughly penetrate the soil at a microscopic scale. Cyanobacteria (bacteria that can photosynthesize) and algae (photosynthetic single-celled eukaryotes) are also present in high numbers in most soils. At its simplest, bioremediation seeks to make use of these indigenous microorganisms.

Indigenous microorganisms are very useful for remediating **hydrocarbons**. Hydrocarbons are organic compounds that are made up from carbon and hydrogen atoms. These atoms can combine in straight chains, branched chains, ring structures, and combinations of these to produce an extraordinary range of compounds varying in structure and molecular weight. Crude oil contains a wealth of different hydrocarbon compounds that include:

- Methane.
- Alkanes such as propane and butane.
- Aromatics; benzene, toluene, ethylbenzene, and xylene, collectively known as BTEX.
- Polycyclic hydrocarbons such as naphthalene.
- Heterocyclic compounds that contain nitrogen, sulphur, and heavy metals.
- Tars and bitumen.

The chemical industry can convert hydrocarbons into a host of very widely used synthetic organic (**xenobiotic**) compounds. These include products like polychlorinated biphenyls (PCBs) (Figure 6.2), chlorinated compounds (e.g. trichloroethene) used widely as solvents, and polycyclic aromatic hydrocarbons (PAHs). Many xenobiotics are complex in structure and slow to degrade in the environment. Many are also highly toxic, particularly those that contain halogens (e.g. chlorine, fluorine, and bromine).

Around 1% of bacterial cells typically present in soil can degrade hydrocarbons (Namkoong et al., 2002) as can some soil-dwelling cyanobacteria and algae. In fact, soil that is contaminated with hydrocarbons can contain more bacteria than non-contaminated soil, because pollutants can act as additional resources for those species able to exploit them (Milcic-Terzic et al., 2001). Thus, with sufficient time, soil polluted with hydrocarbons can be remediated by **indigenous microorganisms**. However, it may be desirable to speed up the process by encouraging the growth of microorganisms (Section 6.4.2) or augmenting them (Section 6.4.3). Hydrocarbon pollution can also have a major effect on marine ecosystems, but even here indigenous microorganisms are present that can degrade hydrocarbons (Case study 6.1) (Head and Swannell, 1999).

The breakdown of any organic compound is primarily affected by its chemical structure. Compounds with simple molecular structures, like alkanes and monocyclic aromatics, are more readily degraded by microorganisms than complex compounds such as PAHs. Organochlorines can be more difficult still with an increase in the number of halogen atoms present increasing a molecule's resistance to degradation (Figure 6.3). Despite the difficulties of breaking down molecules like organochlorine pesticides, there are bacteria naturally occurring in soil that can achieve it (Barragán-Huerta et al., 2007).

6.4.2 Bioremediation by stimulating indigenous microbial growth.

Microorganisms capable of breaking down pollutants may occur naturally in polluted soils and water, but such bacteria usually occur as a relatively tiny component of the microbiota. **Natural selection** will favour those organisms that can tolerate the

Figure 6.2 Chemical pollutants can enter the environment from a variety of sources. For example, polychlorinated biphenyls (PCBS) were commonly used as insulating fluids in electrical transformers, which are significant sources of PCBs in the environment.

Source: Photograph courtesy of Sturmovik/CC BY-SA 3.0.

CASE STUDY 6.1

Bioremediation of marine oil spills

Crude oil is primarily transported around the world using large capacity oil tankers and when such ships run aground or otherwise become holed, the leakage of oil can have considerable negative environmental impact. It is especially harmful if the oil comes ashore, since in the sea most organisms are found close to land and many human activities are also focused there (Figure A).

At sea, booms and barriers can be used to contain the spill and various machines and chemical treatments can be used to collect and disperse the oil. If the oil reaches shore then it can be physically removed (using excavators, cold water flushing, or hot water and high pressure removal) and chemically dispersed, but both physical and chemical techniques have been found to harm flora and fauna to the extent where the treatment may cause greater damage than the oil.

Bioremediation of oil at sea or on the shore seeks to increase the activity of naturally occurring microorganisms that are capable of degrading oil, including *Alcanivorax borkumensis*, which is probably the most important bioremediator of oil-contaminated marine environments (Kasai et al., 2002). The world's oceans have all been shown to contain bacteria capable of degrading oil, but those bacteria function as part of a **community** that is adapted for the particular environment in which they are found. It is mostly because of this community adaption that **bio-augmentation** (Section 6.4.3), seeding the polluted site with species of microorganisms known to degrade oil, has been largely unsuccessful. Much more successful has been the addition of nutrients to stimulate the growth of existing microorganisms (**biostimulation** Section 6.4.2).

Adding slow-release oil-soluble **fertilizers** to shorelines affected by oil spills can result in rapid and near-complete degradation of hydrocarbons (Atlas, 1995). What is more, this approach has no negative impact on the flora and fauna of the beach, unlike spraying with chemical dispersants or the physical disturbance associated with excavation, flushing, or pressure treatments.

Chemical dispersants can be used to accelerate the biodegradation of oil spills. The dispersants break up the crude oil to produce droplets with a greater surface area on which bacteria can act. The dispersants can themselves also act to stimulate growth, and this combined positive effect underpinned the decision to use more than 830,000 gallons of dispersant on the Deepwater Horizon oil spill in the Gulf of Mexico (Biello, 2010). However, despite the benefit, addition of large quantities of dispersants to the environment can have harmful effects. They may be toxic (including potentially to the bacteria being targeted) and could also stimulate bacteria to such an extent as to deplete oxygen levels, especially if used in already oxygen poor environments such as deep sea sediments and marshland.

Figure A Oil spills at sea can have significant environmental effects, harming organisms physically as well as causing toxic chemical effects.

Source: Image courtesy of Jordan Macha/CC BY-ND 2.0.

If oil cannot physically be removed entirely from the sea or seashore (and such removal is practically impossible in most if not all cases) then it will be bacteria in those environments that will ultimately degrade it. Bioremediation of oil spills focuses on enhancing their growth and in making the oil more available to them.

REFERENCES

Atlas, R.M. (1995) Bioremediation of petroleum pollutants. *International Biodeterioration & Biodegradation*, Volume 30, 317–327.

Biello, D. (2010) Slick solution: how microbes will clean up the Deepwater Horizon oil spill. *Scientific American* http://www.scientificamerican.com/article/how-microbes-clean-up-oil-spills/

Kasai, Y., Kishira, H., Sasaki, T., Syutsubo, K., Watanabe, K., & Harayama, S. (2002) Predominant growth of Alcanivorax strains in oil contaminated and nutrient-supplemented sea water. *Environmental Microbiology*, Volume 4, 141–147.

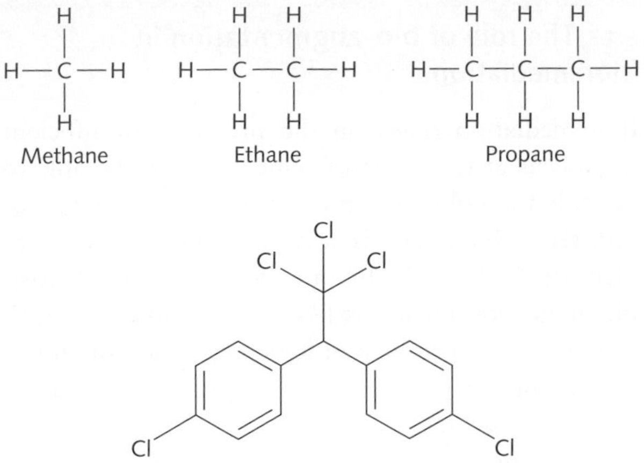

Methane Ethane Propane

Figure 6.3 Simple molecules like the alkanes (top) are more readily degraded than more complex molecules such as polycyclic aromatic hydrocarbons (PAHs) and organochlorines (e.g. the pesticide DDT; bottom).

pollutants or even thrive in their presence, and the **microbial community** will shift to one that can survive in the polluted soil. Over time, microorganisms will break down the pollution, but it will frequently be desirable to speed up this natural process.

One way to achieve a faster result than would occur under natural conditions is to stimulate the growth of indigenous microorganisms through active management of the remediation site. There are a number of physical, chemical, and biological factors that affect the rate at which pollutants can be degraded by microorganisms and in many cases it is possible to enhance conditions to increase the rate of bioremediation. The feasibility of doing this depends both on the parameter being altered (oxygen, pH, etc.) and on the environment. For example, whilst it is possible to increase the temperature of soil remediation systems it is not possible to increase, say, the temperature of large volumes of seawater affected by a crude oil spill. Managing microorganisms for bioremediation is fundamentally no different to managing any other species or community of species. Bioremediation can be enhanced by using one or more of the following approaches:

- **Increasing oxygen:** the speed of degradation of hydrocarbons is closely linked to oxygen availability, because aerobic degradation is far faster than anaerobic degradation (Holliger and Zehnder, 1996). In soils, oxygen is closely related to **soil structure**, which describes the arrangement of the particles of sand, silt, and clay that comprise the soil, the way that these particles aggregate into larger units and the presence of air spaces. An open soil structure has relatively more oxygen available than densely packed soil, while high levels of organic matter or waterlogging lower oxygen availability. Oxygen levels can be increased by improving the soil structure (e.g. or tilling, as in land farming Section 6.4.4) and drainage or by introducing air or oxygen to soil directly (e.g. by bioventing and biosparging (Section 6.4.4)). In water bodies, oxygen is linked to the level of mechanical mixing (so fast flowing water will generally be more oxygen rich) and the amount of dead organic matter. Oxygen levels can be improved by aerating water, perhaps by inducing turbulent flows or pumping air through the water column.
- **Raising temperature**: temperature is an important factor in any chemical reaction, including those that underpin metabolism. Reactions proceed faster, up to a point, as temperature increases, and below a certain temperature biochemical reactions may cease altogether. Temperature can be increased by gathering contaminated soil in heaps and mixing it with composting material such as wood chips or straw. This is principle behind **bioremediation composting** (Section 6.4.4) although fine control will be difficult or impossible in many cases. Temperature can also be enhanced with the **biopile** process (Section 6.4.4) and finely controlled in a **bioreactor** (Section 6.4.4).
- **Optimizing pH:** pH affects the rate of biochemical reactions and different microorganisms are suited to different pH environments. The solubility, and therefore the availability, of compounds to be degraded is also affected by pH. The addition of chemicals to make soil or water more or less acidic can therefore increase the availability of pollutants for degradation by microorganisms and enhance conditions for faster degradation. For example, sodium hydroxide can be added to raise pH, as in the Leviathan Mine bioreactor

(Case Study 6.2). Fine control of pH is achievable with some bioremediation approaches (e.g. a bioreactor, Section 6.4.4) and extremely difficult with others, especially those that are large-scale and open to the environment (e.g. windrow composting, Section 6.4.4.).

- **Availability of the pollutant**: adsorption describes the process whereby molecules adhere to a surface. Clays in soils, or water bodies that have a lot of clay-based sediment, are especially adept at adsorbing molecules, including pollutants. Adsorbed molecules are no longer available to be metabolized by microorganisms. At sea or in larger bodies of water on land, the solubility and miscibility (how readily a molecule mixes within the water column) of pollutants influence how available that pollutant is for bioremediation. Pollutant availability can be enhanced by treating with **surfactants**. Surfactants lower the surface tension between two liquids or between a liquid and a solid. Treating clay soils with a surfactant releases adsorbed molecules, making them available for microorganisms to degrade. Some surfactants act as detergents, **emulsifying** and dispersing otherwise immiscible pollutants. Some microorganisms naturally produce surfactants (**biosurfactants**). These can be added or produced *in situ* by encouraging indigenous biosurfactant bacteria or adding biosurfactant-producing bacterial species. In some cases, the surfactant itself may also be a pollutant that will require removal from the treated soil.
- **Availability of nutrients:** microorganisms able to degrade pollutants may require additional nutrients to survive. Microorganisms requiring additional nutrients can be supplemented with inorganic nitrogen and phosphorous-containing compounds, in effect, fertilizing the bioremediation community. This approach, termed **biostimulation**, can be very successful. For example, the addition of nutrients via fertilizer greatly enhances the bioremediation of oil at sea and on shore (Head and Swannel, 1999; Case Study 6.1) and a similar approach has been successful for soils (e.g. on alpine soils contaminated with hydrocarbons; Margesin and Schinner, 1997).

6.4.3 The role of bio-augmentation in bioremediation.

Bioremediation relies on the presence of sufficient organisms at the point of remediation to be able to degrade the pollutants present. If indigenous microorganism abundance is low then they can be supplemented through bio-augmentation. In theory, bio-augmentation seems like an obvious and simple solution to the problem of low target microorganism population. In practice, however, bio-augmentation has had mixed results.

Bio-augmentation can be effective in the laboratory, but it can be difficult to translate that effectiveness into the field. An inability to compete with indigenous bacteria is probably the most important factor limiting the growth of introduced bacteria. The effects of competition can be more complex and less easy to predict in the real world than in the controlled conditions of the laboratory. For example, in one study, laboratory cultures of bacteria were identified that showed enhanced crude oil degradation, and yet bio-augmentation with these cultures showed no effect on the rate of degradation of diesel fuel in arctic soils (Venosa and Zhu, 2003).

Bacterial species communities are subject to the full range of complex ecological interactions in the field. The physical environment plays a key role in the effectiveness of bioremediating microorganisms and the biological environment is no less important. Factors that can reduce the effectiveness of bio-augmentation include:

- The presence of compounds that limit microorganism growth (these may have caused the limited growth that triggered a bio-augmentation strategy in the first place).
- An insufficient concentration of the pollutant to sustain growth of the introduced microorganism.
- Introduced microorganisms may be metabolizing other substrates and effectively 'ignoring' the pollutant.
- Bacteria may be predated by protists (single-celled eukaryotes), which are common in soil and water.
- The bio-augmentation population may not be able to penetrate the soil sufficiently to access the pollutant.

Not all microorganism-based bio-augmentation involves bacteria. The structure and ecology of fungi makes them particularly effective at penetrating soil and reaching pollutant molecules, and a number of different fungi have been used to bio-augment contaminated soils. One pollutant that can be treated by fungal bio-augmentation is fluorene. Fluorene is a polycyclic aromatic hydrocarbon obtained from coal tar and used in the manufacture of several items, including some anti-malarial drugs. Soil contaminated with fluorine had 47 fungal strains isolated from it and tested for fluorene degradation by Garon et al. (2004). The fungi isolated from the contaminated soil were far more efficient at degrading fluorine-contaminated soil than were strains in a laboratory reference collection. The species best at degrading fluorene, *Absidia cylindrospora*, was added to fluorene-contaminated soil slurry and removed more than 90% of the fluorene present in 288 hours compared with 576 hours without bio-augmentation.

The effectiveness of bio-augmentation can be improved in a number of ways:

- **Adding substances or nutrients to assist the bio-augmenting organisms (biostimulators):** the combined effects of biostimulation through the addition of nutrients and bio-augmentation through the addition of sewage sludge in the bioremediation of creosote-contaminated soil are explored in the Online Case Study for Chapter 6, while the use of fertilizers to enhance the bioremediation of marine oil spills is outlined in Case Study 6.1.
- **Improving the ease of uptake of pollutants:** in the case of fluorene contamination, for example, degradation was enhanced through the use of maltosyl-cyclodextrin (Garon et al., 2004). Cyclodextrins are used in the pharmaceutical industry where they enhance the delivery of drugs across cell membranes, and this property is useful in enhancing the uptake of fluorene for bioremediation.
- **Bio-augmenting with activated populations**: this involves working with populations of microorganisms that have recently been exposed to the pollutant in question *in situ* rather in a laboratory culture. This selects for a community that is better able to degrade the pollutant under field conditions. For example, using activated soil to bio-augment the soil of polluted sites has been shown to be effective in degrading chlorobenzoate in soils (Gentry et al., 2004).
- Genetically engineering superior strains (Hot Topic 6.1).

You can find the online case study at www.oxfordtextbooks.co.uk/orc/goodenough

6.4.4 Techniques for bioremediation using microorganisms.

Making practical use of microorganisms in the environment to biodegrade pollutants, whether using indigenous populations, supplementation, bio-augmentation, or **genetically engineered microorganisms (GEMs)**, will generally require some degree of management if bioremediation is to occur within an acceptable timeframe.

Contamination can be treated on or off site. Off site, or *ex situ*, bioremediation involves the excavation or removal of contaminated material (e.g. soil or water) from the contaminated site and its treatment at another location. The treated material may or may not be returned to the original site. *In situ* bioremediation is, in many ways simpler than *ex situ*, since it does not require expensive and potentially complex excavation, transport, and containment of material at another location. However, *in situ* bioremediation inevitably means that the site remains contaminated until the bioremediation is complete and that the site may continue to be a source of pollution as pollutants leach from soils or otherwise disperse. *Ex situ* bioremediation may be more complex, but is often preferred, particularly in the USA, as it can avoid litigation that might arise from *in situ* remediation if any contamination remains.

The different strategies available for bioremediation are:

- **Land farming:** a simple bioremediation technique that involves mixing contaminated soil, sediment, or sludge with uncontaminated soil by ploughing or tilling. Mechanically processing the land increases aeration and more evenly distributes contaminants and soil microorganisms enhancing natural oxidation and bioremediation activity of indigenous microorganisms. Nutrients and soil bulking agents may be added. A typical land farming operation is shown in Figure 6.4. Land farming can be used for contaminated soil *in situ*, especially if the contamination is shallow and if the contaminated soil can be mixed with uncontaminated soil lying below it. Alternatively, the process can be undertaken *ex situ*. If soil is excavated and treated off site, then the possibility of leaching contaminants into new, unpolluted, ground and groundwater is a major concern. Collecting and monitoring runoff, lining, and damming treatment areas, for example using raised berms, can greatly reduce or prevent this problem, but increases costs and construction may have further environmental implications. Land farming has proved highly successful in treating land contaminated with diesel fuel, fuel oils, oily sludge, pentachlorophenyl (PCP), polyaromatic hydrocarbons (PAHs), creosote (Online Case Study Chapter 6), and some pesticides (e.g. Straube et al., 2003).

- **Bioventing:** works by increasing the oxygen supply to microorganisms within contaminated soil *in situ*. A borehole is drilled through contaminated soil and a vacuum applied. Assuming reasonable aeration, the reduced pressure inside the borehole draws air through the surrounding soil, which is replaced by air drawn in through other boreholes (Figure 6.5). This air flow helps to remove volatile organic compounds as well as increasing the rate of degradation by aerobic microorganisms in the soil. Bioventing is reasonably straightforward and can be a cost-effective way of treating contamination in inaccessible areas (e.g. under buildings), but it is not useful in all sites (e.g. those with low soil permeability or high clay content) and may require treatment of the vented gases. It has been used to treat, among other pollutants, jet fuel contamination of soil at military sites (e.g. Ong et al., 1994).

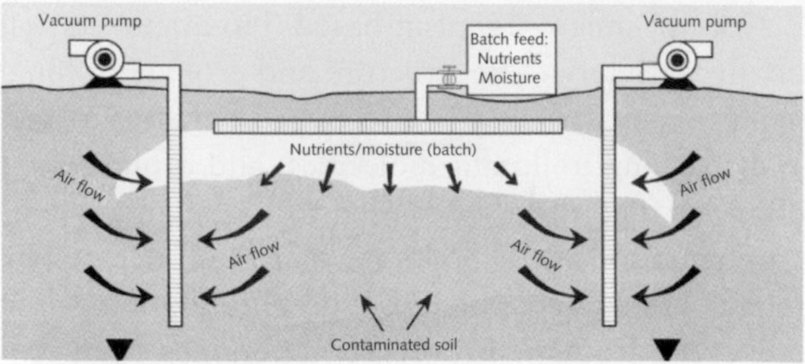

Figure 6.5 Bioventing increases oxygen availability within contaminated soil and enhances microbial degradation. Vacuum pumps draw air out of the soil through boreholes, thereby drawing more air in.

Source: Reprinted with permission from Macdonald, J. A., & Rittmann, B. E. (1993). Performance standards for *in situ* bioremediation. Environmental science & technology, 27, 1974-1979. Copyright 1993 American Chemical Society

- **Biosparging:** like bioventing, biosparging is an *in situ* bioremediation technique that involves increasing the level of oxygen in contaminated soil, but in this case the oxygen or air is actively pumped into the zone where bioremediation is needed. Additional nutrients can be added to enhance microorganism activity, and treatments that enhance the condition of the soil for biodegradation, such as altering pH, can also be carried out. Biosparging has proved to be particularly effective in reducing petroleum products at leaking underground storage tanks, and is most often used at sites contaminated with mid-weight petroleum products like diesel and jet fuel. Biosparging should be carried out with caution, since pumping air into soil can cause further problems by forcing contaminants above ground or into groundwater, basements, and sewers (Johnson et al., 2001).
- **Composting:** if contaminated material is excavated then it can be treated, either on or off site, in ways that accumulate that material and enhance the action of biodegrading microorganisms. Composting works by piling up contaminated soil and mixing it with composting material such as straw, wood chips or bark. Manure may be added, which supplies additional nutrients and microorganisms and chemicals can be used to control pH. Just as in a

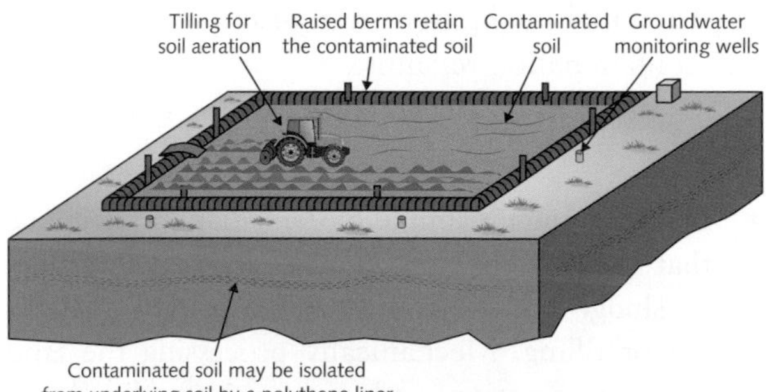

Figure 6.4 A typical land farming operation. Contaminated soil is contained, often within raised berms, and may be isolated from underlying soil by a polythene liner.

Source: Adapted from: Landfarming, United States Environmental Protection Agency (May 2004) www.epa.gov/oust/pubs/tums.htm

Figure 6.6 Large biopiles for treating contaminated soil. The biopiles have been constructed and are in the process of being covered with black plastic to increase their temperature and moisture. The white pipes are connected to a vacuum pump and draw air through the piles.

Source: Image courtesy of Akifer. http://akifer.ca/en/

garden compost heap, the biological activity of decomposing organisms increases the temperature within the pile, sometimes to 70°C or more (Gestel et al., 2003), and enhances the growth and activity of bioremediating microorganisms. Turning of the material within the composting pile increases aeration and further speeds up the process. It is also possible to aerate compost through the use of blowers or by applying a vacuum to the base of the heap. With some contaminants, composting can result in very rapid breakdown. For example, composting diesel contaminated soil resulted in 80% pollution removal within just 10 days (Namkoong et al., 2002). Composting includes windrow composting (basically a linear compost heap that can be relatively easily made, managed, and disassembled once the composting is complete) and biopiles. Biopiles are heaps of contaminated soil (typically between 1–3 metres in height) that are contained in some way (e.g. by lining areas where soil is piled) to prevent leaching (Lei et al., 1994). Polythene is typically used to cover the pile, and moisture and nutrients can be added to the surface (as in Figure 6.6). Biopiles can be additionally aerated by applying a vacuum to the base of them or by forcing air through perforated pipes passing through the pile.

- **Bioreactors:** typically a vessel or contained bed, a bioreactor allows for relatively fine control over many of the factors that are key to the biodegradation of pollutants by microorganisms such as temperature, pH, aeration, moisture, and the presence and abundance of microorganisms, which together enhance the rate and effectiveness of degradation. Bioreactor design varies according to the needs of the system and may include different forms of aeration, stirring, and rotation as well as different methods to expose contaminated material to microorganisms. Some bioreactors are quite simple and may consist merely of a lined pit filled with medium on which bacteria can grow and through which polluted water is induced to flow, either by gravity or pumping. A simple bioreactor filled with woodchips to denitrify field run-off is shown in Figure 6.7. However, other bioreactor designs can be more complex and include ways to control the conditions within them. Common to all bioreactor systems is the principle of control, and of all the techniques used for bioremediation, bioreactors are the most artificial. They can also be expensive and can be prone to malfunction (e.g. of stirring, aeration or pumping equipment), chemical problems (e.g. pH imbalance), and biological contamination (causing, for example, competition between microorganisms). However, bioreactors can still be very effective, as illustrated by Case Study 6.2, which details the sulphate-reducing

Figure 6.7 A simple bioreactor, consisting of a long pit lined with polythene and filled with woodchips being constructed at Southeast Missouri State University's David M. Barton Agriculture Research Center in Gordonville, Missouri, to denitrify the drainage coming from fields. Drainage is diverted through the bioreactor before discharging into local water courses.

Source: Image reproduced by permission of Dr Michael Aide, courtesy of Southeast Missouri State University.

CASE STUDY 6.2 The Leviathan Mine bioreactor

The Leviathan Mine is located in the Sierra Nevada mountain range near the California–Nevada border. Covering 100 ha at an elevation of approximately 2100 m, the mine operated intermittently since mining activities were established in the 1860s. The Anaconda Company purchased the mine in 1951 and extracted sulphur by open pit mining for 10 years from 1952, after which Anaconda ceased operations and no further mining has occurred.

During the open pit period significant pollution was released from the site as snowmelt, rain, and groundwater interacted with waste rock and tailings to create sulphuric acid. Acidified water leached additional contaminants including arsenic, copper, zinc, and chromium from the native rock, resulting in **acid mine drainage** (AMD) that flowed eventually into streams and rivers. The underlying geology of the area caused the drainage from the mine to enter water courses at a number of points, one of which is known as Aspen Seep. In 2003 a sulphate-reducing **bioreactor** was installed at Aspen Seep to treat AMD from the mine workings and surrounding area.

The bioreactor covers 0.3 ha and consists of two connected bioreactor 'pits' (Figure A). The water is first treated with sodium hydroxide to raise the pH from 3.1 to 4, and has ethanol added to provide a carbon source for the sulphate-reducing microorganisms housed in the bioreactors (acting therefore as **biostimulation** Section 6.4.2).

Bioreactor No. 1 is a long pit, lined with high density polyethylene and filled with river rocks. In total the pit has a volume of 350 m^3 and the rocks provide a surface on which bacteria that reduce sulphate to sulphide can grow. Water from Bioreactor No. 1 flows in to Bioreactor No. 2, which is around half the volume and houses bacteria capable of removing metals. From there, the water is treated again with sodium hydroxide to increase its pH to 7. It then passes to a settling pond with a slow but continuous flow. After an average of three days' retention in the pond, water passes out to the environment through a rock-lined channel that promotes degassing of residual hydrogen sulphide. In all, water takes 107 hours to pass through, at a rate of 136 l/minute.

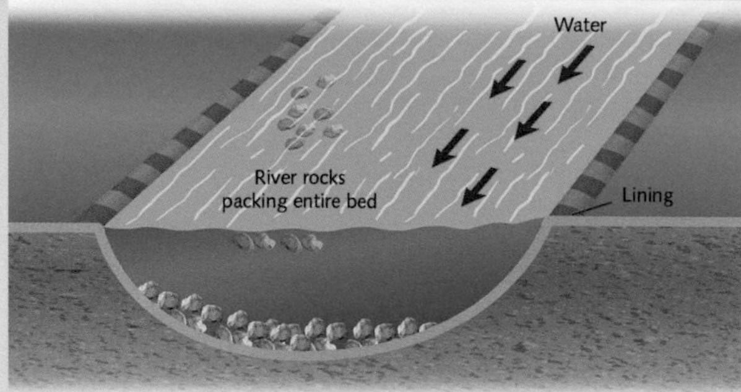

Figure A Schematic of the Aspen Seep bioreactor.

Using relatively simple and cost-effective *in situ* bioreactors, biostimulation (through the addition of ethanol), and sodium hydroxide, the Leviathan Mine bioreactors are able to convert AMD from pH5.3 to pH7 and reduce sulphate concentrations by 17%. They can also reduce target metal concentrations from up to 580 times the allowable standards to more than 43 times lower than those standards.

FURTHER READING

Reviews of acid mine remediation: Johnson, D.B. & Hallberg K.B. (2005) Acid mine drainage remediation options: a review. *Science of the Total Environment*, Volume 338, 3–14.

Reviews of acid mine remediation: Papirio, S., Villa-Gomez, D.K., Esposito, G., Pirozzi, F., & Lens, P.N.L. (2013) Acid mine drainage treatment in fluidized-bed bioreactors by sulphate-reducing bacteria: a critical review. *Critical Reviews in Environmental Science and Technology*, Volume 43, 2545–2580.

bioreactor used at the Leviathan Mine in California, USA.

Bioremediation approaches that enclose pollution, such as bioreactors, reduce the chances of animals and plants coming into contact with the pollution. However, with open-field approaches, such as land farming and windrows, contact between pollution, resident, and transitory flora and fauna is inevitable. Soil-dwelling animals will be directly affected, as will plants in the seed bank within the soil. The mechanical tilling of the soil will attract ground feeding birds and animals that feed on the soil invertebrates such activity exposes, possibly with negative effects for them. Animals are often highly mobile and they may move freely into and out of polluted areas, where they may become prey for others, or die and allow pollutants in their bodies to enter soil at other sites. In reality the wider **ecological risks** of a bioremediation project may be inconsequential. Even if the risks are substantial when balanced against the risks of the untreated pollution, projects involving higher ecological risks may be worth undertaking.

FURTHER READING FOR SECTION

Information compounds and chemistry important in remediation: Clayden, J., Greeves, N. & Warren, S. (2001) *Organic Chemistry*. Oxford University Press, Oxford, UK.

General primer of bacterial biology: Todar, K. Todar's Online Textbook of Bacteriology www.textbookofbacteriology.net

Introduction to bioremediation in practice: Crawford, R.L. & Crawford D.L. (2005) *Bioremediation: Principles and Applications*. Cambridge University Press, Cambridge, UK.

Overview of practicalities, risks and challenges: Gavrilescu, M., Demnerová, K., et al. (2015) Emerging pollutants in the environment: present and future challenges in biomonitoring, ecological risks and bioremediation. *New Biotechnology*, Volume 32, 147–156.

6.5 Phytoremediation: bioremediation using plants.

Phytoremediation is the use of plants to remove pollutants from soil or water, or otherwise render them harmless, including **phytoextraction, phytodegradation, phytostabilization, phytovolatilization,** and processes that take advantage of microorganisms associated with the roots of aquatic plants (**rhizofiltration,** covered in Section 6.5.1). Phytoremediation has a number of advantages over other approaches:

- Plants are more naturalistic than a land farm or bioreactor, and phytoremediation projects can even be made into useful visual landscaping features, e.g. reed beds, Section 6.5.1.
- Plants have a higher degree of public approval and are more readily accepted than microorganisms.
- Minimal disturbance of the topsoil, unlike land farming and some composting approaches.
- Phytoremediation can be effective with low concentrations of contaminants, which can be difficult to treat with microorganisms.
- Plants are easier to monitor than microorganisms.
- Metals can be accumulated in the growing plants and recovered, a process known as phytomining.
- Root penetration of the soil can act to stabilize soils and spoil heaps, preventing the loss of soil to erosion and the production of potentially harmful wind-blown dust. Stabilization with plants also reduces water run-off, thereby reducing the spread of pollution away from the contaminated site.
- It can be 50–80% less expensive than alternative approaches (Pulford & Watson, 2003).

Despite the advantages, there are disadvantages that may mean phytoremediation is not the best bioremediation option. Phytoremediation is relatively slow and can take years rather than months, as plants complete multiple growing seasons. In addition, plants accumulate pollutants, removing them from the soil, but in so doing they might make those pollutants far more readily available to wildlife and thereby enter the food chain.

Phytoextraction is a successful form of phytoremediation Plant roots in the contaminated soil uptake pollutants that are then dispersed through the roots, leaves, and stems. Roots finely penetrate the soil and have a relatively enormous surface area for the uptake of pollutant molecules. For a plant to be useful for phytoextraction it must be tolerant of the pollutant, have a dense root system capable of excellent soil penetration, and be able to develop a sufficiently harvestable biomass quickly enough to remove pollutants within an acceptable time frame. The **bioconversion factor** is the concentration of the pollutant in the plant compared to that in the environment. A bioconversion factor >20 is preferred, but around 400 species of plant have bioconversion factors 50–100 times more than normal plants. These so-called hyperaccumulators have been identified that can accumulate metals including zinc, cadmium, cobalt, nickel, lead, selenium, arsenic, and copper.

In addition to heavy metals such as mercury and cadmium, phytoextraction has been used to remove explosives such as TNT and trichloroethylene. Other pollutants that have been shown to be accumulated in some plants include uranium, which can be accumulated in sunflowers *Helianthus* from contaminated water with a biological conversion of some 30,000 (Meagher, 2000).

An interesting use of phytoextraction is the remediation of land that has been flooded with seawater and has become contaminated with sodium chloride. Barley *Hordeum vulgare* and sugar beets *Beta vulgaris* are commercial crops with moderate salt tolerance and are often grown on such land to extract the salt and render it usable for other less tolerant crops (Raats, 2015).

Trees are a low cost and high biomass option for phytoextraction, and numerous species have been examined and assessed for remediation potential, especially for treating metal contaminated sewage and other sludge. Successful trees must be tolerant of the pollution and if used *in situ* must also be able to grow on the site regardless of any issues such as soil compaction, acidity (especially spoil heaps from mining), salinity, and water and nutrient availability. Species that can grow on poor soil, have a deeply penetrating root system, relatively fast growth and tolerance of metals include willow *Salix*, poplar *Populus*, alder *Alnus*, birch *Betula*, and sycamore *Acer*. Of these, willow is particularly promising as it is fast growing and readily coppiced (Figure 11.7).

The remediation of soil or mine tailings (the material left over after processing ores to remove the economically viable fraction) that contain valuable metals can result in sufficient concentration of those metals accumulating in plants to make phytomining a possibility. The approach has the potential to extract metals from sources where metals are present at concentrations too low to make conventional extraction economically viable. Currently, the high market price of gold make it the most promising candidate metal for phytomining (Sheoran et al., 2013) although the first phytomining experiments focused on nickel accumulated by *Streptanthus polygaloides*. Under the right conditions Indian mustard *Brassica juncea* is capable of taking up gold ions (Au^{3+}) from soils, accumulating gold as elemental nanoparticles (Au) and producing a yield of up to 57 mg/kg of dry plant mass (Anderson et al., 1999).

Unravelling the genetic basis of phytoremediation gives the opportunity to genetically engineer plants for phytoremediation. It is still early days, but some success has been had producing transgenic plants (including *Arabidopsis thaliana* and *Nicotiana glauca*) that express genes crucial for metal tolerance and other bioremediation pathways. Other developments include the development of *Arabidopsis* plants whose roots change colour when they contact the degradation products of landmines (Deyholos, 2006). As with genetically engineered microorganisms (Hot Topic 6.1), the use of engineered plants for bioremediation in the environment is controversial and in practice is heavily regulated or banned because of concerns over environmental safety.

Other approaches to phytoremediation that do not entail the bioaccumulation of untransformed pollutants in plant tissue include:

- **Phytodegradation:** sometimes known as phytotransformation, phytodegradation is the breakdown and rendering safe of pollutants through metabolic processes occurring within plants. Organic molecules including some pesticides, explosives, and solvents can be taken up by plants, partially broken down by the action of enzymes (including peroxidases and esterases) and then

HOT TOPIC 6.1

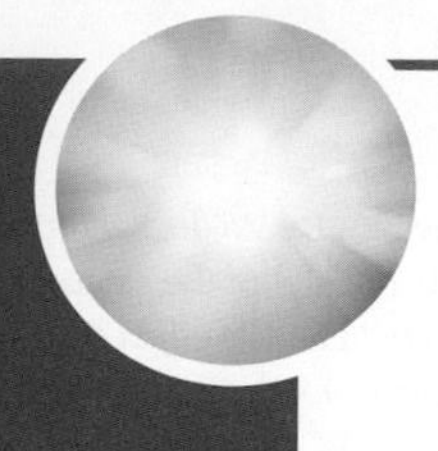

Genetically manipulated organisms in bioremediation

In addition to genetic information encoded in the **DNA** forming their circular **chromosome**, bacterial cells contain smaller DNA molecules that are physically separate from the chromosome. These genetic elements are called **plasmids**. Plasmids contain genes that include those used in antibiotic resistance and the metabolism of more unusual molecules, and can be transferred between individual bacterial cells along with their function.

Bioremediation relies ultimately on the presence of specific genes, and it is possible to make use of the mechanism of plasmid transfer (termed **conjugation**) to introduce genes for degradation pathways directly into polluted water and soil. For example, in one study *Enterobacter agglomerans* containing a plasmid encoding for the degradation of biphenyl was added to soil polluted with biphenyl. The introduced 'donor' bacteria disappeared within three days, but the plasmid, and thereby the ability to degrade the pollutant, was transferred to other bacteria in the soil (de Rore et al., 1994).

Other mobile genetic elements are called transposable elements, or **transposons**. Sometimes called 'jumping genes', transposons are DNA sequences that can change their position within the genome. Class 1 transposons are 'copy and paste' (whereby a copy of the transposon is inserted into another location), while Class 2 transposons are 'cut and paste' (whereby the transposon is removed from one location and inserted instead in another location). Transposons, like plasmids, may code for genes involved in biochemical pathways important for the degradation of xenobiotic pollutants. Examples include Tn5280, a class 1 transposon degrading chlorobenzene, and Tn4651, a class 2 transposon coding for the degradation of toluene and xylene.

Plasmids can be thought of as 'biological software upgrades' for bacteria, and under the right conditions bacteria can thus be 'programmed' to degrade specific molecules simply by inserting the relevant plasmids. Genetically engineering bacteria to produce multi-plasmid bacteria with bioremediation capabilities is far from a new idea. In fact, the first US patent was awarded for a manipulated bacterium in the 1970s. A strain of *Pseudomonas putida* was genetically engineered to contain four plasmids—inserted from four different bacteria—that degrade different hydrocarbons (camphor, octanol, xylene, and naphthalene). The resulting multi-plasmid bacterium is capable of about degrading two thirds of the hydrocarbons present in an oil spill and at a faster rate than the individual species.

Many other examples of genetically engineered bacteria with bioremediation capabilities exist. For example, *Pseudomonas fluorescens* has been engineered to biodegrade 2,4-dinitrotoluene by the insertion of genes encoding for the degradation pathway enzymes that are found naturally on a plasmid within a different, but less robust, bacterial species (Monti et al., 2005). As well as coding for degradation pathways of organic molecules, plasmids are also central to the ability of some bacteria to tolerate heavy metals such as mercury, arsenic, cadmium, and zinc. Resistant cells accumulate heavy metals and, with the right conditions, the cells can be harvested and the pollution removed.

Rapid advances in biomolecular engineering, an enhanced understanding of the bioremediation capabilities of different bacteria, and an ever greater need for bioremediation strategies for a variety of pollutants mean that genetically engineered microorganisms (GEMs) are ever more attractive options. However, the release of any genetically modified organisms (GMOs) into the environment is controversial and public perception of GMOs is generally negative. Containment issues are especially difficult with bacteria. The ability of bacteria to transfer plasmids horizontally between strains and species means that, once in the environment, it can be difficult to control or contain the spread of genes.

Environment concerns and resulting regulatory constraints have greatly limited the *in situ* application of GEMs despite promising results in the laboratory. However, there are biomolecular advances that can limit the ability of GEMs to spread, including plasmid addiction systems and the development of suicidal-GEMs. For example, a bacterium has been genetically engineered to degrade organophosphorus pesticide residue in the environment, that not only fluoresces (making its detection easier), but which can also be induced to die when it detects arabinose, a compound that can be added to soil and thereby allow practitioners to control GEM growth and spread. Such developments may prove instrumental in relaxing the regulatory frameworks that currently constrain the use of GEMs for *in situ* bioremediation.

QUESTIONS

Given that natural selection has resulted in such a wide variety of genes capable of remediating pollutants, are we right to be concerned about genetically engineered microorganisms?

How are concerns over genetically engineered microorganisms different from concerns from genetically modified organisms used in the food chain?

REFERENCES

de Rore, H., Demolder, K., De Wilde, K., Top, E., Houwen, F., & Verstraete, W. (1994) Transfer of the catabolic plasmid RP4:Tn4371 to indigenous soil bacteria and its effect on respiration and biphenyl breakdown. *FEMS Microbiology Ecology*, Volume 15, 71–77.

Monti, M.R., Smania, A.M., Fabro, G., Alvarez, M.E., & Argaraña, C.E. (2005) Engineering Pseudomonas fluorescens for Biodegradation of 2,4-Dinitrotoluene. *Applied and Environmental Microbiology*, Volume 71, 8864–8872.

transformed into less toxic compounds that are then sequestered (isolated) within the plant. Typically, compounds are oxidized within the plant, which makes the molecules more soluble and available for further degradation. The final product may still have some toxicity, and monitoring or containment may be required to ensure that herbivores are not being affected.

- **Phytostabilization:** plants can be very effective physical stabilizers of soil. Their root systems hold soil together and prevent soil erosion. Eroded soils, as well as loosing nutrients, are also prone to excessive runoff and leaching, increasing the dispersion of pollutants. Eroded soils have less plant cover and more prone to drying out, leading to the carrying away of dust (and pollutants) on the wind. Consequently, pollutant tolerant plants can be used to stabilize polluted soils and have a positive effect regardless of any chemical phytoremedation.
- **Phytovolatilization of atmospheric pollutants:** plant bioremediation activity is mostly concerned with soil pollutants and plant–soil–rhizosphere interactions. However, plants also interact with the atmosphere, taking in and expelling gases, and adsorbing particulates. Sometimes this has the potential to have a negative effect as pollutants can be released into the atmosphere, aiding in their dispersal, a process termed phytovolatilization. Poplar *Populus* trees can volatize trichloroethane (Newman et al., 1997), whilst *Eucalyptus* can volatize methyl t-butyl ether (Newman et al., 1999). Tobacco *Nicotiana* can convert methyl mercury to volatile elemental mercury vapour, and mercury ions (Hg^{2+}) in soils can be converted by microorganisms associated with plant roots into elemental mercury (Hg) (Wang et al., 2012). In other cases, phytovolatilization can be useful for remediation. Atmospheric pollutants that occur close to the ground can be taken up by plants through their stomata, tiny pores on the undersides of leaves that are involved in gas exchange. The bioremediation of atmospheric chemical pollutants through uptake, transformation or accumulation is not well developed, but some plants have shown promise in taking up nitrogen dioxide (Lockwood et al., 2008).

6.5.1 Rhizofiltration: reed beds and wetland systems.

Rhizofiltration describes the removal of pollutants from flowing water by the roots and more especially the rhizosphere of plants associated with water and wetlands. Wetland plants such as the common reed *Phragmites* have a high concentration of microorganisms and it is these that are chiefly responsible for the removal, sequestration and transformation of pollutants, including heavy metals and organic compounds.

Wetland plants can also remove nutrients from water flows, making them useful for remediation of water that is contaminated with organic waste, such as manure-contaminated slurry from dairy farms and human sewage. Wetland systems for the treatment of sewage have become popular at eco-friendly visitor attractions, where the reed beds themselves are both attractive to look at and act to attract birds and other wildlife to the site. Given the decline of wetlands, the construction of wetland systems for the treatment of contaminated water has clear ecological benefit, but

AN INTERVIEW WITH AN APPLIED ECOLOGIST 6

Name: Dr Matthew Simpson

Organization: WWT Consulting, consultancy arm of Wildfowl & Wetlands Trust

Role: Associate Director

Nationality and main countries worked: British; worked throughout Europe (UK, Portugal, Spain, Bulgaria), the Americas (US, Brazil, Venezuela, Guyana, Suriname), Africa (Kenya, Botswana), and Asia (India, Sri Lanka, Nepal, South Korea, Japan, China, Vietnam).

What is your day-to-day job?

I work directly with people around the world to manage, and provide advice on managing, the restoration and creation of wetlands for multiple benefits that include water treatment, flood attenuation, carbon storage, biodiversity, green and blue space amenity, and livelihood support.

What is your most interesting recent project and why?

I am currently working on a project in Colombo, Sri Lanka, that involves the management of wetlands within the city to provide essential flood storage (without the wetlands the city would face an annual flood damage of around 1% of Colombo's GDP), to provide water treatment, as many sewers input directly into the wetlands, and to provide a cooling effect on the city (living next to a wetland area can reduce air temperature by 10° compared with living next to a concreted area, which provides a huge electricity saving from reduced air conditioning requirements). The wetlands are threatened by legal and illegal infilling and development, but raising awareness of their importance, training government staff in wetland management, and developing visitor facilities within some of the wetlands to allow public access for recreation and education will maintain the important ecosystem services they provide and protect the impressive range of biodiversity found across the city. The project is a complex mix of technical ecological, hydrological, and socio-economic research and design and extensive engagement with stakeholders with conflicting demands, which makes it very challenging.

What's been best part of these particular projects?

There is a huge sense of satisfaction when a design that you worked on with a local community is constructed and you see people benefitting through improved water quality or through support of sustainable livelihoods, and at the same time that system provides habitat for wildlife and maximizes biodiversity.

What are the main challenges in your field and how can they be overcome?

The main challenge is the acceptance of the use of natural infrastructure such as wetlands by governments, planners, architects, and engineers. There is still an attitude in many parts of the world where hard engineering is seen as the only solution when it has been demonstrated that the use of natural infrastructure not only provides an effective approach to water treatment, for example, but provides multiple benefits for the environment and people that steel and concrete cannot provide. There is still a big educational role that is needed to make sure that this type of understanding becomes mainstream in design and planning.

What next for you, and why?

More of the same—I'm very happy in my current role because I get to travel the world, work with some amazing people and deliver projects that have a benefit to the environment and to people.

Finally, how did you get into your area of work and what advice would to others?

After my first degree I did a master's degree in Environmental Water Management before doing a PhD in Geography on functional wetlands. I then had a number of years doing academic research on wetland management in Vietnam before getting a job in environmental consultancy. My advice to others would be to get as much practical experience as possible whilst also getting a solid academic understanding of a subject. When we recruit we always look at what practical experience people have already got. Unfortunately most people have to volunteer to get that experience, but it is a great way to meet people and make contacts.

it must be realized that the input (or inflow) to such a system is, by definition, polluted. Furthermore, since rhizofiltration relies on flowing water, these filtration systems have an output or effluent that may still contain pollutants. Establishing a monitoring system (Section 3.3.5) to determine the condition of the effluent from the system is necessary to prevent potential downstream pollution, as well as to monitor the effectiveness of the treatment system (Interview with an Applied Ecologist Chapter 6).

Reed bed filtration systems using the common reed *Phragmites* are by far the most common artificial wetland filtration system in use and a typical set-up is shown in Figure 6.8. *Phragmites* can pass oxygen from the leaves to the roots, encouraging the development of a considerable population of aerobic microorganisms on the roots of the reeds, which can degrade pollutants.

Reed beds and other wetland systems are constructed with an impermeable base layer (typically clay or a polypropylene liner) and are contained by walls, with the downstream wall being effectively a dam. These collectively produce a treatment pan in which the reeds or other wetland plants can grow. Contaminated water is fed into the reed bed, under control, passes through the usually water-logged soil from which the reed grow and is taken out of the reed bed under gravity.

Different types of flow can be established through artificial wetlands simply by adjusting the position of the inflow pipe, thereby taking maximum advantage of the plants being employed. Plants may float freely on the surface (such as duckweed *Lemna*) or be completely submerged (including Canadian pondweed *Elodia canadensis*), but surface flow and subsurface flow through the soil are the most commonly used flow patterns. If potentially expansive or invasive plants (including *Lemna* and *Elodia*) are to be used then care must be taken to ensure that they are properly contained. Wetland trees such as willow *Salix* can

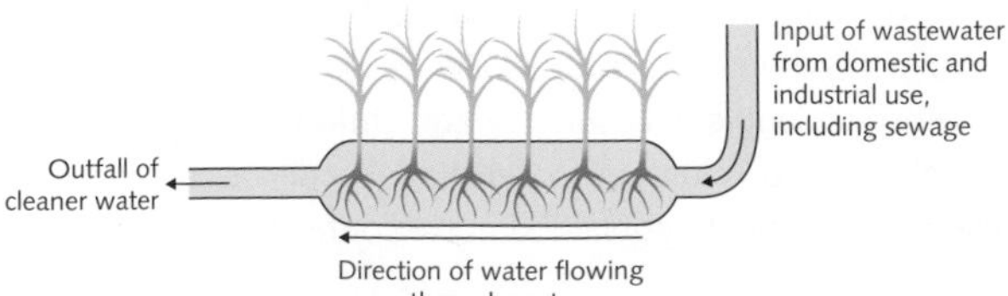

Figure 6.8 Typical reed bed operation. The constructed wetland has an impermeable base and shallow soil is used to root the plants, often the common reed *Phragmites*.

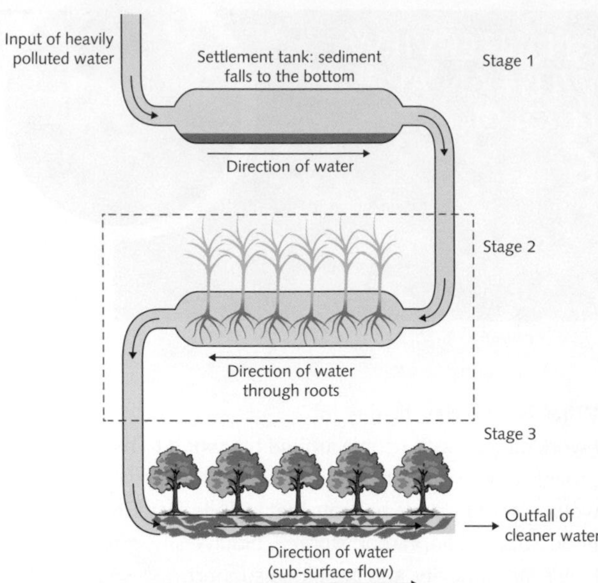

Figure 6.9 Multiple treatment beds can be connected to produce highly 'polished' effluent. Stage 2 can be used in isolation as a single-stage treatment option; see Figure 6.8.

also be incorporated into wetland filtration systems provided that the flow rate through the system is properly controlled.

Multiple artificial wetlands can be joined up to make multiple filtration systems. For example, the effluent from an initial wetland treatment site using cattails *Typha* or reeds *Phragmites* could be fed into a secondary wetland system consisting of planted willows as shown in Figure 6.9.

Rhizofiltration systems using artificial wetlands are commonly used for 'polishing' effluent, a term used to describe the final treatment of wastewater (for example, from sewage treatment works) before discharging it to the environment. Rhizofiltration is ideal for this role, partly because it tends to excel at removing pollutants from water with very low levels of contaminants. There are other reasons why this approach might be taken, including cost effectiveness and landscape amenity, as demonstrated by the Tres Rios Wetlands Project (Case Study 6.3).

FURTHER READING FOR THIS SECTION

Overview of phytoremediation: Wiley, N. (ed.) (2007) Phytoremediation: Methods and Reviews. Methods in Biotechnology v23 Humana Press, Inc., Totowa, NJ, USA.

CASE STUDY 6.3 The Tres Rios wetlands project

Tres Rios ('three rivers') is a **constructed wetland** at the confluence of the Salt, Gila, and Aqua Fria rivers, in the city of Pheonix, Arizona, USA. It demonstrates how wetland projects for treating wastewater can have multiple benefits to local communities and to flora and fauna. Completed in 2010, the project consists of wetlands, riparian corridors (the habitat lining a river or stream), and open water or marshland along a 10 km stretch of the Salt River as it joins with the Gila and Agua Fria Rivers (Figure A). The full scale project involves the construction of almost 200 ha of emergent wetlands in what is essentially the Arizona desert.

The problem with creating wetlands in deserts is the availability of water. The Tres Rios project made use of the water being produced by the 91st Avenue Wastewater Treatment Plant to supply the wetlands, which were also themselves used to treat that water (which has already been treated in the Treatment Plant) before discharge into the river system.

The wetlands use **rhizofiltration** (Section 6.5.1) to treat the influent water, taking advantage of the degrading action of bacteria dwelling on the roots of wetland plants. Wetland trees also play a part in removing contaminants from the already treated incoming water.

What sets Tres Rios apart from a great many other **phytoremediation** projects is firstly its sheer scale and secondly the wealth of interconnected benefits that the project provides, which were integral to the planning of the project. The emergent wetlands provide more than just remediation:

- **Flood protection**: the wetlands help to buffer against sudden water level increases and smooth out the flow of the river.
- **Habitat and biodiversity**: the creation of wetlands in an otherwise arid environment has created a hotspot for birds, animals, and other wildlife.

Figure A The Tres Rios Wetlands provide valuable wetland habitats in a generally arid region. Their creation has benefited wildlife and the community as well as treating wastewater.
Source: Image courtesy of Joyce Cory/CC BY 2.0.

- **Public amenities**: the wetlands provide the local community with multi-use trails and picnic areas as well as bird watching and fishing opportunities. There is also an environmental education centre.

Even smaller and less ambitious wetland remediation projects can provide valuable wetland habitat, especially for birds, can provide the public with amenity space and can act as an education resource.

FURTHER READING

Water Service and Tres Rios website: https://www.phoenix.gov/waterservices/tresrios

6.6 Conclusions.

Pollution is, and will continue to be, a major problem. As new forms of industry and technology develop, so too might novel forms of pollution. For example, nanoparticles and other products of nanotechnology may develop into a significant environmental threat. We may also discover that some molecules not currently considered to be especially harmful might have long-term harmful effects that will only become apparent after sufficient exposure. Ideally we should be finding ways to prevent pollution entering the environment in the first place, and this is indeed the approach being taken in many industries. However, in reality we are also likely to be looking to find new ways to treat polluted soil, water, and air for many years to come. Chemically treating pollutants can be problematic and enhanced legislation means that bioremediation solutions are increasingly attractive. Microorganisms possess a wealth of metabolic pathways that allow many common and even exotic pollutants to be degraded. Techniques ranging from simple land farms to advanced bioreactors are being developed to harness the power of microorganisms in cleaning up pollution. Plants are also powerful pollution remediators, and the use of plants to extract pollutants from soil and water, as well as to clean up polluted water through constructed wetlands, is relatively well advanced.

Future developments will doubtless see more species being identified and employed in remediation, advances in the abiotic components of bioremediation systems (such as in the technology used to convey water through wetland systems), and the use of genetically engineered organisms. Although controversial, the ability to engineer organisms with properties that can lead to targeted pollution remediation is already a reality, and the coming years will likely see legislation, regulation, and public opinion playing catch-up.

ONLINE ACTIVITY

Go to www.oxfordtextbooks.co.uk/orc/goodenough/ to download the activity that accompanies this chapter.

CHAPTER 6 AT A GLANCE: THE BIG PICTURE

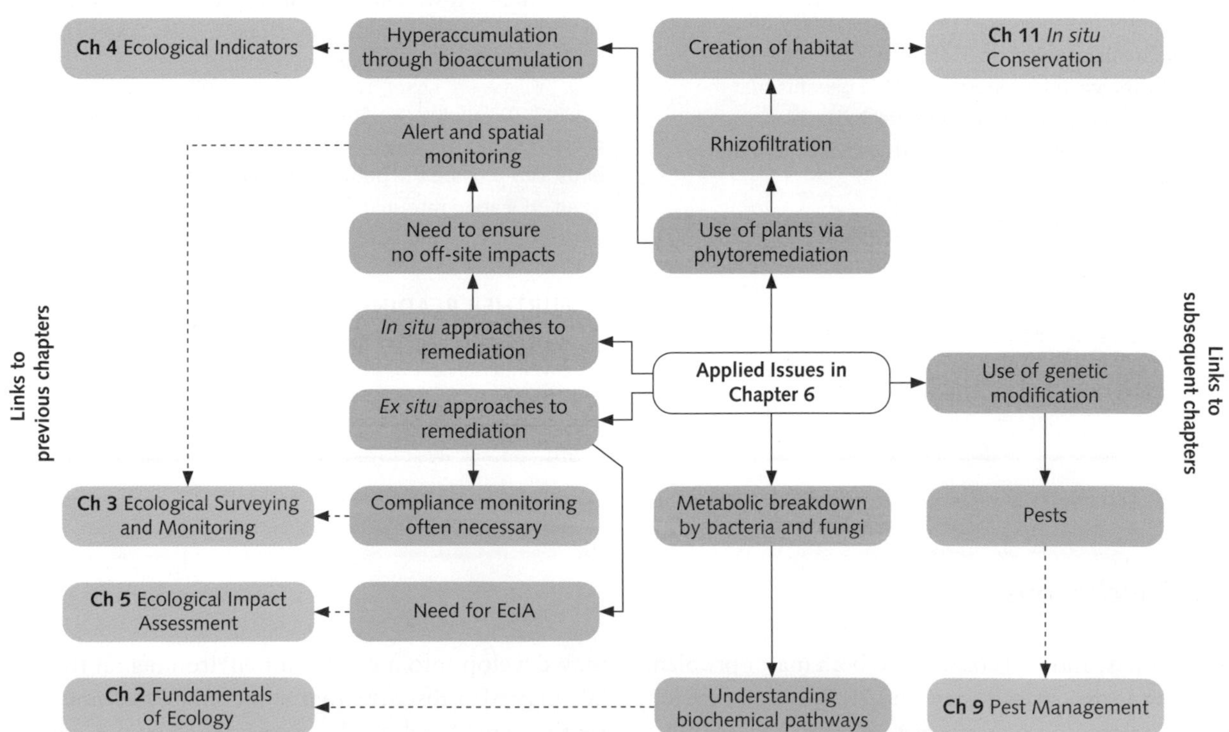

REFERENCES

Anderson, C.W.N., Brooks, R.R., Chiarucci,A., LaCoste, C.J., Leblanc, M., Robinson, B.H., Simcock, R., & Stewart, R.B. (1999) Phytomining for nickel, thallium and gold. *Journal of Geochemical Exploration*, Volume 67, 407–415.

Barragán-Huerta, B.E., Costa-Pérez, C., Peralta-Cruza, J., Barrera-Cortés, J., Esparza-Garcíab, F., & Rodríguez-Vázquez, R. (2007) Biodegradation of organochlorine pesticides by bacteria grown in microniches of the porous structure of green bean coffee. *International Biodeterioration and Biodegradation*, Volume 59, 239–244.

Committee on Toxicity of Chemicals in Foods, Consumer Products and the Environment (2013) Lowermoor Water Pollution Incident. http://cot.food.gov.uk/cotwg/lowermoorsub/draftlowermoorreport/

Deyholos, M., Faust, A.A., Miao, M., Montoya, R., & Donahue, D.A. (2006) *Feasibility of Landmine Detection using Transgenic Plants*. Proceedings of SPIE 6217: Detection and Remediation Technologies for Mines and Minelike Targets XI.

Driscoll, C.T., Mason, R.P., Chan, H.M., Jacob, D.J., & Pirrone, N. (2013) Mercury as a global pollutant: sources, pathways and effects. *Environmental Science and Technology*, Volume 47, 4967–4983

Eckerman, I. (2005) *The Bhopal Saga—Causes and Consequences of the World's Largest Industrial Disaster*. Bhopal, India: Universities Press (India) Pvt. Ltd.

Elkins, J.W. (1999) Chlorofluorocarbons. In: D.E. Alexander & R.W. Fairbridge (Eds) *The Chapman and Hall Encyclopedia of Environmental Science*, pp. 78–80. Boston, MA: Kluwer Academic.

Fierer, N. & Jackson, R.B. (2006) The diversity and biogeography of soil bacterial communities. *Proceedings of the National Academy of Sciences of the United States of America*, Volume 103, 626–631.

Garon, D., Sage, L., Wouessidjewe, D., & Seigle-Murandi, F. (2004) Enhanced degradation of fluorene in soil slurry by Absidia cylindrospora and maltosyl-cyclodextrin. *Chemosphere*, Volume 56, 159–166.

Gentry, T.J., Josephson, K.L., & Pepper, I.L. (2004) Functional establishment of introduced chlorobenzoate degraders following bioaugmentation with newly activated soil: Enhanced contaminant remediation via activated soil bioaugmentation. *Biodegradation*, Volume 1, 67–75.

Gestel, K.V., Mergert, J., Swings, J., Coosemans, J., & Ryckeboer, J. (2003) Bioremediation of diesel oil-contaminated soil by composting with biowaste. *Environmental Pollution*, Volume 125, 361–368.

Harms, H., Schlosser, D., & Wick, L.Y. (2011) Untapped potential: exploiting fungi in bioremediation of hazardous chemicals. *Nature Reviews Microbiology*, Volume 9, 177–192.

Head, I.M. & Swannell, R.P.J. (1999) Bioremediation of petroleum hydrocarbon contaminants in marine habitats. *Current Opinion in Biotechnology*, Volume 10, 234–239.

Holliger, C. & Zehnder, A.J.B. (1996) Anaerobic biodegradation of hydrocarbons. *Current Opinion in Biotechnology*, Volume 7, 326–330.

Johnson, P.C., Johnson, R.L., Bruce, C.L., & Leeson, A. (2001) Advances in *in situ* air sparging/biosparging. *Bioremediation Journal*, Volume 5, 251–266.

Lei, J., Sansregret, J.L., et al. (1994) Biopiles and biofilters combined for soil cleanup. *Pollution Engineering*, Volume 26, 56–58.

Lockwood, A.L., Filley, T.R., Rhodes, D., & Shepson, P.B. (2008) Foliar uptake of atmospheric organic nitrates. *Geophysical Research Letters*, Volume 35, L15809.

Macgregor, C.J., Pocock, M.J.O., Fox, R., & Evans, D.M. (2015) Pollination by nocturnal Lepidoptera, and the effects of light pollution: a review. *Ecological Entomology,* Volume 40, 187–198.

Margesin, R. & Schinner, F. (1997) Bioremediation of diesel-oil contaminated alpine soils at low temperatures. *Applied Microbiology and Biotechnology*, Volume 47, 462–468.

Meagher, R.B. (2000) Phytoremediation of toxic elemental and organic pollutants. *Current Opinion in Plant Biology*, Volume 3, 153–162.

Milic-Terzic, J., Lopezz-Vidal, Y., Vrvic, M.M., & Saval, S. (2001) Detection of catabolic genes in indigenous microbial consortia isolated from diesel-contaminated soil. *Bioresource Technology*, Volume 78, 47–54.

Namkoong W., Hwang, E.Y., Park, J,S., & Choi, J.Y. (2002) Bioremediation of diesel-contaminated soil with composting. *Environmental Pollution*, Volume 119, 23–33.

Newman, L.A., Strand, S.E., Choe, N., Duffy, J., Ekuan, G., Ruszaj, M., Shurtleff, B.B., Wilmoth, J., Heilman, P., & Gordon, M.P. (1997) Uptake and biotransformation of trichloroethylene by hybrid poplars. *Environmental Science and Biotechnology*, Volume 31, 1062–1067.

Newman, L.A., Gordon, M.P., Heilman, P., Cannon, D.L., Lory, E., Miller, K. Osgood, J., & Strand, S.E. (1999) Phytoremediation of MTBE at a California naval site. Soil and Groundwater Cleanup, Feb/March, 42–45.

Ong, S.K., Leeson, A., Hinchee, R.E., Kittel, J., Vogel, C.M., Sayles, G.D., & Miller, R.N. (1994) Cold climate applications of bioventing. In: R.E. Hinchee, B.C. Alleman, R.E. Hoeppel, & R.N. Miller (Ed.) *Hydrocarbon Remediation*, pp. 444–453. Boca Raton, FL: CRC Press Inc.

Pulford, I.D. & Watson, C. (2003) Phytoremediation of heavy metal contaminated land by trees—a review. *Environment International*, Volume 29, 529–540.

Raats, P.A.C. (2015) Salinity management in the coastal region of the Netherlands: A historical perspective. *Agricultural Water Management*, Volume 157, 12–30.

Rascio, N. & Navari-Izzo F. (2011) Heavy metal hyperaccumulating plants: How and why do they do it? And what makes them so interesting? *Plant Science*, Volume 180, 169–181.

Sheoran, V., Sheoran, A.S., & Poonia, P. (2013) Phytomining of gold: a review. *Journal of Geochemical Exploration*, Volume 128, 42–50.

Straube, W.L., Nestler, C.C, Hansen, Ringleberg, D., Pritchard, P.H., & Jones-Meehan, J. (2003) Remediation of Polyaromatic Hydrocarbons (PAHs) through Landfarming with Biostimulation and Bioaugmentation. *Acta Biotechnologica*, Volume 23, 179–196.

Venosa, A.D. & Zhu, X. (2003) Biodegradation of crude oil contaminating marine shorelines and freshwater wetlands. *Spill Science and Technology Bulletin*, Volume 8, 163–178.

Wang, J., Feng, X., Anderson, C.W., Xing, Y., & Shang, L. (2012) Remediation of mercury contaminated sites—a review. *Journal of Hazardous Material*, 221–222, 1–18.

7 Landscape Ecology and Management

7.1 Introduction.

Almost every ecological question has a spatial dimension. Many ecological threats are specifically spatial in origin; for example, habitat loss and habitat fragmentation due to changing land use and infrastructure developments. Even impacts of threats that are not overtly spatial, such as human-accelerated climatic change, have spatial consequences in terms of species range shifts and altered habitat and community structure. As a result, spatial considerations are vitally important in conservation, both in assessing the need for conservation (Chapter 10) and determining the 'where' and 'how' of conservation initiatives across natural and human landscapes (Chapter 11).

Many other aspects of ecological management also have a spatial element. This includes managing non-native species and pests (Chapters 8 and 9) at one end of the spectrum and maintaining ecosystem services at the other. Spatial considerations are also important in the planning of new developments such as housing estates, factories, and transport infrastructure (Chapter 5). All of these are underpinned by spatial monitoring, which is necessary to map species' distributions and understand patterns of dispersal (Chapter 3).

In this chapter we will focus on management of species over space. We will first go through the elements that make up a landscape and discuss how landscapes function and how they can change over time. We then consider the likely effects of such changes on species and their population dynamics, as well as the effects on community structure, habitat characteristics, and environmental conditions. Understanding spatial processes is vital in order for the Applied Ecologist to be able to study landscape-species interactions effectively, predict the effects of landscape change accurately, and implement the landscape-sensitive management that is necessary to buffer the effects of existing problems and prevent future problems.

7.2 Landscape elements.

Landscapes are often complex and dynamic entities, with many different types of land use occurring in complicated spatial arrangements. For example, the city of London, UK, has many different landscape components, as shown in Figure 7.1. Even complicated landscapes such as this, though, can be deconstructed into three fundamental landscape elements:

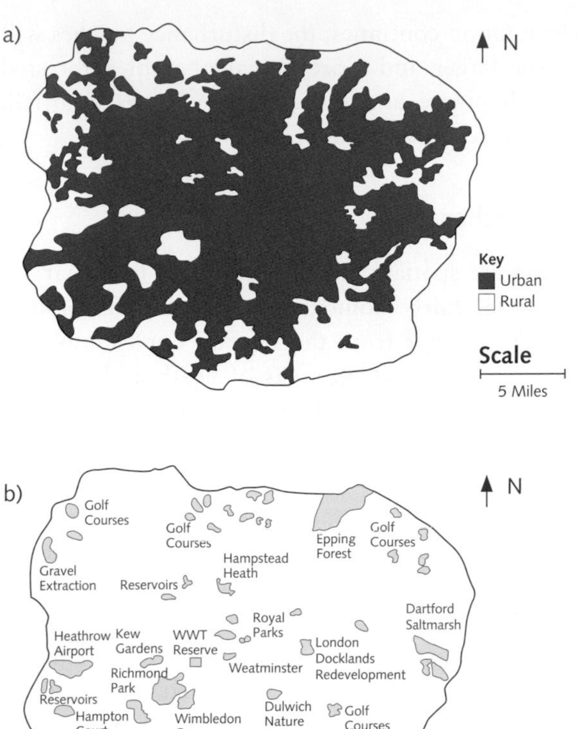

Figure 7.1 The city of London, UK, distilled to its various landscape elements of background matrix of urban land use (a), example patches (b), and major road corridors as example linear features (c). All diagrams use the M25 orbital motorway as the outer boundary.

- **Background matrix:** the dominant land type in a given area at a specific point in time. In the case of London, this is urban land use as shown in Figure 7.1a.
- **Patches:** areas that differ from, and are set within, the background matrix. There are usually multiple patches within a landscape that differ in size and shape, as shown for London in Figure 7.1b.
- **Linear features:** can be natural or anthropogenic and often link patches together. They are sometimes referred to as corridors because of their potential to facilitate the movement of species (rivers, hedgerows) or humans (roads, railways), but given that they can also act as barriers, the term 'linear features' is more accurate, with corridors being a subcategory within that (see section 7.6.2). The main road linear features are shown for London in Figure 7.1c.

As often in Applied Ecology, the question of scale arises here. Figure 7.1 shows the situation for a large-scale complex landscape, but the same deconstruction process can be undertaken for simple landscapes at much smaller spatial scales. For example, Figure 7.2 shows one field of a farm (background matrix), with a small copse of trees (patch), enclosed by hedges and bordered by roads (linear features).

7.2.1 Background matrix.

The type of land use that comprises the background matrix will differ in different areas: where farmland dominates there is an agricultural matrix; in forested areas there is a woodland matrix and so on.

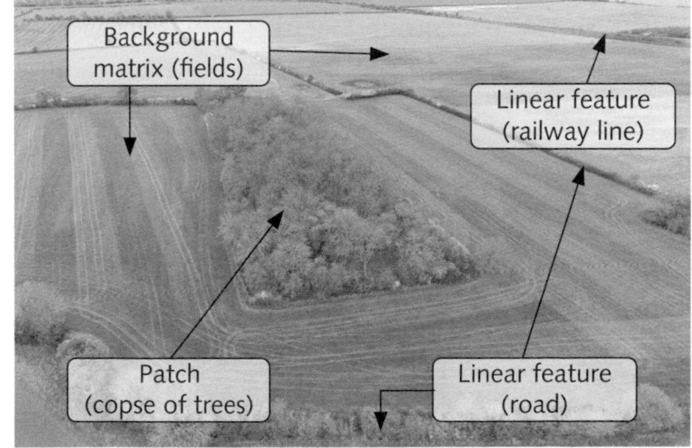

Figure 7.2 Landscape elements at the scale of an individual field near the village of Gotherington, Gloucestershire.

Source: Image courtesy of Will Carpenter, used with kind permission.

This diversity creates a large-scale **mosaic** of dominant land uses. The 'dominant' land use is simply the most prevalent—it need not account for >50% of the landscape. Indeed, in very **heterogeneous** (mixed, diverse) landscapes, the matrix might be fairly uncommon and fragmented.

Sometimes, when the dominant land use is <50%, it can be appropriate to group land uses to describe the matrix more accurately and inclusively using a single land use category. For example, if a landscape is 15% arable farming, 12% livestock farming, 11% orchard, and 9% vineyard, an appropriate background matrix classification might be mixed agriculture accounting for 47%. Grouping habitats in this way is only sensible if—as in this case—the habitats are *broadly* similar, especially in terms of ecological functioning; it would not make sense to combine heathland and urban parkland, nor an estuarine system with woodland.

The background matrix will typically exert a larger influence on how a landscape looks, and the type of ecological processes within it, compared with patches or linear features. For example, a woodland landscape will function, under natural conditions, primarily as a wooded ecosystem with processes such as gap dynamics and competition for light being very important. In contrast, a landscape with a wetland matrix will be influenced by succession (Section 2.8.2), hydrological regimes, and flooding cycles. Thus, if it is hard to define a matrix because different land uses are approximately equal in size, and grouping is not possible, it can be useful to examine which exerts the largest influence on the area's ecology (Forman and Godron, 1986).

Just as the background matrix will change spatially across a landscape mosaic, the matrix at a given point in space is likely to change temporally. This can occur, for instance, if a landscape becomes increasingly urbanized. London's background matrix was 64% urban at the start of the 21st century (Figure 7.1) as a result of 2000 years of urban expansion and a gradual shift from a largely agricultural matrix to one of housing, industry, and retail. Another example would be if a forested landscape is deforested. Initially, cleared areas will be patches within a woodland background matrix, possibly with logging tracks connecting them. Over time, if deforestation continues, the disturbance patches will become larger and more prevalent until deforested areas dominate and the matrix becomes clear-fell, scrub, or secondary woodland.

7.2.2 Patches.

Patches are spatial units or 'parcels' of land that are themselves fairly **homogenous** (similar) in habitat, but which differ from the matrix. For example, in an agricultural matrix, areas of heath and areas of woodland would normally be considered as separate patch types because species composition and dominant ecological processes will differ. This means that just as there is a repeating mosaic of background matrices over a large scale, so there is a repeating mosaic of different patches within each background matrix.

Although patches are usually driven by habitat, they can also be classified into one of six non-mutually exclusive functional groups (classification system after Forman and Godron, 1986).

- **Environmental patches:** these differ from the background matrix due to an abiotic difference. This is often geological: for instance if a patch within a larger area has different geology and soil that can support heathland plants but not trees, a patch of heath might occur in a wooded matrix. Hydrological patches can also occur: a poor draining area becoming a marsh within a grassland matrix, for example.
- **Anthropogenic patches:** patches that result directly from human activity. Examples include gardens, plantations, vineyards, reservoirs, and quarries, as well as (in non-urban landscapes), patches of residential housing, industry, or retail. Anthropogenic patches can be deliberate or accidental in origin, as explained, with examples, in Figure 7.3.
- **Remnant patches:** areas that are now patches within a background matrix, but are actually the habitat of a previous background matrix. For example, a small copse might be a remnant patch in a landscape that used to be wooded but is now agricultural, while a small area of common grassland in an urban area might be the remnant

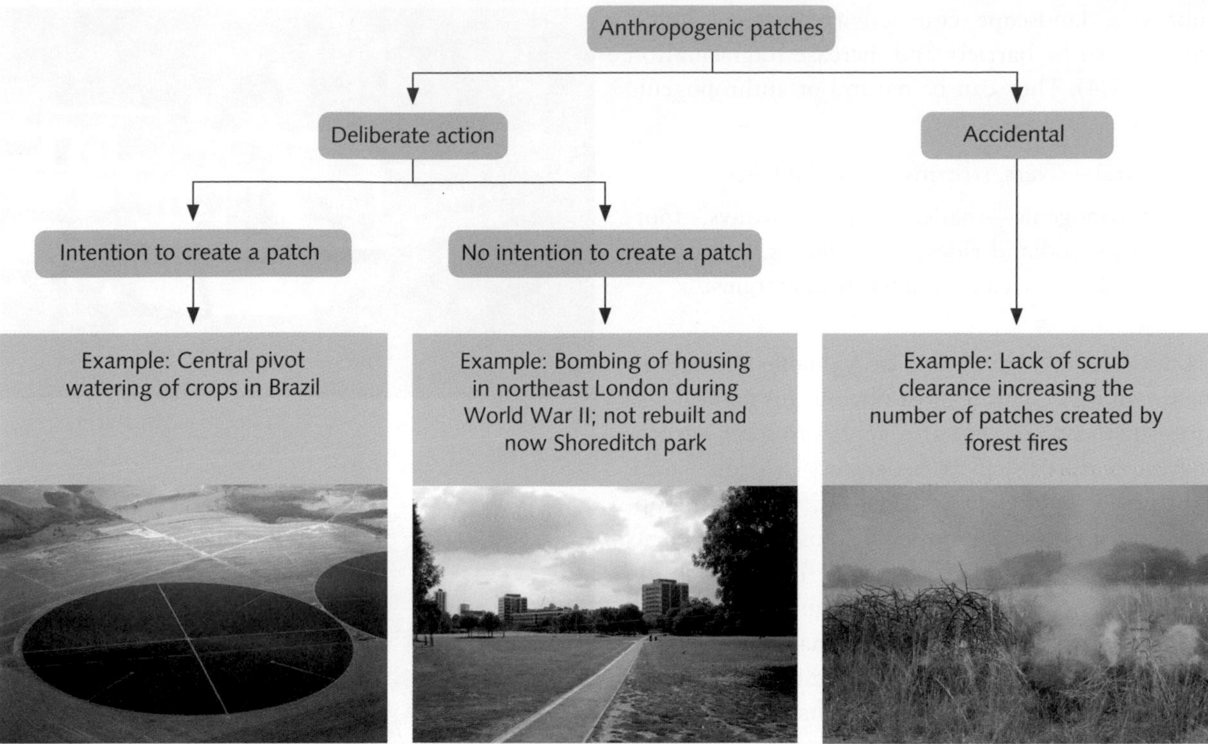

Figure 7.3 Types of anthropogenic patches with examples.

Source: Images (L-R) courtesy of Olearys/CC BY 2.0 ; Ewan Munro/CC BY-SA 2.0; Anne Goodenough.

of a former grassland matrix. Remnant patches can also occur following natural change. For instance, if a change in hydrological regime means an area of semi-desert becomes full desert, land around natural oases (water sources) frequently remain as remnants of the former dominant land type.

- **Disturbance patches:** patches created and/or maintained by disturbance. Disturbance might be human (clear-fell areas, mown grassland) or natural (wind-generated clearings, areas covered by volcanic larva). Some patches are created by **repeated disturbance** while others are created by a **single disturbance** event. In the case of the latter, the persistence of the disturbance patch depends on both the type of disturbance and the location. For example, the loss of a few trees in a tropical rainforest is unlikely to cause a long-lasting patch as other plants rapidly colonize the gap. Loss of a few trees in drought-stressed woodland, however, can have long-lasting effects, as trees might be unable to recolonize as the roots of seedlings are too short to access groundwater supplies.
- **Regeneration patches:** these occur when a factor that was preventing an area from being a different habitat is removed. In the case of disturbance patches, it could be that the factor that is causing the disturbance (grazing, fire, deforestation, etc.) is removed. In this case, the patch will undergo succession to return to a previous land use, which is often the same as the background matrix. There might be human intervention in the regeneration, for instance replanting an area with trees to help with **woodland reversion**, or it might be natural, for example succession from grassland to woodland.

7.2.3 Linear features.

The linear features that criss-cross landscapes have important roles as **corridors** linking patches and improving landscape connectivity (Section 7.6.2), but can also be **barriers** and increase fragmentation (Section 7.4). They can be natural or anthropogenic in origin:

- Natural—rivers, streams, natural ditches.
- Anthropogenic—roads, verges, railways, footpaths, woodland rides, walls, fences, hedgerows, embankments, canals, dikes, field margins.

Linear features can also be valuable habitats in their own right. This is most obvious for natural features such as rivers, but also applies to some anthropogenic features such as hedgerows (Section 7.6.2), road verges, and even walls/fences, which can support important communities of mosses and lichens (Online Case Study Chapter 3). It is worth noting that any linear features can change in form and function over time. For example, a change in a canal from active commercial use for transporting goods to a disused state is likely to alter habitat: the canal is likely to gradually become shallower with an increase in water plants rooted within the canal body, and more enclosed with banks supporting scrub and trees (Figure 7.4).

Figure 7.4 The function and ecological value of a canal varies considerably depending on whether it is being used commercially (top) or is disused (bottom).

Source: Photographs by Anne Goodenough.

7.2.4 Boundaries.

Boundaries between a patch and the matrix, or between two contiguous habitat patches, can be physical or political/administrative. Figure 7.5 shows examples of different boundary types, but the main distinctions are:

- **Physical boundaries:** these have some sort of 'on ground' presence. This might be topographical (e.g. a mountain range or rift valley on a large scale; a ditch or an embankment on a smaller scale) or fluvial (e.g. estuaries and fjords on a large scale; rivers and streams on a smaller scale).
- **Political or administrative boundaries:** these include county or country boundaries, or might be the demarcation between protected areas and the surrounding area. Although they can follow physical boundaries, for instance where boundaries follow rivers, administrative borders can be nothing more than a line on a map. Sometimes these boundaries become noticeable on the ground if the two sides are managed in different ways.

Ecotones

At a small scale, it is rare to find a clearly defined boundary between habitats unless there is substantial human management of that boundary. Far more common is an intermediate zone that has characteristics of the two adjacent habitats—an area of scrub between grassland and woodland, for instance, or an area of marsh between wetland and grassland. These blurred boundaries are referred to as ecotones.

Ecotones are based upon habitat and thus are strongly influenced by the dominant vegetation.

Figure 7.5 Different types of boundaries: (top) Physical boundary between countries (confluence of Iguazú and Paraná rivers, which form the border between Argentina, Paraguay, and Brazil); (middle) original theoretical boundary that now has a physical on-ground presence (Dominican Republic and Haiti); (bottom) theoretical (the majority of the Alaskan border with Canada is a simple line on a map).

Source: Images (top–bottom) courtesy of Stefan Krasowski/CC BY 2.0; NASA; Richard Martin/CC BY 2.0.

However, abiotic parameters such as temperature, moisture, and pH can also be important, and in many cases ecotones are associated with **ecoclines**. Ecoclines can be thought of as substantial gradients in abiotic parameters. A good example of an ecocline-based ecotone is an estuary. This, by its very nature, is a transitional habitat (ecotone) that is based largely on a salinity gradient (ecocline) from freshwater in a river, through a brackish state (a mixture of fresh/saltwater), to a marine ecosystem.

Recent studies have highlighted that ecotones can have important impacts on biodiversity, in some situations driving species richness and within-species genetic diversity. In an example of the former, van Rensburg et al. (2009) studied bird and amphibians in South Africa and Lesotho and found that ecotones held above-average concentrations of endemic, range-limited species. This has important implications for ongoing conservation planning in a biogeographical context. Working in a different part of the African continent, and on mammals, Mitchell et al. (2015) and Sesink Clee et al. (2015) have demonstrated that the population of chimpanzees that inhabit the ecotone regions of Cameroon *Pan troglodytes ellioti* are genetically distinct from the population of the same subspecies that lives in the rainforest region. This genetic distinctiveness at population level within a subspecies is referred to as a deme, and can be thought of as a pre-subspecies stage of distinctiveness which, in the case of chimpanzees, occurred as a result of the two populations splitting about 4000 years ago. This shows how ecotone diversity can link with genetic diversity within an individual species.

FURTHER READING FOR SECTION

Seminal text about landscape ecology: Forman, R.T.T. & Godron, M. (1986) *Landscape Ecology*. Chichester: John Wiley and Sons.

Excellent introduction to landscape elements and landscape ecology: Ingegnoli, V. (2002) *Landscape Ecology: A Widening Foundation*. Amsterdam: Springer.

Useful introduction to landscape elements: Weins, J.A. (2002) Central concepts and issues of landscape ecology. In: K. Gutzwiller (Ed.) *Applying Landscape Ecology in Biological Conservation*, pp. 3–21. Amsterdam: Springer.

7.3 Studying landscape ecology.

In order to study patterns in landscape ecology, and any temporal change in these, three things are necessary:

1. Accurate data on landscape elements (Section 7.2) and how these are organized and arranged.
2. Accurate data on biota, either at species level (Section 7.3.1) or individual level (Section 7.3.2).
3. A suitable framework for bringing these together to quantify patterns, relationships, and interactions, such as that provided by Geographical Information Systems (GIS) (Section 7.3.3).

7.3.1 Spatial patterns of species: quantifying range and distribution.

Studying and mapping species' spatial patterns is vital for managing landscapes in an ecologically robust way. This can be done using numerous metrics, including species range and distributions.

Species range.

At its simplest, the range of a species is the overall area in which it occurs. However, just as a species has fundamental and realized niches (Sections 2.5 and Sections 4.3.1), so it also has a **fundamental range** (spatial area with suitable conditions to support a species and which it could theoretically inhabit; sometimes called a potential range) and a **realized range** (the area it actually does inhabit; sometimes called an actual range; Franklin, 2010). Most species are constrained in location, often by human activity, so the realized range is generally a subset of the fundamental range.

The crudest measure of a species' realized range is its extent of occurrence, which is simply the extremes of the range—these are known as the range margins. This measure is greatly influenced by outliers (perhaps a small population living some distance from main population) and does not give any information on how the species is distributed within that area. However, if used together with the area of occupancy, which is the geographical coverage within that area, the two metrics become rather more meaningful. Indeed, these measures together form an essential part of the International Union for Conservation of Nature (IUCN) species-at-risk lists (Section 10.3.2). Even this combined approach, though, does not give any details of how a species is distributed within that range—for that, other metrics are necessary, as detailed below.

Species distribution

Within its range, a species can have a **continuous distribution** (no substantial gaps in species presence) or a **discontinuous distribution** (notable areas of absence). The difference between these distribution types is displayed in Figure 7.6, which shows the current discontinuous range of the *Pan* genus—chimpanzees and bonobos—and the historic continuous range. In this case, it is apparent that considerable range shrinkage has occurred. With some knowledge of the ecology of the species concerned, it might be decided that this could well be linked to loss of the necessary forest habitat. Examining differences in current and historic distributions in this way can have important links to conservation priorities (Section 10.5) and species reintroduction scenarios (Section 13.5), since one of the key criteria in determining whether a planned reintroduction is sensible—and often whether it is legal—is whether the proposed reintroduction site is within the historic distribution.

When distributions are examined at a very fine scale, very few, if any, will be strictly contentious; there will always be some gaps. The usual criterion for assessing whether gaps are big enough to classify a distribution as being continuous is whether they are greater than the normal dispersal capabilities of that species (Cain, 1944). If gap size exceeds dispersal capacities, distributions are usually considered discontinuous, especially if there are many such gaps. This rule works well in theory, however, the dispersal ecology of many species is poorly known, making classifying ranges in this way problematic in practice.

Figure 7.6 Approximate current discontinuous range of chimpanzees and bonobos (*Pan* genus) and their likely historic continuous range.

Large gaps in **distribution** are called disjunctions. Disjunctions can be caused by differences in climate and habitat; for example, the Alpine marmot *Marmota marmot* occurs at high elevations in the Alps, Pyrenees, and Carpathians and (largely) nowhere else, because low-lying elevations do not have suitable climate/habitat. Humans can also cause disjunctions through severe habitat fragmentation (Section 7.4).

Mapping range and distribution

The easiest way to understand a species' range, and distribution within that range, is by using a map. Mapping is usually undertaken using a computer-based GIS interface. **Dot maps** are often used to show the spatial distribution of a species, with presence indicated by a dot and absence (or more strictly lack of presence—see Chapter 3) indicated by the lack of a dot. An example is the distribution of two different bumblebee species in the UK as shown in Figure 7.7. Dot maps can be made more useful when data on population density or breeding success are combined with occurrence data using differently sized symbols. This is most intuitive when the size of the dot increases with increasing population density or success.

Another common approach is using a **chlorophleth map**, which displays data in blocks and relates species abundance or richness data to shading intensity. Because of the fact that the blocks themselves are often different sizes, chloropleth maps tend to show density rather than an absolute number such as abundance, with darker shading usually indicating higher density (as in Figure 7.8 for global density of fish species related to area). The main issue with chlorophleth mapping is that the area is divided into known blocks and survey data are related to these. Blocks might be countries, counties/provinces, grid squares, or other spatial units (such as ecoregions in Figure 7.8). The use of any spatial unit not only gives a somewhat blocky appearance to the data, but a single record of a species at the extreme edge of a block results in the whole block being coloured, giving a potentially false impression of the extent of a species. This links back to the problems of using political or administrative boundaries (Section 7.2.4).

An alternative is to use **heat mapping.** This takes the data points that would normally be plotted on a dot map and creates a continuous map by interpolating data between points to create a density surface similar to a chloropleth map, but without the 'blocky' appearance (interpolating means inferring new data points between actual data points, in this case to smooth the appearance of the map). Heat mapping can be used to show, say, areas of high density of a species (warm colours, such as red) at one end of the temperature spectrum and areas of low density (cold colours, such as blue) at the other end of the spectrum. The same approach can be used to distinguish geographical areas with high versus low prevalence of a species, high versus low species richness, high versus low nitrate levels, and so on. Because data are interpolated between sites rather than forced into an underlying block pattern, heat maps are often more accurate and potentially more useful. However, this does depend on the original data being good quality and sample sizes reasonably large, otherwise the process of interpolation can make a map with a few data points appear much more complete and accurate than it is in reality.

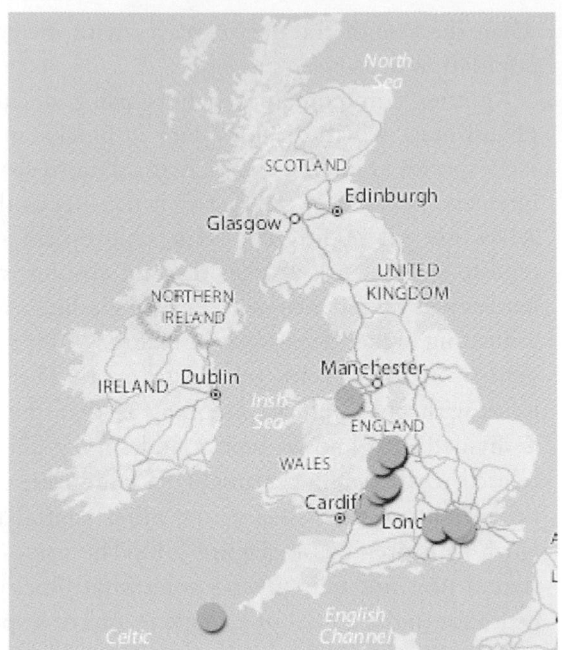

Figure 7.7 Dots showing the distribution of (left) the buff-tailed bumblebee *Bombus terrestris* and (right) the red-tailed bumblebee *Bombus lapidarius* in the UK, as derived from citizen science data based on geo-referenced photographs uploaded to social media sites.

Source: Reproduced from Stafford, R. et al. (2010) 'Eu-social science: the role of internet social networks in the collection of bee biodiversity data'. *PLoS One*, 2010 Dec 17, 5(12)/CC BY.

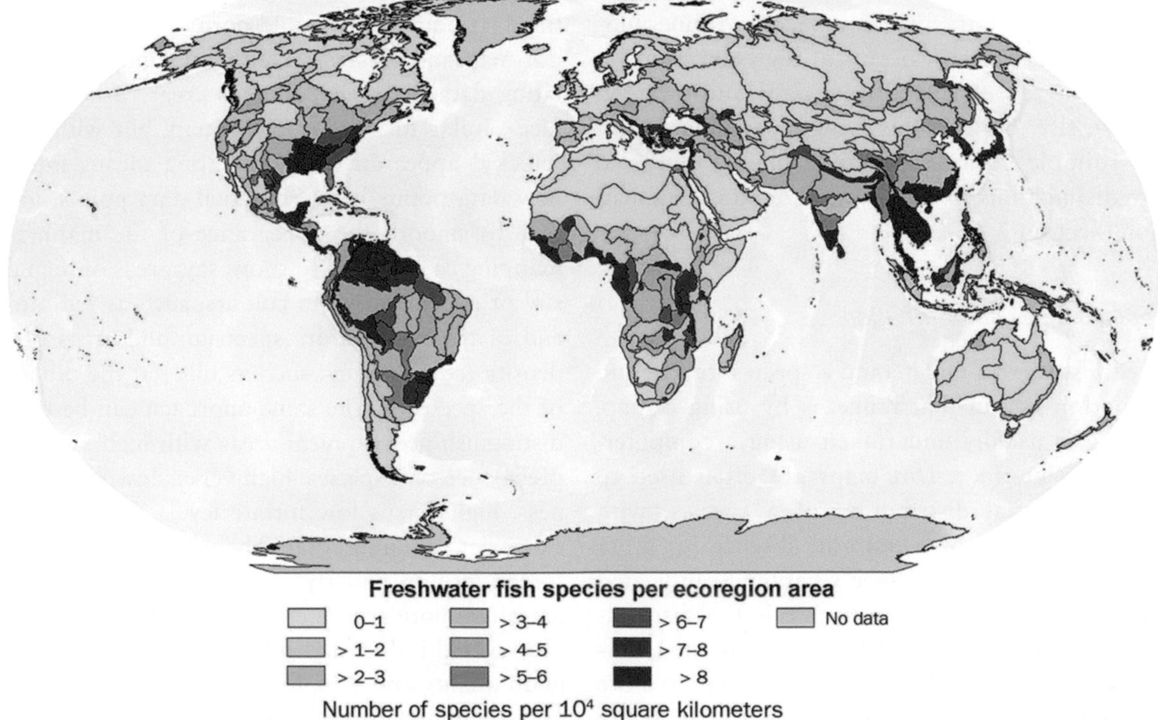

Figure 7.8 Number of freshwater fish species per ecoregion worldwide.

Source: Reproduced from Abell, R. et al. (2008) 'Freshwater ecoregions of the world: a new map of biogeographic units for freshwater biodiversity conservation'. *BioScience*, 58, 403–414. Used with permission of Oxford University Press.

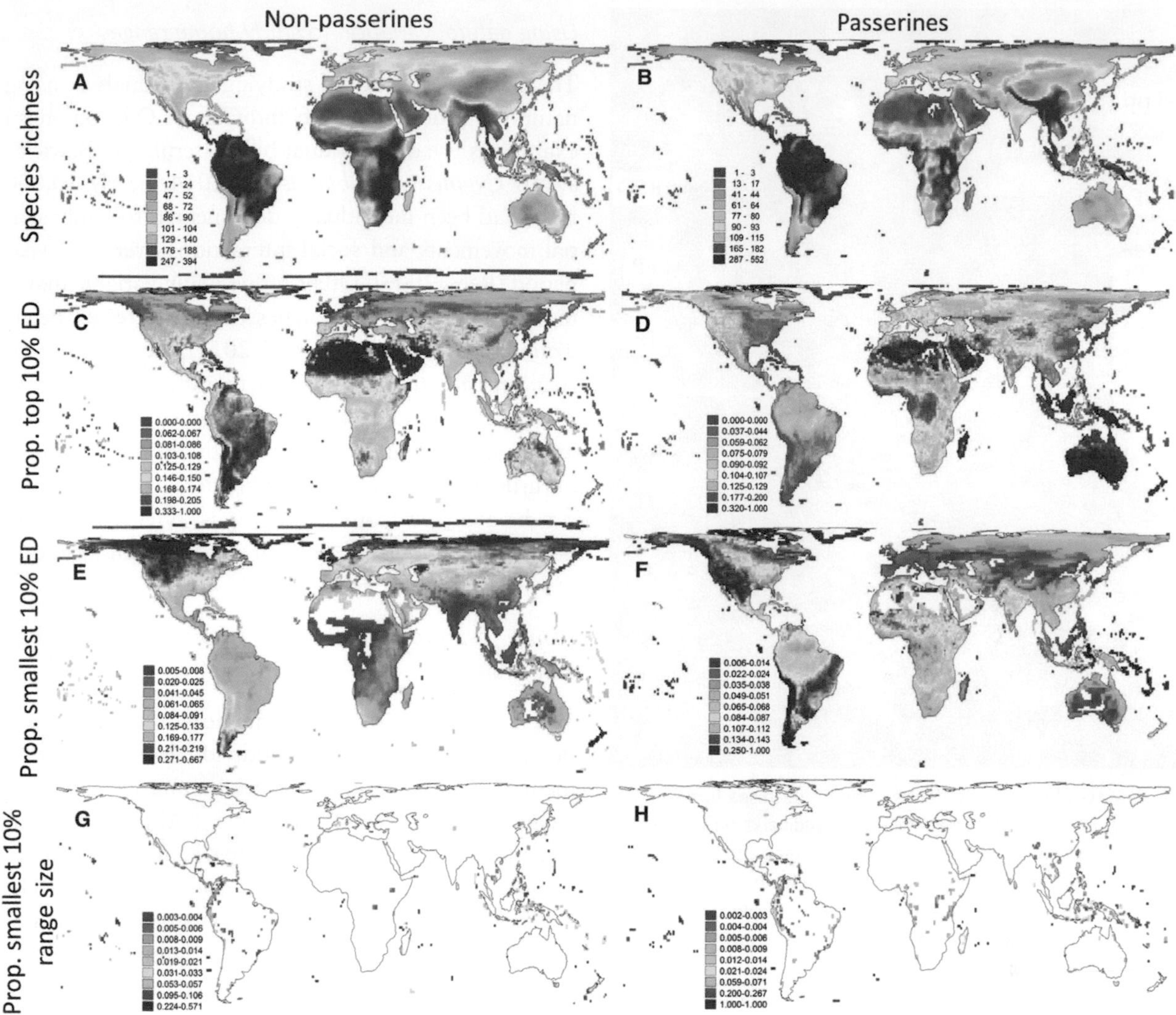

Figure 7.9 Heat map of the number of species of passerine (perching) birds on the right and non-passerine birds on the left. Note that the data categories for each colour differ between the maps, but the overall amount each colour is used remains approximately consistent.

Source: Reproduced from Jetz, W. (2014) 'Global Distribution and Conservation of Evolutionary Distinctness in Birds'. *Current Biology*, Vol. 24, pp. 919–930/CC BY.

An example of a heat map is shown in Figure 7.9, where the number of different bird species has been mapped globally. This was done by assigning the most species rich areas to the 'hottest' colour of red, the next 10% of records to the next hottest colour (dark orange) and so on. In this way, each colour accounted for approximately 10–12% of each map. This relative approach (9 categories that depended on the underlying data) is favoured over fixed class intervals as it ensures that biological data are mapped using an appropriate scale rather than being dominated by 'hot' or 'cool' colours.

7.3.2 Spatial patterns of individuals: quantifying home range and movement.

Sometimes it is useful to focus not on species, but on individuals. Quantifying **home ranges** (the area used by specific individuals) can aid understanding of the area needed to support them, and also helps understanding of natal and breeding dispersal, seasonal movement patterns, and metapopulation dynamics. Tracking individuals is also useful for migratory species in order to understand their entire life cycle.

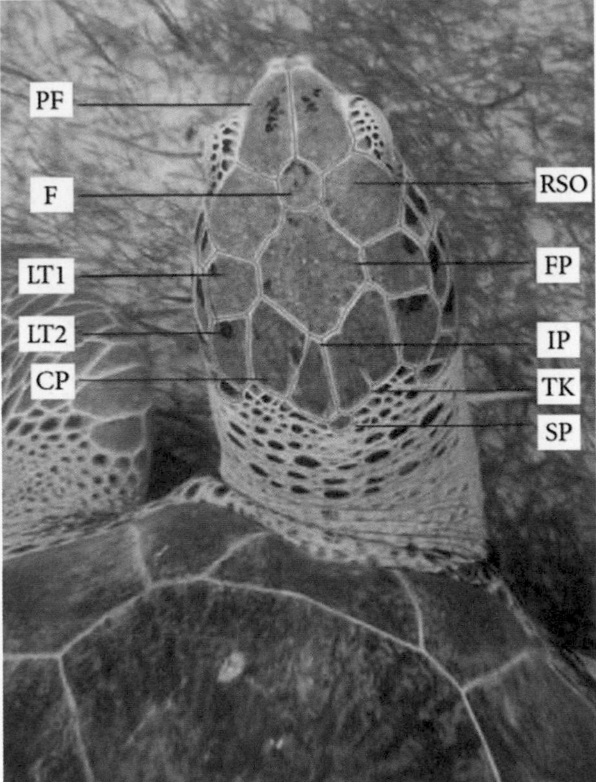

Figure 7.10 Photographic identification of green sea turtles *Chelonia mydas* in Mexico showing key landmarks on the head, specifically the prefrontal, frontal, frontoparietal, and interparietal scales (PF, F, FP, IP), the first and second left temporal scales (LT1, LT2), and the right supraocular scale (RSO). Also noted are locations of ticks and spots (TK and SP).

Source: Reproduced from Lloyd, J. R. et al. (2012) 'Methods of developing user-friendly keys to identify green sea turtles (*Chelonia mydas* L.) from photographs'. *International Journal of Zoology*, Article ID 317568/ CC BY.

Using natural variation to study home ranges

The optimal method of studying individuals is using natural variation to identify individuals. One long-term example is use of individual bill patterning in Bewick's swans *Cygnus columbianus*. By 2014, nearly 10,000 birds had been individually documented to study spatial movements and social interactions over a 50 year period (Rees, 2006). Other individually variable markings include stomach blotches on great crested newts *Triturus cristatus* (Hoque et al., 2011), spots of cheetahs *Acinonyx jubatus* (Kelly, 2001), markings of manta rays *Manta alfredi* and *M. birostris* (Town et al., 2013), and arrangement of scute 'panels' on the heads and necks of turtles (Reisser et al., 2008; Figure 7.10). All of these can be analysed from photographs, often using computation techniques to allow automated identification.

Using capture-mark-resight protocols to study home ranges

When individual markings do not occur, the traditional way of monitoring individuals is to capture them, mark them, and then track them using recapture techniques as discussed in Chapter 3. Alternatively, it is sometimes possible to capture, mark, and resight individuals to reduce handling. This also allows data to be collected over long timeframes since fieldwork is less intensive and less invasive for the animals involved. For birds, this process often uses coloured leg rings. With several colours and the ability to apply rings to either leg and in different orders, multiple unique

Figure 7.11 Birds are often marked using coloured leg rings for capture-mark-resight studies to examine distribution and movement patterns. The colour ringing process is carried out by a trained and licenced ecologist, and often requires an additional licence.

Source: Images courtesy of David Craig/CC BY-SA 2.0 (left) and Anne Goodenough (right).

Table 7.1 Technological approaches to spatial monitoring within landscape ecology.

Method details	Advantages	Limitations
Radio telemetry: animals are fitted with a small radio transmitter. This transmits a 'ping' signal, which is picked up by a receiver aerial. By comparing distance and direction of signals from multiple locations, it is possible to triangulate fixes to estimate location. Radio tags are usually attached using glue, collars, tail mounts or wing mounts. Jennings et al. (2006) used this approach to study Malay civets *Viverra tangalunga* and found that individuals living in undisturbed forest on Sulawesi, Indonesia, needed smaller home ranges than in logged forests on nearby Borneo.	Radio transmitters are small and lightweight and so suitable for many species, including small flying species (e.g. locusts).	Short distance only—normally within 10km. Not automated so fieldworkers need to travel with receiver aerial to get location fixes.
PIT tags: Passive Integrated Transponder tags 'talk' to a receiver unit automatically when in close proximity. Tags can be attached to individuals with receiver units placed wherever appropriate to collect information (e.g. at a fish pass, at breeding ponds for newts, at nest boxes for birds or bats). Tags can be fixed externally or injected under the skin in the same way that pets are micro-chipped. Boarman et al. (1998) used PIT to discover that desert tortoises *Gopherus agassizii* used storm culverts under desert highways to move across the fragmented Mojave desert, California, USA.	No batteries needed in tag so they last a lifetime, unlike tags for other methods; internal tags cannot become detached. Automated data collection.	Only work in close proximity to a receiver so only work for small scale studies such as natal dispersal within a small geographical area.
Satellite trackers: reasonably large and heavy units that communicate with satellites orbiting the earth to provide a location fix like those used by car satellite navigation units. Fixes are transmitted remotely to computers, allowing tracking anywhere in the world in real time. Rahimi & Owen-Smith (2007) used satellite collars on sable antelope *Hippotragus niger* in Kruger National Park, South Africa, to show differences in herd home range size (65–118 km^2) depending on local environment.	The 'gold standard' for tracking; data are very accurate, global, and collected/transmitted automatically.	Units are heavy, so unsuitable for small species, especially those that fly (4% of body weight usually used as upper limit). Expensive.
Geolocators: these little devices house a light detector, battery and data storage unit with an internal calendar/clock. They record and store light level data. Latitude can be determined by assessing hours of daylight in relation to date; longitude can be determined by assessing time of sunrise/sunset. Stutchbury et al. (2009) used geolocators to map migration of wood thrushes *Hylocichla mustelina* and showed birds from the same breeding population in the USA occupied the same wintering sites in central America, suggesting populations remain intact across the globe.	Very small/light; suitable for small species such as many migrant birds that cannot be studied using satellite tracking. Can be used globally to study movement patterns.	Data not transmitted so recapture necessary to recover units; only possible for species that return to the same place each year. Many more units deployed than recovered.

combinations are possible (Figure 7.11). For larger species, colour rings can also include a short alphanumeric code to reduce the number of rings needed.

Using technology to study home ranges

The issue with all techniques that depend on (re) sighting records is that data are essentially a series of 'snapshots' rather than a continuous moving image. In many cases, there are not enough data points to really understand movement patterns and factors driving those—much like it being hard to understand the storyline of a film from a few still images. There are several ways of using technology to obtain a more continuous record of movement patterns, each with its own advantages and limitations as outlined in Table 7.1:

Despite the range of different technologies listed above to track individuals and species over space, there are occasions when none are appropriate and alternatives are needed. At a small scale, the role of environmental DNA (eDNA) was discussed in Hot Topic 3.3; this is especially important for species that are hard to survey using traditional techniques. At a large scale, establishing the wintering ranges of small migratory songbirds is problematic as the spatial scale is too large for radio/PIT tagging and the birds are too small for satellite transmitters. Recent

HOT TOPIC 7.1

Using proxies to determine species range and movement

Many bird species migrate between summer and winter ranges. Improving knowledge of such migrants throughout their entire annual cycle is important for the development of appropriate conservation strategies. It is normally possible to study species' ranges and movements directly using spatial monitoring approaches (primary observations or individual-based capture-mark-recapture: Section 3.6.1) or continuous automated monitoring (satellite transmitters and geolocators: Table 7.1). However, for many small bird species there is very poor ringing recovery data and this, together with the current impracticality of using satellite trackers on small passerines, makes studying migrant species' ecology rather challenging.

Recently, **stable isotope analysis** of feathers has been used to provide valuable insights. Isotopes are 'types' of a chemical element that differ in their number of neutrons. In many chemical elements, such as carbon, nitrogen, and oxygen, there are several 'types' that are stable (i.e. do not change through radioactive decay). Stable isotope analysis involves examining the ratios of different stable types of the same element in a sample; for example, quantifying how much carbon-type-13 (^{13}C) there is in relation to carbon-type-12 (^{12}C).

Unlike living tissue, feathers are keratinized and thus inert once grown. This means feather stable isotope composition reflects the bird's diet and environment (habitat and location) at the time of growth. Depending on moult strategy, it can be possible to undertake isotope profiling on feathers grown on a bird's wintering grounds when that bird is captured thousands of kilometres away on its breeding ground (or vice-versa).

The dominant control of carbon isotopes within the feathers of a bird is the relative proportion of C3 and C4 plants in the environment the bird was in at the time of feather growth. Different plants use different biochemical photosynthesis pathways: C4 plants, such as many grasses, use water-efficient C4 carbon fixation whereas C3 plants, including most trees, use water-inefficient C3 carbon fixation. As a result, C3-dominated habitats tend to be wetter while C4-dominated habitats tend to be drier. Analysing the carbon isotopes in feathers, therefore, can allow insights into habitat type and location. Similarly it is possible to compare relative mass of hydrogen isotopes to ascertain the relative amount of deuterium (^{2}H) to protium (^{1}H) and thus the likely geographical location based on precipitation (Figure A).

Stable isotope analysis has allowed valuable insights into the wintering or stopover locations of a range of

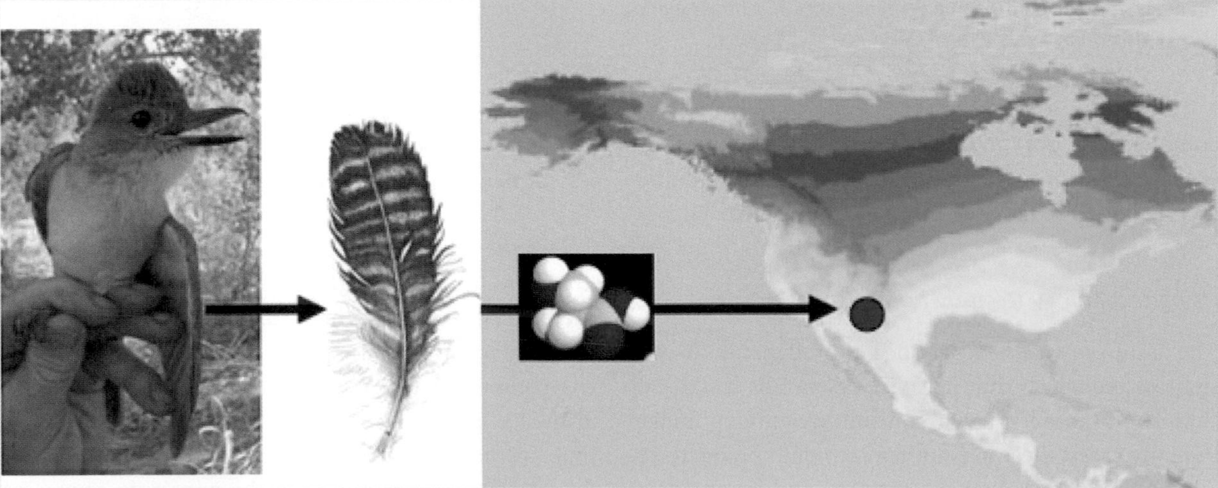

Figure A Use of stable isotope analysis of bird feathers to construct an isomap.

Source: Image courtesy of Jeff Kelly, University of Oklahoma, and Gabriel Bowen, University of Utah, used with kind permission of waterisotopes.org

birds, including several birds that migrate between North and South America and between Europe and Africa (e.g. songbirds such as whitethroat *Sylvia communis* and willow warblers *Phylloscopus trochilus*) (Yohannes et al., 2005; Morrison et al., 2013). The majority of these studies have focused on establishing likely wintering grounds or migratory stop-over areas (and thus flyway routes) and it is noteworthy that most have focused on species that are endangered and/or declining, emphasizing the conservation importance of such work.

QUESTIONS

The examples above relate to bird migration; what other applications are there for stable isotope analysis within ecology?

There are many different chemical elements used in stable isotope analysis including carbon, nitrogen, hydrogen, and oxygen. What are the strengths and weaknesses of each of these when tracking movement?

REFERENCES

Morrison, C.A., Robinson, R.A., Clark, J.A., Marca, A.D., Newton, J., & Gill, J.A. (2013) Using stable isotopes to link breeding population trends to winter ecology in Willow Warblers, *Phylloscopus trochilus*. *Bird Study*, Volume 60, 211–220.

Yohannes, E., Hobson, K.A., Pearson, D.J., & Wassenaar, L.I. (2005) Stable isotope analyses of feathers help identify autumn stopover sites of three long-distance migrants in northeastern Africa. *Journal of Avian Biology*, Volume 36, 235–241.

research breakthroughs have utilized a new approach, **stable isotope analysis**, which is discussed in detail in Hot Topic 7.1.

7.3.3 Spatial analysis framework: Geographical Information Systems.

Spatial analysis is primarily undertaken with a Geographical Information Systems (GIS) approach. This involves digitizing ecological data into coordinates representing latitude and longitude. Data can then be displayed graphically together with other spatial information such as geology, soils, river networks, habitat, management, and conservation designations. These can be shown in layers to create a flat 'map stack', as shown in Figure 7.12. If elevation is important, 3D models can be constructed. Physical environment layers are often derived from remotely sensed data such as satellite images (often from NASA Landsat), electromagnetic-reflectance and light-reflectance images (**radar** and **LiDAR**, respectively) and, in aquatic and marine environments, sound reflectance (**sonar**).

GIS is an extremely powerful technique, and with increasingly user-friendly and freeware GIS packages coming online, GIS as a tool for data display and analysis is within the grasp of Applied Ecologists rather than being solely the preserve of specialist consultants (see the Interview with an Applied Ecologist for this chapter). As a result, the number of ecological studies using GIS for mapping habitats and studying

Figure 7.12 Geographical Information Systems allow Applied Ecologists to add "layers" of information about land type, topography, fluvial species, habitats, species distributions, and so on, to create a composite computer map.

Source: Image courtesy of the Naugatuck Valley Council of Governments.

AN INTERVIEW WITH AN APPLIED ECOLOGIST 7

Name: Peter Fretwell

Organisation: British Antarctic Survey

Role: Geographical Information Officer

Nationality and main countries worked: British; worked in Antarctica

What is your day-to-day job?

I've been lucky to be at the forefront of what is becoming a revolution in Applied Ecology: using remotely-sensed images, such as satellite images, to study the distribution and movement of individual species. The use of satellite remote sensing of wildlife has developed rapidly over the past five years, mainly because of the ease of access to imagery and the increasing resolution of satellite censors.

What is your most interesting recent project and why?

Show most people a picture of an emperor penguin and the chances are good that they will recognize it. The species is charismatic, unique, and well-known in popular culture. However, that familiarity belies the fact that we actually know very little about the species' ecology. It breeds in some of the most inhospitable places on earth—inside the Antarctic Circle—at one of the most extreme times of the year when the Antarctic winter means that temperatures drops to –50°C and the continent remains in total darkness.

In 2008 I was working for the British Antarctic Survey in a cartography (mapping) role. One of the things we did was to create route maps for the aircraft to avoid dangerous areas and no-fly zones, which included penguin colonies. For the colonies of smaller penguins, such as Adélies and chinstrap, that was no problem. However, the data were very poor for emperor penguins. For example, we knew that there were two emperor penguin colonies on the Brunt Ice Shelf, but we had no idea where on the ice shelf they were. As each colony might only occupy a few hundred square metres, and the ice shelf is the size of Lincolnshire, this was a problem. When looking at the Landsat satellite images, I saw a large brown stain on the sea-ice. I started to put ideas together and wondered whether this brown strain was evidence of one of the missing penguin colonies. I looked at the other possible penguin site, and, sure enough, there was another brown stain. At that time, there were 25 known colonies and a few others that were considered probable. Helped by the fact that Landsat had just made images freely available, we remotely surveyed around 90% of the coastline. At the end of our scrutiny in 2009, we had found 38 colonies through these brown stains, which turned out to be guano (bird poo). We were literately sensing penguins from space.

We then tasked a very high resolution (VHR) satellite, Quickbird 2, to fly over the potential colonies later in 2009. The pixels on the images covered by this satellite cover an area of just 60cm by 60cm, but it has to be specifically sent to a target area. On the VHR images, we could actually count individual penguins when they were travelling between the colony and the sea. When they were in a huddle this was much harder, but we developed a computer technique to look at the density of the penguin huddle by digitally analysing the 'penguin pixels' relative to the background of snow, ice, and guano, and we estimated numbers from that. The world population estimate for emperor penguins went from 310,000 to 595,000 in one year by virtue of the world's first satellite census of any species, anywhere.

What's been the best part of this particular project?

Well, it's quite nice to be personally responsible for finding half of the world's emperor penguins! More seriously, though, some quite unexpected things have come out of this project. For example, we had another missing penguin colony. On his exploration on Antarctica in 1893, Carl Anton Larson documented numerous emperor penguins near the Jason Peninsula. We couldn't find the colony until we looked not just at the sea-ice, but at the ice shelf beyond it … and there were our missing penguins. Nobody knew penguins could even get onto continental Antarctica, so we discovered new breeding behaviour using a satellite. Pretty amazing.

What do you see as the main challenges in your field and how can they be overcome?

Ground truthing is an issue; traditional ground based or aerial surveys are done over a period of time, but with satellites you get a snapshot and we need to understand the implications of this when assessing populations. The datasets can be very large so there are problems of data storage and processing. Often region assessments require thousands of square kilometres of imagery to be assessed. Looking for signs of animals that might only be 30cm across is something that takes too long manually, so most of the work is aiming towards automated analysis, developing machine learning intelligent search protocols that will automatically count wildlife.

Crystal ball time: mapping technology and data possibilities have changed a lot over the last 5–10 years. What technological advances do you think the next 5–10 years will bring?

Even higher resolution imagery meaning we can look at smaller organisms, thermal imagery, crowd sourcing, and cheaper images for conservation purposes are all on the horizon. A new even higher resolution satellite, giving us imagery at 30cm resolution, has gone up recently, so now we should be able to see even more species. I hope that we can extend the technology out to monitoring many more types of animal, not just in the polar regions, but in any non-forested areas where there is a conservation need around the world.

What's next for you, and why?
We need to assess the penguin population trends over initially a 10-year period and ideally more in the future. We have a snapshot, but we now need to continue this work to see how this changes over time. More generally, we are also expanding this technique to assess the distribution and population of other species.

Finally, how did you get into mapping and spatial analysis and what advice would you give to others?
Luck mainly, although you have to have the right skill set and the attitude to take the opportunities when they turn up. As for how I got into mapping and spatial analysis, in my case I did a degree in Geography then went on to study Quaternary Geomorphology. That mainly involved mapping and spatial analysis, and from there I took a very low paid temporary job at BAS, which basically was just looking after their map collection. I did a few maps for them and they liked what I produced and I managed to get a job as a junior cartographer. I started applying my science background to a few projects and became an expert at GIS and spatial analysis, mainly through on-the-job experience. Then the penguins turned up and since then I have been concentrating on remote sensing of wildlife, publishing scientific research, and exploring new applications. I am now a senior scientist at British Antarctic Survey with around 50 published papers and over 1000 citations.

species–habitat relationships has increased markedly over the last 25 years.

As well as being used to show data, GIS can be used for analysis. For example, species distribution can be compared with habitat to provide insights into species–habitat interactions that are vital for conservation. The importance of species–habitat interaction will be discussed in more detail in Section 11.7, but one specific use is important to mention here: **habitat suitability mapping.**

Habitat suitability mapping

Once habitat requirements of a species are known, and a landscape has been mapped, it is possible to combine this knowledge predictively to map habitat

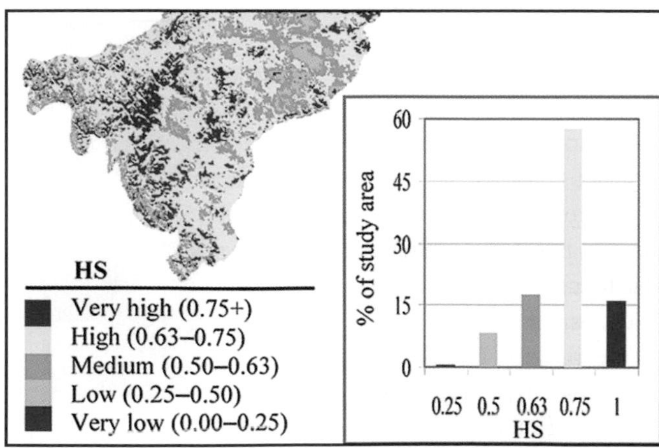

Figure 7.13 The suitability of habitat in New Brunswick, Canada, for supporting balsam fir *Abies balsamea* expressed as a proportion from 0 (completely unsuitable) to 1.

Source: Reproduced from Hassan, Q. K., and Bourque, C. P. A. (2009) 'Potential species distribution of balsam fir based on the integration of biophysical variables derived with remote sensing and process-based methods'. *Remote Sensing*, 1(3), 393–407/CC BY.

suitability for a specific species in a specific landscape. This approach was utilized by Hassan and Bourque (2009) to map habitat suitability for balsam fir, *Abies balsamea*, in Canada. This used climate measures and other biophysical parameters all derived from remote sensing data from satellite images to create a map showing areas that were highly and very highly suitable versus those with medium, low, or very low suitability (Figure 7.13).

FURTHER READING FOR SECTION

Introduction to considerations when mapping species: Franklin, J. (2010) *Mapping Species Distributions: Spatial Inference and Prediction*. Cambridge: Cambridge University Press.

Excellent guide to the use of GIS and remote sensing in Applied Ecology: Horning, N., Robinson, J., Sterling, E., Turner, W., & Spector, S. (2010) *Remote Sensing for Ecology and Conservation*. Oxford: Oxford University Press.

7.4 Effects of landscape processes: habitat loss and fragmentation.

Although landscape elements and the species that use them can be mapped (Section 7.3) it is important to remember that a landscape is a dynamic entity of processes and interactions. It is these processes and interactions, and temporal change therein, that are responsible for the spatial complexity inherent in ecological systems. Two of the most important processes are habitat loss and habitat fragmentation, which are important threats to global biodiversity (Gurevitch & Padilla, 2004).

Habitat loss usually refers to the process of converting land from habitats of high ecological value to highly human-constructed habitats, generally with hard infrastructure and substantial human management such as housing, industry, or mineral extraction; agriculture can be included especially if it is intensive. Such land uses generally have lower ecological value—sometimes substantially so—than the one they replaced. It should be noted that although the term 'habitat loss' is common parlance, it is not strictly accurate since most land uses are habitats of some description. Urban areas, for example, usually contain gardens; industrial areas often have trees that support a whole range of invertebrates; agricultural land generally has hedges, ditches, streams, or ponds. Even walls, fences, and gravestones are habitats for mosses and lichens.

Habitat fragmentation is the process by which a landscape becomes increasingly patchy. Imagine two landscapes, one almost entirely forested and a second that was originally entirely forested, but now has many 'interruptions', perhaps being cross-crossed with roads and punctuated by multiple urban and agricultural patches. Whereas the first landscape is almost entirely one habitat (a background matrix that is coherent, unified, and integrated), the second landscape is very patchy as a result of fragmentation

There are different types of fragmentation, as shown in Figure 7.14. It should be noted that although this figure shows fragmentation of a background matrix, the same processes can operate at a patch level if a large patch is fragmented into smaller sections. The main landscape-level result of fragmentation is that large habitat areas are split into two or more smaller areas that, by their very nature, are not directly connected. Fragmentation thus encompasses two concepts:

1. **The amount of habitat lost:** fragmentation of a matrix or patch invariably leads to a loss of the specific habitat comprising that matrix or patch. This can be minimal—for example a narrow road bisecting a habitat might not use much space—but where

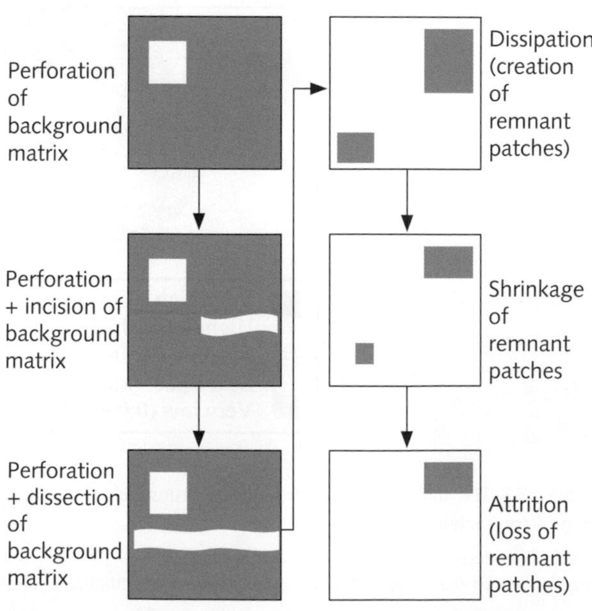

Figure 7.14 Different types of fragmentation, listed in order of severity, whereby blue denotes the original background matrix.

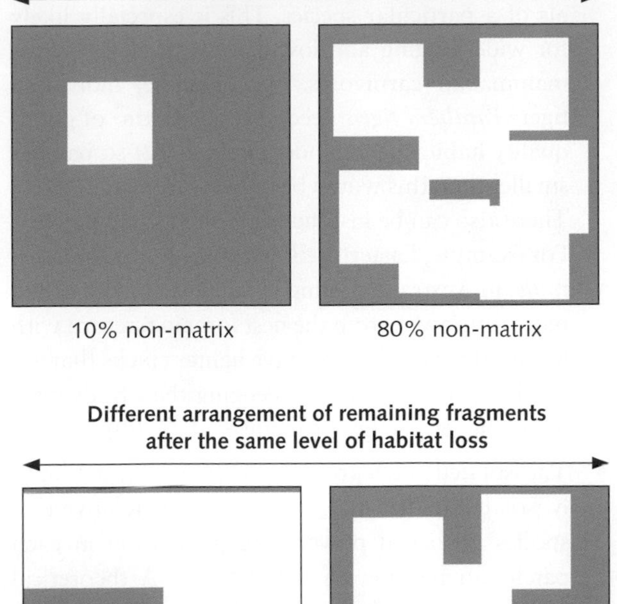

Figure 7.15 The two factors that influence the severity of fragmentation: the amount of habitat lost (top) and the spatial arrangement of the remaining habitat (bottom), whereby blue denotes the original background matrix.

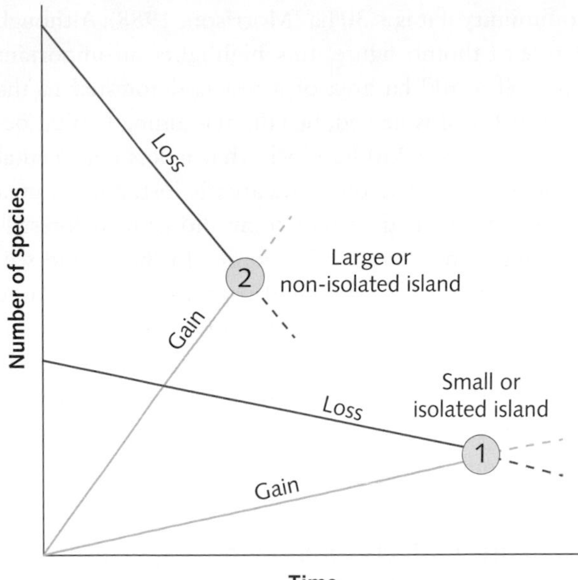

Figure 7.16 The island equilibrium model of biogeography where species arrival and loss occur at the same rate and so the total number of species remains the same. There are two possible equilibrium points: (1) small or isolated islands, (2) large or non-isolated islands.

fragmentation involves a change in land use that takes up a large area, habitat loss can be substantial.

2. **The spatial arrangement of the remaining habitat:** habitat fragmentation is more of a problem when the remaining habitat is very widely distributed so that there is a substantial distance between each patch. This is especially true when there is little connectivity between these patches (e.g. patches of woodland not linked by hedgerows). Such landscapes are likely to be more ecologically compromised compared with loss of the same physical area in one or two larger blocks. This is shown graphically in the fragmentation severity hierarchy in Figure 7.15.

There are strong parallels between fragmentation and **island biogeography** (MacArthur and Wilson, 1967) as small and isolated patches are effectively virtual islands surrounded by a 'sea' of other land uses. When a new oceanic island is formed as a result of volcanic or coralline activity, the number of species starts at zero and will gradually increase, firstly due to immigration and then, given sufficient time, immigration plus speciation. Over time, some colonists become extirpated through competition, predation, and disease so that eventually the rate of increase should equal the rate of loss. This is referred to as island equilibrium and is shown graphically in Figure 7.16. The factors that determine at what point equilibrium occurs are island size (large islands generally support more species and are colonized more rapidly than small islands) and island isolation (isolated islands have slower colonization rates than islands close to other landmasses).

The same principles apply to islands that were once connected to continental landmass and subsequently isolated, such as Madagascar (Case Study 10.3) and Tasmania, as well as the landscape patches discussed in this chapter. The only difference is that such 'islands' come into existence with a pre-existing species assemblage, so extirpation/extinction processes can, and do, occur immediately.

The effect that patch size and arrangement can have is exemplified by fragmentation of old-growth forest in British Columbia, Canada. Here, a patch generally ceases to be able to support a typical woodland community if it is <30 ha (Morrison, 1988). Although a rule-of-thumb figure, this highlights an important point. If a 500 ha area of forest is deforested so the area of forest is halved, but the remaining habitat occurs in a single 250 ha block, that is less detrimental ecologically than if the same area is lost, but the area is also fragmented so remaining old-growth forest is split into ten separate 25 ha units. In the second scenario, none of the individual patches would be large enough to sustain old-growth forest. This is an example of fragmentation having substantial direct effects on individual species (Section 7.4.1) and indirect effects on wider ecology (habitat functioning, ecological interactions, species communities; Section 7.4.2).

7.4.1 Direct effects on individual species.

Just as fragmentation splits up large habitat units into multiple smaller units, so it can also split large populations into multiple smaller populations. The effect this has will depend on the **severity** of the fragmentation (how much habitat is lost and the distance between the remaining blocks) and the **magnitude** of the spatial area involved. Also important are:

1. The **vulnerability** of the species concerned.
2. The **resilience** of the landscape to fragmentation.

Species' vulnerability

Species are not equal in their vulnerability to landscape fragmentation progresses. Species traits that are associated with species that are vulnerable and resilient to fragmentation are highlighted in Table 7.2.

Fragmentation severity and landscape resilience

Minor fragmentation is not necessarily a problem if patches are large and close to one another, especially for species that are more resilient to fragmentation. Severe fragmentation, however, is likely to have substantial ecological implications if patches are either small and/or highly isolated:

- **The small patch problem:** severe fragmentation not only reduces overall habitat area, but can mean what is left is split into areas that are too small to retain their original ecological function. Patch size could be smaller than the **home range** of individuals of a particular species. This is especially likely for wide-ranging and low-density species, such as mammalian carnivores. For example, individual tigers *Panthera tigris* need around 50 km^2 of good-quality habitat (Majumder et al., 2012) so patches smaller than this would be unlikely to sustain them. There also can be insufficient food in small patches. For example, Eastern yellow robins *Eopsaltria australis*, in Australia nesting in smaller patches spend more time away from the nest foraging, return with less food for chicks, and have lighter chicks that are less likely to survive post-fledging than birds nesting in bigger patches (Zanette et al., 2000).
- **The isolated patch problem:** when patches are highly isolated (either in actual distance or relative to a species' dispersal power), the population in each patch can become an isolated unit. A theoretical benefit of the isolation caused by severe fragmentation is isolation-induced speciation. However, in reality this is unlikely, as discussed in Hot Topic 7.2.

Fragmentation can also alter species' behaviour. For example, bobcats *Lynx rufus* and coyotes *Canis latrans* in California have higher-than-normal nightly activity levels in home ranges fragmented by urban development (Riley et al., 2003). This is partly due to fragmented home ranges being significantly larger overall (due to the amount of non-natural habitat) and partly to individuals becoming more nocturnal to reduce disturbance risks.

Metapopulations

Real islands often occur as groups known as archipelagos. Volcanic archipelagos include the Galápagos Islands in Ecuador and the Canary islands off West Africa, while coralline archipelagos include the Maldives. In natural archipelagos, specific species typically occur on multiple islands and mixing occurs between populations due to inter-island movement. This collection of semi-discrete populations is a **metapopulation**, a concept developed by mathematical ecologist Richard Levins (1969), who defined it as 'a population of populations'. An example of an oceanic archipelago metapopulation is Mariana's fruit bats *Pteropus mariannus* on the Mariana Islands of the Western Pacific, where individuals frequently move between islands (Wiles & Glass, 1990).

Table 7.2 Traits that are linked to the vulnerability of species to the effect of landscape fragmentation.

Species trait	Highly vulnerable to fragmentation	Reasonably resilient to fragmentation
Species' mobility through locomotion (the ability of an *individual organism* to move)	Non-mobile species are vulnerable. This includes static marine species such as sea anemones, corals, and oysters and most plants (the birdcage plant *Oenothera deltoides* of the Californian desert is an exception as whole plants move). Also vulnerable are slow-moving species or species where individuals can only travel a short distance per day.	Highly mobile species such as many birds; also some insects (especially butterflies and moths). Species that fly have the added benefit of being able to move over inhospitable terrain or boundaries, rather than being forced to move through or round them.
Species' philopatry (the tendency for a species to remain in its place of birth—natal philopatry—or previous breeding events—breeding philopatry)	Philopatry is the tendency for individuals of a species to move rather than its physical ability to do so. Highly philopatric species are most vulnerable. This includes birds, such as Mauritius kestrels (Case Study 7.1), many newts, and some mammals such as the neotropical bat *Thyroptera tricolor* (Chaverri & Kunz, 2011).	Species that are less philopatric might have individuals that are less well adapted to specific environments (this is one of the theories for the evolution of philopatry), but they are better able to move—subject to having the physical ability to do so—when the landscape changes.
Species' dispersal power (the ability of a species to move to new areas *over successive generations*)	Species with low dispersal power are vulnerable. In plants, this includes species reliant on animals for seed dispersal rather than wind. K-strategist plants (Section 2.7.1) with long generation times have a reduced amount of distance that can be covered per time period.	Species with high dispersal power, such as wind-dispersed species and winged faunal species such as birds, bats, and insects. Species that are r-strategist such as many annuals (Section 2.7.1) are also less vulnerable due to rapid generation times.
Species' specialism	Specialist species closely tied to one habitat might be unable to cross other habitats to disperse to other suitable areas. Likewise species that are highly dependent on other species through specialist mutualistic, grazing, predatory, or parasitic interactions are tied to locations where those species occur.	Species that are generalists are usually less affected by fragmentation, either because they can carry on living in the same area even if that area changes or because they are better able to move over across non-ideal terrain to reach new area.
Species' home range size (the size of area that any one individual uses)	Although species with large ranges might be more mobile, the very fact they have large ranges makes then susceptible to large-scale landscape issues such as fragmentation. Examples include apex predators (top predators in the food chain) such as bears and wolves.	Species are less vulnerable to fragmentation if the remaining patches of their habitat are bigger than the average home range size; this favours species with smaller home ranges.
Species' nativeness	Native species vary in their vulnerability to fragmentation based on attributes noted above.	Non-native species are often generalists, which, by their very nature, are able to survive in novel ecosystems and often have high mobility and/or dispersal power (Section 8.3).

Metapopulations can occur through inter-patch movement in fragmented landscapes. This is most likely when patches are close together and for species that are mobile and are not philopatric. The presence of well-functioning metapopulations decreases fragmentation problems as gene flow is maintained, as for the northern goshawk *Accipiter gentilis* in Canada and Alaska. However, the danger of assuming that fragmentation is unlikely to affect flying species is demonstrated by another raptor—the Mauritius kestrel *Falco punctatus*. The different fate of these bird species in response to similar landscape processes is discussed in Case Study 7.1 and serves to demonstrate the inherent complexity of ecological systems.

Source-sink population dynamics

Movement between islands/patches in a metapopulation framework is not necessarily equal or reciprocal. Patches usually vary in habitat quality, food sources, risk of disease, predation rates, and so on. Together, this means that populations in different patches usually have different productivity rates and mortality rates.

If we imagine a simple two-patch model where one patch has high-quality habitat and optimal conditions

HOT TOPIC 7.2

Fragmentation: a driver of extinction or speciation?

Isolation is a key driver for the evolution of new species through the process of speciation. The type of speciation that arises from geographical isolation is called allopatric speciation and it occurs because separate populations, which cannot mix, adapt to their own specific environment. If geographic isolation persists for long time and/or the environments (and thus the selection pressures) are very different, evolution takes the different populations down a different pathway until they become genetically distinct from one another and are classed as separate species. This process is illustrated here, for ease, for two isolated populations, but multiple populations can be involved, each forming a different species over time.

Islands often support a high number of **endemic species** and these add to the sum of the world's biodiversity. They are also home to highly unusual species, including those that are very small (a new species of chameleon on Madagascar described in 2012, *Brookesia micra*, grows to a maximum length of just 29 mm) and very large such as the Galápagos giant tortoise *Chelonoidis nigra*. This has led to islands being called 'natural laboratories for extravagant evolutionary experiments' (Quammen, 1996).

So why, then, doesn't fragmentation-induced isolation (normally) benefit biodiversity through speciation? It could certainly be argued that human-induced fragmentation will ultimately produce some new species through **isolation** and **founder effects**. However, population genetics predict that most fragmentation events caused by human activities will facilitate not speciation, but extirpation (Templeton et al., 2001).

Part of the reason is that multiple other stressors generally occur at the same time as human-induced fragmentation, not least the habitat loss that often goes hand-in-hand with fragmentation reducing overall carrying capacity. In many cases, artificial habitat fragmentation affects numerous ecological processes as well as 'just' reducing or preventing gene flow—for example, by disrupting other species interactions and changing the environmental conditions and habitat characteristics through edge effects. While founder events have played an important role in the macroevolution of certain groups, this is when ecological opportunities are expanding rather than contracting (Templeton et al., 2001). The other issue to consider is timescale. Although speciation can occur reasonably quickly in some circumstances, especially for species with short generation times, this is unusual. Speciation arising from isolation tends to be a long-term process, but fragmentation can occur very rapidly.

Although it is possible that fragmentation-induced speciation is occurring undetected, the negative impacts of habitat fragmentation generally outweigh any potential speciation-related positive impacts and conservation efforts are essential to decrease habitat fragmentation and loss in the face of human population growth and climate change.

QUESTIONS

How has isolation historically led to allopatric speciation?

Are there any examples of recent habitat fragmentation causing speciation?

REFERENCES

Quammen, D. (1996) Song of the Dodo: Island Biogeography in an Age of Extinctions. Scribner, New York, USA.

Templeton, A.R., Robertson, R.J., Brisson, J., & Strasburg, J. (2001) Disrupting evolutionary processes: the effect of habitat fragmentation on collared lizards in the Missouri Ozarks. *Proceedings of the National Academy of Sciences of the United States of America*, Volume 98, 5426–5432.

for a particular species, and the other is low-quality habitat with suboptimal conditions for that same species, it is likely that the population will increase in the former and decrease in the latter. The second patch is a **sink** patch and might not, long term, be able to sustain a population. However, the first patch is a source patch with a growing population. If individuals do not move from that patch, intraspecific

CASE STUDY 7.1

Differing fates of raptors in fragmented landscapes

The northern goshawk *Accipiter gentilis* breeds in Canada and Alaska in mainland forest patches and near-shore islands. Recent research on population dynamics by Sonsthagen et al. (2012) has examined gene flow between individual populations using mitochondrial DNA (DNA passed down the female line) as a proxy for movement of individuals over time.

This analysis demonstrates that there is a well-functioning metapopulation framework with long-term gene flow between all different populations (Figure A). However, it also shows gene flow is often asymmetric with some populations having more gene flow out (**source populations**) and some having more gene flow in (**sink populations**). Moreover, the symmetry of gene flow rates has changed over time; for example the Admiralty Islands are now a source, but used to be a sink, while coastal British Columbia was a strong source population historically, but this has waned.

In stark contrast is the case of the Mauritius kestrel *Falco punctatus*, a falcon endemic to the island of Mauritius in the Indian Ocean. This once-common species declined to just four birds in 1974. Decline was due to numerous interacting factors, including introduction of non-native species such as small Indian mongoose *Herpestes javanicus* (Chapter 8), and use of pesticides to control mosquitoes (Chapter 9). The most important factor, however, was habitat loss and extreme fragmentation.

By the 1950s, habitat loss was extreme and many patches were becoming too small to support birds, with each patch being vulnerable to chance extinctions due to nest predation and food shortages. By 1974 there was less than 3% of original forest left. Worse still, this was highly fragmented into three virtual islands in the Black River Gorge in the south of the island (Figure B). Although the kestrels can and do fly, individuals are very philopatric and so do not disperse far from their natal site, possibly because of the fitness costs associated in so doing (Nevoux et al., 2013). As a result, even where populations did survive, they were very small and genetically isolated: metapopulation dynamics simply did not operate.

These **population bottlenecks** caused severe inbreeding so that by 1974 not only was the population size extremely low at just 2 breeding pairs, so too was the overall gene pool (Groombridge et al., 2001). Conservation action, including provision of supplemental food and safe nest boxes, has worked in the short term. Although still listed as Vulnerable by IUCN, the population is now around 250–400 birds. However, the fundamental problems of the fragmented landscape remain and the long-term effects of

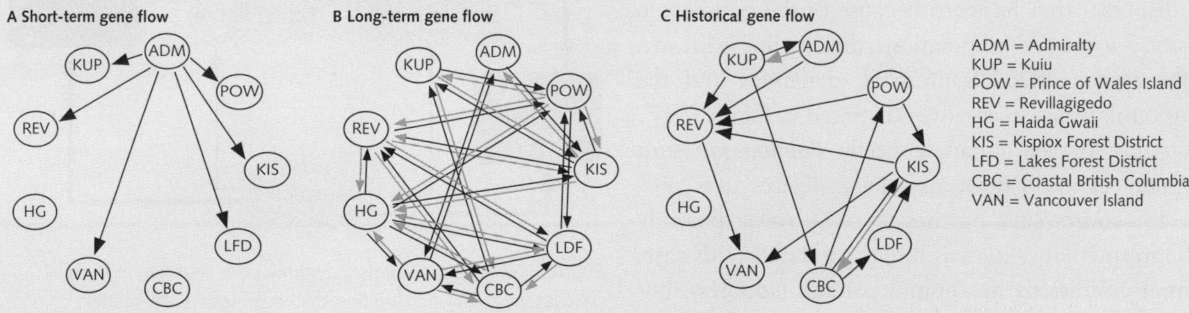

Figure A Metapopulation dynamics in northern goshawks *Accipiter gentilis* in Alaska, USA, and British Columbia, Canada, showing gene flow direction (arrow) and strength (thickness of line) between different populations within a metapopulation framework. The current, short-term, situation shows ADM is a vital source population, but over the longer term, and historically, there is a well-functioning metapopulation.

Source: Reproduced from Sonsthagen, S. A. et al. (2012) 'Identification of metapopulation dynamics among northern goshawks of the Alexander Archipelago, Alaska, and Coastal British Columbia'. *Conservation Genetics*, 13(4), 1045–1057; with permission from Springer.

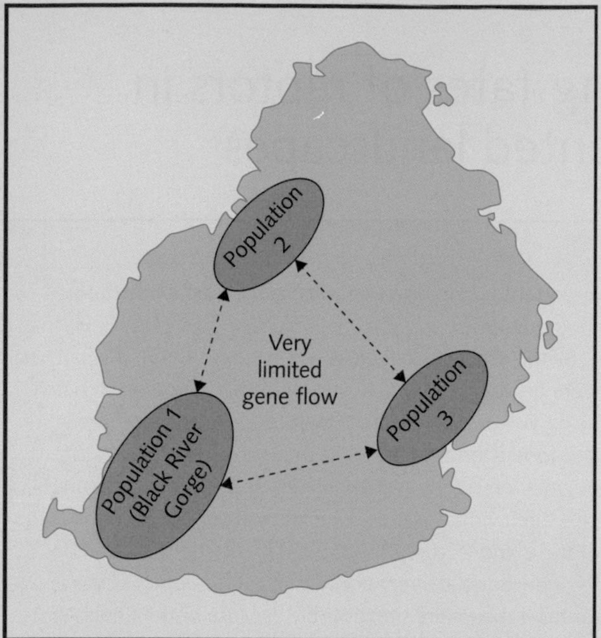

Figure B Map of Mauritius showing Mauritius kestrel *Falco punctatus*, distribution.

potential **inbreeding depression** (reduced biological fitness in a given population as a result of inbreeding; Section 12.4.3) remain to be seen. The moral of this story is not to assume that because a species can physically move through a fragmented landscape that it will do so: well-functioning metapopulations do not always arise automatically to buffer the effects of patchiness.

REFERENCES

Groombridge, J.J., Bruford, M.W., Jones, C.G., & Nichols, R.A. (2001) Evaluating the severity of the population bottleneck in the Mauritius kestrel *Falco punctatus* from ringing records using MCMC estimation. *Journal of Animal Ecology*, Volume 70, 401–409.

Sonsthagen, S.A., McClaren, E.L., Doyle, F.I., Titus, K., Sage, G.K., Wilson, R.E., Gust, J.R., & Talbot, S.L. (2012) Identification of metapopulation dynamics among northern goshawks of the Alexander Archipelago, Alaska, and Coastal British Columbia. *Conservation Genetics*, Volume 13, 1045–1057.

Nevoux, M., Arlt, D., Nicoll, M., Jones, C., & Norris, K. (2013) The short-and long-term fitness consequences of natal dispersal in a wild bird population. *Ecology Letters*, Volume 16, 438–445.

competition (Section 2.4.1) will increase and carrying capacity might be exceeded, which will itself have an impact on productivity and mortality. Accordingly, some individuals are likely to emigrate from the high-quality (source) patch to the low-quality (sink) patch in what is referred to density-dependent dispersal; literally dispersal that happens because of the population density. For the sink populations, this life-line is known as the **rescue effect**. Source/sink dynamics and the metapopulation concept are illustrated in Figure 7.17.

A study of the Japanese beetle *Popillia japonica* (Regniere et al., 1983), also demonstrates it is possible for source/sink patches to swap roles, possibly with intermediate stages as refuge patches. In this case, optimal conditions are humid patches (sources), but this is also where most predators occur. Thus, when reproduction rates are high and predator populations increase, beetles spread to less favourable drier patches (sinks), which do not support permanent predator populations. In years when predation in humid patches is intense, beetle survival can be higher in drier patches, which become refuges rather than sinks. Beetles can

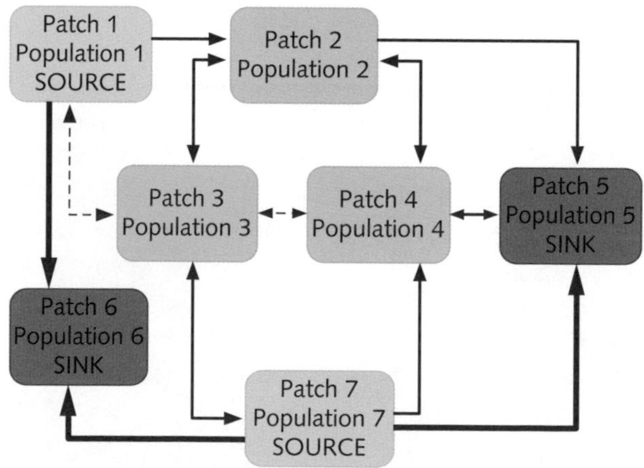

Figure 7.17 Metapopulation dynamics for seven hypothetical patches/populations. The line thickness shows the relative number of individuals moving between the patches and the arrows the direction of movement. In some cases, arrows are bidirectional because individuals move in both directions. Source populations are shown in green; more individuals leave such patches than move to them. Sink patches are shown in red; more individuals move to such patches than leave them. Some patches have approximately equal movement and are shown here in blue.

then reinvade humid patches when predator numbers decline, such that the original sink is now a source.

When sink populations are isolated because metapopulations are not viable, extirpation is the likely long-term outcome. Even for population/patch combinations viable long-term, there is potential for inbreeding due to the small number of individuals and a small gene pool (Section 12.4.3). Isolated populations are also at a greater risk of chance extinctions through **stochastic processes**—random events such as disease outbreaks (Section 10.3.1).

7.4.2 Effects on wider ecology.

The effect of fragmentation on wider ecology through its influence on environmental conditions, habitat, and overall species community is largely driven by **edge effects**.

Edge effects

Environmental parameters are often different at the edge of a habitat compared with the **interior** or **core**. A good terrestrial example is woodland. The edge of a woodland tends to have more light, be more susceptible to wind and precipitation, have more temperature variability, and be more prone to human disturbance, compared with the interior of the same wood. In the aquatic environment, the edge of a lake is usually shallower (and thus warmer, lighter, and with more rooted aquatic plants) than the middle of the lake.

As a result of core:edge habitat differences, species composition can be dramatically different. As a general rule, interior-dwelling species include those that are susceptible to disturbance, whereas edge species are often more disturbance-tolerant (or even thrive in disturbed habitats). Thus the interior community often includes more **K-strategist** species while the edge community includes more **r-strategist** species (Section 2.7.1). The edge might also include a higher proportion of non-native species as it is prone to invasion by disturbance-tolerant species. In terms of plants, there are likely to be more pioneer species at the edge than the interior. Indeed, in Brazilian Atlantic forest 83% of plants were pioneers in the 100-m edge zone compared with the 37% in the interior (Oliveria et al., 2004)

It is sometimes thought that interior species are more likely to be specialist and edge species more likely to be generalist, but this is an over-simplification. If we return to the examples above, woodland edge species in Northern Europe include flowering plants such as bramble *Rubus fruticosus*, dog rose *Rosa canina*, and honeysuckle *Lonicera periclymenum*. These flower after leaf-burst of deciduous trees when the canopy is well developed and so thrive at the edges where there is more light. Such plants are vital for butterflies, including the woodland specialist silver-washed fritillary *Argynnis paphia*, while honeysuckle is the primary nectar plant for another woodland specialist, white admiral *Limenitis camilla*. In lakes and ponds, dragonfly larvae such as the broad bodied chaser *Libellula depressa*, benefit from shallower and warmer water at the water's edge, while newts Pleurodelinae depend on edge emergent vegetation for egg-laying. Accordingly, it is more correct to say that the distribution of some species, at a local scale, might be driven by distance to habitat edge with some species increasing in abundance (**edge specialists**) and others decreasing (**interior specialists**).

Ecological processes, such as predation, can also differ between edge and interior. For example, there is a strong edge effect for moorland-breeding wading birds in Scotland where natural moorland habitat has been changed by planting pine plantations. Some wading birds, such as dunlin *Calidris alpina* and golden plover *Pluvialis apricaria*, avoid moorland:woodland edges, probably due to higher abundance of nest predators near woodland edges, especially hooded crows *Corvus cornix*, red foxes *Vulpes vulpes*, and pine martens *Martes martes* (Wilson et al., 2014).

Finally, the role of **ecosystem engineers** can also differ between edge and interior. For example, leafcutting ants *Atta cephalotes* and *A. sexdens* (Figure 2.8), can be ecosystem engineers by reducing above-ground biomass of plant species, which, through a reduction in leaf litter in foraging areas, can reduce soil fertility in those areas. There is a difference between edge and interior in density of *A. sexdens* (6–8 times more nests in edge habitats) and the herbivory rate of *A. cephalotes*, which could have long-term effects on forest structure and regeneration, especially in secondary forest (Wirth et al., 2007; Urbas et al., 2007; Meyer et al., 2013).

7.4.3 Fragmentation metrics and mapping.

Various analyses and metrics can be used to quantify fragmentation of the background matrix or patches of a specific habitat using maps or remotely sensed data (Section 7.3.3). Common metrics include:

- Total habitat area and area of intact core.
- Number and size of patches (and variation in this—standard deviation).
- Number of patches with core areas.
- Mean core area per patch (and variation in this—standard deviation).
- Total edge length.

Another very useful metric is **nearest neighbour.** This is the distance from one patch of a habitat to the nearest patch of the same habitat. When the nearest neighbour is calculated for multiple patches in a landscape, it is also possible to calculate the standard deviation as a measure of variability. Generally, nearest neighbour distances increase with fragmentation. Variability is not affected by the amount of fragmentation, but by its spatial arrangement—values are low when fragmentation is consistent spatially and high when fragmentation itself is patchy. More detail can be gained using a specific nearest neighbour analysis formula to classify the spatial arrangements of the patches. As with all formulae, it can look confusing at first, but is actually very simple:

$$Rn = \frac{\overline{D}(Obs)}{0.5\sqrt{\frac{a}{n}}}$$

where Rn is the nearest neighbor value, $\overline{D}$(Obs) is the mean observed nearest neighbor distance, a is the size of area under study (use same unit of measurement as for nearest neighbour distances; usually m^2 or km^2), and n is the total number of patches measured.

Rn values close to 0 indicate patches are highly clustered, values near 1 indicate patches are regularly distributed (as might occur for plantation patches, for example), and values near or over 2 indicate patches are arranged randomly across the landscape. It is worth noting that although the nearest neighbour statistic has been applied here to patches, it can be used to describe the spatial distribution of individuals of a species too.

Friction maps and connectivity maps.

Friction maps show the ease with which a species can traverse a landscape and can help predict impacts of fragmentation on dispersal (Section 7.4.1). For example, Joly et al. (2003) created a friction map to model movement potential for common toads *Bufo bufo* in the Rhône floodplain, France, whereby each habitat was given a resistance coefficient. The lowest values were given to 'toad friendly' habitat, while higher values were given to open areas (due to desiccation effects) and habitats where substratum unevenness involves high movement costs. Roads were given high values due to the risk of mortality while crossing.

An alternative to friction mapping is connectivity mapping. In some senses the two approaches are similar, but whereas friction mapping focuses on barriers to dispersal, connectivity mapping focuses on dispersal pathways. An example is the connectivity mapping work undertaken for otters *Lutra lutra*, in Italy (Carranza et al., 2012).

FURTHER READING FOR SECTION

Habitat fragmentation importance in forested ecosystems worldwide: Haddad, N.M., Brudvig, L.A., Clobert, J., Davies, K.F., Gonzalez, A., Holt, R.D., Lovejoy, T.E., Sexton, J.O., Austin, M.P., Collins, C.D., & Cook, W.M. (2015) Habitat fragmentation and its lasting impact on Earth's ecosystems. *Science Advances*, Volume 1, e1500,052.

Studying the effects of habitat fragmentation on species—examples with primates: Arroyo-Rodríguez, V., Cuesta-del Moral, E., Mandujano, S., Chapman, C.A., Reyna-Hurtado, R., & Fahrig, L. (2013) Assessing habitat fragmentation effects on primates: the importance of evaluating questions at the correct scale. In: L.K. Marsh and C. Chapman (Eds) *Primates in Fragments Complexity and Resilience*. New York, NY: Springer.

Effect of patch size and edge effects on vegetation: Gonzalez, M., Ladet, S., Deconchat, M., Cabanettes, A., Alard, D., & Balent, G. (2010) Relative contribution of edge and interior zones to patch size effect on species richness: An example for woody plants. *Forest Ecology and Management*, Volume 259, 266–274.

Habitat suitability mapping for at-risk species: Razgour, O., Hanmer, J., & Jones, G., 2011. Using multi-scale modelling to predict habitat suitability for species of conservation concern: the grey long-eared bat as a case study. *Biological Conservation*, Volume 144, 2922–2930.

The effect of road building on species' dispersal—example of a friction mapping approach: Vuilleumier, S. & Prélaz-Droux, R., 2002. Map of ecological networks for landscape planning. *Landscape and Urban Planning*, Volume 58, 157–170.

7.5 Effects of non-landscape processes on landscape ecology.

In addition to change in landscape ecology driven by change in the landscape itself, there are non-landscape drivers of change in species distribution (and thus species interactions and community composition):

- **Introduction of non-native species** automatically alters the range of the introduced species, but there can be substantial impacts on the distribution of other species too (Chapter 8).
- **Population change in ecosystem engineers** has a profound effect on local environments and can alter ecology substantially. This has happened for the southern white rhino *Ceratotherium simum simum*, a megaherbivore with a high herbivory rate that is highly selective over which species are consumed. This species has considerable influence on habitat structure in areas such as Kruger National Park, South Africa (Cromsigt & te Beest, 2014). Reduction in population size through poaching has changed grass communities and there are cascade ('knock on') effects for populations of other grazers, which again has landscape-level impacts.
- **Disease** that affects keystone or dominant species can have a noticeable effect on the landscape, as seen in the case of ash dieback, a serious disease of ash trees *Fraxinus excelsior*, caused by the fungus *Chalara fraxinea*. Disease-induced tree mortality over a large scale has landscape-level effects, especially where the tree was dominant. This is explored further in Case Study 7.2.
- **Climate change** is a considerable influence on species range and, whether natural or human-accelerated, can alter species distributions substantially (Section 7.5.1).

7.5.1 Climate change.

As noted in Section 4.3.1, species have tolerance ranges for abiotic parameters such as temperature and precipitation. This means that the range of a given species is partly determined by climatic factors both directly and through climate influences on other species with which they have mutualistic or feeding interactions. This means that change in climate often results in change in range. This is not a new phenomenon, but has normally been a gradual process as climate change itself has been slow. Human-accelerated climate change, as the name suggests, is occurring much faster than the historical norm, which is increasing the distances species are moving within a short period. Such climatic change is likely to alter many species' ranges substantially by the end of the 21st century.

There are several general types of temporal range change, which are explained in Figure 7.18. These

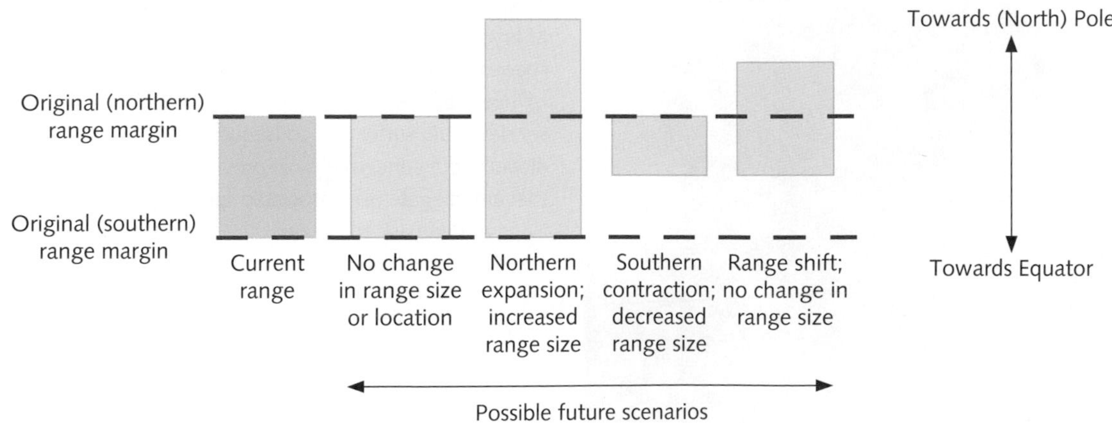

Figure 7.18 Possible range change scenarios over time (orange) relative to the current situation (green) in response to an external stimulus, in this case climate change, for species in the northern hemisphere.

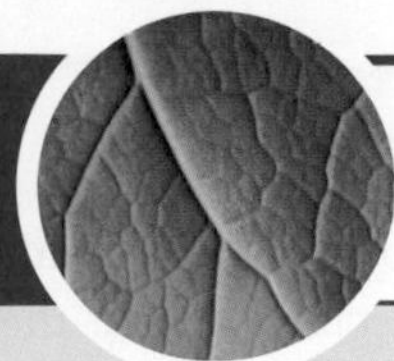

CASE STUDY 7.2 Ash dieback

Ash trees (especially European ash *Fraxinus excelsior*) (Figure A) are prone to the disease referred to as Chalara dieback, ash dieback, or simply Chalara. This is caused by a fungus now known as *Hymenoscyphus fraxineus*, but previously known as *Chalara fraxinea*. Ash trees affected by Chalara experience leaf loss, dieback of the crown, and distinctive bark lesions (Figure B). European ash trees first started dying in large numbers from Chalara dieback in the early 1990s in Poland, and within a few years symptoms were also reported in Latvia, Lithuania, and Estonia. By 2012 the disease was well established across Western Europe, reaching as far as mainland Britain. Infected trees usually die, either from the direct effects of the fungus or because infection makes them more susceptible to other pathogens, especially honey fungus *Armillaria*.

The disease has, had and continues to have, substantial effects across Europe. Where ash is an abundant species, ash dieback has the potential to have significant landscape-level effects. In the UK, ash is the third most common tree species and is widely found in woodlands, hedgerows, and forestry plantations. Consequently, there was widespread concern over the effects of ash dieback when it first appeared. In response to these concerns, the Joint Nature Conservation Committee (JNCC) published a report in 2014 summarizing the potential effects of ash dieback on UK biodiversity (Lawrence & Cheffings, 2014).

Figure A Ash trees, *Fraxinus excelsior*, are a familiar part of the European landscape.

Source: Courtesy of Jim Champion/CC BY-SA 2.0.

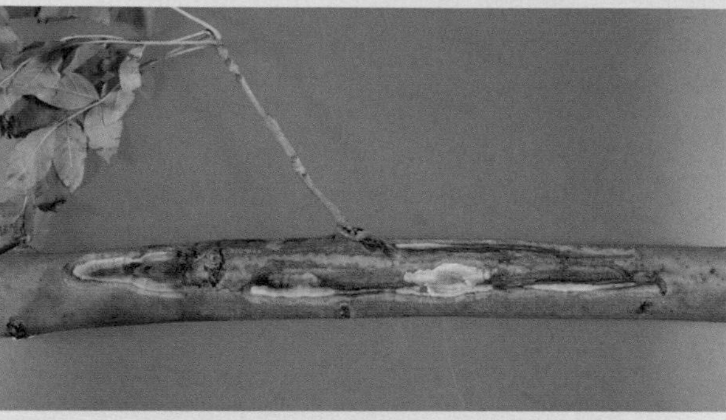

Figure B Distinctive bark lesions are a symptom of ash dieback.

Source: Image courtesy of The Food and Environment Research Agency (FERA), Crown Copyright, used under Open Government Licence.

Ash leaf litter is nutrient-rich and rapidly degradable, whilst the canopy has high light penetration. Together, these factors raise the pH of soil, lead to fast nutrient turnover and create a light understorey encouraging the growth of light-tolerant plants in the shrub layer. These conditions favour specific assemblages of species. In total, 1058 species have been identified as being associated with ash trees (Mitchell et al., 2014), all of which either make use of the ash trees themselves or the habitat created by them. They include 12 species of birds, 55 species of mammals, 58 species of bryophytes (mosses, liverworts, and hornworts), 68 species of fungi, 78 species of vascular plants, 239 species of invertebrates, and 546 species of lichens. Many of these species could suffer as a consequence of the spread of ash dieback, but concern is most pressing for the 44 species that are obligate ash-associated species (i.e. only found on living or dead ash trees) and the 62 additional species highly associated with ash (including fungi, lichen, bryophyte, and invertebrate species).

Where woodland comprises $<10\%$ ash, the gaps created by ash dieback will likely be filled by canopy expansion of surrounding species, which in turn will lead to a growth of shade-tolerant plants in the shrub layer. In woodland with $>20\%$ ash, the gaps created will recruit other species and sycamore *Acer pseudopanatus* is likely to become particularly dominant.

Overall, the widespread death of ash trees within Europe has caused, and will continue to cause, significant changes in the proportion of different tree species in woodlands. The knock-on effects of ash decline will likely also have a negative impact on species that use ash, especially those that do so exclusively.

REFERENCES

Lawrence, R. & Cheffings, C.M. (2014) A summary of the impacts of ash dieback on UK biodiversity, including the potential for long-term monitoring and further research on management scenarios. JNCC Report No.501.

Mitchell, R.J., Bailey, S., Beaton, J.K., Bellamy, P.E., Brooker, R.W., Broome, A., Chetcuti, J., Eaton, S., Ellis, C.J., Farren, J., Gimona, A., Goldberg, E., Hall, J., Harmer, R., Hester, A.J., Hewison, R.L., Hodgetts, N.G., Hooper, R.J., Howe, L., Iason, G.R., Kerr, G., Littlewood, N.A., Morgan, V., Newey, S., Potts, J.M., Pozsgai, G., Ray, D., Sim, D.A., Stockan, J.A., Taylor, A.F.S., & Woodward, S. 2014. The potential ecological impact of ash dieback in the UK. JNCC Report No. 483.

are applicable not only to climate-induced changes, but also other distribution shifts. In the case of climate warming, range shifts tend to be polewards and thus north in the northern hemisphere and south in the southern hemisphere. Some species are also moving altitudinally, with mountain tops being the altitudinal equivalent of polar regions.

Climate envelope modelling.

An important method of predicting change for individual species (and, by extension, species communities) is Climate Envelope Modelling (CEM). This uses a species' current range to infer climatic requirements for that species (the 'climatic envelope'). Future distribution can then be predicted by establishing envelope movement under climate change scenarios. CEM is crucial for understanding long-term implications of climate change and the best guide available for informing climate-related conservation (Goodenough and Hart, 2013). One study that has used CEM to good effect is the UK multi-taxon MONARCH project discussed in the Online Case Study for Chapter 7. However, CEM is necessarily a simplification and does have limitations:

- The premise is that species will not adapt to different climates and climate-altered habitat.
- It assumes species can move freely to areas with favourable climate, whereas actually dispersal power might restrict movement (e.g. the Scottish crossbill *Loxia scotica* is predicted to move to Iceland, which seems unlikely given the distance involved and the absence of land between current and predicted ranges).
- Climate is not the only factor that influences range—food sources, competitors, predators, and parasites are all important and are only loosely associated with climate (Davis et al., 1998).

You can find the online case study at www.oxfordtextbooks.co.uk/orc/goodenough

FURTHER READING FOR SECTION

Example of approaches to climate envelope modelling: Huntley, B., Green, R.E., Collingham, Y.C., & Willis, S.G. (2007) *A Climatic Atlas of European Breeding Birds*. Barcelona: Lynx Editions.

Effect of climate change on butterflies in an already-fragmented landscape: Wilson, R.J., Davies, Z.G., & Thomas, C.D. (2010) Linking habitat use to range expansion rates in fragmented landscapes: a metapopulation approach. *Ecography*, Volume 33, 73–82.

Links between landscape ecology, climate change and conservation: Hannah, L., Midgley, G., Andelman, S., Araújo, M., Hughes, G., Martinez-Mayer, E., Pearson, R., & Williams, P. (2007) Protected area needs in a changing climate. *Frontiers in Ecology and the Environment*, Volume 5, 131–138.

7.6 Landscape ecology management.

This chapter has highlighted that while many changes in landscape ecology are natural, a considerable number of current changes are driven by human change to those landscapes, including habitat fragmentation (Section 7.4), as well as broader changes such as climate change and spread of disease (Section 7.5). Understanding such processes and the impact they have on species/habitats allows Applied Ecologists to design management to reduce existing problems and, in some cases, prevent problems occurring in the first place. Such management usually falls under one of the following headings:

- **Site-level management:**
  - Increasing heterogeneity (Section 7.6.1 and Section 11.3.1).
  - Habitat creation and management (Section 11.3.1).
- **Species-level management:**
  - Facilitating movement:
    - improving connectivity (Section 7.6.2);
    - moving species physically (translocation) (Chapter 13).
  - Preventing species moving to, or aggregating in, 'high risk' areas (Section 7.6.3).
- **Landscape-scale approaches:**
  - Eco-regions and 'living landscapes' (Section 7.6.4).
  - Agri-environment schemes (Section 11.5.3).

7.6.1 Heterogeneity.

A heterogeneous environment is one that contains variation (hetero meaning mixed; the opposite is homogenous). This might be obvious variation at a landscape scale, with different habitats forming a mosaic. Heterogeneity might also be more subtle, for example, variation within a grassland environment driven by changes in slope angle and aspect, different geologies, or different grazing pressures. Variation can also occur at even smaller scales, such as differences in depth and substrate in one small area of a stream. The multiple interdependent causes of heterogeneity, both natural and anthropogenic, are shown in Figure 7.19 along with their ecological effects.

Creating heterogeneity is an important part of *in situ* conservation management. Examples include cutting, grazing, or burn management that is applied to subsections of an area or nature reserve in rotation to create the habitat mosaic affect mentioned above. Such approaches will be discussed in more detail in Section 11.3.1.

In other cases, effective conservation management requires understanding and maintaining natural heterogeneity-causing processes. An example of the consequences of failure to do this is the historic fire suppression policy in several American National Parks, including Yosemite created in 1890. The lack of heterogeneity, change in habitat through the process of natural ecological succession, and the lack of growth of species that depend on fire to germinate, such as giant redwoods *Sequoiadendron giganteum* led to this policy being revised in the 1960s (including through the Wilderness Act of 1964) so that fires are now allowed to burn unless they are threatening humans or livestock.

There are few 'one size fits all' landscape ecology rules and good management sometimes involves flouting the guiding principles discussed above. One example is creating woodland rides—linear strips of open habitat in woodland—which are often actively encouraged to increase flowering plant diversity. This involves fragmenting woodland (albeit on a small scale) to create more 'edge'. Because rides are linear, they can be used as **corridors** (Section 7.6.2), but they are important habitats in their own right. In the same way, field margins—uncultivated areas at the edge of fields—can be a vital habitat for species. In both cases, these edge habitats are important because they create heterogeneity in the landscape, often having a very different structure complexity and species assemblage to the surrounding area. This also underlines the fact that edges are not necessarily 'bad': they can constitute valuable habitat and are sometimes created deliberately for the purposes of conservation.

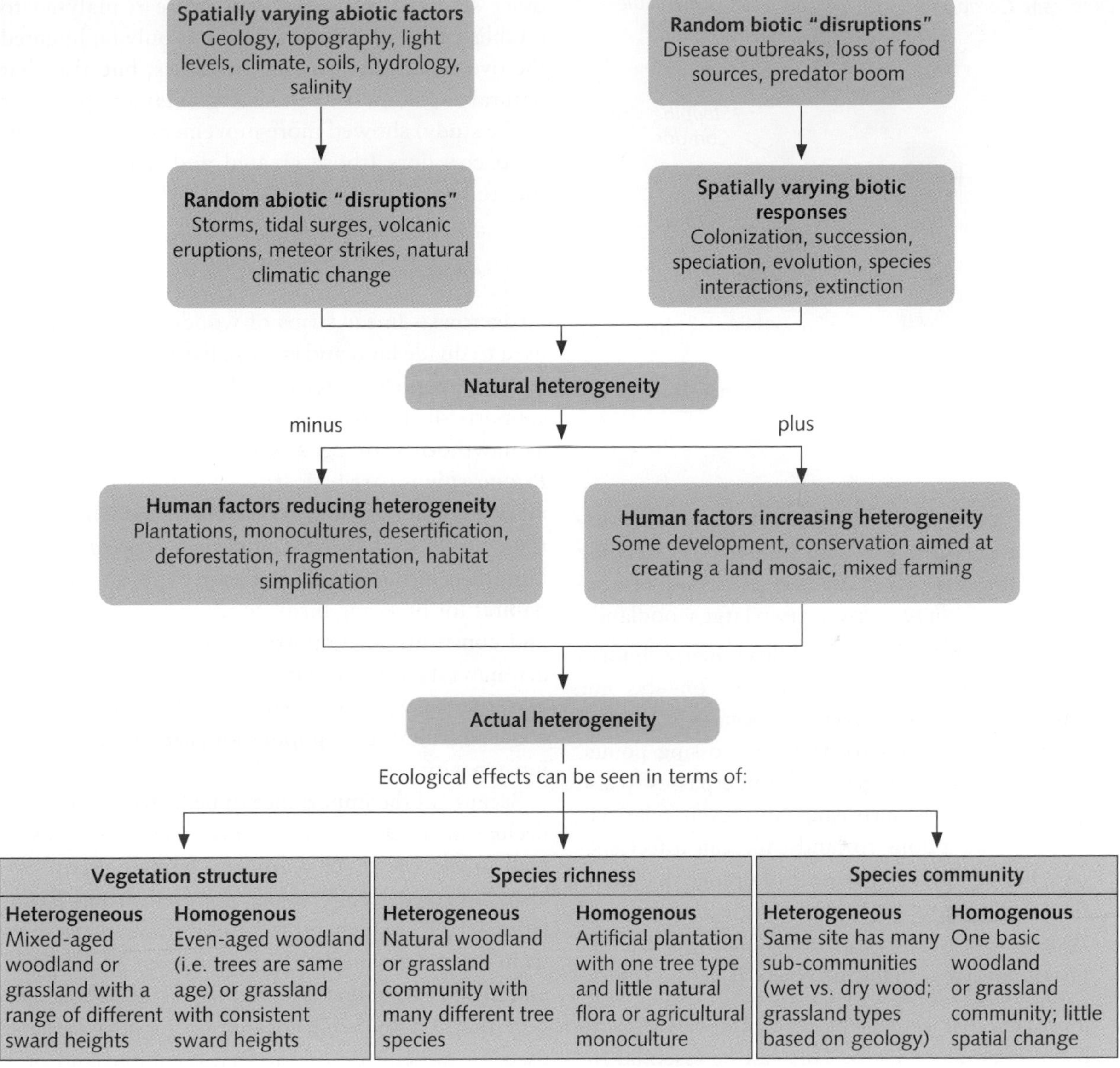

Vegetation structure		Species richness		Species community	
Heterogeneous	**Homogenous**	**Heterogeneous**	**Homogenous**	**Heterogeneous**	**Homogenous**
Mixed-aged woodland or grassland with a range of different sward heights	Even-aged woodland (i.e. trees are same age) or grassland with consistent sward heights	Natural woodland or grassland community with many different tree species	Artificial plantation with one tree type and little natural flora or agricultural monoculture	Same site has many sub-communities (wet vs. dry wood; grassland types based on geology)	One basic woodland or grassland community; little spatial change

Figure 7.19 Sources of habitat heterogeneity within the landscape, and the effects of heterogeneity on ecological complexity and value.

7.6.2 Connectivity.

The ecological effects of landscape fragmentation (Section 7.4) are related to landscape connectivity—the ease with which species can move from one patch to another. One vital part of managing fragmented landscapes, therefore, is to improve connectivity. This can be done in four main ways as listed below in order of desirability (Gilbert-Norton et al., 2010). The first three are shown graphically in Figure 7.20.

1. Creating **linear corridors:** constructing specific linear habitats such as hedgerows or woodland rides along which species can move.
2. Creating **landscape corridors:** ensuring that patch elements are managed so there are 'pathways' across the landscape—such as a contiguous line of fields between two grassland patches.
3. Creating **stepping stone islands:** forming small patches of land that species use in a series of

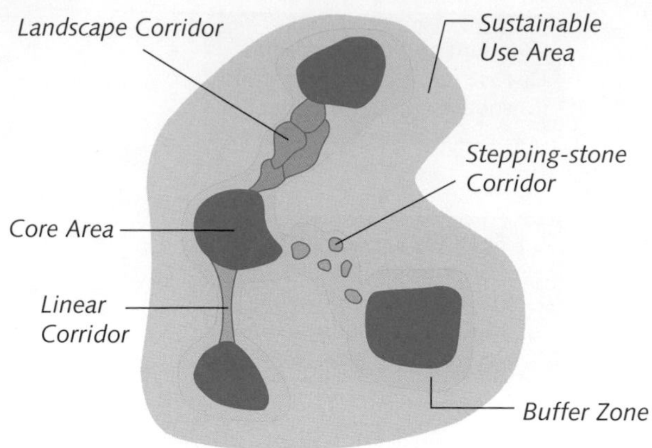

Figure 7.20 Approaches to improve connectivity in the landscape.

'jumps' as they move between larger patches, exactly in the same way that humans use stepping stones to cross a river. An example would be a series of small copses between two large woodlands.

4. Improving **barrier permeability**: many human corridors—roads, railways and so on—become barriers for species movement (Section 7.2.3) and it can be important to provide 'crossing points' such as green bridges and underpasses (Case Study 5.2). It is surprising how even relatively narrow disruptions to landscape can affect species. For example, movement of orange tip butterflies *Anthocharis cardamines* is greatly restricted by motorways, and most individuals that attempt cross-motorway flight turn back (Dennis, 1986).

Improving connectivity basically allows species to 'join the dots' more easily. This can be essential for movement and metapopulation dynamics, especially when patches are isolated and/or species have small home ranges and low dispersal power. Landscapes with high connectivity have more resilient ecological networks, especially given the need for species movement in response to climate change (Section 7.5.1). From a purely pragmatic point of view, it may be more cost-effective to promote and enhance the existing network of corridors and stepping stones (by appropriate management, and by widening them) than to create new ones (Lawton, 2010).

A comprehensive study of the effectiveness of corridors for different taxa has been undertaken by Gilbert-Norton et al. (2010). This used a meta-analysis approach whereby the results of lots of other studies (35 in this case) were reanalysed to establish general patterns. This not only highlighted the overall effectiveness of corridors, but also that natural corridors (those existing in landscapes prior to the study) showed more movement than manipulated corridors (those created and maintained for the study).

Hedgerows

Hedgerows—linear strips of wooded scrub that are used to divide land and enclose livestock—are a particularly important terrestrial corridor, which also supports an entire ecosystem. Indeed, plants such as hawthorn *Crataegus monogyna* and blackthorn *Prunus spinosa* are hedgerow specialists and are primary larval food plants for black-veined white *Aporia crataegi* and brown hairstreak *Thecla betulae* butterflies in Europe. They also provide vital nesting habitat for breeding birds, food for wintering birds, and constitute an important 'overflow' or secondary habitats for the hazel dormouse *Muscardinus avellanarius* when preferred woodland habitat becomes unsuitable or population size exceeds carrying capacity.

Because of the importance of hedgerows, and their decline in much of Europe, hedgerows are considered in the UK to be a priority habitat. Many are also protected legally under the Hedgerows Regulations 1997, which protects important hedgerows from being destroyed if they are at least 30 years old and meet at least one criterion in Part II of Schedule 1 of that Act by being located on or next to: land used for agriculture or forestry, common land or a village green, or a Site of Special Scientific Interest or a legally protected nature reserve. Further details can be found at http://www.legislation.gov.uk/uksi/1997/1160

Ecological problems of enhanced connectivity

Although enhancing connectivity has many benefits, not all movement is desirable. **Non-native species** (Chapter 8) often use corridors to spread through a new ecosystem with surprising speed. One of the best examples is species spread from the Red Sea to the Mediterranean Sea via the Suez Canal. The opening of the canal in 1869 also facilitates the movement of more than 400 non-native

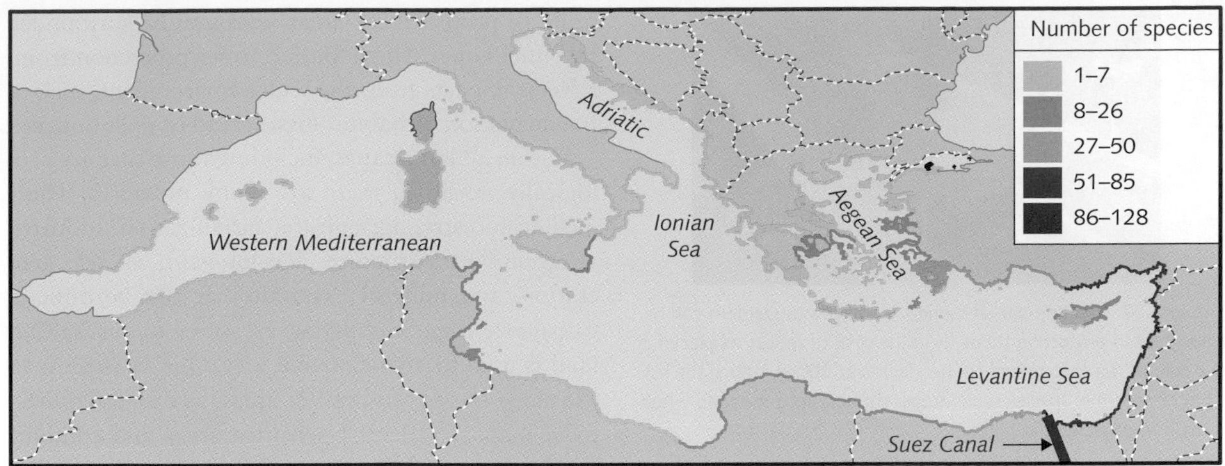

Figure 7.21 Distribution of non-native species in the Mediterranean Sea that have entered from the Red Sea via the Suez Canal.

Source: Reproduced from Katsanevakis, S. et al. (2014) 'Invading the Mediterranean Sea: biodiversity patterns shaped by human activities'. *Frontiers in Marine Science*/CC BY.

species (Katsanevakis et al., 2014). Movement tends to be unidirectional since the Red Sea is extremely salty and nutrient-poor, and is thus a more challenging environment for Mediterranean species, whereas the Red Sea species are able to survive and thrive in the nutrient-rich waters of the Mediterranean and have spread considerable distances around the coastline (Figure 7.21). Such species movement is termed Lessepsian migration, after Ferdinand de Lesseps, who managed the Canal's construction.

There can be other disadvantages of corridors too:

- **Spread of disease:** a range of human diseases vectored by mosquitos, including malaria, spread along the Trans-Amazon highway in Brazil (indeed, it killed many of the road's construction workers in the 1970s), while bubonic plague, spread by fleas associated with rats, followed early trade routes, such as the 'silk road', to spread from Asia to Europe in the 1300s.
- **Spread of fire:** there have been long-standing concerns about fire spreading along corridors (e.g. Simberloff & Cox 1987), although a recent experimental study in open herbaceous habitat in South Carolina, USA, suggests that, in some landscapes at least, the corridor-fire interaction is not one that increases fire spread, but increases fire temperature (and thus its effects on the landscape) (Brudvig et al., 2012).

7.6.3 Management to prevent species movement.

Sometimes it is important to prevent species entering or aggregating in an area either to protect them or to prevent them endangering human activity. Examples in Section 5.6 highlighted use of temporary fencing to stop amphibians straying into foundation trenches and permanent fencing to stop mammals straying onto roads. 'Virtual fences' can also be important. To reduce deer-vehicle collisions, two virtual fence solutions have been trialled: chemical fences that emit a smell that the target species finds unpleasant (Germany) and light reflectors (UK). The chemical fences performed well in trials (repellent chemicals encapsulated in slow release organic foam at six Bavarian test sites showed 60% of the animals encountering the treated areas withdrew and crossed the road beyond the 'scent fence' at an untreated section and a further 20% crossed very rapidly without delay: Staines et al., 2001). The reflectors (explained diagrammatically in Figure 7.22) appear to work initially, but can then lose effect as animals become used to them through the process of habituation.

Habituation can also be an issue for methods used to disperse species from high risk aggregation areas. For example, in California, submerging acoustic 'pingers' near fisheries eliminates entanglement of Cuvier's beaked whales *Ziphius cavirostris* and Hubb's beaked whales *Mesoplodon carlhubbsi* (Carretta et al., 2008), but such interventions have been found to be ineffective in the long term.

Figure 7.22 Creating partial barriers to species movement can be important in protecting them, as in the case of reflectors placed at the side of roads to reflect car headlights at 90° to reduce the risk of light-sensitive species such as deer running into the road when a car is approaching at night.

Source: 'Zerokill', designed by Sungi Kim & Hozin Song. Image courtesy of Yanko Design (2009) http://www.yankodesign.com/2009/12/01/save-the-animals-and-yourself/

7.6.4 Ecoregions and 'living landscapes'.

Although human activity is influenced by administrative units such as countries and states/provinces, species are not governed by these theoretical boundaries (Section 7.2.4). Accordingly, there is an increasing drive to manage areas as **ecoregions**, which are geographically distinct areas with a common species community that shares similar environmental conditions and ecological dynamics.

Managing areas as ecoregions is a robust form of management; however, it is not without its challenges. For example, many ecoregions span multiple countries and thus **multilateral agreements** (agreements between more than one country) are needed even for a management plan to be agreed. Different countries have different priorities, planning laws, conservation bodies, wildlife legislation, protected area frameworks, and funding networks and for an ecoregion to work, all of these must be reconciled. One example of ecoregion-level management is the Yellowstone to Yukon or Y2Y initiative between Canada and the USA (Case Study 7.3).

The designation of areas as **core areas** and **buffer zones** has a role to play in both ecoregion management and management of smaller scale landscapes. Core areas have high nature conservation value and form the heart of a landscape's ecological network. They contain habitats that are rare or important because of the wildlife they support or the ecosystem services that they provide. They generally have the highest concentrations of species or support rare species. In order to protect these areas, they can be surrounded by buffer zones. These 'buffer' (offer protection from) adverse impacts from the wider environment, such as fragmentation or habitat loss, spread of pollution, etc.

Within all landscapes, including those that are ecologically sensitive, there are many pressures. These include forestry, agriculture, urbanization, industrialization, infrastructure development, energy generation, and mineral extraction. It can be difficult to balance these competing pressures to ensure that land is used in a sustainable way. One strategy is to use maps of core and buffer areas, as well as connectivity maps, to identify sensitive areas and combine this landscape ecology approach with an Ecological Impact Assessment approach (Chapter 5). This is essentially a **zonation** approach, and can be used at any scale from managing people versus wildlife conflicts in a small nature reserve to a 3200 km long ecoregion such as Y2Y (Case Study 7.3).

Many areas are also developing **'living landscapes'** approaches. Again the motivating factor here is to balance use of the landscape with conservation needs, but the focus tends to be on smaller-scale landscapes than ecoregions—for example, a river catchment area. Living landscape projects also seek to identify and protect core areas and enhance connectivity, but they also have a more overt social focus, aiming to reconnect people with the natural world and promote the benefits it provides.

• FURTHER READING FOR SECTION

Creating corridors between key sites—the case for Kenyan elephants: Nyaligu, M.O. & Weeks, S. (2013) An Elephant Corridor in a fragmented conservation landscape: preventing the isolation of Mount Kenya National Park and National Reserve. *Parks*, Volume 19, 91–101.

Review of corridor effectiveness: Gilbert-Norton, L., Wilson, R., Stevens, J.R., & Beard, K.H. (2010) A meta-analytic review of corridor effectiveness. *Conservation Biology*, Volume 24, 660–668.

Landscape ecology assessments—an example from the US: Theobald, D.M. (2013) A general model to quantify ecological integrity for landscape assessments and US application. *Landscape Ecology*, Volume 28, 1859–1874.

Managing landscapes—an example framework from the UK: Lawton, J. (2010) *Making Space for Nature: A review of England's Wildlife Sites and Ecological Network*. London: DEFRA report.

CASE STUDY
7.3

Yellowstone to Yukon: an ecoregion approach

The Yellowstone to Yukon or Y2Y region is a mountainous area that is 3200 km long by around 600km wide. It spans both the USA (encompassing five states—Montana, Idaho, Wyoming, Oregon, and Washington), and Canada (covering two Canadian provinces—Alberta and British Columbia, and two territories—Yukon and Northwest Territories) (Figure A). There are many threats to the Y2Y region, including the direct threats of fragmentation, urbanization, logging and mineral extraction, and external threats such as climate change that are much harder to mange at a local level. Approximately 10% of the Y2Y region has some form of protected area status, such as wildlife refuges, wilderness areas, and state and provincial parks (Figure A).

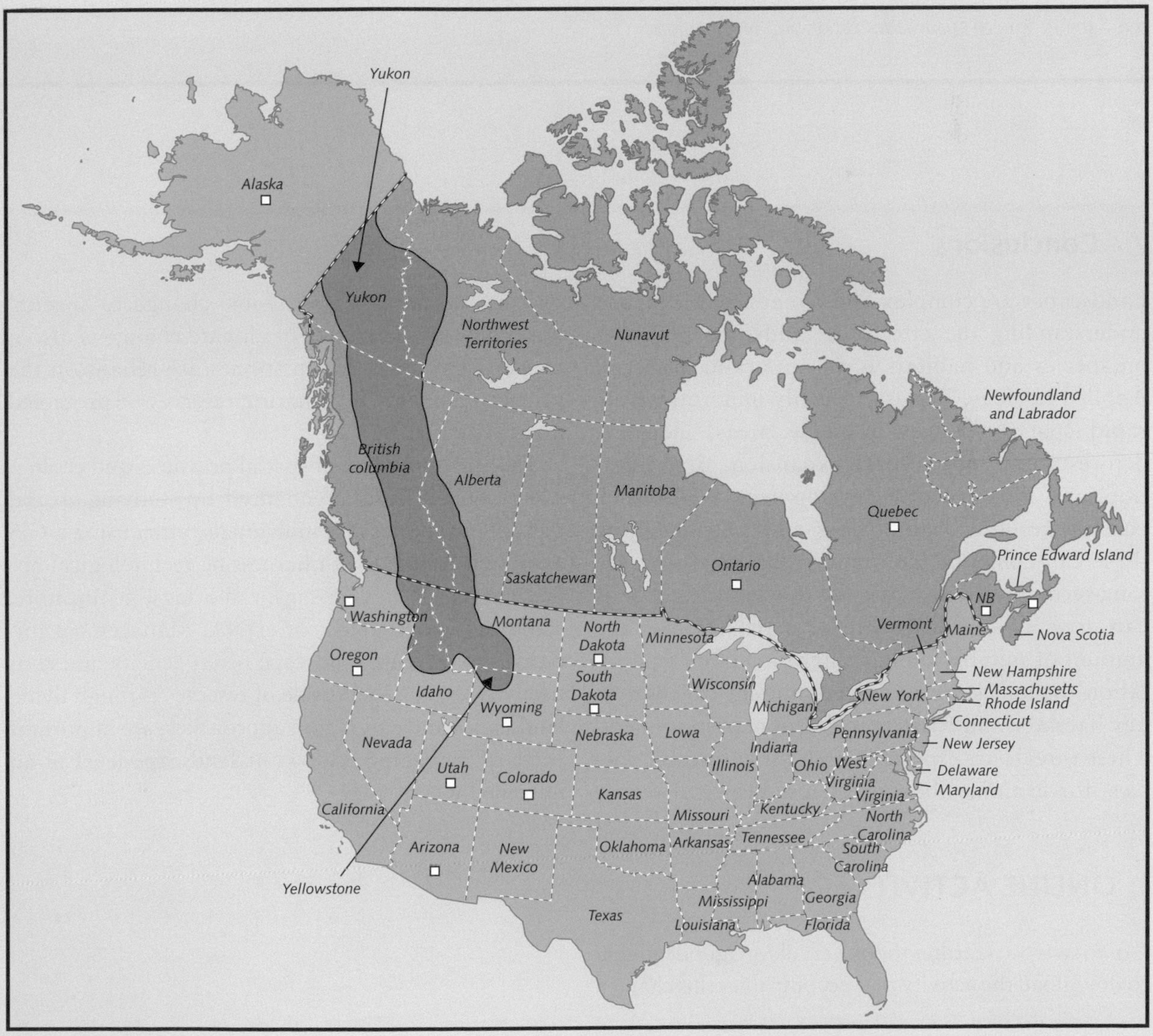

Figure A The Y2Y region.

The aim of Y2Y is that the entire Yellowstone to Yukon region is managed as one interconnecting unit. The emphasis is on identifying and protecting core areas (known as priority areas in Y2Y parlance) via appropriate conservation designation and use of buffer zones, and ensuring high levels of connectivity between core areas through corridor networks and stepping stone islands. **Gap analysis** has been undertaken to identify where there are spatial gaps in corridors and/or lack of designation for core areas. For example, the Flathead Valley of southeast British Columbia has been identified as a gap—lacking both corridors and protected land—and it is part of the Y2Y vision that this area be incorporated into Waterton Lakes National Park. Regeneration of areas through remediation, restoration, and rewilding initiatives (Chapters 6 and 13) is also an important strategy.

Because Y2Y is so big, numerous distinct projects occur under the Y2Y umbrella (Figure A). One of these was the Cabinet-Purcell Mountain Corridor (CPMC) project, which started in 2005 with the aim of reconnecting the isolated grizzly bear *Ursus arctos* population in the Greater Yellowstone Ecosystem with populations in Canada, Idaho, and Montana to create a metapopulation. This involves managing a highly fragmented landscape and the conflicts between people and wildlife, especially the attraction of bears to poorly managed refuse centres. The CPMC project has a partner network of more than 60 different organizations.

So far, the work has focused on purchasing and then managing land in linkage zones, large-scale habitat restoration efforts in the Clearwater National Forest, research into bear movement, and community educational Bear-Aware programs to help reduce human–wildlife conflicts.

FURTHER READING

The Y2Y Ecoregion: http://y2y.net

7.7 Conclusions.

Landscapes are complex and dynamic entities, and understanding the effects of landscape processes on species and habitats is an important aspect of Applied Ecology. This is especially important given rapid spatial changes in many areas, including deforestation, agricultural expansion, desertification, and infrastructure development. Where such changes result in habitat loss and fragmentation, these can constitute substantial threats to species' long-term persistence within the landscape. This can arise as a consequence of a reduction in the amount of suitable habitat, alteration of metapopulation dynamics, an increased amount of 'edge' in key habitats, and barriers to freedom of movement. These threats are especially high for vulnerable species, for example those with low dispersal power or large home ranges. Possible change to species' distributions as a result of climate change is also a concern, especially when species are reliant on the protection afforded by nature reserves or protected areas (Chapter 8).

The understanding of spatial pressures and ecological responses can be enhanced by studying spatial patterns of species and individuals, often using a GIS approach coupled with harnessing technological approaches such as GPS tags, radio tags, geolocators, stable isotope analysis, or eDNA. Management initiatives to increase landscape heterogeneity and connectivity, especially outside of reserves through living landscapes and eco-region approaches, are important ways of managing ecology at landscape level in an informed way.

ONLINE ACTIVITY

Go to www.oxfordtextbooks.co.uk/orc/goodenough/ to download the activity that accompanies this chapter.

CHAPTER 7 AT A GLANCE: THE BIG PICTURE

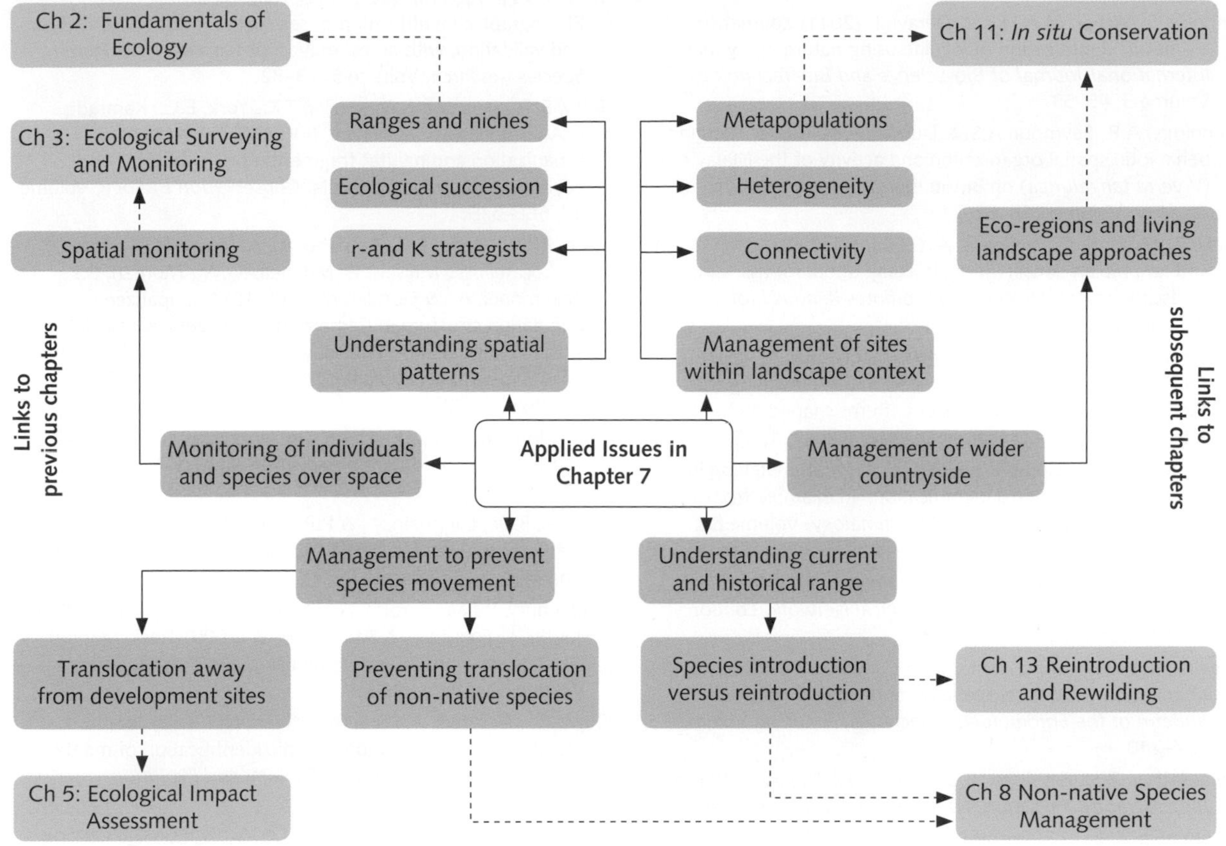

REFERENCES

Boarman, W.I., Beigel, M.L., Goodlett, G.C., & Sazaki, M. (1998) A passive integrated transponder system for tracking animal movements. *Wildlife Society Bulletin*, Volume 26, 886–891.

Brudvig, L.A., Wagner, S. A., & Damschen, E.I. (2012) Corridors promote fire via connectivity and edge effects. *Ecological Applications*, Volume 22, 937–946.

Cain, S.A. (1944) *Foundations of Plant Geography*. New York, NY: Harper and Brothers.

Carranza, M.L., D'Alessandro, E., Saura, S., & Loy, A. (2012) Connectivity providers for semi-aquatic vertebrates: the case of the endangered otter in Italy. *Landscape Ecology*, Volume 27, 281–290.

Carretta, J.M, Barlow, J. & Enriques, L. (2008) Acoustic pingers eliminate beaked whale bycatch in a gillnet fishery. *Marine Mammal Science*. Volume 24, 956–961.

Chaverri, G. & Kunz, T.H. (2011) All-offspring natal philopatry in a neotropical bat. *Animal Behaviour*, Volume 82, 1127–1133.

Cromsigt, J.P. & te Beest, M. (2014) Restoration of a megaherbivore: landscape-level impacts of white rhinoceros in Kruger National Park, South Africa. *Journal of Ecology*, Volume 102, 566–575.

Davis, A.J., Jenkinson, L.S., Lawton, J.L., Shorrocks, B., & Wood S. (1998) Making mistakes when predicting shifts in species range in response to global warming. *Nature*, Volume 391, 783–786.

Dennis, R.L.H. (1986) Motorways and cross-movements. An insect's 'mental map' of the M56 in Cheshire. *Bulletin of the Amateur Entomologists' Society*, Volume 45, 228–243.

Forman, R.T.T. & Godron, M. (1986) *Landscape Ecology*. John Wiley and Sons, Chichester, UK.

Franklin, J. (2010) *Mapping Species Distributions: Spatial Inference and Prediction*. Cambridge: Cambridge University Press.

Fuhr, B. & Demarchi, D.A. (1990) *A Methodology for Grizzly Bear Habitat Assessment in British Columbia*. Vancouver: Ministry of Environment.

Gilbert-Norton, L., Wilson, R., Stevens, J.R., & Beard, K.H. (2010) A meta-analytic review of corridor effectiveness. *Conservation Biology*, Volume 24, 660–668.

Goodenough, A.E. & Hart, A.G. (2013) Correlates of vulnerability to climate-induced distribution changes in European avifauna: habitat, migration and endemism. *Climatic Change*, Volume 118, 659–669.

Gurevitch, J. & Padilla, D.K. (2004) Are invasive species a major cause of extinctions? *Trends in Ecology & Evolution*, Volume 19, 470–474.

Hoque, S., Azhar, M.A.H.B. & Deravi, F. (2011) Zoometrics: biometric identification of wildlife using natural body marks. *International Journal of Bio-Science and Bio-Technology*, Volume 3, 45–53.

Jennings, A.P., Seymour, A.S. & Dunstone, N. (2006) Ranging behaviour, spatial organization and activity of the Malay civet (*Viverra tangalunga*) on Buton Island, Sulawesi. *Journal of Zoology*, Volume 268, 63–71.

Joly, P., Morand, C., & Cohas, A. (2003) Habitat fragmentation and amphibian conservation: building a tool for assessing landscape matrix connectivity. *Comptes Rendus Biologies*, Volume 326, 132–139.

Katsanevakis, S., Coll, M., Piroddi, C., Steenbeek, J., Lasram, F.B.R., Zenetos, A., & Cardoso, A.C. (2014) Invading the Mediterranean Sea: biodiversity patterns shaped by human activities. *Frontiers in Marine Science*, Volume 1, 32.

Kelly, M.J. (2001) Computer-aided photograph matching in studies using individual identification: an example from Serengeti cheetahs. *Journal of Mammalogy*, Volume 82, 440–449.

Lawton, J. (2010) *Making Space for Nature: A review of England's Wildlife Sites and Ecological Network*. London: DEFRA report.

Levins, R. (1969) Some demographic and genetic consequences of environmental heterogeneity for biological control. *Bulletin of the Entomological Society of America*, Volume 15, 237–240.

MacArthur, R.H. & Wilson, E.O. (1967) *The Theory of Island Biogeography. Princeton*, NJ: Princeton University Press.

Majumder A, Basu S, Sankar K, Qureshi Q, Jhala Y.V., & Nigam, P. (2012) Home ranges of Bengal tiger (*Panthera tigris tigris* L.) in Pench Tiger Reserve, Madhya Pradesh, Central India. *Wildlife Biology in Practice*, Volume 8, 36–49.

Meyer, S. T., Neubauer, M., Sayer, E.J., Leal, I.R., Tabarelli, M., & Wirth, R. (2013) Leaf-cutting ants as ecosystem engineers: topsoil and litter perturbations around Atta cephalotes nests reduce nutrient availability. *Ecological Entomology*, Volume 38, 497–504.

Mitchell, M.W., Locatelli, S., Ghobrial, L., Pokempner, A.A., Clee, P.R.S., Abwe, E.E., Nicholas, A., Nkembi, L., Anthony, N.M., Morgan, B.J., & Fotso, R. (2015) The population genetics of wild chimpanzees in Cameroon and Nigeria suggests a positive role for selection in the evolution of chimpanzee subspecies. *BMC Evolutionary Biology*, Volume 15, 1–15.

Morrison, P.H. (1988) *Old growth in the Pacific Northwest: a Status Report*. Alexandra, VA: Wilderness Society.

Oliveira, M.A., Grilloa, A.S., & Tabarelli, M. (2004) Forest edge in the Brazilian Atlantic forest: drastic changes in tree species assemblages. *Oryx*, Volume 38, 1–6.

Rahimi, S. & Owen-Smith, N. (2007) Movement patterns of sable antelope in the Kruger National Park from GPS/GSM collars: a preliminary assessment. *South African Journal of Wildlife Research*, Volume 37, 143–151.

Rees, E.C. (2006) *Bewick's Swan*. Calton: T and A.D. Poyser.

Regniere, J., Rabb, R., & Stinner, R. (1983) *Popillia japonica* (Coleoptera: Scarabaeidae) distribution and movement of adults in heterogeneous environments. *The Canadian Entomologist*, Volume 115, 287–294.

Reisser, J.W., Proietti, M.C., Kinas, P.G., & Sazima, I. (2008) Photographic identification of sea turtles: method description and validation, with an estimation of tag loss. *Endangered Species Research*, Volume 5, 73–82.

Riley, S. P., Sauvajot, R.M., Fuller, T.K., York, E.C., Kamradt, D.A., Bromley, C., & Wayne, R.K. (2003) Effects of urbanization and habitat fragmentation on bobcats and coyotes in southern California. *Conservation Biology*, Volume 17, 566–576.

Clee, P.R.S., Abwe, E.E., Ambahe, R., Anthony, N.M., Fotso, R., Locatelli, S., Maisels, F., Mitchell, M.W., Morgan, B.J., Pokempner, A., & Gonder, M.K. (2015) Chimpanzee population structure in Cameroon and Nigeria is associated with habitat variation that may be lost under climate change. *BMC Evolutionary Biology*, Volume 15, DOI 10.1196/s12862-014-0275-z

Simberloff, D. & Cox, J. (1987) Consequences and costs of conservation corridors. *Conservation Biology*, Volume 1, 63–71.

Staines, B.W., Langbein, J., & Putman, R.J. (2001) *Road Traffic Accidents and Deer in Scotland*, Report to the Deer Commission. Edinburgh: Deer Commission.

Stutchbury, B.J.M., Tarof, S.A., Done, T., Gow, E., Kramer, P.M., Tautin, J., Fox, J.W., & Afanasyev, V. (2009) Tracking long-distance songbird migration by using geolocators. *Science*, Volume 323, 896–896.

Town, C., Marshall, A., & Sethasathien, N. (2013) Manta Matcher: automated photographic identification of manta rays using keypoint features. *Ecology and Evolution*, Volume 3, 1902–1914.

Urbas, P., Araújo, M.V. Jr, Leal, I.R., & Wirth R (2007) Cutting more from cut forests—edge effects on foraging and herbivory of leaf-cutting ants. *Biotropica*, Volume 39, 489–495.

van Rensburg, B.J., Levin, N., & Kark S. (2009) Spatial congruence between ecotones and range-restricted species: implications for conservation biogeography at the sub-continental scale. *Diversity and Distributions*, Volume 15, 379–89.

Wiles, G.J. & Glass, P.O. (1990) *Interisland movements of fruit bats* (Pteropus mariannus) *in the Mariana Islands*. Washington, DC: Smithsonian Institution.

Wilson, J.D., Anderson, R., Bailey, S., Chetcuti, J., Cowie, N.R., Hancock, M.H., Quine, C.P., Russell, N., Stephen, L., & Thompson, D.B.A. (2014) Modelling edge effects of mature forest plantations on peatland waders informs landscape-scale conservation. *Journal of Applied Ecology*, Volume 51, 204–213.

Wirth, R., Meyer, S.T., Almeida, W.R., Araújo, M.V. Jr, Barbosa, V.S., & Leal, I.R. (2007) Increasing densities of leaf-cutting ants (*Atta spp.*) with proximity to the edge in a Brazilian Atlantic forest. *Journal of Tropical Ecology*, Volume 23, 501–505.

Zanette, L., Doyle, P., & Trémont, S. M. (2000) Food shortage in small fragments: evidence from an area-sensitive passerine. *Ecology*, Volume 81, 1654–1666.

8 Non-native Species Management

8.1 Introduction.

There might seem little to connect Japanese knotweed *Fallopia japonica* in Canada, grey squirrels *Sciurus carolinensis* in the UK, Asian gypsy moths *Lymantria dispar* in the USA, and cane toads *Rhinella marina*, in Australia, but these species exemplify an increasingly common phenomenon: the introduction of **non-native species** through human actions. Movement of species from their **native range** to novel environments, often thousands of kilometres from where they evolved, is becoming more frequent year-on-year as the world becomes ever more interconnected.

In cases where non-native species not only arrive, but go on to survive and thrive in their new environment, this can fundamentally alter that area's ecological interactions. Sometimes introductions have negative impacts on native species that can turn out to be catastrophic, as is exemplified by the multiple extinctions that have resulted from the introduction of black rats *Rattus rattus* especially to islands. In other circumstances, however, non-native species can have beneficial effects by providing valuable new food sources or hosts for rare native species. The frequency of species introductions and the broad spectrum of possible impacts mean that few topics in Applied Ecology are so important, and so inherently complex, as is the management of non-native species. Perhaps this complexity goes some way to explaining the biggest criticism of the field of invasion ecology: the lack of effective translation of scientific research to applied management (Hulme, 2003).

This chapter introduces non-native species and considers why they are such an important part of Applied Ecology, before discussing impacts (positive and negative), monitoring, and management. Numerous examples are given throughout and areas of current debate, such as when management of non-native species might be counter-productive, are discussed. Finally, the various problems that this topic creates for Applied Ecologists and ways of optimizing links between research and management are considered.

8.2 Non-native species: key concepts and questions.

Ecosystems comprise collections of interacting populations of different species called communities. Species within such communities are often categorized as native or non-native. Native species are typically regarded as those species that occur naturally within a specific geographical area, whereas non-native species are those that have been introduced artificially, either deliberately or accidentally as a by-product of human activity and trade, through a process called **translocation** (Section 8.3).

In reality, splitting species into 'native' and 'non-native' is rarely this clear-cut. Species that are colonizing new areas naturally, such as the tree bumblebee *Bombus hypnorum* in the UK, are considered by some as 'invaders' (e.g. Jones & Brown, 2014). This is especially true when geographical spread is facilitated by human-induced climatic change or habitat alteration. Conversely, species that arrived in a new area many years ago, such as honey bees *Apis mellifera* in North America 500 years ago, or even the dingo *Canis lupus dingo* in Australia 4000 years ago, are often seen as a valued part of an area's 'natural' biodiversity (e.g. Schlaepfer et al., 2011). This is particularly likely when a species becomes **naturalized** so that it occurs in wild self-sustaining populations comparatively quickly, or if it provides a valuable ecosystem service, such as pollination in the case of the honey bee. Hot Topic 8.1 discusses some of the terminology and perspectives relating to non-native species.

8.2.1 'Nativeness': an academic debate with applied implications.

Although the concept of nativeness—whether a species occurs naturally within a given ecological community—feels like an intuitive one, in practice there is a lack of consensus about the demarcation between 'native' and 'non-native'. The definition of a species as either non-native or native depends not only on its geographical history, but how that history fits with a specific date. To add yet another level of complexity, there are also cases when the history of a species in an area is unclear, as in the case of the European perch *Perca fluviatilis* in Ireland; such species cannot easily be defined as native or non-native. As there are often very different laws and policies for native and non-native species, this is far from simply being an academic debate.

Temporal scale

One key question when considering nativeness is how long a species should be present in an area before it is regarded as native. Within the many possibilities are two common suggestions:

- **The last glacial maximum:** species located in an area before the deglaciation at the end of the last ice age around 8000–10,000 years ago are regarded as native; post-ice age arrivals are considered non-native.
- **The year 1500:** species that occurred in an area before this date are considered native (**archæophytic** species), while those arriving after this are considered non-native (**neophytic** species).

The first definition encompasses the concept that species move around the globe naturally, and especially in response to changes in climate and **geographical barriers**. For example, the United Kingdom was joined to mainland Europe by a temporary land bridge, called Doggerland, in the last ice age, while Siberia was joined to Northern America via the Beringia land bridge across the Bering Straight. This result of sea level change meant land-dwelling species, such as amphibians, could move between major landmasses without restriction (Case Study 8.1).

The second definition seems to use an arbitrary date, but in fact 1500 AD is around the time when substantial global movement of humans (mostly Europeans) first occurred. Accordingly, this marks the division in ecological history where the start of the era of trade, exploration, and widespread human movement began the process of accidental and deliberate movement of species much more widely, and in much higher numbers, than previously. It is worth noting that this date can vary between locations, depending on human activity.

HOT TOPIC 8.1

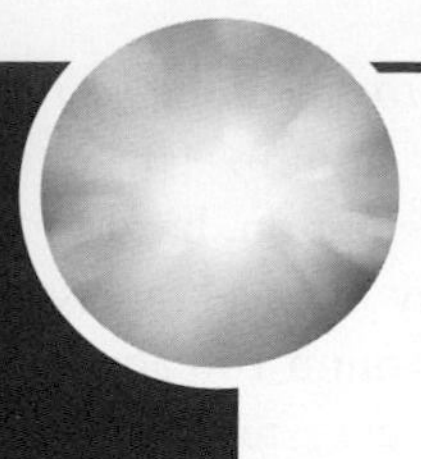

Invasion terminology and perception of non-native species

There are many terms for 'non-native species' that are essentially synonymous: 'alien', 'non-indigenous', 'introduced', and 'exotic' are especially common. However, one common term, 'invasive', has a very different meaning. It refers to a non-native species that has spread widely and rapidly within its introduced range, increased substantially in abundance, and is now regarded as a pest because of negative interactions with non-native species and/or damage to the wider environment. Accordingly, there are many non-native species that are not invasive, and these terms should not be confounded. In practice, and especially in the media, they are used interchangeably and this can cause problems in managing these species (and in managing scientific and public perceptions of them).

The damage that non-native species can cause in their new introduced range is evident by invasion nomenclature; for example, species translocation is sometimes referred to as 'biological pollution' or being a 'biological threat'. While there is no doubt that that the introduction of non-native species can have a significant detrimental effects on native biota (Section 8.4), there are potential benefits and occasions when translocation becomes an ecological opportunity. The pejorative nature of the word 'invasion' can, in part, help to mask the more complex aspects of the invading species' ecology.

Consideration of the entire spectrum of impacts is a necessary prerequisite in formulating objective and justifiable policies and management initiatives. There needs to be a move away from a sole focus on negative impacts and biotic resistance to recognize that facilitative interactions can also exist and a reworking of classification systems where only 'risk' is considered—for example, where species are graded according to their impact from one (mildly negative) to five (severely negative) (Fuller, 1991).

QUESTIONS

How important is language in ecology, both for non-native species and generally?

REFERENCES

Fuller, J.L. (1991) *The Threat of Invasive Plants to Natural Ecosystems*. MPhil thesis, University of Cambridge, UK.

Spatial scale

In addition to the question of the time period a species has to be in an area to be classified as native, there is also the issue of spatial scale. If a species' range begins to extend into a contiguous area, perhaps because of climate change, loss of a geographical barrier, or the creation of a corridor allowing movement (Section 7.6.2), then this movement is typically regarded as range change. Range change can be the consequence of natural processes or it can be human-mediated, for example, as a result of habitat alteration.

Distinguishing range changes from introductions can be straightforward. The movement of the brown tree snake *Boiga irregularis* from Australia to the Island of Guam in the late 1940s or early 1950s is a clear example of a non-native species introduction. In this case, individuals were accidentally transported several thousand kilometres over water on cargo planes. To take a completely different example, the arrival of collared doves *Streptopelia decaocto* in the UK from mainland Europe and the Middle East, which occurred at about the same period, is typically regarded an example of range change with individual birds moving only a short distance.

Between these two clear ends of the continuum, there is a complex set of examples that ask a difficult question of ecologists—how far does a species have to move before it is considered to be non-native in its new environment? What is certain is that a species does not necessarily have to come from another country to be considered non-native. For example, lake trout *Salvelinus namaycush* are native to the Great Lakes, but considered non-native in Yellowstone Lake in Wyoming, while hedgehogs *Erinaceus europaeus* are native to mainland Scotland, but viewed as non-native on offshore Scottish islands.

Ecological divergence and naïve behaviours

One way of circumventing the debates regarding 'how long is too long' and 'how far is too far' is to consider ecological divergence, rather than temporal or spatial scales. This shifts the focus onto:

- The difference between the new species and the native ecological community in terms of taxonomy, ecology, and behaviour.
- The similarity of the novel environment the species now inhabits to its original (native) range (Hettinger, 2001).

Ecological divergence summed up by comparing community and environment is arguably more useful in categorizing non-nativeness in an applied sense than some arbitrary rule-of-thumb based on the distance between the native and new ranges.

An alternative strategy is to look at species interactions. One of the reasons that non-native predators can have such a profound effect in a new ecosystem is because their prey is naive: individuals simply do not know how to avoid predation because they have not evolved suitable anti-predation strategies (Section 8.4). Carthey & Banks (2012) have taken the interesting position of flipping this concept to argue that if native prey species have learnt tactics to reduce predation by a non-native predator, that predator is no longer new enough to be considered non-native.

Cryptogenic species

When a species is deliberately introduced into a new area, or escapes from captivity, the date of release is usually fairly well known. Such species can thus be defined as non-native with relative ease. In some cases, however, it is not clear whether a species is native to a geographical area or whether it has been introduced. Such species are called **cryptogenic** species (crypto from the Greek *kryptos* meaning secret, and *genic* from the Latin meaning origin). The number of cryptogenic species is, by definition, hard to quantify. However, a systematic study in San Francisco Bay (USA) suggests that there are around 100 aquatic plants, animals, and protists in that ecosystem that are cryptogenic, in comparison with around 300 species that are known to have been introduced (Carlton, 1996).

Cryptogenic species usually occur as a result of one of the following scenarios:

1. A species is known to be 'new' in an area, but it is unclear whether it is a natural colonist or has been translocated by humans. For example, it is not known whether the grapsoid crab, *Brachynotus sexdentatus*, in France is expanding naturally from the Mediterranean or has been introduced.
2. A species is known or suspected to have been introduced by humans to a new area, but the timescale is unclear (such that it could be classified as native or non-native). An example is the sycamore tree *Acer pseudoplatanus*, which could have been introduced to the UK before 1500 AD (and regarded as native if that date is used as the demarcation between native and non-native; see 'Temporal scale' section) or soon afterwards (such that it would be regarded as non-native).
3. Individuals of a species are known or suspected to have been introduced by humans to a specific location, but it is unclear whether the species occurred there historically (effectively an unofficial reintroduction: Figure 13.1) or has never been in the area before (a straightforward introduction of a non-native species). An example of this is the European eagle owl *Bubo bubo* in the UK: the individuals currently breeding in northern England are likely to be captive escapees, but there is some evidence in the fossil record that the species was once native (Stewart, 2007).
4. There is simply uncertainty as to the provenance of a particular species in an area, with no evidence of an introduction, but equally no evidence of long-term presence in the fossil record (e.g. common periwinkle *Littorina littorea* on the eastern seaboard of North America).

As there are very different attitudes and policies to native and non-native species, it can be important to try to resolve cryptogenic status. Until then, the way such species are viewed from management and legal standpoints is complex. For example, the European eagle owl is protected under UK law where it is breeding in a natural state (Wildlife and Countryside Act, 1981), but the species is not included on the British Ornithologists' Union list of native (or naturalized) birds. Moreover, if it was classified as native, the small population size would probably afford it special protection and potentially make it a target for conservation measures, whereas if it was classified non-native it might be subject to population control measures. Some of the other considerations associated with cryptogenic species are considered for pool frogs *Pelophylax lessonae*, as explored in Case Study 8.1.

8.2.2 The importance of non-native species in Applied Ecology.

There are several reasons why invasion ecology is an important aspect of Applied Ecology:

1. **Prevalence and abundance:** non-native species inhabit all continents and often occur in substantial numbers. Worldwide, rates of invasion are climbing each year. The effect of this on the cumulative number of non-native species is shown in Figure 8.1. This process is referred to as **biological homogenization**, as non-native species are fast becoming the rule rather than the exception. For example, using the year 1500AD to demarcate native and non-native species, it is estimated that 28% of Canadian flora and 47% of New Zealand flora is non-native (Heywood, 1989; Green, 2000).
2. **Diversity:** non-native species are extremely diverse, both in terms of taxonomic group (microbes, plankton, algae, plants, fungi, insects, fish, amphibians, reptiles, mammals, and birds are all represented, see Figure 8.2 for examples) and also in terms of their ecological niche. This means that invasion ecology is complex theoretically and even more challenging in the field.
3. **Risk:** not all geographical locations or habitats are equally at risk of invasion, and not all non-native species are equally likely to create ecological problems. Identifying high-risk sites and high-risk species requires ecological knowledge and insights.
4. **Impacts:** non-native species interact with native species in complex and sometimes unpredictable ways. Invasive non-native species are recognized as a major threat to biodiversity (Gurevitch & Padilla, 2004) and can have a profound impact on the environment (Pimentel et al., 2005).
5. **Preventing problems:** preventing introductions typically requires understanding of how species spread, while preventing problems immediately post-colonization requires routine monitoring and quick effective action to prevent spread when non-native species are discovered.
6. **Management:** control of non-native species can be necessary to stop their geographical spread, to reduce or eradicate populations, or to reduce or buffer impacts. Such mitigating action requires detailed knowledge of species' ecology and ecological interactions if it is to be maximally effective.

FURTHER READING FOR THIS SECTION

Defining nativeness: Carlton, J.T. (1996) Biological invasions and cryptogenic species. *Ecology*, Volume 77, 1653–1655.

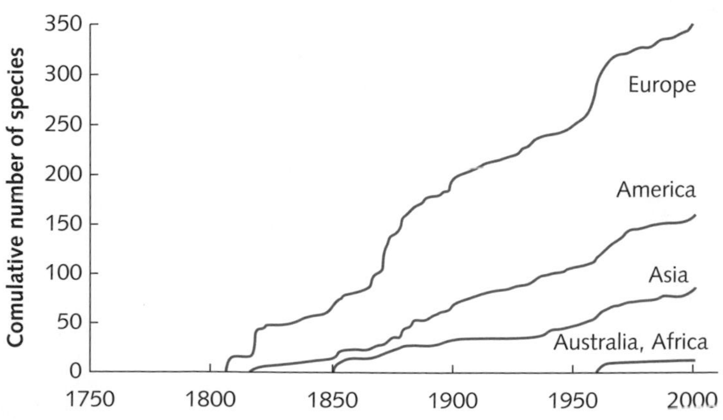

Figure 8.1 The cumulative increase in the number of non-native plants in one country (Czech Republic) from different continents.

Source: Reproduced from Pyšek, P. et al. (2003) 'Czech alien flora and the historical pattern of its formation: what came first to Central Europe?' *Oecologia*, 135, 122–130, with permission from Springer-Verlag.

CASE STUDY 8.1

To cull or to conserve? Practical implications when species are 'of unknown origin'

The northern pool frog *Pelophylax* (formerly *Rana*) *lessonae* (Figure A) has a cosmopolitan distribution across Europe, but until recently was considered to be non-native in the UK. A few individuals occurred in isolated areas in the early 1900s, but were considered either to be atypical native common frogs *Rana temporaria*, or non-native individuals from other non-native species. This was not illogical since other so-called 'green' or 'water' frogs were known to have been introduced; for example, edible frogs *Pelophylax* (formerly *Rana*) *esculentus* were released in Norfolk in 1837.

Pelophylax frogs were thought to have become **extirpated** (locally extinct) in the UK, but individuals were rediscovered in Norfolk in the mid-1960s. These were considered to be descended from non-native populations and, importantly from an applied perspective, were candidates for listing under Schedule 9 of the UK's Wildlife and Countryside Act (1981). This schedule lists non-native species that are legally permissible to cull as a pest. However, there was sufficient uncertainty as to the species' status to investigate further. Accordingly, as part of the UK's **Biodiversity Action Plan (BAP)** initiative, research was undertaken to establish nativeness or non-nativeness definitively.

Figure A Northern pool frog, Pelophylax lessonae.

Source: Image courtesy of Viridiflavus/CC BY-SA 3.0.

The investigation was prompted by historical documents, which suggested that *Pelophylax* frogs were already widespread in Cambridgeshire and Norfolk before the earliest known introductions in the 1830s. These records were cross-referenced with museum specimens, which were found, upon re-examination, to be pool frogs. In addition, the researchers:

- found archaeozoological evidence suggesting pool frogs were present in the UK 1000 years ago;
- identified other isolated native populations of pool frogs in Europe (e.g. Uppsala, Sweden);
- compared the genetics and call acoustics of UK individuals with those of different European populations, which showed UK pool frogs were strongly related to Scandinavian pool frogs rather than the central and southern European populations known to be the source of the original UK pool frog introductions

One of the most compelling pieces of evidence was the UK and Scandinavian populations having a genetic separation equating to around 10,000 years' isolation. This opened the possibility of invasion across the temporary land bridge (called Doggerland), which was present during and immediately after the last ice age, with populations in Scandinavia and the UK then being divided geographically by the rising North Sea.

As a result of this analysis, the pool frog was redefined as a native species in the UK using the end of the last ice age as the demarcation between native and non-native. Crucially, that meant the UK/Scandinavian individuals were part of a clade (population group) that was highly threatened. Far from being listed as a pest on Schedule 9, then, it was considered a conservation priority. By this time, however, the species had become extirpated, and the protection that would otherwise have been afforded by Wildlife and Countryside Act (1981) could not be invoked. A species **reintroduction strategy** co-ordinated by the statutory regulator Natural England (then English Nature) (Section 13.3) was undertaken and there are now small, legally protected populations of the species in the UK.

It is interesting to reflect that the species itself and its ecological interactions (positive and negative) remained constant throughout this process. The decision as to whether Applied Ecologists had the task of eradicating or conserving was dictated purely by which side of the native/non-native boundary the species was assessed to be. Moreover, the legal protection given to pool frogs has only been given to individuals of the northern clade, despite those individuals being synonymous ecologically with non-native pool frogs.

FURTHER READING

Pool frog nativeness debate: Beebee, T.J.C., Buckley, J., Evans, I., Foster, J.P., Gent, A.H., Gleed-Owen, C.P., Kelly, G., Rowe, G., Snell, C., Wycherley, J.T., & Zeisset, I. (2005) Neglected native or undesirable alien. Resolution of a conservation dilemma concerning the pool frog *Rana lessonae*. *Biodiversity and Conservation*, Volume 14, 1607–1626.

Pool frog reintroduction plan: Buckley, J. & Foster, J. (2005) *Reintroduction strategy for the pool frog Rana lessonae in England*. Peterborough: English Nature Research Report 642.

Natural England pool frog species action plan: Available at: http://www.arc-trust.org/pool-frog

8.3 Translocation.

Translocation refers to the process by which a species is introduced to a new environment. Although non-human-mediated introductions do occur as the result of storms or sea currents, most introductions (especially post 1500AD) are human-mediated. These can be subdivided into accidental or intentional, as illustrated in Figure 8.2.

8.3.1 Invasion pathways.

There are many ways that species are introduced across the globe, but these can be broadly divided into six principal invasion pathways (Hulme et al., 2008):

1. **Unaided:** a species introduces itself to a new area via wind or sea currents.
2. **Release:** a species is deliberately released into a new environment either from captivity (planned or unplanned (Section 12.4.4)) or within a biological pest control programme (Section 9.4.3).
3. **Escape:** a species escapes from captivity or cultivation.
4. **Contaminant:** a species is introduced inadvertently as contaminant within a product such as timber.
5. **Stowaway:** similar to contaminant, but whereas that requires some kind of host species, a stowaway can be transported inside or attached to a container or within packaging (e.g. rats on board ships).
6. **Corridor:** a species moves along a linear feature, such as a river or road (Section 7.6.2).

The relative importance of these pathways varies between taxa: vertebrate pathways tend to be deliberate releases, invertebrates as contaminants or stowaways, and plants as escapes. Pathogenic microorganisms and fungi are generally introduced as contaminants of their host (which may be a vertebrate, an invertebrate, or a plant).

Trade and the movement of goods across the globe provides many opportunities for species to **'hitch-hike'**, potentially over great distances. This includes species shipped as goods themselves (garden plants, species for the pet trade, etc.) or species shipped within goods, such as within the soil around plant roots. Another common way non-native species can invade a new area inadvertently is through the importation of timber, as happened in the case of Dutch elm disease caused by the beetle *Ophiostoma*, and more recently, ash dieback caused by the fungus *Hymenoscyphus fraxineus* in Europe (Case Study 7.2) and the Asian longhorned beetle *Anoplophora glabripennis* in the USA. Transporting vehicles are another primary vector, as in the case of ship ballast, discussed in more detail in Hot Topic 8.2.

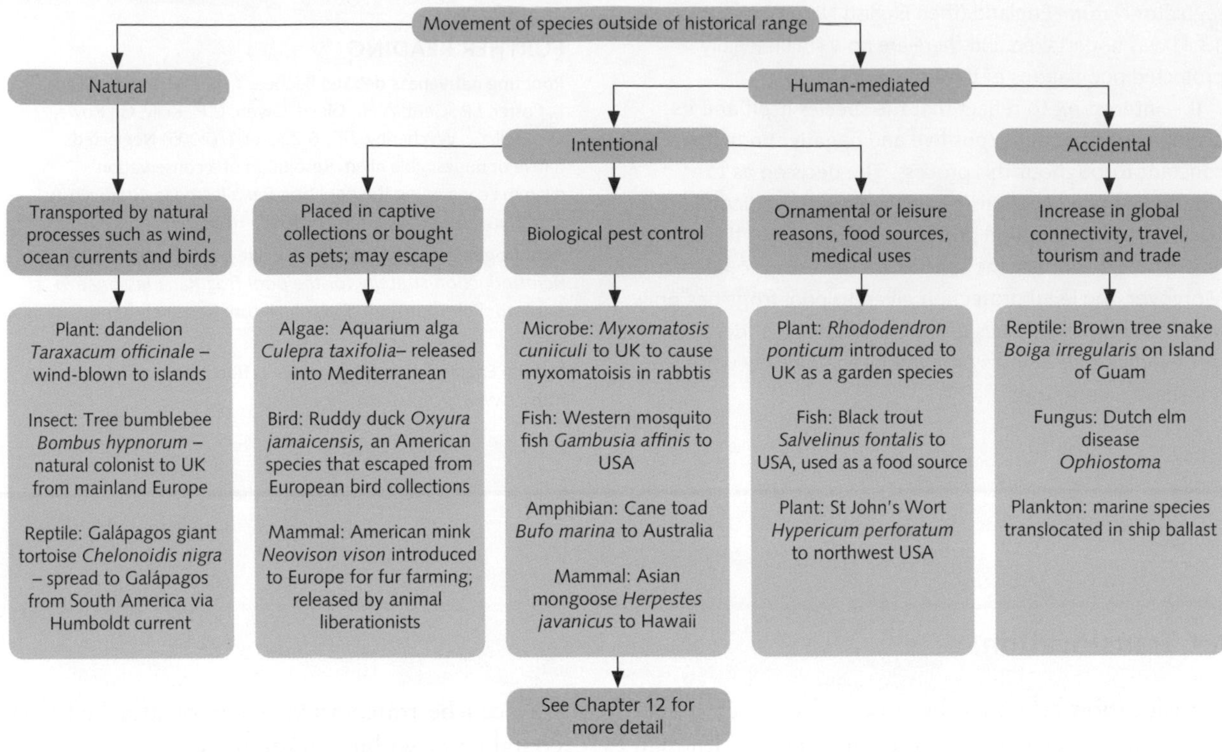

Figure 8.2 Classification of species translocation – motivations and pathways.

While **commercial trade** is, globally, the primary driver of increasing biotic homogenization, especially given the rising volume of air and ship transport, movement of people for tourism is another way in which non-native species are spread, for example, on footwear. It is also possible for species to move through military mobilizations, as happened when the brown tree snake *Boiga irregularis* was introduced to Guam, a key USA military hub, via ship or airplane cargo.

8.3.2 Likelihood of establishment.

A non-native species has three major obstacles to overcome if it is to be a successful 'invader'.

1. First, it must **arrive** through the process of translocation along a suitable pathway via a suitable vector.
2. It needs then to **survive** the initial introduction stage.
3. Finally, it needs to attain enough of a foothold in the new environment to **thrive.**

According to the **tens rule** of Williamson & Fitter (1996), only one in ten imported species appears in the wild and only one in ten of those become self-sustaining. Although the tens rule has been challenged as being too conservative and under-estimating the number of introduced species that do establish (e.g. Jeschke & Strayer 2005), it serves as a reminder that establishment in the wild is the exception rather than the rule. Figure 8.3 shows the different possible outcomes for a species following translocation.

Fundamentally, for a species to become established in a new area, there needs to be an exploitable niche. This means that although abiotic conditions, including climate, topography, and soil properties might not be identical to those in the native range, they have to be within the species' tolerance range (Section 4.3.1). Biotic conditions also have to be appropriate: there must be suitable food resources available, and predation and disease levels must not be prohibitive.

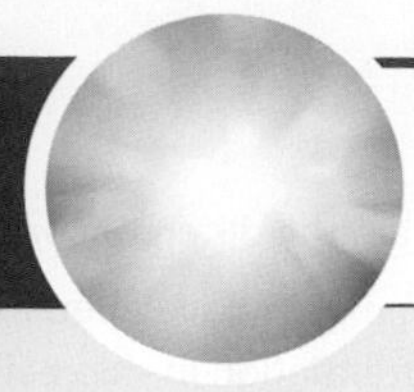

HOT TOPIC 8.2

Hitch-hiking in ship ballast

Translocation of species often occurs via 'hitch-hiking'—unintentional transport of species through movement of goods and people (Section 8.3), especially along shipping routes. In the marine environment, one primary **vector** of species is shipping, either through species attaching to ship hulls or, more commonly, through ballast water.

Because ships are engineered to carry substantial loads, they sit high in the water when empty. This makes them unstable and prone to capsize, especially in heavy seas.

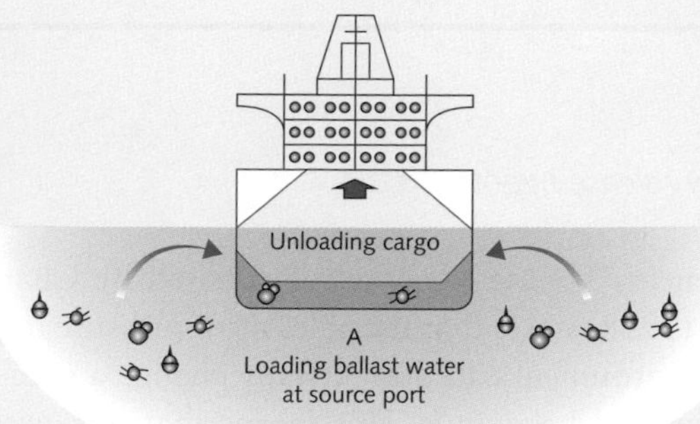

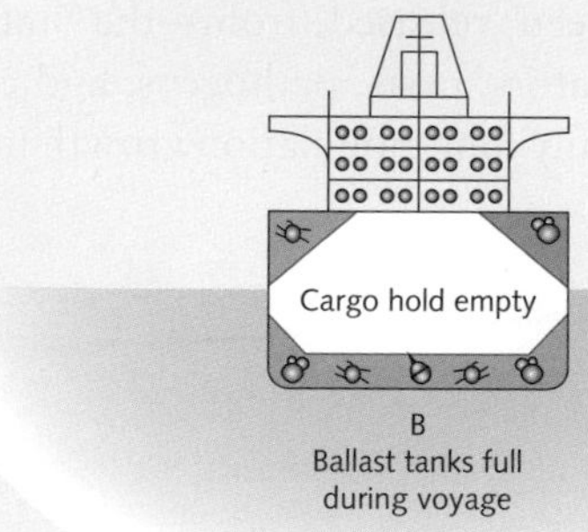

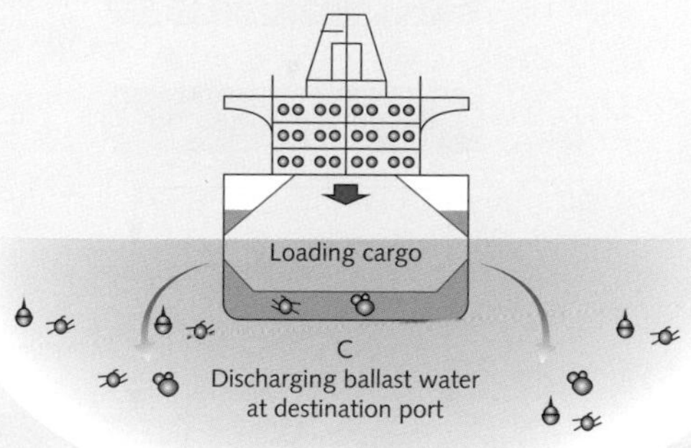

Figure A Translocation of marine species via ship ballast.

Source: Image adapted from Global ballast water management programme.

To prevent such problems, ships often pump sea water into voids in the hull around the (empty) cargo holds when travelling unladen or only partially laden. Because water is usually pumped into ships directly from the sea (usually in shallow ports or onshore anchorages), aquatic organisms usually enter along with the water. When the water is discharged at the destination port—often a substantial distance from the port of origin—these organisms are likewise discharged into a new environment, where, over time, they have the potential to become invasive (Figure A). The same process can happen in large freshwater environments, such as the Great Lakes in the USA/Canada.

The International Convention for the Control and Management of Ships' Ballast Water and Sediments was adopted on 13 February 2004 and came into force 12 months later after ratification by 30 countries (representing 35% of world merchant shipping tonnage) with the remit to prevent, minimize, and ultimately eliminate the transfer of harmful aquatic organisms and pathogens through the control and management of ships' ballast water.

There are several methods to prevent or reduce the likelihood of organism translocation through ship ballast:

- **Physical filtration**: a filter is placed over the ballast water pipe inlet. This stops larger organisms entering the ballast tanks but not microorganisms or, potentially, very small pieces of aquatic plant. The filtration slows down the ballast-loading process and filters require maintenance.
- **Centrifugal separators**: these are drums that spin fast horizontally around a central point to separate sediment (and associated species) from the water using centrifugal (outwards) force. As with physical filtration, this does not prevent microorganisms or small water-borne species entering the tanks
- **UV light**: ballast water is subject to UV light to kill microorganisms or small water-borne species—often used in conjunction with centrifugal separators.
- **Heat**: pioneered in Australia, this method involves rerouting the sea water that is heated as part of the engine cooling process to the ballast tanks. It is a feasible solution, but requires engineering changes, and there are no data on the long-term effects on the integrity of ballast tanks by carrying hot water, which has a greater capacity to cause corrosion.

- **Chemical control:** Germany is developing a biodegradable ballast water treatment chemical to kill hitch-hiking species in the ballast tanks, but cost could well be the prohibitive factor here.

A complementary scheme is the GloBallast programme, which focuses on enhancing developing countries' capacity to reduce the environmental and socio-economic risks associated with the transfer of aquatic invasive species in ship ballast water. This has established best practice and stimulated innovative ballast water management solutions in six sites that were chosen as representative of the six main developing regions of the world—South America, East Asia, South Asia, Arab Countries/Persian Gulf, Africa, and Eastern Europe. The project also developed mechanisms for compliance monitoring and enforcement.

QUESTIONS

How effective are the different control techniques?

How could techniques developed for minimizing the risk of species translation in ship ballast be applied to other translocation pathways?

FURTHER READING

International Convention for the Control and Management of Ships' Ballast Water and Sediments: http://www.imo.org/en/OurWork/Environment/BallastWaterManagement/Pages/Default.aspx

Globallast programme: http://globallast.imo.org

For an established non-native species to become invasive (Hot Topic 8.1 discusses the difference between non-native and invasive), its population has to grow and spread rapidly. This means that it needs to be able to out-compete native species in similar ecological niches to become dominant in the ecosystem. This is most likely for **generalist species** that can flourish in a range of conditions and/or r-strategist species with rapid population growth.

Enemy release hypothesis

The enemy release hypothesis proposed by Charles Elton (1958) states that the 'release' of pressure in the new environment can dramatically change a species' ecology and behaviour. This means that non-native species often thrive in novel environments because they have lost—been released from—the natural 'enemies' (i.e. predators, pests, pathogens, and competitors) that typically limit population growth in the

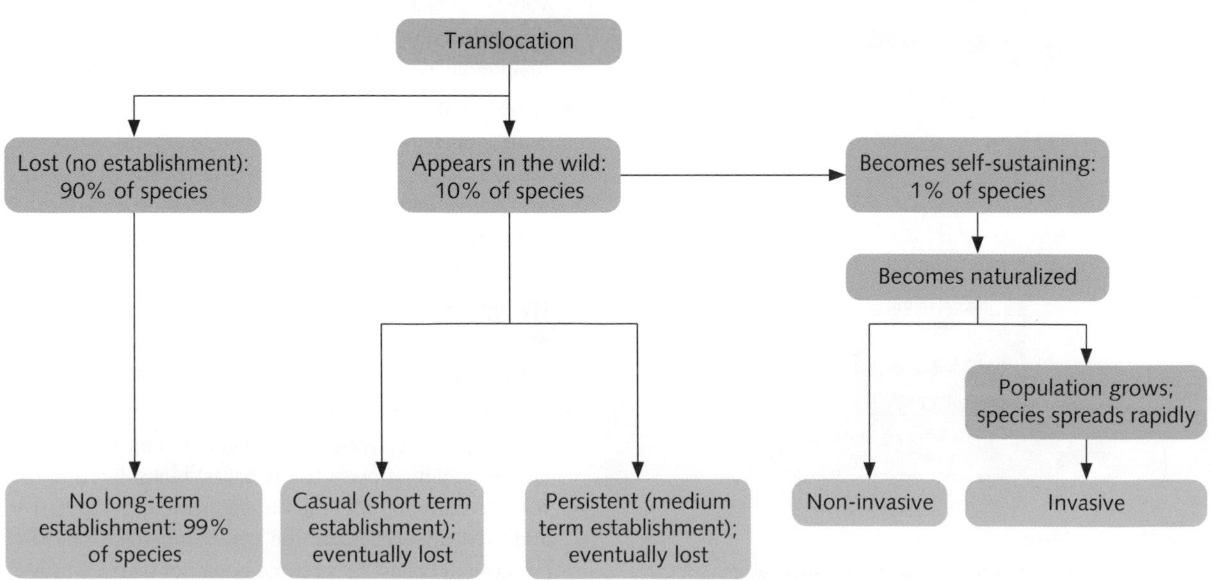

Figure 8.3 Possible outcomes for a species following translocation.

native range. However, while it is undeniably true that species will have lost most negative interactions (unless their enemies' range is large enough to cover both native and introduced ranges), this does not automatically mean species will thrive, since there will be plenty of other species in the new range to take on these roles (Section 8.4).

The snowball effect: invasion meltdown

Arrival of a non-native species can sometimes facilitate the arrival of other non-natives, either by changing the habitat of its new environment such that the conditions are more suited to new invaders (ecosystem engineering; Section 8.4.1), or through close relationships with other species from the original native range (the concept of symbiosis: Section 2.3). This snowball effect of one invasion making subsequent invasions more likely has been termed **'invasional meltdown'** by Simberloff & von Holle (1999).

One example is caused by parasitic interactions, such that when a new host species is introduced its parasites can also be introduced. In some cases, a non-native species that has no measurable impact on native species could facilitate the arrival of another non-native species that has considerable impact. An example is the Pacific oyster *Crassostrea gigas*, which had no measureable impact when it was introduced as a non-native in the Mediterranean directly, but its arrival facilitated invasion by two parasitic copepods, *Mytilicola orientalis* and *Myicola ostreae*, which have caused other native Crustacea to decline (Torchin et al., 2002, and references therein). Another example is avian malaria in Hawaii, when both parasite and vector were introduced with catastrophic consequences, as discussed in the Online Case Study for Chapter 8.

You can find the online case study at www.oxfordtextbooks.co.uk/orc/goodenough

8.3.3 Invasion risk.

Invasion risk may be thought of as the probability of non-native species becoming established in a given area. It should be recognized that different geographical locations, and habitats, will have different invasion risks. Invasion risk is generally the culmination of three different factors:

1. **Prevalence of translocation opportunities:** how many species/individuals have the *potential* to be released. Locations that are trade or tourism hubs, or near areas where non-native species have been introduced in captivity, have a higher invasion risk simply because there is more potential for invasion (Hot Topic 8.2).
2. **Propagule pressure:** how many species/individuals *are* released. This is a composite measure of the number of individuals of a species released into a new environment. It is based upon three interlinked components—the absolute number of individuals in any one release event (**propagule size**), the number of discrete release events (**propagule number**), and the number of taxa or genotypes introduced (**propagule richness**). Propagule richness usually relates to the diversity of pathways. For example, an area that is an air and shipping cargo hub, that is frequented by tourists from many different geographical areas, is likely to receive a greater range of non-native species, purely as a result of the diversity of translocation pathways and origins. Propagule size and number, by contrast, tend to relate to vector prevalence (the number of tourists, the amount of shipping, etc.).
3. **Environmental vulnerability:** how vulnerable or resilient an area's environment is to invasion.

Because invasion risk comprises several different factors, and each factor is itself multifaceted, invasion risk can be a hard topic to comprehend (and, indeed, hard to model for any given species). A schematic of invasion risk is given in Figure 8.4, to clarify the role of each factor.

Environmental vulnerability

Environmental vulnerability is determined by how resilient the native community is to invasion; in other words, how good it is at buffering invasions. There are several traits that make some environments especially vulnerable to invasion:

- **Island ecosystems:** islands are often cited as being more prone to invasion than mainland areas, although this in not the case for all taxa: birds invade islands and mainland areas equally (Sol, 2000). The general pattern could link to the num-

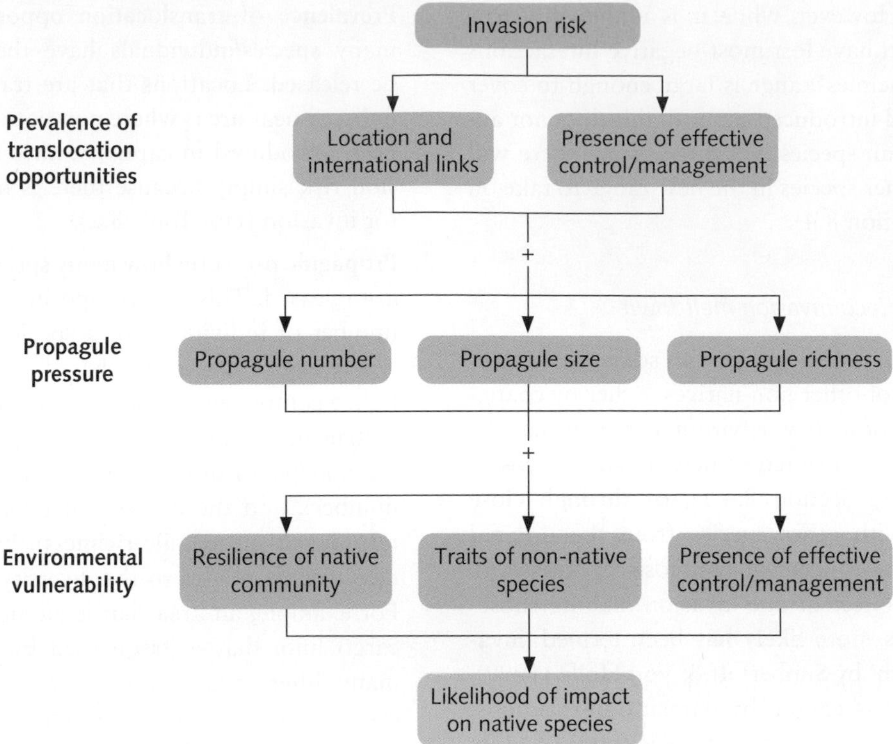

Figure 8.4 Schematic of invasion risk.

ber of empty niches as island ecosystems tend to be less species-saturated than mainland environments at the same latitude. This is especially true for **oceanic islands** (islands that have never been connected to the mainland).

- **Disturbance**: disturbed, nutrient-enriched, or polluted ecosystems are highly susceptible to invasion. Non-native species are often better able to tolerate disturbance due to their generalist ecology and their high adaptability. It is worth noting that this could itself affect assessment of the damage caused by non-native species, since these environmental traits are also associated with native species decline. Thus, the dominance of non-natives within a species community may be, initially at least, a consequence of ecosystem disturbance, not the driving force behind it (Chabrerie et al., 2008).
- **Type of habitat and connectivity**: the number of non-native plants and insects is highest in urban areas and riparian habitats (i.e. the marginal land between aquatic and terrestrial ecosystems, including river beds). This is probably because of escapees from urban gardens and the ease of spreading through a riparian ecosystem. (For example, the seeds of Himalayan balsam *Impatiens glandulifera* are readily carried downstream, so the species spreads along river banks very rapidly.) Such examples highlight the importance of the study of landscape ecology from an applied perspective (Chapter 7), and that wildlife corridors, although important for conservation, can cause problems by facilitating the spread of non-native species.

Risk modelling

Understanding translocation pathways is the first step in being able to model risk in the future. When combined with understanding of the areas that are especially vulnerable to both initial establishment and the impacts of non-native species (Section 8.4), Applied Ecologists can start to predict and model risk in a way that combines both **invasion risk** (probability of invasion) and **impact risk** (magnitude of likely effects).

Areas with especially high impact risk include ecosystems that are finely balanced and that have few native predators. For instance, oceanic islands are often home to a highly specialized biodiversity, including many species that have lost anti-predator defences. The most obvious example of loss of anti-predator defence is in flightless birds on islands with no ground predators. Extant examples include kiwi *Apteryx* in New Zealand, while extinct examples include the Saint Helena hoopoe *Upupa antaios* (extinction largely due to introduction of black rats *Rattus rattus*), and the Stephens Island wren *Xenicus lyalli* (extinction largely due to the introduction of cats—probably not just the lighthouse keeper's cat, as legend has it, but certainly a very few individuals). One of the other reasons why Applied Ecologists tend to be so protective of oceanic islands is that they are often centres of endemism, containing a high number of species found nowhere else. This means that if a species is driven to extinction on that one oceanic island, it becomes extinct (at least in the wild) in the biosphere as a whole.

Risk modelling can be extended using a map-based approach called **species distribution modelling**, which is similar to the Climate Envelope Modelling (CEM) approach outlined in Section 7.5.1. Briefly, CEM involves mapping where a species currently is, establishing what the climate is like in that area, and then running a series of simulations to where the species would move under climate change. A similar approach can be applied to non-native species by considering where they are now (native range and current introduced range, if any), considering what the abiotic conditions are like, and then mapping other locations with similar abiotic factors. These would be potential risk areas. The level of risk for each area can be weighted according to proximity and connectivity with areas in the species' current range. In this way, if a map was being devised for the invasive European plant *Rhododendron ponticum*, the highest risk areas would be those with similar abiotic characteristics to where rhododendron is already found, that are close to the current range, and that have strong trade, transport, or infrastructure links with areas in the current range. The approach can be powerful, but is limited in that it requires good-quality data to be available, and it takes time, effort, and knowledge to undertake the modelling.

At a more local scale, areas where the establishment of non-native species is likely to have a more than proportionate effect include those that:

- **Support rare species:** conservation implications are likely.
- **Are very small:** it is then more likely that native species have small populations, so any negative impacts of non-native species are more likely to cause native species to dip below the minimum number that is needed for that population to continue (the **minimum viable population** concept discussed in Section 10.3.1).
- **Are very isolated:** for example, a wood surrounded by agricultural land or an isolated lake. Such locations are often termed virtual islands—they are not surrounded by the sea but are effectively 'islands' in a 'sea' of other land uses (Section 7.2.2). If native species become extirpated (locally extinct) in that area, they are less likely to be able to recolonize that area at a later time.
- **Are relatively species-poor:** likely to be less stable under the diversity–stability hypothesis.

Challenging the diversity–stability hypothesis

The concept that communities with more empty niches are more prone to invasion dates back to independent work in the 1950s by two of ecology's most influential researchers, Eugene Odum (*Fundamentals of Ecology*, 1953) and Charles Elton (*Ecology of Invasions by Animals and Plants*, 1958). Both scientists found that simple communities were more easily 'upset' than richer ones, being more sensitive to oscillations in populations of native species and more vulnerable to invasion by non-native species. This is called the diversity–stability hypothesis.

The fact that species-poor environments have a higher risk of being successfully invaded than species-rich environments makes intuitive sense, since there will be fewer potential competitor species present in the new location (fewer species overall = fewer potential competitors = greater chance of

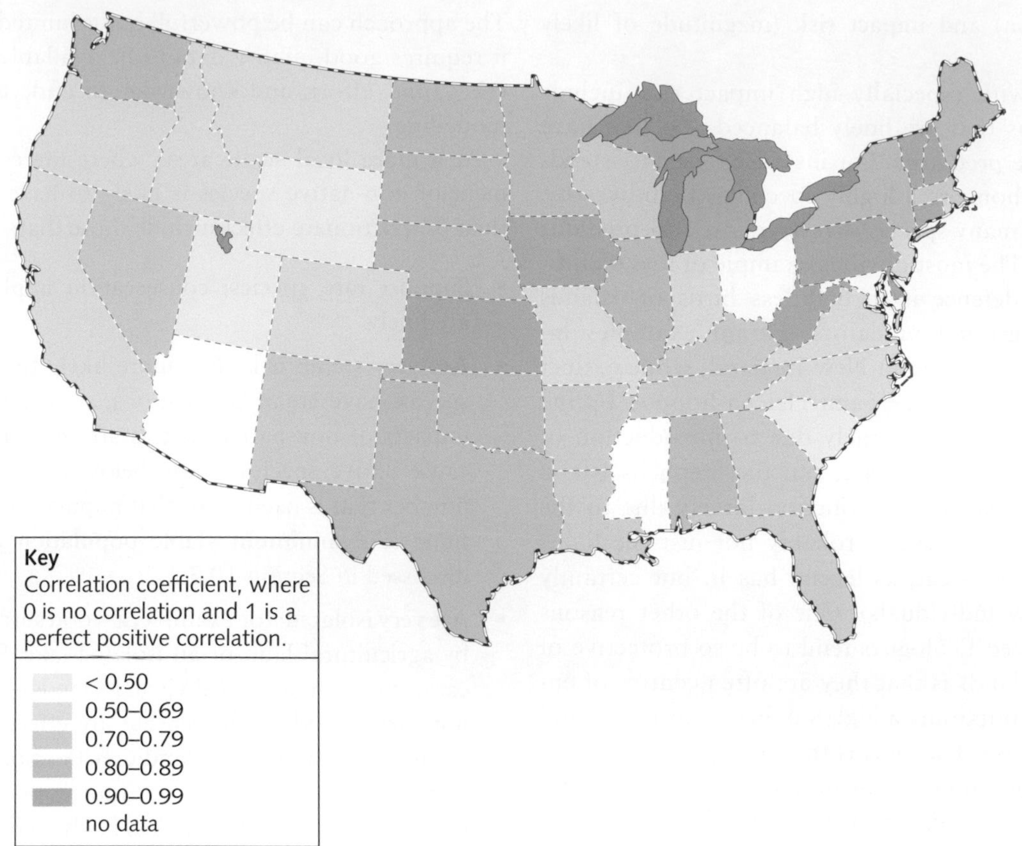

Figure 8.5 The correlation between native plant species richness and non-native plant species richness across the USA on a state-by-state basis.

Source: Data from Stohlgren, T. J., Barnett, D. T., & Kartesz, J. T. (2003) 'The rich get richer: patterns of plant invasions in the United States'. *Frontiers in Ecology and the Environment*, 1, 11–14.

successful invasion). This rule was based on repeated observations that simplified terrestrial communities were subject to reduced **species turnover** (i.e. fewer new species entering the community and fewer existing species being out-competed to the point of local extinction) relative to more diverse communities. However, although the diversity–stability hypothesis can hold on a small scale, this is not always true on large scale. For example, a study of plant richness across the USA found that the number of non-native plants increased as the number of native plants increased (Stohlgren et al., 2003; Figure 8.5). This is termed a positive correlation and is the opposite of the negative relationship the diversity–stability hypothesis would predict.

The diversity–stability hypothesis is still widely considered to be 'fact'. This might have real implications in terms of applied ecological management, as it specifically states that areas with high species richness, such as **biodiversity hotspots** (Section 10.5.3), might be relatively safe from invasion. First, this is not necessarily true, as some such locations can have a high invasion risk. Secondly, and more importantly, sites with high species richness are frequently considered to be **conservation priorities**. Together, these factors might mean that the risks to these areas and the species communities they support are being continually underestimated.

8.3.4 Understanding translocation: preventing species introduction.

Understanding translocation pathways and invasion risk is fundamental to undertaking the most basic (and most effective) form of non-native species control—preventing introductions.

In many countries there are restrictions on release of non-native species. In the UK, for example, Section 14(1) of the Wildlife and Countryside Act (1981) makes it an offence to release or allow to escape into the wild any animal of a kind (generally taken to mean species or subspecies) not ordinarily resident in, and not a regular visitor to, Great Britain in a wild state. Offences can carry an unlimited fine commensurate with the offence and/or two years' imprisonment. Within Europe more generally, Article 22(b) of the Habitats Directive requires member states to ensure that deliberate introduction of non-native species is regulated or, if necessary, prohibited, so as not to damage natural habitats or wild native flora and fauna, while Article 11 of the EU Directive on the Conservation of Wild Birds 1979 (79/409/EEC; the 'Birds Directive') requires member states to ensure that any introduction of a bird species which does not occur naturally in a wild state within their territory does not prejudice the local flora and fauna. Many other countries or country groups have similar policies, usually equating to a 'general no-release' approach.

While such legislation covers purposeful introductions, accidental introductions are, by their very nature, harder to prevent. Some countries, especially those affected by non-native species in the past or that are especially vulnerable (e.g. island ecosystems), have strict protocols at entry points. Examples of restrictions include:

- Limitations on goods tourists can bring in (foodstuffs, products made from rattan or wood etc.).
- Regulations on shipping of items such as soil.
- Quarantine for imported animals and sometimes plants to reduce risk of non-native species hitch-hiking.
- A 'clean footwear' policy for travellers (can also be extended to items such as tents or fishing equipment) to reduce spread of seeds, spores, or plant fragments.
- Use of sniffer dogs at port/airports.
- Visual inspection of luggage and shipping containers.
- Fumigation of cargo and aircraft cabins/holds, especially for aircraft coming from tropical areas where mosquito-vectored diseases are common.

FURTHER READING FOR THIS SECTION

Invasion success: Colautti, R.I., Ricciardi, A., Grigorovich, I.A., & MacIsaac, H.J. (2004) Is invasion success explained by the enemy release hypothesis? *Ecology Letters*, Volume 7, 721–733.

General overview of non-native species introductions: Lockwood, J.L., Hoopes, M.F., & Marchetti, M.P. (2013) *Invasion Ecology*. Chichester: John Wiley and Sons.

General overview of non-native species introductions: Simberloff, D. (2013) *Non-native Species: What Everyone Needs to Know*. New York, NY: Oxford University Press.

8.4 Impacts of non-native species.

Significant ecological impacts (positive or negative) are only likely to occur if non-native species become established and self-sustaining in a novel environment, which is far from guaranteed (Section 8.3). Even where establishment does occur, impacts on native biota are not inevitable. Indeed, many native communities are not species-saturated (Sax et al., 2007) and new species can sometimes thrive without significant problems.

Co-existence between non-native and native species is most likely when there is little competition between them. This usually happens when an invader exploits an unoccupied niche, a situation that is most common when non-natives have few ecological and behavioural similarities with native species. Alternatively, native species may be unaffected by non-natives as a result of pre-existing adaptations, such as the generalist anti-predator defences of many toad (*Bufo*) species that secrete a toxin that is unpalatable or poisonous to would-be predators. The likelihood of native species being substantially affected by the arrival of non-native species is also decreased when:

- the non-native species has not travelled far (more likely to have **co-evolved** with similar species);
- the non-native species is in a lower trophic level (fewer negative interactions such as predation).

8.4.1 Ecological impacts: negative and positive.

It does not necessarily follow that a non-native species will affect native species in its new environment. However, impacts do frequently occur and can do so at a variety of scales and at different magnitudes. At the smallest scale are **gene-level** interactions, such as hybridization. Moving up in scale, there are **individual-level** interactions; for example, a non-native mediated change in the behaviour of a native species. At a larger scale still, non-native species can have **population-level** effects, such as competition with a native species, and even **community-level** impacts, such as herbivory-driven changes in native vegetation.

There are numerous ecological mechanisms through which a non-native species could have an effect on native species. These include predation, competition, vectoring disease or parasites, and hybridization. These different mechanisms, together with examples, are outlined in Figure 8.6. However, while impacts can be negative, and indeed often are negative, it is vital that potential positive interactions are also considered. Non-native species can act as food sources, hosts, pollinators, and seed-dispersers for native species. Examples of all these possibilities (and more) are given in Figure 8.7. Consideration of all potential impacts, both positive and negative, is vital if Applied Ecologists are to formulate objective and justifiable policies and management initiatives (Goodenough, 2010).

Although not directly relevant to Applied Ecology, it is important to note that non-native species can have profound effects on the wider environment. For example, 39–64% of non-native plant species affect hydrology and geomorphology in their new environments (Gordon, 1998); nitrogen-fixing species can also alter soil biogeochemistry. There can also be effects on human health, especially through non-native disease vectors: there is no example that is more infamous than non-native black rats *Rattus rattus*, their fleas, and the bubonic plague that those fleas vectored.

Ecosystem engineering

Non-native species can cause substantial direct modification of a new environment through alteration of the physical environment: the process of ecosystem engineering (Jones et al., 1997). In some cases, this can benefit native species. For example, the occurrence of Australian pines *Casuarina equisetifolia* behind Florida beaches seems to increase nesting by native loggerhead turtles *Caretta caretta* by blocking light pollution (Salmon et al., 1995), while non-native hornsnails *Batillaria attramentaria* have stabilized the Northeast Pacific ocean floor and facilitated colonization by native hermit crabs *Pagurus* (Wonham et al., 2005). In other instances, effects can be negative. For instance, in California, USA, soil salinization by the ice plant *Mesembryanthemum crystallinum* is inhibiting native non-halophytic plants.

In some cases, a non-native species does not affect native species through physical alteration of an area via ecosystem engineering, but simply by virtue of its presence. *For* example, on Robben Island off the coast of South Africa, stands of red-eyed wattle *Acacia cyclops* provide nest sites for African penguins *Spheniscus demersus*, a species classified as vulnerable and declining by the IUCN (Crawford et al., 1995).

8.4.2 Impact complexity.

As useful as it might be to split impacts of non-native species into 'positive' and 'negative', in reality this is usually over-simplistic. One non-native species could easily have a beneficial interaction (mutualistic or commensual relationship: Section 2.3.2) with one native species and a negative effect on another. The two examples below show how complex interactions between non-native and native species can be, and the difficulties this can cause for management strategies. Even if it is possible and

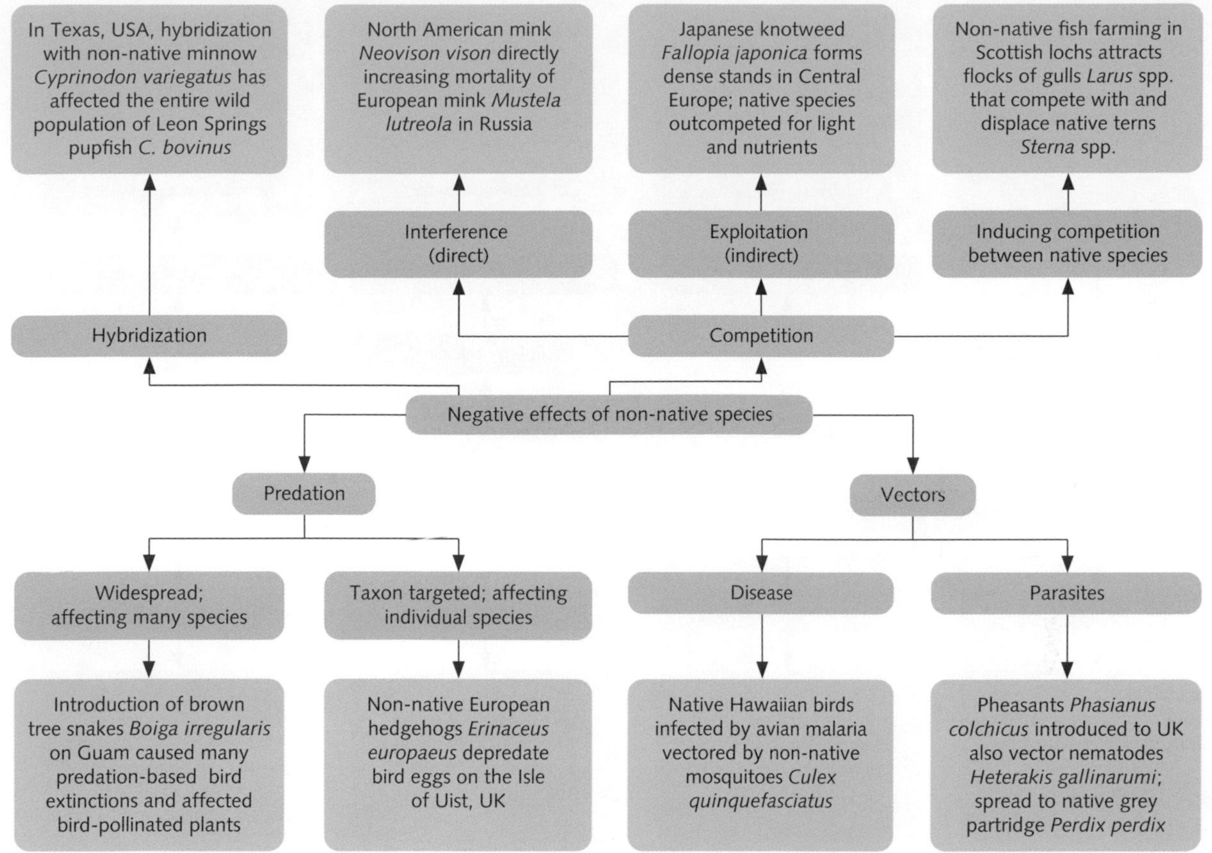

Figure 8.6 Potential negative effects of non-native species on native species with examples.

desirable to remove a non-native species, the impacts on native species can be mixed. This type of management quandary is discussed in more detail in Section 8.6.

1. In Europe, the sugar-rich nectar produced by the non-native plant Himalayan balsam *Impatiens glandulifera*, is used by native bumblebees *Bombus*, and bee abundance can increase as a result. However, in utilizing this non-native plant, bee visitation rates to native plants, such as marsh woundwort *Stachys palustris* decrease. This so-called 'pollinator switching' means that native plants receive fewer pollinator visits and have significantly lower rates of seed set (Chittka & Schürkens, 2001). The negative impact on native flora is not offset by increasing bee numbers, at least at current population levels, because Himalayan balsam can produce sugar very rapidly. This means that increasing bee populations can be supported without bees reverting to native species for their nectar needs.
2. In Canada, introducing the crested wheat grass *Agropyron cristatum* to the prairies affected natives in different ways. Of 33 natives, 31 species were negatively influenced (population size decreased) due to competition (Heidinga & Wilson, 2002). Of the remaining two species, one, *Bouteloua gracilis*, was unaffected while the other, *Poa sandbergii*, was positively affected due to competitor release (i.e. one of the grasses that normally outcompeted it was itself outcompeted by non-native crested wheat grass). This final example is sometimes referred to as the 'an enemy of my enemy is my friend' scenario.

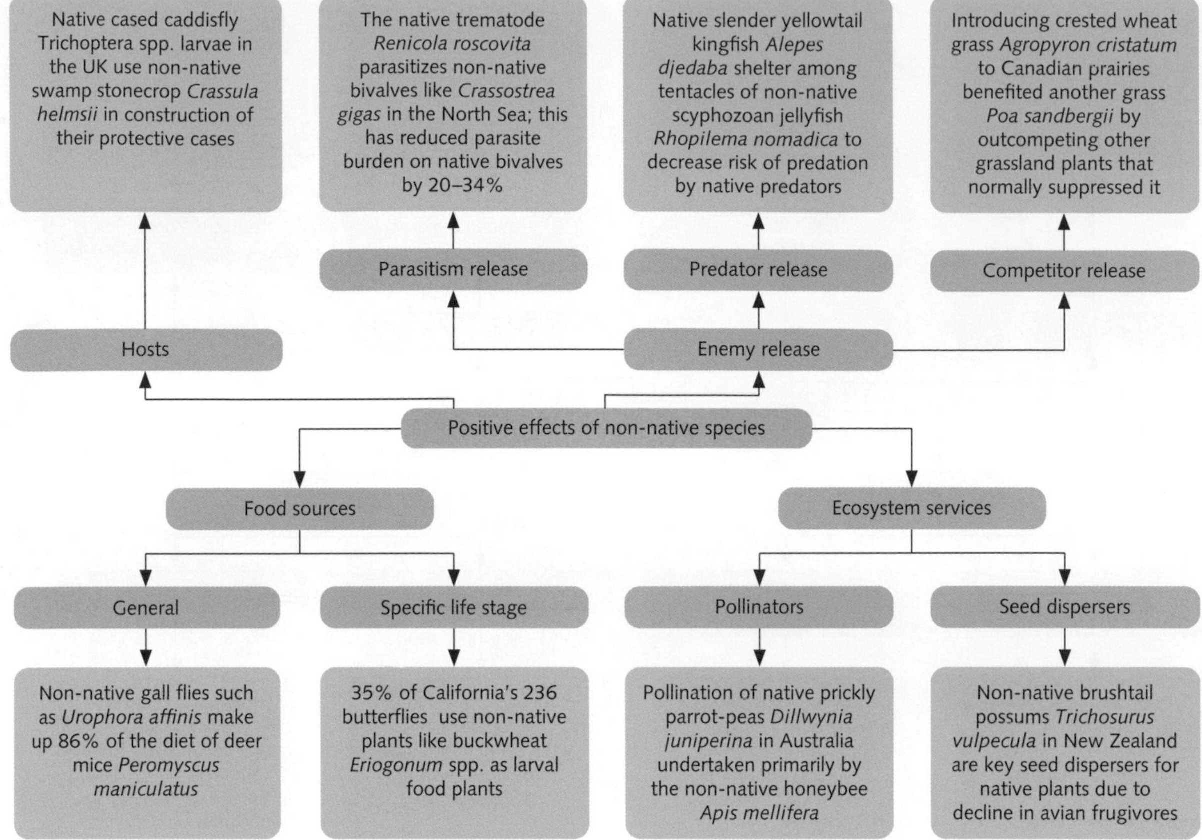

Figure 8.7 Potential positive effects of non-native species on native species with examples.

One other factor adds to the complexity of invasion ecology: it can be difficult to quantify the effects of non-native species independently of other factors. This is demonstrated by the decline of the European otter *Lutra lutra* in the UK. In this case, population decline in the 1970s was mainly caused by pollution, but coincided with a rapid increase in the non-native American mink *Neovison vison*. If these events were linked, then the directionality of the link is not in the expected direction. Rather than mink out-competing otters (a theory commonly suggested at the time and that still persists to some extent), the increase in mink numbers is more likely to have been facilitated, at least in part, by **competitive release** as a result of otter decline.

8.4.3 Responses of native species to non-native species.

Native species are not just the passive recipients of the impacts, positive or negative, associated with non-native species. In many cases there is a period of adjustment followed by native species altering behaviour, or adapting physically (changes in size, shape, form, etc.), or physiologically (changes in biological processes, such as digestion or reproduction). The potential for **native response** to non-native species is evidenced by cane toads *Rhinella marina* which secrete a toxin to deter predators. Initially, the Australian native amphibian-eating red-bellied black snake *Pseudechis porphyriacus* was affected by this toxin, but as exposure increased there was a genetic increase in toxin resistance and the negative impact (poisoning) of the native snakes consuming the non-native toads was negated (Phillips & Shine, 2006).

8.4.4 Expansive native species.

In the same way that non-native species can become invasive (i.e. spread and reproduce quickly to the point they can become pests), native species can also spread and reproduce quickly, and cause similar problems by doing so. The term 'invasive' is reserved for non-native species; the term for similar behaviour in native species is 'expansive'. Native brambles *Rubus fruticosus* in British woodland often become expansive by forming dense thickets that reduce light levels and restrict ground flora, which can reduce the diversity of native flowering plants. This, in turn, can affect butterfly diversity because the majority of butterfly species lay eggs on spring-flowering host plants, such as violets *Viola* which are especially prone to being outcompeted.

Thus, while rapid increases in population or distribution are common in non-native species (due to a lack of predators, pathogens, and parasites, a concept summarized in the **enemy release hypothesis**), the real issue is often not non-native status itself, but the presence of biological traits that encourage invasiveness/expansiveness.

FURTHER READING FOR THIS SECTION

Overview of ecological impacts: Goodenough, A.E. (2010) Are the ecological impacts of alien species misrepresented? A review of the 'native good, alien bad' philosophy. *Community Ecology*, Volume 11, 13–21.

Overview of environmental impacts: Pimentel, D., Zuniga, R., & Morisson, D. (2005) Update on the environmental and economic costs associated with alien-invasive species in the United States. *Ecological Economics*, Volume 52, 273–288.

8.5 Importance of monitoring.

As discussed in Chapter 3, monitoring is an extremely important part of Applied Ecology, and this is certainly true for non-native species. Identification of initial introduction can be an unintentional result of any ecological surveying (e.g. **baseline monitoring**: Section 3.2.1) or can result from deliberate and targeted **alert monitoring** (Section 3.2.4).

Not only can surveys identify the arrival of a non-native species, the longitudinal data that result from **temporal monitoring** (Section 3.3.3) and repeated **spatial monitoring** (Section 3.2.2) can also be used to identify and quantify species' spread, and any change in abundance. Monitoring of co-occurring native species is also the first step in establishing if a non-native species is having substantial impacts (and this, in turn, often dictates the priorities for management). Finally, if non-native species in an area are subject to management to eradicate, contain, or control non-native species (Section 8.6), monitoring also becomes vital to establish the effectiveness (or otherwise) of that management. Such data are essential information for ecologists and continue to optimize management methods.

In order for non-native species monitoring to be maximally effective, surveys should follow the key surveying principles outlined in Section 3.6.4.

8.5.1 The role of citizen science.

To complement detailed scientific surveys, it is also possible to harness the power of citizen science and ask members of the public to record data. This was explored in Section 3.3.4 for Applied Ecology in general, but it should be noted that this approach can be especially useful for monitoring many non-native species.

Citizen science surveys of non-native species can allow very wide spatial coverage (country- or even continent-wide) and also have the useful additional benefit of making the general public aware of non-native species and overall biodiversity. They tend to work best for species that are charismatic, to get people interested, and relatively easy to identify so that the data are as robust as possible. Examples of non-native species citizen science projects include the Ontario Reptile and Amphibian Atlas non-native species recording system in the USA and the Centre for Ecology and Hydrology's harlequin ladybird *Harmonia axyridis* survey in the UK (see the Interview with an Applied Ecologist for this chapter).

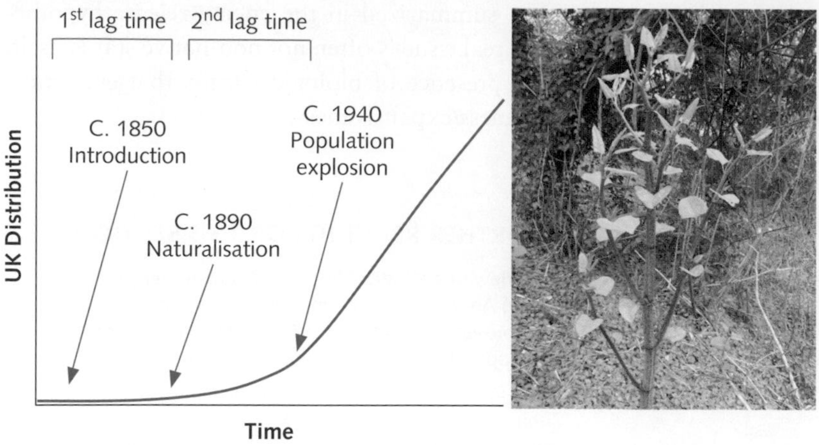

Figure 8.8 Lag time between Japanese knotweed, *Fallopia japonica*, introduction into Europe, and its eventual spread and population explosion.

Source: Graph based on information in Child, L. and Wade, M. (2000) *The Japanese Knotweed Manual*. Packard Publishing Limited, Chichester. Photograph by Anne Goodenough.

8.5.2 Lag effects.

The impacts of a non-native species on its new environment and native species can change temporally. There is usually an initial lag time between arrival and establishment, as well as a second, less-reported, lag between establishment and population growth. These lag phases can be substantial, for example, as shown in Figure 8.8, when Japanese knotweed *Fallopia japonica* was introduced to Europe in the 1850s, the first and second lag phases combined lasted almost 100 years, but when population growth started it progressed very rapidly.

The existence of lag effects means that long-term temporal monitoring is vital, both to ensure that any impacts of non-native species are recorded and to establish how resilient the ecosystem is in responding to new interactions (or, conversely, where active management is needed).

Some situations can be especially complex. For example, in Puerto Rico, a considerable amount of native forest was deforested to make way for agriculture. Those trying to restore deserted farmland and recreate woodland (**rewilding** Section 13.6) found that it was hard to re-establish native trees as there were few pioneers able to survive. Instead, non-native trees such as the African tulip tree *Spathodea campanulata* colonized the landscape. However, although the new forests remain dominated by non-native trees for their first three or four decades, native trees are increasingly able to recolonize as the environment becomes more stable and nutrient rich (i.e. as the environment moves from early- to mid-successional stages).

FURTHER READING FOR THIS SECTION

Overview of ecological impacts: monitoring networks: Graham, J. Newman, G., Jarnevich, C., Shory, R., & Stohlgren, T.J. (2007) A global organism detection and monitoring system for non-native species. *Ecological Informatics*, Volume 2, 177–183.

8.6 Management.

The need for management of non-native species depends on the magnitude and scale of negative impacts of non-native species. If management is necessary, the immediacy, magnitude, and spatial and temporal extent of that management is often dictated by resourcing constraints, especially those directly relating to funding.

Because resources for non-native species management are not infinite, management resources need to be prioritized in a similar way to conservation resources, as discussed in Section 10.5. The problem with this approach is that it is usually reactive. Most resources are spent tackling the 'major

AN INTERVIEW WITH AN APPLIED ECOLOGIST 8

Name: Professor Helen Roy

Position: Research Ecologist

Organisation: Biological Records Centre, Centre for Ecology & Hydrology UK

Role: Consultant Ecologist

Nationality and main countries worked: British; worked across Europe

What is your day-to-day job?

I am a full-time research scientist focusing on community ecology. I work both on long-term and large-scale datasets derived from the inspiring contributions of volunteers and through experimental work to assess the effects of environmental change on wildlife. My day-to-day work involves collaborating with the volunteer organizers of many surveys and recording schemes to gather the vast datasets (comprising biological records: essentially occurrence data) that are critical for addressing ecological questions in relation to environmental change. I have worked on broad inventories of all non-native species in Britain and across Europe, but also in depth on one particular invasive non-native species, the harlequin ladybird *Harmonia axyridis* at global scales.

I am a passionate natural historian and feel privileged to work with so many amazing natural historians through the volunteer schemes and societies. My work involves collaborating with large teams around the world to make predictions and test hypotheses on invasive non-native species. Some work has a direct link to policy and I have been delighted to lead projects that ultimately inform decision-making, for example, through the new EU Regulation on invasive non-native species. I also enjoy the opportunity to share my enthusiasm for Applied Ecology—science communication is critical to engage people in understanding the issues of environmental change.

What are the most interesting recent or currents projects you have been working on within both the broad and specific parts of your role, and why?

Broad: I have been involved in developing so called 'horizon scanning' approaches to enable predictions about which future invasive non-native species likely to arrive, establish, spread, and ultimately adversely affect biodiversity and ecosystem function. Such approaches are challenging—the pool of potential species is vast—but it is incredible what can be achieved with a group of experts working collaboratively to reach consensus.

Specific: In 2005 I helped establish an online survey for people to report sightings of harlequins and all species of ladybird through the UK Ladybird Survey. Although never intentionally introduced into the UK, the harlequin ladybird was first reported in Autumn 2004 in the southeast of the UK, from where it has spread rapidly north and west. Contributions from volunteer recorders through the UK Ladybird Survey (www.ladybird-survey.org) have provided an amazing resource for studying invasion of habitats across the UK by the harlequin ladybird. By simultaneously monitoring native ladybirds (Coccinellidae) through the UK Ladybird Survey, we have been able to assess the effect of the harlequin ladybird on changes in the distribution of native species.

What's been best part of these particular projects?

Broad: I have been so privileged in working with incredible teams of scientists around the world, which has been tremendously rewarding. It has also been fascinating to work across a range of different ecosystems, from arctic tundra to Mediterranean regions via temperate woodland and alpine ecosystems.

Specific: It has been fascinating to see how new technologies have changed the way that people can get involved in scientific research. Certainly online recording and the development of the smartphone application have increased participation in the UK Ladybird Survey. Around 17,000 people have been involved with the survey and collectively contribute around 25,000 records each year. It's also been really interesting working with partner organizations such as the BBC, the British Science Association, the Royal Society of Biology, and many more.

What are the main challenges in monitoring and managing non-native species, and how can they be overcome?

The number of first records of non-native species being recorded in the UK (and, for that matter, elsewhere) is increasing rapidly year on year. We are still in the growth phase of invasion biology both in terms of new arrivals and knowledge. Moreover, because the cumulative number of non-native species in most regions is still growing, the proportion of those non-native species which have an ecological impact is also still growing.

What next for you, and why?

Perhaps the most threatening species on the British horizon is the Asian hornet *Vespa velutina*. This species has recently appeared in the UK. This predatory species consumes a range of insects, but is commonly seen feeding on honeybees. We are putting a huge amount of effort into publicity, especially for bee keepers, to raise awareness of the species to encourage people to report sightings of concern through an online alert system for non-native species.

Finally, how did you get into community ecology and what advice would you give to others?

I have had a passion for natural history for as long as I can remember. I got involved in a number of wildlife projects (particularly monitoring small mammals) as a teenager so it was natural for me to study biology at University. My advice for others is simple: look for opportunities to get involved as a volunteer, and use your enthusiasm to contribute to projects and learn new skills. Collaborate and communicate—it is fun, rewarding, and ultimately will almost certainly lead to many other opportunities. Sharing your enthusiasm and knowledge through blogs or magazine pieces, or offering talks to local groups, national organizations, and even academic conferences is a great way to put you in contact with potential employers and collaborators. Collaborations and working in partnership with organizations can be a very powerful way of breaking into the job market.

problems'—controlling, or attempting to control, headline invasive species such as Japanese knotweed. Other non-native species are frequently ignored until and unless they have a detrimental effect on native species, by which time they have often spread and are much harder to control. The old adage of 'a stitch in time ...' is apt: proactive management, especially on key sites such as nature reserves, can have huge benefits by allowing problems to be prevented or solved, rather than simply managed.

8.6.1 Non-native species databases.

Until comparatively recently, there were few accessible central resources for management of non-native species. With the digital age, this has changed and there are now major online databases that are accessible to all, from which management protocols can be obtained. Three of the most important are the worldwide GISB (Global Invasive Species Database http://www.iucngisd.org/gisd/), the North American Center for Invasive Species and Ecosystem Health (http://www.invasive.org/) and the European DAISIE (Delivering Alien Invasive Species Inventories for Europe http://www.europe-aliens.org/default.do).

8.6.2 Aims of non-native species management.

The aim of non-native species management depends not only on what species and locations are involved, but also on the time the non-native species has been present in its novel environment. The best time to manage non-native species is as soon as they are introduced into (or escape into) a natural ecosystem—or at least as soon as they are discovered. When populations are small and the non-native range is small, **eradication** can be feasible. However, if suitable management is not undertaken promptly, the species can spread and population sizes can increase (as shown for Japanese knotweed in Figure 8.8), often becoming invasive and passing the point where eradication is feasible. Early in this process, it can be possible to **contain** the species spatially and/or **control** population levels (and therefore likely impact). If extensive spread occurs, though, the management focus will usually switch to **buffering impacts** of invasive species. These different management aims and phases are summarized in Figure 8.9.

Eradication

Eradication, the complete removal of a non-native species in a given area, is the 'gold standard' in management of non-native species. Eradication can be

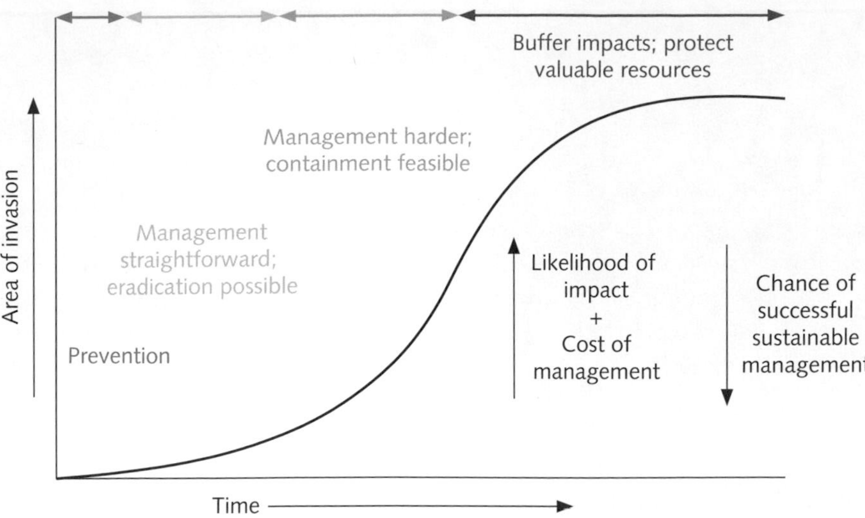

Figure 8.9 Phases of non-native species management in relation to time since invasion and spatial spread.

feasible very soon after translocation, but the opportunity can quickly be lost as there is a strong link between the size of the invaded area and the success of any eradication programme, both in terms of overall result and the time involved.

Eradication must involve removing all individuals (and, in the case of plants, all yet-to-germinate seeds) to be effective. This is imperative, since even leaving one individual behind (a pregnant female, a viable seed, or even an unmated female in the case of species such as aphids that reduce parthenogenetically without fertilization) could allow the population to re-establish. Similarly, successful eradication is only possible if there is a low chance of the species reinvading. This means that unless the target population is isolated, it can be particularly hard to eradicate, especially in the case of mobile species, such as birds.

Because of the greater success of eradication schemes for isolated populations, eradication has historically been most successful on islands where even well-established species can be successfully removed if the island is relatively small. Indeed, in Europe, 90% of successful eradication programmes have occurred on islands. These include successful removal of rats *Rattus rattus/norvegicus* from French and Italian islands, American mink *Neovison vison* from the Estonian island Hiiumaa, and cats *Felis catus* from several Spanish islands.

An example of a missed eradication opportunity is afforded by the marine aquarium alga *Culepra taxifolia*. This was first discovered in its non-native range in 1984 outside an aquarium in Monaco. The affected area was 1 m^2. By 2008, the distribution covered much of the Mediterranean coast with secondary spread occurring through shipping, fishing activity, and natural currents—the spread is shown in Figure 8.10. *Culepra* could probably have been easily and quickly eradicated if prompt action had been taken. The lack of such action is arguably more to blame for the substantial problems subsequently caused in much of the Mediterranean (including out-competition of other photosynthesizing organisms) than the original accidental introduction. In this case, the decision on whether to start an eradication programme was delayed because of political sensitivities regarding the source of the problem and a lack of consensus on who should undertake the eradication, especially since the species quickly spread along the coast of more than one country (spreading from Monaco to France and Italy).

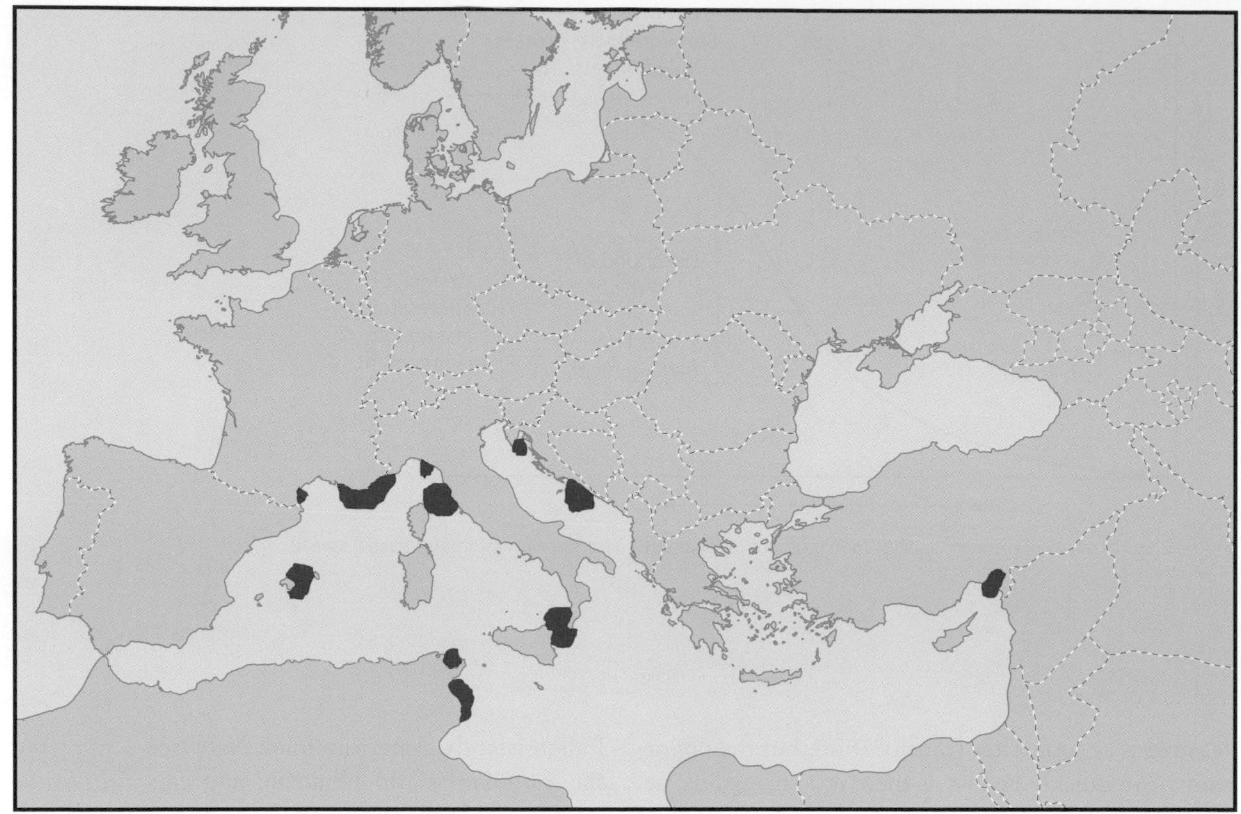

Figure 8.10 The distribution of *Culepra taxifolia* along the Mediterranean coast.

Source: Information from the DAISIE (Delivering Alien Invasive Species Inventories for Europe) database complied by Galil, B.S. (2006) and available from http://www.europe-aliens.org/pdf/Caulerpa_taxifolia.pdf

Unfortunately, such stories of missed opportunities are all too common for invading species. Missed opportunities can also arise as a result of:

- Lack of monitoring with species not spotted in time (Section 8.5).
- Lack of understanding of impacts (no motivation to manage).
- Lack of money and/or labour to undertake the management at that juncture, even though it is likely to be more cost-effective in the long term.

Of course, with the benefit of hindsight, it is very easy to look back at missed opportunities to control infamous invasive species. On the other hand, however, applying the **precautionary principle** and implementing eradication programmes for all non-native species would have significant resourcing implications, especially remembering that only one species in every 100 is likely to establish long-term.

Spatial containment

Species can spread very quickly in a novel environment, as evidenced by the spread of starlings *Sturnus vulgaris* in the USA as shown in Figure 8.11. The measure of the speed of spread of a non-native in a new area without direct human aid is called its **secondary dispersal power**. Species with high secondary dispersal power are more likely to become invasive than species with low dispersal power. This includes plants that reproduce vegetatively (e.g. Japanese knotweed *Fallopia japonica*) or produce numerous seeds (e.g. Himalayan balsam *Impatiens glandulifera*) and animals that are highly mobile. Species that have lost all or most of their 'natural enemies' (Section 8.3) are also more likely to be able to spread rapidly, especially if they are also able to tolerate multiple different habitats and environmental conditions.

In many countries, national **legislation** includes provision both to reduce the risk of accidental human-mediated spread and to punish deliberate spread. For example, in the UK, the Wildlife and Countryside Act lists many non-native species within Schedule 9. One such species is Japanese knotweed and it is an offence under Section 14(1) of

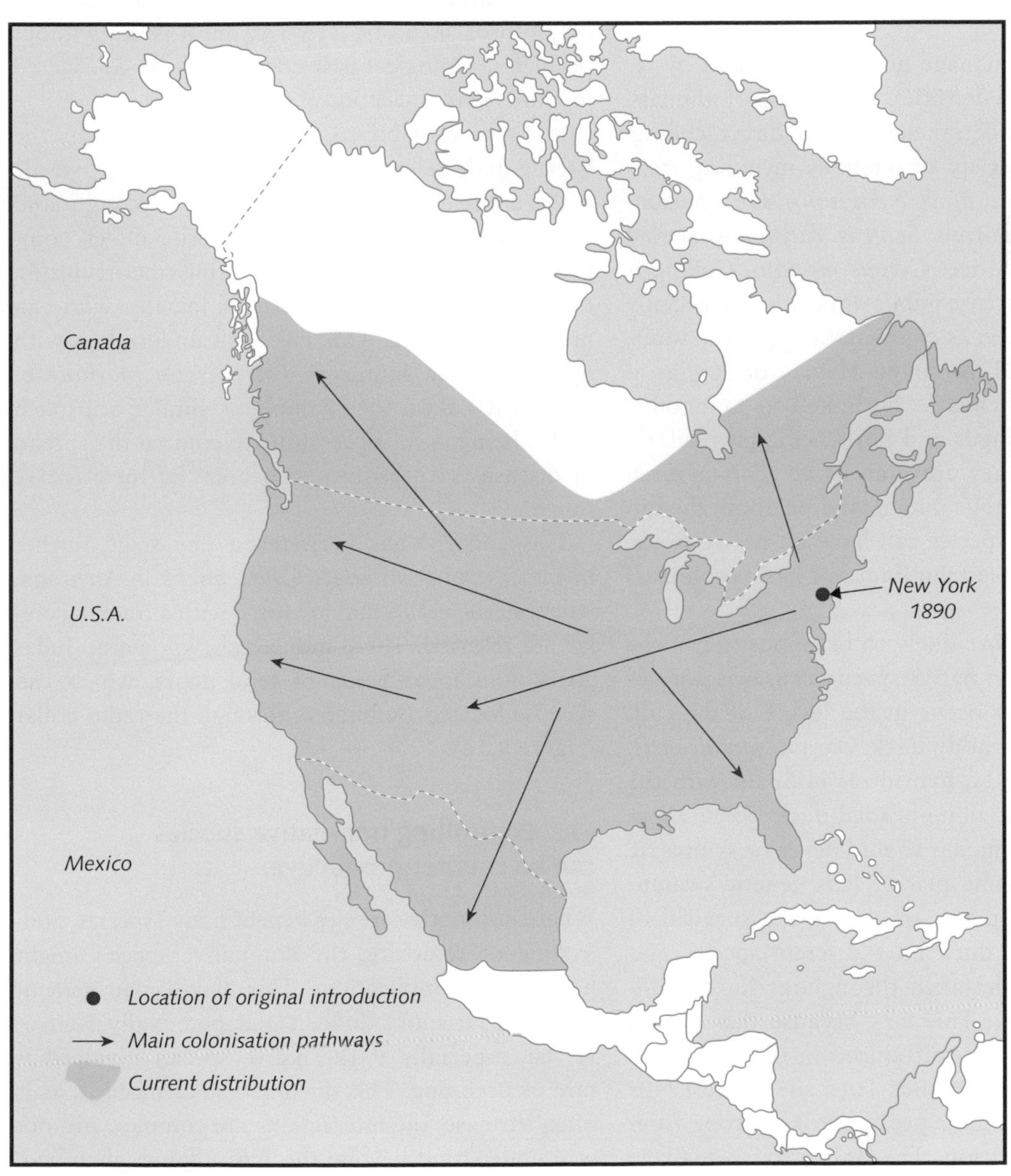

Figure 8.11 Spatial spread of European starlings *Sturnus vulgaris* after introduction to New York in 1890.

the Act to spread it. Any excavated soil from areas where Japanese knotweed has established must be disposed of at a licensed landfill site and not reused in further construction or landscaping. Moreover, Section 34 of the Environmental Protection Act 1990 places a duty of care on all waste producers to ensure that any wastes are disposed of safely, and that a written description of the wastes, and any specific harmful properties, is provided to the landfill operator.

Population control and buffering impacts

It is common to manage non-native species by reducing population densities. In the case of animals, this is usually undertaken through direct **culling**. This can use a variety of methods including trapping (e.g. American mink *Neovison vison*), poisoning (e.g. grey squirrels *Sciurus carolinensis*) and shooting (e.g. sika deer *Cervus nippon*). Culling is often subject to intense public opposition, especially if the species is seen as charismatic, e.g. grey squirrels in the UK and the coypu *Myocastor coypus* in Sicily. Charisma tends to be linked to taxonomic group, with mammals and birds being particularly popular even if they are non-native. This is often due to a lack of public understanding about the impacts non-native species can have on native biota, and more widespread ignorance of basic ecological concepts.

There is also more likely to be opposition if the problem that a non-native species causes is not obvious, or does not occur in the locale of the cull. For example, the ruddy duck *Oxyura jamaicensis*, an American species, hybridizes in Spain with the rare and declining white-headed duck *Oxyura leucocephala*, with which it should never be sympatric (co-exist in the same place). This **genetic swamping** is one of the major threats to the survival of the white-headed duck as a discrete species and so a cull was undertaken throughout Europe, including in the UK. This was because the species, which was originally introduced in the UK at the first Wildfowl and Wetlands Trust site in the 1940s by Sir Peter Scott, had spread across Europe from this source population. The only way to manage the problem long-term was considered to be a Europe-wide eradication programme (culling only in Spain would almost certainly lead to reinvasion). The cull has been subject to considerable criticism, especially as the species was having little if any impact in the UK.

Another method of controlling non-native species is by releasing 'enemies' from the natural range within a **biological pest management** framework. This could involve releasing pathogens, diseases, parasites (with vectors if appropriate), predators, or herbivores. In a sense this might be seen to be fighting fire with fire, and just as in the control of bush fires that approach can be effective, but it is not without problems. Biological pest control will be considered in more detail in Section 9.4.3.

It is also possible to control populations by the 'sterile-male-release' technique, whereby sterile males are released into a population (or caught and re-released) on the basis that sterile males compete with 'normal' males for mating opportunities, thereby reducing the number of females who can produce offspring. This has been undertaken with non-native sea lampreys *Petromyzon marinus* in the Great Lakes, for example. A similar approach is also being used successfully to combat the spread of disease via mosquitoes in areas of their native range.

This concept has been taken one stage further in the case of feral goats *Capra hircus* in Australia, where sterile males that are fitted with a tracking collar are released. These individuals, known as Judas goats, join local herds of feral goats, which can then be located by hunters through the radio collar (Figure 8.12).

8.6.3 Controlling non-native species can be counter-productive.

Where non-native species benefit native species, controlling or removing the non-native species might have serious implications. This makes management decisions far more difficult and potentially controversial, especially if the native species involved is rare or declining. This dilemma is enhanced if, as is often the case, the interactions are complex and not well understood beyond the **'native good, alien bad' philosophy**.

Figure 8.12 Judas goats in New Zealand with individual tags and tracking collars are used to alert authorities to the location of feral goat populations.

Source: Image courtesy of www.flyafrica.info/CC BY 3.0.

Such issues are most likely to occur when there are producer–herbivore or predator–prey interactions. This can occur when native predators preferentially depredate (or graze) non-native species, possibly because they lack co-evolved deterrent mechanisms. Similarly, there are examples of situations where non-natives are important, even vital, in the reproduction of native plants. For example, the red-whiskered bulbul bird *Pycnonotus jocosus* is now the sole pollinator of the rare Mauritian endemic plant, *Nesocodon mauritianus*, while native Hawaiian ie'ie vines *Freycinetia arborea* rely almost exclusively on pollination by non-native Japanese white-eye birds *Zosterops japonicus*.

Non-native species can also benefit native species of key conservation importance through ecosystem engineering (Section 8.4.1). For instance, eradication of the non-native saltcedar tree *Tamarix* in substantial parts of its introduced range in the USA has been repeatedly delayed because it provides important nesting habitat for the endangered southwestern willow flycatcher *Empidonax traillii extimus*. In both cases, control of the invader, even if justified for other reasons, could cause the extinction of a rare indigenous species: an example of just how complex formulating effective conservation policies can become.

When a non-native species is depredated by two (or more) other species that themselves have a predator–prey interaction, elimination of the non-native top predator can lead to the decline of native prey populations through a process called mesopredator release. This occurred on Little Barrier Island, New Zealand, where removal of non-native top predator feral cats *Felis catus* led to a substantial increase in the mesopredator black rat *Rattus rattus*. As the black rat population grew as a result of lower mortality rates, this then caused a substantial decline in native Cook's petrels *Pterodroma cookii* (Rayner et al., 2007). This example, together with the case of rabbits and large blue butterflies in the UK (Case Study 8.2), serves as a reminder that applied management of non-native species must be carefully planned.

Threatened non-native species

Very occasionally, species are translocated for the purpose of conserving them. This usually involves taking a species into captivity, rather than simply translocating that species to another area and releasing individuals into the wild, but sometimes seminatural refugia are used. For example, the Devil's Hole pupfish *Cyprinodon diabolis* was translocated from its only known natural habitat, Devil's Hole in Death Valley, USA, to a refugia pool near the Hoover Dam because of decreasing water quality in Devil's Hole. With such translocations, there is always potential for escape into a novel environment. In these cases, the potential threat to native species needs to be weighed carefully against the threat to the escapee in its natural range.

FURTHER READING FOR THIS SECTION

Conservation of non-natives: Schlaepfer, M.A., Sax, D.F., & Olden, J.D. (2011) The potential conservation value of non-native species. *Conservation Biology*, Volume 25, 428–437.

Weighing up non-native species management: Zavaleta, E.S., Hobbs, R.J., & Mooney, H.A. (2001) Viewing invasive species removal in a whole-ecosystem context. *Trends in Ecology & Evolution*, Volume 16, 454–459.

CASE STUDY 8.2

Non-native species management with conservation implications

The European rabbit *Oryctolagus cuniculus* is thought to have been introduced to the UK from its native range in southwestern Europe and northwest Africa by the Romans around 100–400 AD. However, the species seems to have been kept under domestic conditions and only started to occur in the wild following the Norman invasion of Britain in the 11th century. After this time, rabbits became established in the wild and populations expanded rapidly to the point that they have been seen, in many areas, as pests that needed controlling.

One method of control was the release of Myxoma virus, which causes myxomatosis, a disease inducing substantial **mortality** and **morbidity**. The disease was introduced to Scottish offshore islands (without governmental knowledge, finance, or permission) in the early 1950s, but the introduction failed because of a lack of insect vectors. The disease was, however, introduced in Switzerland and from there spread to the UK (either though movement of insect vectors or deliberate introduction; the translocation pathway is unclear). After myxomatosis reached the UK in 1953, rabbit numbers declined by 99% in 3 years across mainland Britain. This decline, together with concurrent changes in agricultural practice, resulted in grassland ecosystems changing and a dramatic increase in sward height through lack of grazing.

What was, at this time, unknown was the impact this habitat change would have on three oddly interconnected native species: two red ant species, *Myrmica sabuleti* and *M. scabrinodis*, and the charismatic butterfly species the large blue *Phengaris arion* (Figure A). The habitat change increased shading in grassland ecosystems and so ants, which need warm soil for their nests, declined. The large blue butterfly has a complex life cycle. After the female lays eggs, the eggs hatch, the caterpillars fall to the ground, and there secrete pheromones in a **chemical mimicry** of ant larvae. Worker ants carry the 'ant imposters' into the nest, where the caterpillars use **sound mimicry** to sustain the deception. In a mixture of predation and brood parasitism interactions, each caterpillar depredates ant larvae and obtains food from worker ants. Each caterpillar pupates and, ultimately, the adult emerges from cocoon, emerges from the ant nest, finds a mate, and the cycle starts all over again (Figure B).

In this example there was a beneficial relationship of a non-native herbivore (the rabbit) creating an optimal

Figure A Large blue *Phengaris arion*.
Source: Photograph by Anne Goodenough.

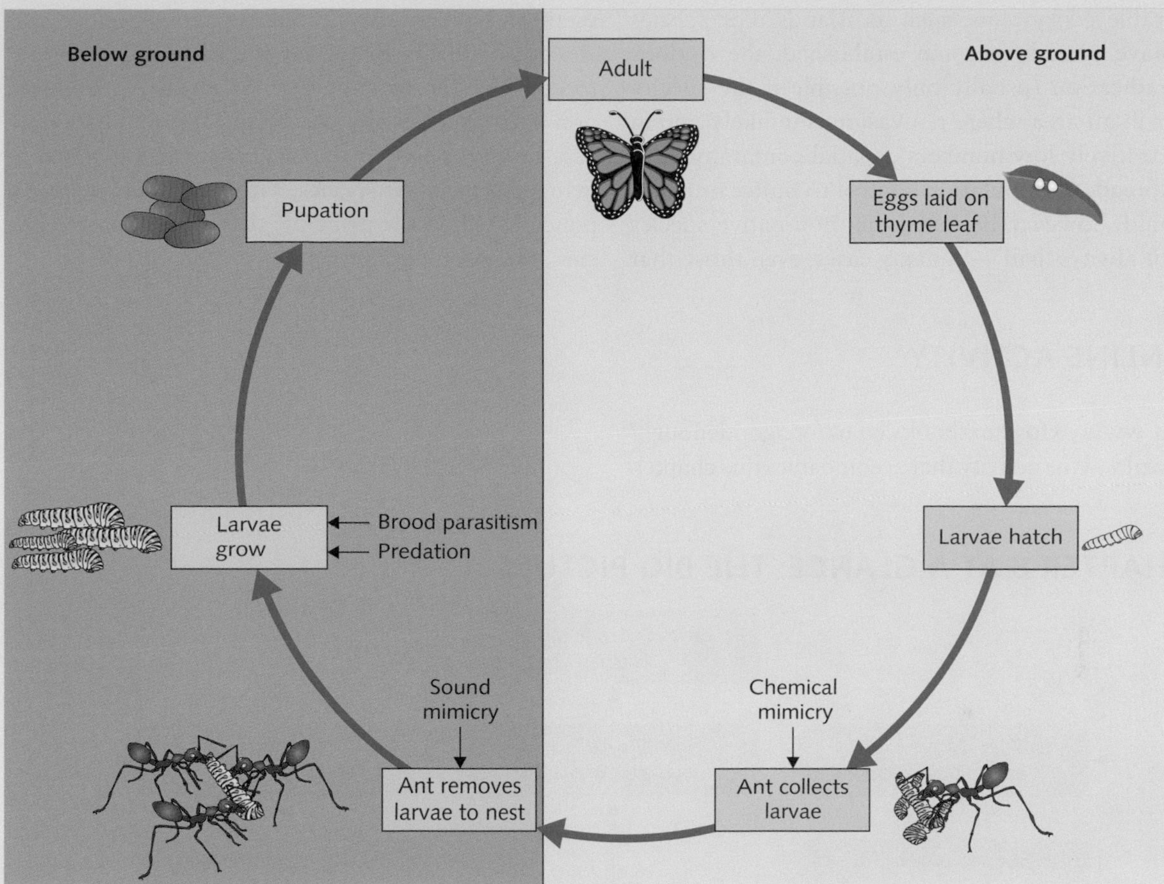

Figure B Life cycle of the large blue butterfly and the importance of ants.

landscape for native ants. Introduction of the Myxoma virus had a detrimental effect not only on the target non-native species (the rabbit) but also ants and, through a cascade effect, the large blue. As a result, the large blue became extirpated in the UK in 1979 (but has now been successfully reintroduced).

FURTHER READING

Large blue butterfly review: Hayes, M.P. (2015) The biology and ecology of the large blue butterfly *Phengaris (Maculinea) arion*: a review. *Journal of Insect Conservation*, Volume 19, 1037–1051.

8.7 Conclusions.

With the increase in movement of people and goods around the world, the introduction of non-native species has increased substantially over the past few hundred years. This increase is not only continuing in the 21st century, but is actually accelerating. Although many non-native species do not become established in their new environment, those that do can be a substantial threat to native species through interactions such as predation, competition, parasitism, and hybridization.

The most ecologically efficient and cost-effective method of addressing the problems of non-native species is to prevent introduction in the first place. This requires an understanding of translocation pathways and vigilance at key points of entry, such as ports and airports. This is especially important for vulnerable ecosystems, such as islands. For species that have already become established, the options are eradication (usually only possible if the species occurs in an area where reinvasion is unlikely and in comparatively low numbers), spatial containment to stop spread, or population control to buffer impacts. It should, however, be noted that non-native species are not always 'bad'—in many cases, even those that establish have no substantial impacts on native species or can even be beneficial. It is therefore important to consider the management of already introduced non-native species that occur in high numbers on a case-by-case basis so that resources are targeted at managing *problem* species, rather than employing a policy based on the principle that all non-native species are a problem.

ONLINE ACTIVITY

Go to www.oxfordtextbooks.co.uk/orc/goodenough/ to download the activity that accompanies this chapter.

CHAPTER 8 AT A GLANCE: THE BIG PICTURE

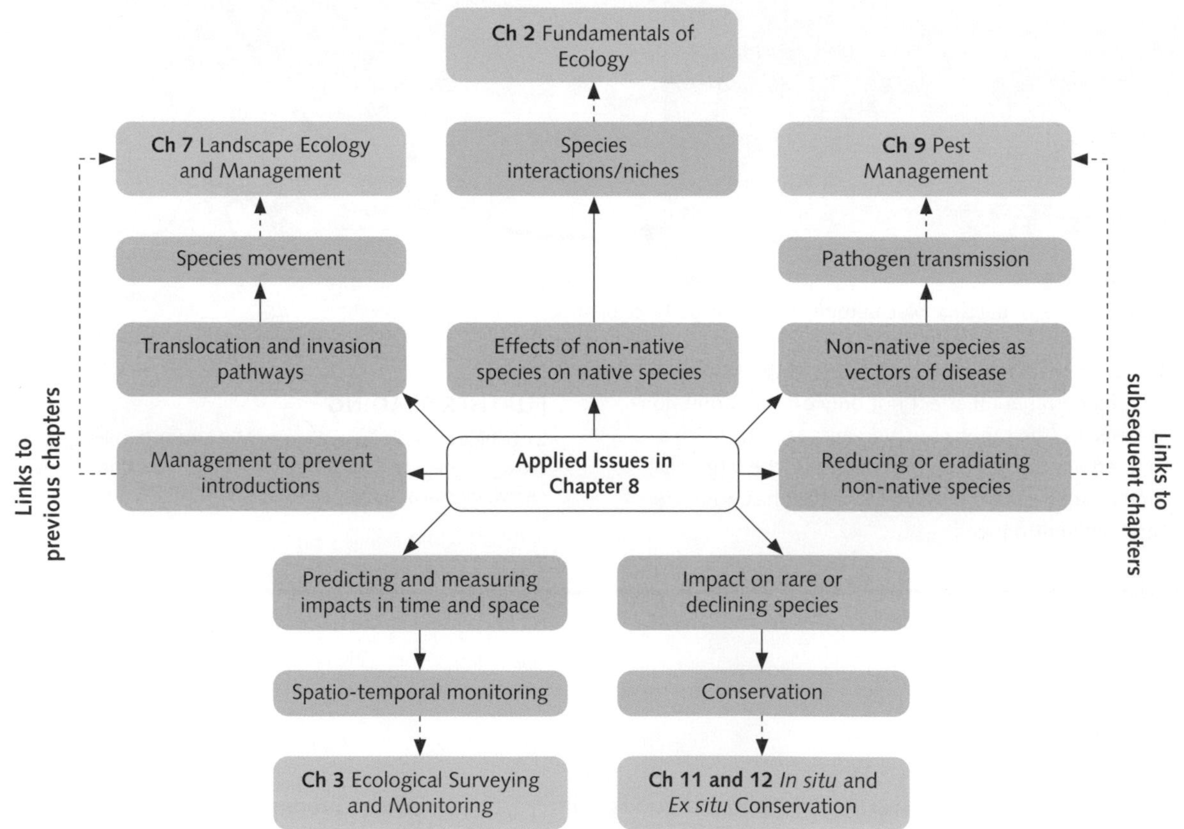

REFERENCES

Carlton, J.T. (1996) Biological invasions and cryptogenic species. *Ecology*, Volume 77, 1653–1655.

Carthey, A.J. & Banks, P.B. (2012) When does an alien become a native species? A vulnerable native mammal recognizes and responds to its long-term alien predator. *PLoS One*, Volume 7, e31,804.

Chabrerie, O., Verheyen, K., Saguez, R., & Decocq, G. (2008) Disentangling relationships between habitat conditions, disturbance history, plant diversity, and American black cherry (*Prunus serotina* Ehrh.) invasion in a European temperate forest. *Diversity and Distributions*, Volume 14, 204–212.

Chittka, L. & Schürkens, S. (2001) Successful invasion of a floral market. *Nature*, Volume 411, 653.

Crawford, R.J.M., Boonstra, H.G.v.D., Dyer, B.M., & Upfold, L. (1995) Recolonization of Robben Island by African penguins, 1938–1992. In: P. Dann, I. Norman, I. P. Reilly (Eds) *The Penguins: Ecology and Management*, pp. 333–363. Chipping Norton: Beatty.

Elton, C. (1958) *Ecology of Invasions by Animals and Plants*. Chicago, IL: University of Chicago Press.

Green, W. (2000) *Biosecurity Threats to Indigenous Biodiversity in New Zealand: an Analysis of Key Issues and Future Options*. Parliamentary Commissioner for the Environment, New Zealand Government, Auckland.

Goodenough, A.E. (2010) Are the ecological impacts of alien species misrepresented? A review of the 'native good, alien bad' philosophy. *Community Ecology*, Volume 11, 13–21.

Gordon, D.R. (1998) Effects of invasive, non-indigenous plant species on ecosystem processes: lessons from Florida. *Ecological Applications*, Volume 8, 975–989.

Gurevitch, J. & Padilla, D.K. (2004) Are invasive species a major cause of extinctions? *Trends in Ecology & Evolution*, Volume 19, 470–474.

Heidinga, L. & Wilson, S.D. (2002) The impact of an invading grass (*Agropyron cristatum*) on species turnover in native prairie. *Diversity and Distributions*, Volume 8, 249–258.

Hettinger, N. (2001) Exotic species, naturalisation, and biological nativism. *Environmental Values*, Volume 10, 193–224.

Heywood, V.H. (1989) Patterns, modes and extents of invasions by terrestrial plants. In: J. Drake (Ed.) *Biological Invasions: A Global Perspective*, pp. 31–51. New York, NY: John Wiley and Sons.

Hulme, P.E. (2003) Biological invasions: winning the science battles but losing the conservation war? *Oryx*, Volume 37, 178–193.

Hulme, P.E., Bacher, S., Kenis, M., Klotz, S., Kühn, I., Minchin, D., Nentwig, W., Olenin, S., Panov, V., Pergl, J., Pyšek, P., Roques, A., Sol, D., Solarz, W., & Vilà, M. (2008) Grasping at the routes of biological invasions: a framework for integrating pathways into policy. *Journal of Applied Ecology*, Volume 45, 403–414.

Jeschke, J.M. & D.L. Strayer. (2005) Invasion success of vertebrates in Europe and North America. *Proceedings of the National Academy of Sciences of the United States of America*, Volume 102, 7198–7202.

Jones, C.M. & Brown, M.J.F. (2014) Parasites and genetic diversity in an invasive bumblebee. *Journal of Animal Ecology*, Volume 83, 1428–1440.

Jones, C.G., Lawton, J.H., & Shachak, M. (1997) Positive and negative effects of organisms as physical ecosystem engineers. *Ecology*, Volume 78, 1946–1957.

Odum, E. (1953) *Fundamentals of Ecology*. Philadelphia, PA: Saunders.

Phillips, B.L. & Shine, R. (2006) An invasive species induces rapid adaptive change in a native predator: cane toads and black snakes in Australia. *Proceedings of the Royal Society B*, Volume 273, 1545–1550.

Pimentel, D., Zuniga, R., & Morrison, D. (2005) Update on the environmental and economic costs associated with alien-invasive species in the United States. *Ecological Economics*, Volume 52, 273–288.

Rayner, M.J., Clout, M.N., Stamp, R.K., Imber, M.J., Brunton, D.H., & Hauber, M.E. (2007) Predictive habitat modelling for the population census of a burrowing seabird: a study of the endangered Cook's petrel. *Biological Conservation*, Volume 138, 235–247.

Salmon, M., Reiners, R., Lavin, C., & Wyneken, J. (1995) Behavior of loggerhead sea turtles on an urban beach: correlates of nest placement. *Journal of Herpetology*, Volume 29, 560–567.

Sax, D.F., Stachowicz, J.J., Brown, J.H., Bruno, J.F., Dawson, M.N., Gaines, S.D., Grosberg, R.K., Hastings, A., Holt, R.D., Mayfield, M.M., O'Connor, M.I., & Rice, W.R. (2007) Ecological and evolutionary insights from species invasions. *Trends in Ecology & Evolution*, Volume 22, 465–471.

Schlaepfer, M.A., Sax, D.F., & Olden, J.D. (2011) The potential conservation value of non-native species. *Conservation Biology*, Volume 25, 428–437.

Simberloff, D. & von Holle, B. (1999) Positive interactions of nonindigenous species: invasional meltdown? *Biological Invasions*, Volume 1, 21–32.

Sol, D. (2000) Are islands more susceptible to be invaded than continents? Birds say no. *Ecography*, Volume 23, 687–692.

Stewart, J.R. (2007) The fossil and archaeological record of the Eagle Owl in Britain. *British Birds*, Volume 100, 481–486.

Stohlgren, T.J., Barnett, D.T., & Kartesz, J.T. (2003) The rich get richer: patterns of plant invasions in the United States. *Frontiers in Ecology and the Environment*, Volume 1, 11–14.

Torchin, M.E., Lafferty, K.D., & Kuris, A.M. (2002) Parasites and marine invasions. *Parasitology*, Volume 124, 137–151.

Williamson, M. & Fitter, A. (1996) The varying success of invaders. *Ecology*, Volume 77, 1661–1666.

Wonham, M.J., O'Connor, M., & Harley, C.D.G. (2005) Positive effects of a dominant invader on introduced and native mudflat species. *Marine Ecology Progress Series*, Volume 289, 109–116.

Pest Management

9.1 Introduction.

A field of wheat, a forest plantation or a hay meadow are potential income for a farmer or forester. They are also key landscape components (Chapter 7) and constitute important habitats that support a range of species, some of which may be of conservation concern (Chapter 10). Likewise, parks, gardens, and other managed ecosystems may be home to a great many species, some of which use such land without any kind of negative impact on human activity or productivity. Indeed, in some cases (such as pollinators), species can provide very valuable ecosystem services. However, such ecosystems also present a considerable opportunity for organisms that are able to exploit the species being managed within those ecosystems (such as crops and livestock), or to compete with them for resources. If those exploiting or competing species are detrimental, perhaps by reducing crop yields or damaging property, then they become **pests**.

In this chapter we consider the definition of a pest, the diversity of pests and pest management strategies, and the theoretical and practical factors that dictate pest management decisions. Physical, chemical, and biological pest control strategies are discussed, and the limitations and problems of these approaches are examined from an Applied Ecology perspective. Integrated Pest Management, where multiple approaches are combined, is discussed as a way to solve some of the problems that single approaches to pest control can cause.

9.2 What are pests?

The wide ranging nature of human activities results in a remarkable **diversity** of organisms that may be regarded as pests. Few taxonomic groups are exempt. Within the animals, insect species feature heavily and are drawn from a great many insect taxa (Hill, 1996). These include:

- **Coleoptera:** beetles with both adult and larval forms eating crops.
- **Hymenoptera:** wasps, ants, and sawflies (bees are also in this order, but are rarely considered pests).
- **Lepidoptera:** butterflies and moths, whose larval caterpillars consume large quantities of plant material before undergoing metamorphosis.
- **Orthoptera:** crickets, locusts, and grasshoppers, whose juvenile and adult stages are often voracious herbivores.

- **Isoptera:** termites, whose wood eating habits results in structural damage to wooden structures.
- **Diptera:** true flies including mosquitoes that **vector** (spread and transmit) yellow fever, malaria, and dengue fever (Applied Ecologist, Chapter 9), sand flies (that vector leishmaniasis), and Tsetse flies (that vector sleeping sickness).
- **Hemiptera:** the true bugs, mostly herbivorous and including the aphids, but also including the bed bug.
- **Phthiraptera and Siphonaptera:** lice and fleas can cause intense skin irritation and in some cases transmit disease (Figure 9.1).

Other invertebrate groups can also be pests. The best known arachnids, the spiders and scorpions, may become pests in human settlements especially if venomous species are present. Ticks and mites, also arachnids, can have a large cost in terms of disease transmission in humans, plants (where they vector viruses), and animals. Nematode worms are pests of both plants and animals (though they can also be used as an effective means of **biological pest management**; Section 9.4.3), terrestrial molluscs (slugs and snails) affect horticulture and agriculture, and aquatic snails feed on rice and are intermediate hosts of organisms causing schistosomiasis.

Many vertebrate species can become pests if their activities impinge upon our lives in a negative way. Seed eating birds can devastate crops; rats and mice can consume grain stores; introduced grey squirrels *Sciurus carolinensis* have had a detrimental effect on native red squirrels *Sciurus vulgaris* in the UK; rabbits *Oryctolagus cuniculus* have caused declines of native plants in Australia (Case Study 8.2); gulls *Larus* cause noise disturbance and foul cars and buildings with their excrement; and woodpeckers Picidae can cause structural damage to timber-framed homes.

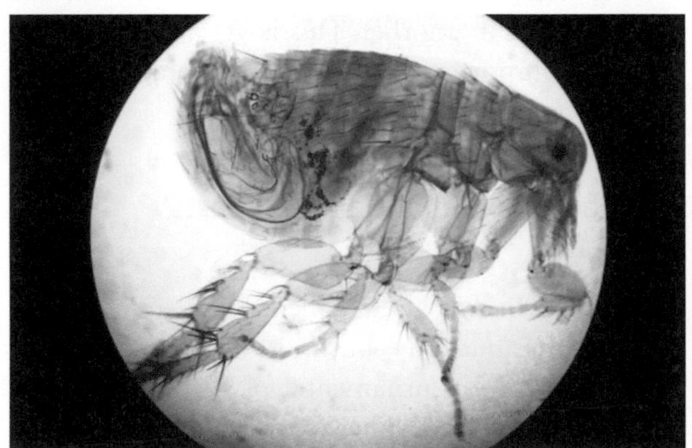

Figure 9.1 Fleas are pests that can cause harm ranging from mild skin irritation to death if the flea, like this rat flea *Xenopsylla cheopis* transmits *Yersinia pestis*, the bacterium that causes plague.

Source: Image courtesy of Fay Celestial/CC BY-ND 2.0.

Plant pests have a similar diversity both taxonomically and in terms of their mode of action. Often called 'weeds', a plant becomes a pest if it grows where it is not wanted and if, by growing in such a place, it causes harm. Plants may be undesirable in a specific location for a variety of reasons, some of which are ecological:

- They may compete with desired plants for sunlight, nutrients, and water, leading to a diminished yield of the desired plants, e.g. Downy brome *Bromus tectorum*. A pest of crops in North America, this plant has a shallow and extensive root system that can extract significant amounts of soil water from fields in which it grows.
- They may provide shelter and food for animal pest species (such as seed-eating birds) that allow these species to survive for longer and cause more damage to desired plants.
- They can act as hosts and then sources for plant pathogens and parasites.

Other reasons why plants may be undesirable in certain locations are because of the direct effect they have on us and our constructions:

- They may be cosmetically undesirable, for example in formal planting schemes in parks and gardens, e.g. chickweed *Stellaria media*, a common weed in parks and gardens in the UK.
- Some species may cause skin irritation by contact (e.g. poison ivy *Toxicodendron radicans* in North America and Asia), or physical injury because of thorns (e.g. bramble *Rubus fruticosus*).
- Their roots may cause damage to pavements, roads, walls, and other structures, e.g. ivy *Hedera helix*. Although not capable of creating cracks in masonry, the roots and stems of ivy growing on walls can widen existing cracks and weaken the structure.

- They may block waterways, e.g. creeping water primrose *Ludwigia peploides* in Europe.

Fungal pests can cause also considerable damage. Many plant diseases, such as white rust *Albugo candida* (affecting brassicas), powdery mildews (affecting many plants and especially common in greenhouse crops), and cavity spot *Pythium sulcatum* (affecting carrots), are caused by fungi, and in such cases the pest fungi are acting both as parasites and pathogens (disease-causing organisms). Fungi are readily spread by wind, water, and contaminated soil, machinery, clothing, and tools, and can easily enter plants through natural openings (such as stomata on the underside of leaves) and through wounds caused by pruning, herbivory, or mechanical damage. Animals are also affected by diseases caused by pest fungi. For example, the fungus *Batrachochytrium dendrobatidis* is causing dramatic population declines in amphibians worldwide, and in captive-bred populations it is carefully monitored and managed (Case Study 12.2).

9.2.1 The definition of 'pest'.

The crucial factor in determining whether an organism is a pest is its **impact**. Consider a female *Anopheles* mosquito taking a blood meal. She removes such a small amount of blood that she cannot be considered to be having any measurable negative impact in terms of her grazing (Section 2.2). However, mosquitos are considered a pest to humans in some parts of the world and this is not because they take blood, but because in so doing they can also vector potentially fatal diseases such as malaria, yellow fever, and dengue fever. Mosquito bites can also cause intense irritation (an impact on our comfort levels) and subsequent scratching can cause bites to become infected, potentially leading to serious medical consequences.

The cost and impact of mosquitoes is considerable, and consequently considerable effort goes into their control by draining marshes and ditches, spraying insecticide, developing infertile males, and other increasingly sophisticated methods. It is only because of the scale of their impact on humans that they are considered a pest, and the scale of the attempts to control them is directly related to that impact.

Pests and **pest management** as subtopics within Applied Ecology have tended, for obvious economic reasons, to focus on pests within agricultural systems. Pests have an enormous economic impact on agriculture, estimated as more than 40% of potential food production annually (Pimentel, 2009). However, if pests are defined as any organism with a detrimental effect on a system of interest, then pests are neither limited to agricultural systems nor to those organisms causing a direct, measurable, economic cost. For example, animals causing excessive noise in urban areas (such as roosting birds) can have a negative impact on people's ability to sleep and on housing value. Annual increases in social wasp numbers lead to increased stings and can be problematic at attractions open during the peak months. In principle, it may be possible to quantify such costs (poor sleep results in lowered productivity, for example), but it is not always easy to find a robust method for doing so.

Of course, it is possible to go too far with our definition of pest. People that suffer from hay fever might consider all pollen-producing grasses to be a pest, and from their perspective they are right. If hay fever sufferers were in the majority and the symptoms were serious enough that they caused people to miss large numbers of working days, then it is entirely feasible that 'pollen control measures' would be actively sought. Being considered a pest or not is, like beauty, very much in the eye of the beholder, and this means that it is entirely possible for a species to be a pest in one situation, or at one time, and not in another. This is yet another example of the importance of spatial and temporal scale in Applied Ecology.

9.2.2 The problem of defining 'pests' in practice.

Buddleia davidii, often called buddleia, summer lilac or the butterfly bush, is a vigorous shrub widely used as an ornamental plant because of its relatively large size, long green leaves, and highly attractive flowers (Figure 9.2). The flowers are nectar-rich and attract large numbers of nectar-feeding insects, especially butterflies, leading to its common name. Native to central China and Japan, it was cultivated within gardens and parks in Europe from

Figure 9.2 *Buddleia davidii* is a common sight in gardens. Known as 'butterfly bush', it attracts large number of insects to its flowers but is now considered problematic in many regions where it is found.

Source: Photograph by Anne Goodenough.

the late 19th century onwards. As it grew in popularity, it escaped from cultivated sites and spread rapidly. It is now found growing wild throughout much of central and southern Europe (including as far north and west as the UK), in Australia, New Zealand, and much of the USA (Tallent-Halsell & Watt, 2009).

In the UK in the 1970s and 1980s it was commonplace to hear buddleia being recommended as a plant to grow to encourage butterflies. Now, it is considered an invasive pest although the full extent of its impact is still unknown in many cases. It is defined as 'problematic' in north-western and north-eastern USA and Canada, throughout New Zealand, and in central Europe where it out-competes native, agricultural, and forestry species (Invasive Species Compendium, 2015). In some states of the USA, for example Oregon, it is considered a 'noxious weed' because it has the capacity to dominate open habitats and change the ecology of dry-land meadows, open slopes, and dunes (Oregon Department of Agriculture, 2014). It can also invade reforested sites and have a substantial negative impact on the growth of tree species used in forestry plantations by restricting light availability (Richardson et al., 1996). Although overall there is no full-scale analysis of the costs caused by the plant's negative impact, some figures are available for local cases: for example, it has been estimated that in New Zealand, buddleia costs the forestry industry between NZD$0.5 and 2.9 million annually in control costs and loss of production (Kriticos, 2007).

In both space and time, buddleia varies in its status:

- In its native range it is simply a component of the ecosystem.
- In late 19th century and early 20th century Europe it was viewed as a highly desirable ornamental shrub.
- In the 1970s and 1980s in the UK and other countries in Europe it was seen as a butterfly-friendly addition to parks and gardens, and could be considered to have been part of a wider insect conservation agenda.
- Now, in those non-native areas, Applied Ecologists have started to gain an appreciation of the impact it has on both human activities and on ecosystems and it is viewed variously as problematic, a pest, an invasive species, or a noxious weed.

What this highlights is that it is not possible to answer the question 'is buddleia (or any species) a pest?' without first knowing the location and the system being considered. Organisms become pests when they cause what we define as a **negative impact,** but even defining 'we' can be problematic. Tigers *Pantheria tigris* taking domestic cattle as prey in parts of India are undoubtedly a pest to rural villagers, who suffer a demonstrable economic loss that could be reduced or eliminated if tigers were removed from the land. To the wider global community, however, tigers are a conservation icon to be conserved at almost any price. Just as in the case of human values playing a key role in deciding conservation priorities (Section 10.5), it is important to realize that different human value judgements can also play a crucial and sometimes contradictory part in the complex process of deciding what is, or is not, a pest and in deciding which control methods should be used.

9.2.3 Pest management.

Pests are organisms involved in sometimes complex individual interactions with both their target organisms and the wider ecosystem. Understanding these interactions, their effects on pest and host population dynamics, and their wider effects on the biotic and abiotic environment, requires us to take an Applied Ecological approach, and such an approach also assists in designing effective management strategies. In pest management, knowledge is power.

FURTHER READING FOR THIS SECTION

Insect pests: Onstad, D.W. (Ed.) (2009) *Insect Resistance Management: Biology, Economics, and Prediction*. Academic Press, California, USA.

Buddleia as a pest: Tallent-Halsell, N.G. & Watt, M.S. (2009) The invasive *Buddleja davidii* (butterfly bush). *Botanical Review*, Volume 75, 292–325.

9.3 The theory of pest management.

It is tempting to think that the best way to control a pest is to remove it completely. If a field of cotton *Gossypium* is being attacked by cotton boll weevils *Anthonomus grandis*, then why not simply spray the field intensively with the most effective pesticide currently (and legally) available and continue to do so until the last weevil is killed? If mosquitoes are a problem around a village then why not drain all natural and artificial water bodies, contain the rivers and streams in enclosed pipes, impose and enforce regulations on standing water within and around the village, and treat the area with insecticide regularly?

The problems with such eradication approaches are three-fold.

1. **In many cases they cannot work, even in the short term.** Many successful pest organisms (such as flying insects) are adept at dispersal, and removing pests from one location will not stop them reinvading from other sites. Indeed, removing potential competitors from the treated site will likely make that site even more attractive to subsequent waves of invaders, since there will be less **intraspecific competition.**
2. **It may not be technically or practically possible.** Natural environments are full of places where pests can be protected from chemical or physical treatment, especially if, as is often the case, the pest species is small. However, even larger species can benefit from natural cover (Case Study 9.1).
3. **It is unlikely to be economically sensible.** Even if it was practically and ecologically possible to eliminate a pest entirely from a given location it is quite possible that the cost of doing so will far outstrip the cost that the pest imposes, especially if the pest is well established. If it costs more to eliminate a pest than it does to put up with it, then it makes no sense to pursue elimination.

Where elimination tactics can be effective is if a pest management strategy can be implemented very soon after a pest emerges in an area, when the area invaded tends to be small and constrained and population size is low. In non-native species terms, this is often within the lag phase (Figure 8.10), which is frequently before any impacts become obvious.

9.3.1 The Economic Injury Level.

Of course, in practice it is not a simple choice between 'do nothing' and 'total elimination'. In the middle ground there may be a level of pest control that reduces the population of the pest, and the damage it causes, to an acceptable level with an implementation cost that is **financially sensible.** This balance, between the harm a pest causes and the cost of controlling it, is the theoretical underpinning of modern pest control.

Engaging in any form of pest control has a cost. For example, treating a crop by applying an insecticide has the obvious cost of buying the necessary chemicals, but there are other factors that contribute. Equipment for delivering the chemicals will be required and that equipment may be bought (imposing an initial capital cost, a maintenance cost, and a depreciation cost as the equipment ages) or hired. Personal protective equipment (PPE) will likely be needed for the people applying the insecticide, who

CASE STUDY 9.1

Burmese pythons in the Florida Everglades

The Burmese python *Python bivittatus* is one of the longest snakes in the world, with larger specimens growing to more than 5.5 m in length (Figure A). These snakes are constrictors, using their bodies to coil around their prey to apply pressure that prevents breathing and eventually causes death. They are native to South East Asia and are found in a wide range of habitats including wetlands, swamps, forests, and river valleys. They are excellent swimmers, climbers, and hunters taking small, medium, and sometimes large mammals and birds as prey.

In the Florida Everglades non-native Burmese pythons, presumably released as unwanted pets, have increased in numbers. First sighted in the 1980s, the population has increased dramatically since 2000 and estimates range from 30,000 to 300,000 snakes (Snow et al., 2007a).

Although they clearly do not 'belong' in the Everglades this fact alone does not make them a pest. Moreover, they do not cause any financial detriment since the Everglades is not an agricultural system. However, some studies have indicated that they might be causing problems for the native fauna. Severe declines in medium-sized mammals previously common in the Everglades, including raccoons *Procyon lotor*, Virginia opossums *Didelphis virginiana*, and cottontail rabbits *Sylvilagus*, coincide with the increase in the Burmese python population. There are also reported declines in larger mammals including white-tailed deer *Odocoileus virginianus* and competing predators such as bobcats *Lynx rufus*. These declines were reported by Dorcas et al., (2012), but disputed by others: Barker & Barker (2012) provide a particularly robust rebuttal, which illustrates the contentious, and sometimes personal, arguments that can develop in scientific literature.

Figure A A fully grown Burmese python *Python bivittatus* can grow to more than 5m in length.

Source: Photograph by U.S. Fish and Wildlife Service/CC BY 2.0.

Management in this case is driven not by the philosophy of EIL, but by one of eradication achieved through physical removal and destruction. However, despite their large size and abundance, pythons are difficult to find and to capture, so much so that a hunt was organized in 2013 (*The 2013 Python Challenge*) that attracted 1600 participants. These individuals hunted the pythons for a month, with each dead snake attracting a financial reward. Despite the large number of hunters, the duration of the hunting period and the lure of the bounty placed on the snakes, only 68 individuals were recorded as captures. Managing to be both ineffective and hugely inefficient, the *Python Challenge* illustrates the difficulty of controlling a species once it has become a pest.

The complexity of the physical and biotic environment and the secretive, cryptic nature of the python's ecology combine to make the jump from theoretical pest management (culminating in the organization of the hunt) to useful applied pest management especially difficult. The fact that there is even disagreement over whether or not the animals are a pest serves to illustrate the fundamental problem of quantifying the impact of an organism on an ecosystem. Although an extreme case, the Burmese python in the Everglades is a lesson to the Applied Ecologist, who should be ever-aware of the difficulties that can arise in defining pests and in putting pest control plans into practice (see also Case Study 8.1 on pool frogs in Europe).

REFERENCES

Barker, D.G. & Barker, T.M. (2012) A Review of: Dorcas et al (2012) Severe mammal declines coincide with proliferation of invasive Burmese pythons in everglades national park. *Bulletin of the Chicago Herpetological Society*, Volume 47, 45–50.

Dorcas, M.E., Wilson, J.D., Reed, R.N., Snow, R.W., Rochford, M.R., Miller, M.A., Meshaka, W.E. Jr, Andreadis, P.T., Mazzotti, F.J., Romagosa, C.M. & Hart, K.M. (2012) Severe mammal declines coincide with proliferation of invasive Burmese pythons in Everglades National Park. *Proceedings of the National Academy of Sciences of the United States of America*, Volume 109, 2418–2422.

Snow, R.W., Krysko, K.L, Enge, K.M., Oberhofer, L., Warren-Bradley, A. & Wilkins, L. (2007a) Introduced populations of *Boa constrictor* (Boidae) and *Python molurus bivittatus* (Pythonidae) in southern Florida. In: R.W. Henderson and R. Powell (Eds.) *Biology of the Boas and Pythons*, pp. 416–438. Eagle Mountain Press, USA, 416-438.

will also need to be paid. Vehicles for transporting personnel and chemicals need fuel and maintenance. Warning signs may need to be erected, personnel may need specific training, licenses may need to be obtained and pest and damage levels may need to be monitored during and after treatment (Chapter 3).

While such costs can be complex they are, nonetheless, calculable. It is also possible, though sometimes difficult in practice, to arrive at a figure for the damage that a pest is causing, perhaps in terms of decreased yield or quality of a crop. With the costs quantified it becomes possible to undertake a cost:benefit analysis to determine a level of harm caused by the pest:

- Below: it costs more to treat than you gain from the increased revenue the treatment would provide; it is thus not worthwhile to treat.
- Above: the costs of the treatment are outweighed by the gains in revenue and it becomes worthwhile to treat.

This threshold value of harm is termed the **Economic Injury Level** (**EIL**; e.g. Pedigo et al., 1986).

If the population of a pest species (which is assumed to be proportional to the damage it causes) is plotted over time then such a plot might resemble Figure 9.3. As an example, consider an insect pest such as an aphid and a target such as a crop, although the principles outlined hold true for any pest scenario. Initially, a small number of aphids (or even just a single aphid since females can reproduce asexually) arrives and begins to consume the crop. From this founding population, the population of the pest begins to increase. In the initial growth phase the pest has a low population and the level of damage caused to the crop may be so small as to be practically undetectable. Indeed, the pest itself may be present in such low numbers and on so few plants that sampling techniques to find insect pests are unlikely to find it. However, assuming that other pests are not present in high numbers then the absence of strong interspecific competition and the low level of intraspecific competition coupled with abundant food plants at high density allows the aphids to reproduce rapidly.

As the pest population increases, so too does the damage it causes. At some critical population size the level of damage, or injury level, caused to the crop exceeds the EIL. At that point it makes economic sense to control the aphids. Without control the

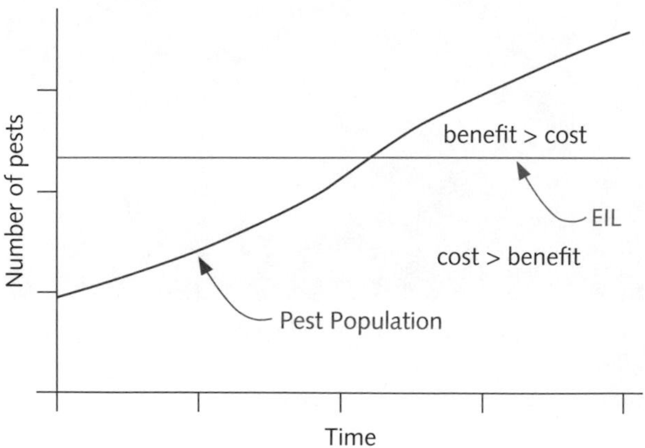

Figure 9.3 When pest levels exceed the Economic Injury Level (EIL) then it becomes cost effective to control them. Below that level it costs more to control the pests than is gained from doing so.

population will continue to increase until it reaches its **equilibrium abundance**, which is the population level that can be supported in that environment at that time.

If the EIL is above the equilibrium abundance then the species under consideration has a population that is naturally self-limiting to a level below the EIL. It is clearly causing damage, but it is doing so at a level so low that it would cost more to control than to ignore. However, it is possible that the control of other pests, or some other aspect of management, could relax those population constraints and result in the focal species becoming a pest. If birds are a natural predator and remove large numbers of the potential pest, but also cause problems for people by roosting in occupied buildings, then bird-scarers or control by shooting could remove these natural predators. Changing the trophic relationships within the environment could have a knock-on effect on population dynamics and shift the focal species from being a potential to an actual pest. It could also be that the focal species is being out-competed by another species whose population is then reduced by pest control methods. The removal of the competitor species results in **competitor release**, leading to an increase in the focal species population and its subsequent classification as a pest.

EIL is most commonly applied to the management of pests on agricultural crops. In these systems the economic injury is calculable and the costs of control may be known precisely. However, the general principle of EIL (that there is a threshold below which it is not worth the cost to control the pest) is one that applies to any pest control situation. In practice though there are problems with the approach regardless of the system to which it is applied.

9.3.2 Problems with the EIL approach.

EIL maximizes 'benefit minus cost' and is a theoretically sound approach to defining and managing pests. However, those seeking to control pests in the field come up against a number of practical problems. These problems do not undermine the underlying theory or diminish the value of the EIL principle, but they have led to the need for operational procedures to bridge the 'theory–practice divide'.

Heterogeneity

The first problem relates to some basic ecological principles outlined in Chapter 2, principally spatial and temporal heterogeneity. For example, species might be pests in one field or house but not another, EILs change as market forces change (adding economic complexity to the mix), pests and their targets have complex interconnected life cycles, and the pest's capacity for causing harm might not remain constant over time. In addition, analysing ecological relationships as interacting pairs (hunter–hunted; consumer–consumed) is a gross simplification, since both pest and target may have significant ecological relationships with other species that have knock-on effects for the trophic relationship between food and pest.

The difficulty of quantification

A second problem concerns the practicality of quantifying or measuring some of the basic parameters required to set the EIL in the first place. While it is entirely possible to measure the damage caused by pests, it is far from easy to do so in a field setting. Also, environmental heterogeneity and variation within species are such that values obtained in one field infested with one population of a pest might be wholly different to those from a neighbouring field infested with a population of the same pest arising from a different founding event.

A related problem is that the EIL approach requires that the pest population is known. Since counting every pest individual is not practical, then a **sampling regime** must be employed and sampling carries with it a sampling error. For example, there may be many aphids on plants in the corner of a field, but the environmental conditions in the rest of the field might not favour aphids so they are almost non-existent away from the sampled corner. It is possible to improve both the sampling regime (how many samples are taken and how they are selected) and the sampling technique to reduce that error, but the best way to reduce sampling error is to increase the number of samples until it reaches its logical conclusion whereby the 'sample' is simply the whole field and every individual is counted. The incontrovertible conclusion in most cases is that to get better estimates requires better sampling, which requires more effort, which incurs greater cost. EIL is a modelling approach, and a model is only as good as the data fed into it. If fundamental values like initial population or population growth rates are not reliably known then it can be difficult to convert the theory into practice.

Lag time

A final problem is that in most cases pests do not respond instantly to whatever control measures are employed against them. This is referred to as lag time, and is similar in concept to the lag time between a non-native species being introduced to a novel environment and any effects on that system (Section 8.5.2). If a pest population is sampled and the population is at the EIL, then by the time the measures are implemented on the ground and have taken effect, it is likely that the actual pest population will have continued to rise, exceeding the EIL and, by definition, causing economic damage. Ideally the lag phase will be known (and be constant) and the dynamics of the pest population around the EIL will be sufficiently understood so that control measures can be put in place at just the right time so that they take effect just as the EIL is reached.

Such mathematical and biological balancing is difficult, but practical pest control can make effective decisions on the ground based on the **economic threshold** (ET), also known as the **control action threshold** (CAT) or simply the action threshold (Pedigo et al., 1986).

9.3.3 The economic threshold.

The economic threshold is a pest population below the EIL that can be used rather like an early warning. When the ET is reached, the density of the pest has reached the level at which action should be taken to prevent the pest from reaching the EIL. The ET is based on detailed ecological knowledge of the pest–food interaction, often from studies of previous outbreaks, as well as knowledge of the EIL and climatic records. Armed with this information, pest managers can formulate rules that can vary depending on the time of the year and that set levels of observable pest density and activity that, if breached, should trigger control measures (Case Study 9.2 and Figure 3.7 on soybean rust outbreaks in southeast USA).

However, if the EIL is subject to change depending on season, local environment, commodity prices and other factors, then so too are any other thresholds set below that level. Just as with EILs, setting effective ETs relies on a sound ecological understanding of the pest and its associated species, as well as an accurate and frequent monitoring programme (Chapter 3).

FURTHER READING FOR THIS SECTION

The concept of EIL: Pedigo, L.P., Hutchins, S.H., and Higley, L.G. (1986) Economic injury levels in theory and practice. *Annual Review of Entomology*, Volume 31, 341–368.

9.4 Pest management in practice.

Humans have always competed with other species for food. In our earliest hunter-gatherer stages (and indeed still in some indigenous cultures in the modern world) we would have competed with birds for newly ripe berries and with large predatory and scavenging mammals at kill sites. Such competing organisms would have surely been regarded as 'pests' by our ancestors, who would have also provided nutrition for a suite of invertebrates like fleas, lice, mites, and ticks. As we settled into more stable and larger communities and developed agriculture, then the resulting aggregation of resources for pests to exploit would have resulted in greater numbers of pests associated with human settlement, including new associations with pests making use of novel resources such as stored grain or preserved meat.

Early attempts to control pests included **physical** measures (such as rat-proof grain stores), **chemical** measures (Homer refers to the use of sulphur in pest control in 1000BC), and **biological** measures (with Chinese farmers establishing predatory weaver ants in citrus trees to control aphids in 300AD). In principle, pest control

CASE STUDY 9.2

Alfalfa weevil in alfalfa crops in Colorado

Alfalfa *Medicago sativa* is a clover-like plant that is used as fodder for livestock. It is an economically important crop that is widely grown although it is especially significant in North America. The alfalfa weevil *Hypera postica* is a beetle that can cause substantial damage, with heavy infestations reducing yield and quality through herbivory (Figure A). Adults feed on plants, but it is the larvae that cause most damage. An adult female can lay up to 4000 eggs in a lifetime and the emerging larvae feed voraciously on leaves.

Alfalfa weevil control becomes essential only when the EIL (Section 9.3.1) is exceeded, in other words, it is usually only sensible to invest in control when doing so is better economically than doing nothing. In practice, the economic threshold (ET) is used to indicate when pest control is worthwhile (Section 9.3.3).

Figure A The alfalfa weevil *Hypera postica* is a beetle that causes substantial damage to alfalfa *Medicago sativa* crops.

Source: Clemson University - USDA Cooperative Extension Slide Series, Bugwood.org

In the USA, farmers can consult State-based Extension Offices for advice on a range of agricultural issues including pest control. Run from universities, the network of Extension offices offers advice tailored to the local situation. The Colorado State University Extension offers advice for farmers seeking to deal with alfalfa weevil using insecticides. Note that the indication to treat is not usually based on knowing the actual pest population, but is instead based on a proxy measure of population that is possible to obtain with far less time, effort, and expense. This is a key feature of an effective ET. In this case, the proxy measure involves:

- Assessing the severity of an infestation by calculating the percentage of damaged terminals [a type of leaf], by counting the number of larvae per stem, or by counting the number of larvae captured with a standard 15-inch diameter sweep net used in a 180 degree sweep.
- Counting stem infestations by gently cutting several groups of 20 stems per field and shaking the larvae into a bucket or pan for counting. Determine larvae per sweep by averaging the counts from several sets of 25 sweeps.

If 30–50% of terminals are damaged, if larval counts average 1.5 to 2 per stem, or if larvae average 20 per 180° sweep, advice is often to make an insecticide treatment or cut the crop immediately.

FURTHER READING

Details of alfalfa weevil control in Colorado: Available at: http://www.ext.colostate.edu/pubs/insect/05500.html

The very comprehensive alfalfa weevil control programme in Iowa: Available at: http://www.extension.iastate.edu/

measures have not changed a great deal, with modern pest control also being framed around physical, chemical, and biological action. However, in practice these measures have become far more sophisticated.

9.4.1 Physical pest management.

Physical pest control involves removing pests or setting up barriers that prevent them from becoming a pest. Typically low-tech solutions, physical control methods can work well for some pests in some situations, but when pest numbers and the area of land requiring control increase, physical methods often become less economically sensible.

Simple physical removal of pest organisms can be effective in some cases. A gardener weeding a window box is undertaking effective pest control, but scaling up the problem to a large garden or park might mean that weeding is no longer the preferred option. **Physical removal** of animals is possible when their numbers are relatively low and when they are capable of being physically removed. Hence, it is entirely feasible to cull deer by shooting where they are causing damage to young trees in woodland, but removing aphids from a rose garden is far more likely to involve chemical control.

Trapping is another example of physical pest control when it is done for this purpose rather than for fur or food. Rodent traps, including the traditional spring-loaded mouse traps, are effective physical pest control where rodent numbers are relatively low and the trapping area manageable. Fly paper and other sticky traps are similarly effective for small-scale insect removal.

Barrier methods work by physically preventing pests from coming into contact with whatever resource is being defended. Netting can prevent birds from taking growing berries and cloches (or row covers) can help to reduce insect pests. Preventing pests from entering buildings that store food or resources attractive to pests can be achieved by designing pest-proof measures into the buildings themselves. Grain stores for example might be raised off the ground on rodent-proof stands (Figure 9.4), including the first recorded example in 13BC designed by the Roman architect Marcus Pollio.

Physical pest control measures can be highly cost effective, especially if they involve design features (like rodent proofing) that work without any further input. However, removal methods like trapping, culling, and weeding are generally labour-intensive and do not scale to larger applications. For these, the pest controller generally has to turn to chemical (Section 9.4.2), biological (Section 9.4.3), or Integrated Pest Management (Section 9.5).

Figure 9.4 Physical pest control can be simple and cost-effective. In this case, a grain store in Zermatt, Switzerland, is raised off the ground on mushroom shaped staddle stones to protect the grain from rodents.

Source: Photograph by Anne Goodenough.

9.4.2 Chemical pest management.

Pesticides are simply chemicals that kill pests and these can be subdivided into herbicides (killing plants), insecticides (killing insects, although often affecting other arthropods), acarides (targeting the acari, or mites and ticks), nematicides (acting against nematodes), fungicides (targeting fungi), and so on.

The development of pesticides into a large-scale global industry started after the Second World War. During the War there had been a rush to develop effective insecticides to control insects vectoring organisms causing diseases in the tropics. Agriculture also required effective insecticides and in 1939 the insecticidal action of dichlorodiphenyltrichloroethane (DDT) was first identified. The discovery of DDT won Paul Hermann Müller the Nobel Prize in Physiology or Medicine in 1948 in recognition of the enormous benefits that this so-called organochloride provided in controlling the vectors of yellow fever, malaria, and other insect-vectored disease. After some years of use the dangers of DDT to the wider environment were identified and its use was banned (Hot Topic 4.1). DDT was followed by other organochlorides, organophosphates, and then the carbamates, giving a trio of synthetic organic (chemicals based around carbon) insecticides that, after the war, were rapidly employed in pest control in agriculture. These insecticides were cheap, easy to apply, **broad spectrum** (which is to say they were capable of killing a wide range of pests) and, most importantly, effective.

At more-or-less the same time that this insecticide revolution was occurring, similar leaps forward were being made in the development of chemical herbicides. Controlling competitive plants is a more difficult prospect than controlling insects because the biology of the pest is often so similar to the biology of the associated species. Broad-spectrum herbicides are likely to be just as effective against the crop as it is against the weed. However, research into plant growth regulatory chemicals led, in the 1930s and 1940s, to the development of the first selective herbicides, which were used initially to control weeds in cereal crops. The ever-increasing use of chemical agents in the control of pests has led to the period from the end of the Second World War being described as the 'Golden Age of Pesticides' (Casida & Quistad, 1998).

It is useful for an Applied Ecologist to have a broad understanding of the different types of pesticides in use historically and today. It is also valuable to understand their modes of action since this underpins their effectiveness in practice. However, pest control and its regulation are areas in which developments can happen relatively quickly and changes can have serious legal and ecological implications. For practitioners working in pest control it is clear that the use of pesticides requires extreme care, up-to-date knowledge, and good understanding. Partly, this knowledge and understanding is ecological and involves an understanding of EILs and ETs as well an ability to plan, implement, and understand monitoring programmes for pests and their impact. However, the complex legal landscape making the use of certain pesticides illegal or subject to stringent regulations means that practitioners also need to keep up to date with the current regulations regarding pesticide use in their region.

It should also be noted that pesticides bought legally can sometimes be stored for long periods before use and during that period regulations concerning their use can change. Such regulations could include the introduction of blanket bans, regulations concerning their use that restrict applications to specific doses, times of the year or crops. Regulations might also include changes in Personal Protective Equipment (PPE) or level of training required by those applying the pesticides.

For a more detailed examination of their chemistry, mode of action, indicative use, and application then it is wise to consult a text dedicated to pest management in conjunction with up-to-date resources covering their regulations. It is also desirable, and in many countries essential, to undertake relevant **specialist training** and to gain formal qualifications before undertaking chemical pest control. Details of such requirements can be found through **regulatory bodies**. Even if no such formal regulation is required or enforced, the harmful nature of many pesticides to human health and to the environment means that a responsible Applied Ecologist should seek to follow best practice regardless of the territory in which they are operating. Overall, pest control should be a highly professional activity carried out by skilled and knowledgeable practitioners.

The aim of the following sections is not to provide a 'how to' manual of pest control or to give an exhaustive summary of the pesticides available. For a single, definitive source covering pesticides, including details of their modes of action and application, consult Krämer et al. (2011). The following sections provide an overview of some of the main processes and procedures used, giving a foundation on which more specialist knowledge can be built should it be required.

Common insecticides and how they work

An insecticide is a chemical that is effective at killing insects, and they are widely used in pest control including agricultural control (to kill insects consuming crops), domestic control (to kill cockroaches and flies in kitchens for example), and medical and veterinary control (to kill fleas and lice). Insecticides include chemicals effective against different life stages and can include specialized ovicides (eggs) and larvicides, as well as chemicals affective against adults.

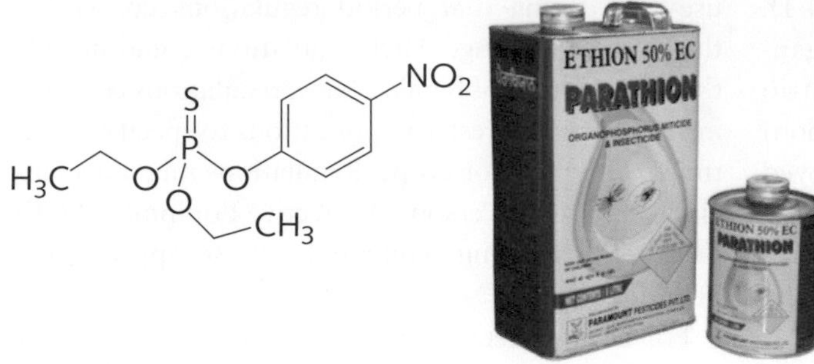

Figure 9.5 Parathion was a widely used organophosphate insecticide but it has now been banned in some countries because of its potential to harm humans.

There are different ways to classify insecticides but the broadest classification is between inorganic and organic insecticides.

Inorganic insecticides are commonly compounds containing copper, arsenic, and other metals, or sulphur-based compounds. Inorganic insecticides are generally stomach poisons and work by being applied directly to plants where they can come into contact with insects. Consequently, they are known as **contact insecticides**. Residues of metallic compounds persist in the environment and are broadly toxic and for these reasons they are seldom used in modern pest control.

Most insecticides are organic compounds containing multiple carbon atoms linked in a wide variety of different ways, and associated with range of other elements. The majority of these are contact insecticides. In some cases (e.g. neonicotinoids) insecticides can be incorporated in seeds or applied to soil such that the plant takes up the compound and incorporates it into its tissues, transferring it to the pest as the pest feeds. Compounds that are incorporated in to the body of the plant in this way are known as **systemic** rather than contact insecticides. Insecticides have a number of different **modes of action** (for an overview see Coats, 1982):

- **Organochlorides** including aldrin, chlordane, dieldrin, endrin, and DDT target insects' nervous systems. Such a mode of action is broad-spectrum and DDT and other organochlorides have deleterious effects on organisms other than target insects. They are also long-lived in the environment, accumulate in organisms within the food chain, and many are classified as **persistent organic pollutants** (POPs). As a consequence of their broad toxicity a great many organochlorides have been banned in different territories, although they were used extensively during the mid-20th century.
- **Organophosphates** include parathion (Figure 9.5) and malathion. Like organochlorides, they target the nervous system and this makes organophosphates effective against organisms other than the target insects. In fact some organophosphates such as sarin have been developed as nerve agents against humans. Organophosphates degrade far more rapidly than organochlorides in natural conditions. They are used widely in agriculture, residential gardening, park management, and in the control of disease-vectoring insects. Despite this widespread use, a number of countries have banned some organophosphates, most notably parathion although other organophosphates such as dichlorvos and azinphos-methyl have also been banned or heavily restricted. For an up-to-date list of current restrictions it is essential to consult local organizations overseeing pesticide usage.
- **Carbamates**, such as aldicarb (the active substance in the insecticide Temik®) are similar in their mode of action to organophosphates. As well as being effective against arthropods, aldicarb is extremely effective against nematode plant pests and it is widely used in controlling such pests on potato crops. It is also used when pests have become **resistant** to organophosphates (see Section 9.4.4). Most carbamates target the nervous system, but fenoxycarb acts as an insect

growth regulator, mimicking an important insect hormone called juvenile hormone, which prevents immature forms of insects becoming adults. Insecticides that work this way are sometimes called juvenoids. One of the most widely used juvenoids is methoprene, which is approved by the WHO for addition directly to water cisterns to control mosquitoes (and consequently to control the diseases they transmit).

Plant secondary compounds inspire pyrethroids and neonicotinoids

Plants produce a wealth of so-called **secondary compounds**. These are chemicals produced by plants that are not essential to their functioning, but that play some otherwise important role such as defence against herbivores. In some cases plants provide natural 'on-board' protection against insect predators through secondary compounds. The development of organic chemistry means that versions of these compounds can be synthesized by chemists and used for large-scale pesticide applications. Two of the best known classes of such bio-inspired compounds are pyrethroids and neonicotinoids.

Pyrethroids are compounds that mimic pyrethrins, one of which, pyrethrum, is a naturally occurring insecticide found in the flowers of *Chrysanthemum* plants. Pyrethroids target the nervous system of insects and are the active compounds in most domestic insecticides as well as sometimes being used in medical pest control. Permethrin for example is the primary treatment for the mites that cause scabies in humans. It is on the World Health Organization's List of Essential Medicines (World Health Organization, 2013), and it is also used as an agricultural insecticide, as an insect repellent, as a frontline defence against bedbugs and fleas and as veterinary insecticide in both agricultural and domestic settings. First generation pyrethroids, developed in the 1960s (such as bioallethrin and resmethrin), were not especially stable in the environment, being broken down within a few days under natural field conditions. In one sense this lack of environmental persistence is a desirable quality since it prevents the build-up of insecticide pollutants in the soil and groundwater, but such a lack of persistence makes it difficult to apply agriculturally. In the 1970s more stable second generation pyrethroids (including permethrin) were developed and used agriculturally, but with the trade-off that these compounds were substantially more toxic to mammals. Pyrethroids are also toxic to many aquatic invertebrates even at extremely low levels, to non-target insects, and to fish.

Nicotine, like pyrethrin, is a naturally occurring insecticide produced as a secondary compound by a plant, in this case *Nicotiana* or the tobacco plant, a member of the nightshade group of plants and a relative of the infamous deadly nightshade (Figure 9.6). Nicotine itself is an alkaloid, a group of compounds that includes caffeine, cocaine, morphine, and quinine. In low doses, such as those delivered by smoking tobacco, nicotine is a stimulant, but in higher doses it is highly toxic and more so to vertebrates than insects. Nicotine itself has been used as an insecticide, but chemists in the late 20th century seeking effective insecticides with lower environmental toxicity (and with the potential for lucrative commercial

Figure 9.6 The tobacco plant *Nicotiana* produces nicotine, which is an effective insecticide. Neonicotinoids are insecticides with a molecular structure similar to nicotine but with lower vertebrate toxicity.

Source: Image courtesy of Carl Lewis/CC BY 2.0.

patents) began to seek molecules similar to nicotine with lower vertebrate toxicity. Neonicotinoids (commonly referred to as neonics) were a result of this search. They work by targeting the nervous system and a number of different neonicotinoids have been synthesized and licensed for use.

Imidacloprid, a neonicotinoid developed by Bayer, is now the most widely used pesticide in the world. As well as being used agriculturally, neonicotinoids are used in domestic gardening, the control of pests in domestic animals and for treating timber against insect pests. It seemed for a while that neonicotinoids were 'wonder' insecticides (a term also used by some to describe DDT in the 1940s: Metcalfe, 1980). They are highly effective against a range of insect pests in different applications, they have relatively low environmental toxicity, and they have low toxicity against vertebrates. However, in the 2000s they began to be implicated in the decline of honeybees (and other vital pollinators) and these initial implications were soon supported with scientific studies that showed them to be having important **sub-lethal** effects. This has led to restrictions in use in a number of territories (Hot Topic 9.1).

Plant-incorporated protectants

A relatively recent development in the battle against pests is to incorporate into the DNA of crop plants genes that code for pesticidal proteins that are naturally present in other species. These donor species could be from any taxonomic group since it is the gene, or more correctly the function of the protein that it codes for, that is important. The plant that results from this genetic engineering is a **genetically modified organism**, or GMO, that has the ability to produce built-in protectant molecules. Genetic engineering of this type is controversial, but can also be very effective (see Hot Topic 9.2; also Section 6.8 and Hot Topic 6.1).

The complex trade-offs between target-toxicity, **non-target effects**, environmental persistence, effectiveness, and range of application has led to a succession of chemical insecticides being developed throughout the 20th and 21st centuries. Early insecticides were highly effective, but had high environmental toxicity and subsequently each new generation of insecticide has attempted to reduce non-desirable harm while maintaining effectiveness. However, fundamentally, insecticides target some aspect of insect physiology and insect physiology tends to be conserved (i.e. is similar) between diverse insect orders. This means that both contact and, to a lesser extent, systemic pesticides always have the capacity to have detrimental effects on non-target insects and other organisms.

Common herbicides and how they work

Effective chemical control (herbicide) for plant pests within cereal crops (a common application of herbicides) requires that the chemical is selective in its action, targeting dicotyledons, broadleaf plant pests, but sparing monocotyledons, the cereal crop. Ideally, such a chemical should also be non-toxic to humans. The first widely and successfully used chemical herbicides were 2,4-dichlorophenoxyacetic acid (2,4-D) and 2-methyl-4-chlorophenoxyacetic acid (MCPA) and were selective in their action. Plant growth is regulated by plant hormones of which the most important are auxins. 2,4-D and MCPA mimic an auxin, causing abnormal and unsustainable growth that results in wilting and death. Broadleaf plants are more susceptible to this effect than monocots, hence the selective action of 2,4-D and MCPA (Song, 2014).

The commercial release of 2,4-D and MCPA in the mid to late 1940s transformed global agriculture and their success led to the development of more herbicides:

- **Atrazine** was developed in the late 1950s and is one of the most widely used herbicides, especially in the USA and Australian agricultural systems. It has found favour among farmers growing maize *Zea mays* and sugarcane *Saccharum*, and is also widely used as a herbicide on lawns including golf courses, roadway verges, and residential lawns. It works by binding with a protein essential to photosynthesis, leading to plant death. There are some concerns about its effect on human health (TEACH, 2007) and on the health of animals in ecosystems exposed to atrazine and these led to a ban in the European Union in 2003 (Sass & Colangelo, 2006). Atrazine continues to be legal in other territories.
- **Paraquat** was first identified as a herbicide in 1955. By 1962 it was being synthesized and mar-

HOT TOPIC
9.1

Neonicotinoids

Introduced in the 1990s, neonicotinoids are a family of highly effective and commercially extremely successful insecticides used globally to control insect pests on a wide range of agricultural crops. Indeed, imidacloprid is the most widely used insecticide in the world (ChEBI, 2014). Neonicotinoids are systemic insecticides that are absorbed by treated plants, entering into insects that subsequently feed on those plants. They are toxic to non-insect animals, but overall their toxicity is considered to be low under normal use.

Despite being effective and seemingly low in toxicity, neonicotinoid use has been restricted in the European Union for two years from December 1st 2013. The reason behind that ban is the detrimental effect that neonicotinoids have been shown to have on pollinators.

Many crops require insects to transfer pollen from male flowers to female flowers, thereby pollinating the flower and allowing it to set fruit. If pollinators decline then this will have knock-on effects to productivity in agricultural systems and wider ecological effects to the ecosystem in general. Bees (which include solitary bees as well as the better known honeybees and bumblebees that live socially) are important insect pollinators in many systems and have tended to hog the limelight when it comes to pollination, but a great many other insects also undertake pollination.

The plight of pollinators became news in the 2000s, when honeybees *Apis mellifera* in the USA were hit by Colony Collapse Disorder (CCD). In these cases, honeybees left hives, but failed to return. This caused massive losses to beekeepers and sparked an international media interest in bee health that continues to the present day. CCD is still not fully understood, but many beekeepers believed that pesticides had a role in CCD as well as causing other problems for honeybee health (Stankus, 2014). This period saw a growing number of calls for a ban on the use of neonicotinoids.

The scientific evidence for harmful 'in-the-field' effects of neonicotinoids on honeybees has been difficult to gather, partly because natural doses are typically likely to be tiny (bees contact systemic pesticides principally through pollen and nectar) and partly because of the problems controlling for the many variables that affect natural systems. However, despite these problems, there is a growing evidence base to support the fact that the normal use of neonicotinoids has a detrimental effect on pollinators (Fairbrother et al., 2014) and it was a consequence of this evidence base that the EU decision was taken to impose a two year ban on their use.

The effect of neonicotinoids on bees (which have borne the brunt of the experimental work) is more subtle than simply causing death. For example, it has been shown that exposing bumblebee *Bombus* foragers impairs their ability to learn how to forage (Gill & Raine, 2014). This, in turn, reduces the productivity of the nest, which may result in nest failure. As more evidence mounts, additional territories are considering bans or reductions in use. However, it remains to be seen whether the temporary ban in the EU will have a beneficial effect and the debate continues to be highly polarized.

QUESTIONS

Why is it so difficult to quantify the effects of neonicotinoids on natural populations in the field?

Should different users of pesticides (e.g. farmers and gardeners) be subject to different legislation regarding pesticides?

REFERENCES

ChEBI (2014) *The Database and Ontology of Chemical Entities of Biological Interest: Imidacloprid* http://www.ebi.ac.uk/chebi/searchId.do?chebiId=CHEBI:5870

Fairbrother, A., Purdy, J., Anderson, T. & Fell, R. (2014) Risks of neonicotinoid insecticides to honeybees. *Environmental Toxicology and Chemistry*, Volume 33, 719–731.

Gill, R.J. & Raine, N.E. (2014) Chronic impairment of bumblebee natural foraging behaviour induced by sublethal pesticide exposure. *Functional Ecology*, Volume 28, 1459–1471.

Stankus, T. (2014) Reviews of Science for Science Librarians: an update on honeybee colony collapse disorder. *Science and Technology Libraries*, Volume 33, 228–260.

HOT TOPIC 9.2

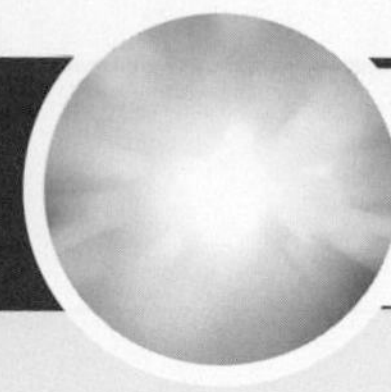

GMO

One way to control insect pests is to expose them to pathogens. *Bacillus thuringiensis* is a bacterium that kills insects if they ingest it and consequently it has been used as a microbial insecticide. Handily, it has different **pathotypes** that are specific against different types of insects, giving some level of specificity as well as high toxicity to the target and lack of toxicity to other organisms.

Genetic engineering techniques allow scientists to incorporate 'useful' genes from a donor organism (in this case, the gene encoding insecticidal crystal protein Cry from *Bacillus thuringiensis*) into a recipient organism. If the recipient organism is a plant then the resulting 'transgenic plant' produces the compound naturally as it grows.

So-called plant-incorporated protectants are systemic insecticides, and the first such use was in 1996 with the development of *Bt* corn. This **transgenic** (genetically modified or GM) variety of maize had been genetically altered to express the Cry protein from the bacterium *Bacillus thuringiensis* (hence *Bt*). This protein is produced throughout the plant, and when ingested by vulnerable insects (for example, the larvae of the European corn borer moth *Ostrinia nubilalis*) it is converted in the alkaline environment of the stomach to a toxin that rapidly paralyses the insect's digestive system and forms holes in the gut wall, leading to starvation and death.

It is now possible to genetically alter crop plants to produce on-board insecticides and *Bt* genes have been incorporated into commercially important crops particularly maize, soybeans, and cotton. However, it is also possible to incorporate herbicide resistance genes into crop plants. This means that very effective, but broad-spectrum, herbicides like glyphosate can be used with impunity to eradicate weeds from crops genetically modified to resist its actions. Of course, it is also possible to combine insecticide and herbicide properties in the same transgenic plant.

Triple-stack corn developed by Monsanto is a plant that has been modified to protect against corn borer (an insect pest) and corn root worm (despite its name, a beetle larvae pest), as well as to provide herbicide resistance to Roundup, the Monsanto manufactured glyphosate herbicide. Such crops are termed Roundup Ready and the advantages for pest management are clear.

Genetically modified crops, or biotech crops, are widely grown especially in the developing world. Their use has increased year on year with 79% of soybean production and 32% of maize being from GM crops in 2013. The leading nations in GM crop use by field area in 2013 were the USA (70.1 million hectares), Brazil (40.3), Argentina (24.4), India (11.0), Canada (10.8), and China (4.2). In stark contrast, there are concerns over the effect of GM crops on ecosystems and human health in the EU, and GM plants are highly restricted (although recent legal changes might mean an increase in their uptake).

QUESTIONS

In Hot Topic 6.1 the use of genetically engineered microorganisms in remediation was discussed. What differences are there between GEMs and crops modified genetically to resist herbicides?

Genetically modified organisms can be 'owned' by companies. Why are some people concerned about this?

FURTHER READING

Overview of GMOs: Available at: http://www.gmo-compass.org/eng/agri_biotechnology/gmo_planting/257.global_gm_planting_2013.html

Developments in EU legislation: Available at: http://www.bbc.co.uk/news/world-europe-30794256/

Introduction to the GM debate and its scientific background: Halford, G.M. (2003) *Genetically Modified Crops*. London: Imperial College Press.

Impact of GM crops: Ferry, N. & Gatehouse, A.M.R. (2009) *Environmental Impact of Genetically Modified Crops.* Oxford: CABI Publishing.

keted by ICI, and became the most commonly used herbicide globally. It is a non-selective and quick acting herbicide that is highly effective at killing plants on contact. It is rapidly absorbed into the plant, within a few minutes of application, which means that it is not washed off and rendered ineffective if it rains shortly after application. These properties make it ideal for clearing land both agricultural and residentially. It is marketed under names including Gramoxone®, Weedol®, and Pathclear® (Figure 9.7). The fact that it does not harm mature tree bark has led to its widespread use for weed control in fruit orchards, plantation crops (such as coffee, cocoa, and oil palm), in ornamental trees and shrubs and in forestry. Although used as a herbicide, paraquat is toxic to all organisms and is fatal to humans even in small doses. In most developed countries there are now strict controls over the use of paraquat that include quantities, location, and methods of application. Some countries (including Finland, Norway, and Sweden, none of which have much reliance on agricultural exports) have banned paraquat completely, others have restrictions on purchase (requiring certification for people wanting to obtain it), on use, and on both purchase and use. However, despite restrictions it continues to be one of the most widely used herbicides.

- **Glyphosate** was developed as a herbicide by Monsanto and brought to market in the 1970s under the trade name Roundup®. It is a broad-spectrum herbicide that works by inhibiting an enzyme required by actively growing plants to synthesize amino acids essential for their growth. Glyphosate is not deemed to pose a health risk to humans (Williams et al., 2000) and its environmental toxicity is considered to be low. However, glyphosate use is controversial in some sectors and arguments are often conflated with issues surrounding the development of glyphosate-resistant GM crops and wider concerns over GMO (Hot Topic 9.2; also Hot Topic 6.1).

Figure 9.7 Paraquat is a highly effective and widely used herbicide sold under a variety of brand names.

Source: Colin Underhill / Alamy Stock Photo

Other pesticides

The diversity of organisms considered as pests has inevitably led to a range of chemical treatments available to control them, of which herbicides and insecticides represent the largest classes. Other pesticides include:

- **Metaldehyde**, used to control slugs and snails (a molluscicide).
- **Methiocarb**, a carbamate insecticide also used as an acaricide, a molluscicide, and as a seed dressing to repel birds.
- **Rotenone**, used an unselective piscicide to control fish, as an acaricide to treat infested humans and livestock and as short-lived broad-spectrum pesticide for arthropods, especially as pesticide dust in gardens.
- **4-Aminopyridine and starlicide**, used as an avicide to control birds.
- **Strychnine**, used to control birds and small vertebrates, especially rodents (a rodenticide).
- **Warfarin and pindone**, used as rodenticides.
- **Sulphur**, commonly used as a fungicide.

All pesticides should be handled according to best practice guidelines, but substances that are targeted at vertebrates should be handled especially carefully. Strychnine and warfarin, for example, are potent poisons and misuse could have serious consequences.

9.4.3 Biological pest management.

Physical pest control is cost effective in some applications (e.g. grain stores designed to protect against pests) and direct pest removal can be effective at small scale. At larger scales agricultural techniques like tilling or weeding can also work. Chemical pest control often provides widespread and effective coverage coupled with proven commercial advantage. In some cases, inspiration for these chemical controls has been taken from biology with plant secondary compounds acting as models for insecticides, for example. A logical next step is to move from bio-inspired chemical control to purely biological control, taking the approach that 'an enemy of my enemy is my friend'.

Biological pest control combines ecological processes like herbivory, predation, and parasitism with active human management. The principle is simple. If a crop is being attacked, for example by sawfly larvae, and there is a naturally occurring parasite or parasitoid that targets this insect species, then processes that introduce or encourage those natural enemies should result in a reduction in the damage caused by the pest.

Regardless of the species employed, biological pest control can occur in three ways (O'Neil & Obrycki, 2009):

- **Introduced** control aims to bring natural enemies of pests to an area where they do not naturally occur. This is often used for pests that are themselves introduced (Chapter 8) when that introduction occurred in isolation from species that regulate its population (through predation, parasitism, etc.) in its native range. Although controversial at times (it is, after all, trying to combat one problematic introduced species by introducing another) this 'fighting fire with fire' approach has been highly successful in many cases. An early example of success concerns cottony cushion scale insect *Icerya purchasi*, detailed below.
- **Augmentation** of natural enemies acts to boost existing natural populations of species targeting the pest. Augmentation can involve small numbers of released organisms (innoculative release, such as the release of the parasitic wasp, *Encarsia formosa*, for the control of greenhouse whitefly *Trialeurodes vaporariorum:* see Online Case Study for Chapter 9), or very large numbers (inundative release, such as the use of nematodes for the suppression of black vine weevils *Otiorhynchus sulcatus*) (O'Neil & Obrycki, 2009).
- **Conservation** of naturally occurring enemies of pests involves land management techniques that are sympathetic to natural enemies (including predators like birds), thereby increasing their population and their effect on pest populations. Conservation approaches require good knowledge of the pest enemy's ecology so that suitable refugia, nesting sites and food can be provided.

You can find the online case study at www.oxfordtextbooks.co.uk/orc/goodenough

Any potential natural predator, pathogen, parasite, or parasitoid is a potential biological pest control agent and consequently the organisms that have been used for control are diverse. They include insects (parasitoid wasps feature prominently, as well as beetles and various species of fly), nematodes, bacteria, fungi, amphibians, and plants. An unusual form of biological pest control, using mosquitos infected with a bacterium that can spread through a population and prevent mosquitoes vectoring some diseases, is discussed in the Interview with an Applied Ecologist for this chapter.

An excellent, and early, example of the value of biological pest control is provided by the cottony cushion scale insect *Icerya purchasi* shown in Figure 9.8. This tiny insect feeds directly on the sap of plants and was originally associated with *Acacia* trees from

Figure 9.8 Cottony cushion scale *Icerya purchasi* is a pest of citrus trees but can be controlled biologically by introducing a natural predator, the cardinal ladybird *Rodolia cardinalis*, shown here attacking a scale insect.

Source: Image courtesy of Mark S. Hoddle, Department of Entomology, University of California Riverside.

AN INTERVIEW WITH AN APPLIED ECOLOGIST 9

Name: Luciano A. Moreira
Organisation: Oswaldo Cruz Foundation, Brazil
Role: Senior Researcher
Nationality and main countries worked: Brazilian; worked in Brazil, The Netherlands, United States, and Australia

Why is pest management an important aspect of Applied Ecology?
Mosquito-borne diseases are a source of great suffering and finding new ways to deal with them has never been more pressing. Insecticides have proved useful in many cases, but we urgently need to think about different approaches. Understanding the ecology of the pest you are dealing with gives you the opportunity to develop novel ways to manage them.

What is your day-to-day job?
I coordinate the Eliminate Dengue Project: Brazil Challenge. I manage two large teams and coordinate their research efforts. Keeping up-to-date with their work requires me to attend a great many meetings!

What are your most interesting recent projects and why?
Dengue fever is a mosquito-borne viral disease that causes sudden fever and acute joint pain. I am the lead Scientist of the Eliminate Dengue Project (ED), which intends to release *Aedes aegypti* mosquitoes in urban settings. These mosquitoes will have been infected with a commonly found bacterium called *Wolbachia* and by breeding with wild mosquitoes will spread *Wolbachia* through the population. The presence of *Wolbachia* appears to prevent the dengue virus from replicating in the mosquito and prevents its transmission.

What's been the best part of these particular projects?
The best part of being involved with ED is very simple—I am involved in a project that has the clear potential to save many lives. It is also rewarding to work with communities and to get their support in making a difference.

What are the main challenges in your field and how can they be overcome?
Aedes aegypti is an introduced species in Brazil, but once it was here it found great conditions to survive. It has subsequently spread all over the country. Undoubtedly the biggest challenge we face in the field of mosquito pest management is finding ways to control this species (and others) and reduce or eliminate the transmission of diseases to humans.

What next for you, and why?
I will be expanding the Eliminate Dengue project to a big city and evaluating the impact of this approaching in reducing mosquito-borne diseases in general and dengue fever in particular.

Finally, how did you get into your area of work and what advice would you give to others?
I was fortunate to work in different institutions in different countries and my advice would be not to be afraid of changing your direction and working on different subjects. This, in the end, will broaden your way of thinking and make you a better ecologist.

Australia. In the latter part of the 19th century, these plants were imported to California. The cottony cushion scale insect, transported with the acacias in a hitch-hiking scenario (Section 8.3.1), found new plants on which to feed in the citrus groves of California. In less than a decade, and without the control of natural predators, parasites, or pathogens, cottony cushion scale became a major pest of citrus to which no native control solution could be found.

In its natural range, where it was not a pest, the cottony cushion scale had interactions with a predator and a parasitoid that helped to regulate its numbers and prevent it breaching the EIL. The predator was the cardinal ladybird *Rodolia cardinalis*, a beetle of the family Coccinellidae shown in Figure 9.8, and the parasitoid was a dipteran fly *Cryptochaetum*. These insects were imported to California as part of a biological pest control strategy, starting in 1888. Within a couple of years, a population explosion of beetles led to complete control of scale insects. In the longer term, the combined effect of both control agents led to a successful biological control strategy that has been used across the world's citrus producing countries (DeBach &

Rosen, 1991). Some other examples of successful biological control include:

- The use of the velvet bean *Mucuna pruriens* as a biocontrol in West African agriculture (Ceballos et al., 2012). Planted on clear-cut fields the bean both outcompetes and suppresses weeds that would typically grow on the cleared fields and, as a nitrogen-fixing legume, acts as a green manure when it is ploughed in.
- Nematodes are extensively used as biocontrol agents against a wide variety of invertebrate pests because of their ecological role as parasites (Grewal et al., 2005). For example:

  Steinernema feltiae used to control fungus gnat larvae. Fungus gnat is a general term for small flies from a number of families within the Diptera (the true flies), the larvae of which can cause widespread damage to seedlings and to houseplants.

  Deladenus siricidicola is used to control the woodwasp, *Sirex noctilio*, an invasive insect pest of pine trees in many parts of the world.

  Phasmarhabditis hermaphrodita is increasingly used in domestic, municipal, and agricultural systems as a biocontrol agent of slugs.
- Green lacewing larvae (order Neuroptera) are voracious predators on aphids and mites and are released as biocontrol agents in agriculture and gardens. Gardeners often encourage ladybird beetles (family Coccinellidae) with the aim of naturally controlling aphids.
- *Encarsia formosa* is a parasitoid (Section 2.2.3) wasp that lays its eggs within whitefly (Hemiptera; an aphid relative) with the resulting larvae ultimately killing the host. It is used as a biocontrol agent in greenhouses.
- *Bacillus thuringiensis* (or Bt) is a bacterium widely used as a biocontrol agent, especially of Lepidoptera (butterfly and moth) caterpillars. The genes coding for the protein that kills insects have been incorporated into some plants, an example of genetic engineering explored in Hot Topic 9.2.
- *Beauveria bassiana* is a fungus that parasitizes arthropods and is used as a biocontrol agent of pests including termites, thrips, beetles, and aphids.

9.4.4 Problems with pest management.

We have already seen that pest control can be problematic. Pesticides might kill species other than the target pest or cause **environmental contamination**. Biological control can cause unintended knock-on ecological consequences. Effective physical control can be severely limited by scale. There are other biological and ecological effects that may be difficult to predict.

Target pest resurgence

Pest populations can sometimes, counter-intuitively, increase after the application of a pesticide. Such an increase is termed a target pest resurgence and is caused by the non-selective action of the pesticide killing the predators, parasites, or competitors of the pest as well as the pest itself. This is known as the enemy release hypothesis (Section 8.3.2) and can occur in pest management with insecticides, since many natural enemies of pest insects are themselves insects, such as parasitoid wasps. It can also happen when **bioaccumulation** means that pesticides have non-target effects on other species in the ecosystem, including species in other trophic levels (Section 2.2). Any pest individuals that survive, and any individuals that move into the area, are released from the pressures of natural enemies and competitors, and can thrive.

Secondary pests

The release from natural enemies that can lead to target pest resurgence can also lead to species that were not considered pests before the pest control treatment becoming pests after the treatment. If species that were kept to populations below the EIL through ecological interactions with natural competitors or predators are released from these pressures, they can thrive. The classic example of secondary pest outbreaks concerns cotton crops in Central America. Insecticide use began in earnest in the 1950s. At that time there were two pests of cotton in that region: the cotton leafworm (the caterpillars of the moth *Alabama argillacea*) and the boll weevil *Anthonomus grandis*, a type of beetle that feeds on cotton buds and flowers. Initial application of organochlorine and organophosphate insecticides achieved impressive improvements in yield, but by the mid-1950s three new, secondary, pests had emerged. Increasing pesticide use reduced two of these new pests, but caused the outbreak of five further species. This pattern continued for a decade until there were eight pest species and a staggering 28 applications a year, compared to fewer than five applications initially (Smith, 1998).

Resistance

A further serious problem with chemical control is that pest organisms can evolve resistance to pesticides. Evolution depends fundamentally on variation (Section 2.6). In the case of a pest evolving resistance to a pesticide, the variation of interest is the ability to survive when the pesticide is applied. Many pesticides rely on targeting specific pathways or molecules within the target pest, and some individuals within the pest population may, as a consequence of random mutations, have slightly different but still functional pathways. It is also possible that, in some cases, these mutations allow the individual possessing them to withstand the effect of the pesticide. The mutation may not endow the individual with complete resistance and the pesticide may still have an effect, but that effect is no longer lethal.

In some cases, these mutant individuals might have had no advantage before the application of the pesticide. Indeed, it is even possible that mutant individuals are at a disadvantage compared with normal individuals (termed the 'wild type' in genetic studies) under natural conditions. Alternatively, they may have had an advantage against some of the natural analogues of pesticides that in some cases have been the inspiration for the development of the synthetic version. Whatever the situation under natural conditions, in an environment where the pesticide becomes a strong selection factor, individuals bearing genes that confer resistance are at an enormous advantage: while wild type individuals die, resistant individuals survive and are able to have offspring.

Provided that the ability to withstand the effects of the pesticide is genetic, then the offspring of the mutant plants may also carry the gene conferring the ability to withstand the pesticide. Thus, the frequency of **pesticide resistant genes** in the pest population increases and the population itself moves from the wild type, susceptible to the pesticide, to a pesticide resistant population (Insecticide Resistance Action Committee, 2007).

If pesticides remain in the environment then they will also continue to act as a selection factor even if they have not been recently applied. Furthermore, pests often have a short generation time and are capable of large potential reproductive output. Since each generation is effectively a selection event, many pests are able to purge populations of susceptible types and develop a resistant population relatively rapidly (Figure 9.9).

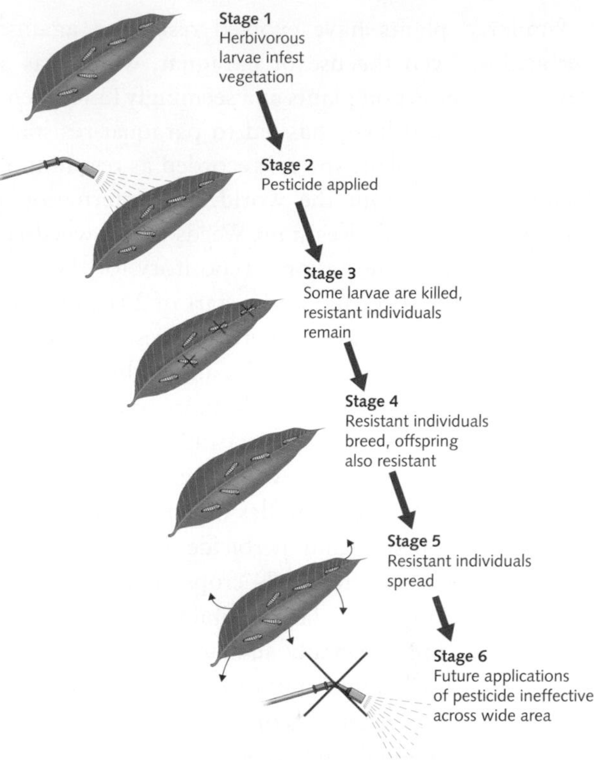

Figure 9.9 The use of pesticides leads to the selection of individuals able to tolerate them. Over time this selection leads to the evolution of pesticide resistant populations.

Of course, the mutations that cause resistance are likely to be rare. Furthermore, in the face of some pesticides it may be biologically very difficult to resist, especially if their action affects multiple pathways or involves fundamental biochemical processes that cannot work in mutated forms. However, despite these apparent obstacles, resistance has evolved in multiple pests against multiple pesticides. Indeed, it is virtually certain that a new pesticide will lead to resistance in target pests, often in just a few years. Whalon et al. (2008) reported 'more than 7747 cases of resistance with more than 331 insecticide compounds involved', and that 553 of the estimated 10,000 arthropod pest species are reported to have resistance to insecticides.

Similarly, plants have evolved resistance against herbicides. Even the use of paraquat, which has a devastating effect on plants at a seemingly fundamental physiological level, has led to paraquat-resistant plants, with 57 plant species recorded as resistant in locations throughout the world. The International Survey of Herbicide Resistant Weeds (www.weedscience.org) is the international repository for data on herbicide resistance, and at the start of 2015 they reported 437 unique cases of herbicide resistant weeds (a unique pairing of both species and herbicide mode of action), with 238 species split more or less equally between monocots (100 species) and dicots (138 species). Overall, weeds have evolved resistance to 22 of the 25 known herbicide modes of action and to 155 different herbicides, and herbicide resistant weeds have now been reported in 84 crops in 66 countries.

Pesticide resistance, like antibiotic resistance in medicine (a fundamentally similar evolutionary scenario), is a predictable and serious problem. Evolution by natural selection acting on genetic, heritable, variation is powerful, and pesticide manufacturers are ultimately in an arms race against natural processes that can only be won by keeping one step ahead of evolution. The fact that chemical pesticides still work effectively in many cases is why they are still so widely used, but the evolution of resistance, as well as valid environmental and human health concerns, leaves the door open for alternative methods of pest control.

Knock-on ecological effects

Doing one thing to a complex system can have knock-on effects that can be difficult or impossible to predict. Biological pest control, by its very nature, exploits ecological interactions. Frequently, such interactions are complex, extend well beyond the focal organisms, and are poorly understood. It is because of this that, as well as the potential for great success, biological pest control has the potential for unintended, and serious, ecological consequences, of which the introduction of the cane toad *Bufo marinus* into Australia is a famous example.

Cane toads were introduced into Australia from Hawaii in 1935 in an attempt to control the cane beetle *Dermolepida albohirtum*, which was causing serious harm to sugar cane crops, largely through the action of its larvae feeding on plant roots. Toads bred well in captivity and later that year over 100 were introduced into the cane fields around Queensland. Further introductions occurred in different regions and they fared extremely well in their new habitat. In fact, their population and their range increased dramatically until by the late 1970s there were millions extending throughout Queensland.

The Australian cane toad population now is estimated to be in the hundreds of millions and the species causes serious environmental harm. The toads spread diseases to the native amphibian fauna, they are voracious predators, and they poison animals trying to consume them through the toxins in the glands of their warty skin. Their sheer numbers, and the need to conserve native amphibians who may be affected by control measures, make control extremely difficult, and to date, largely unsuccessful. Control methods using sterile males (who would mate with females to produce spawn that does not develop) and females that produce only male offspring have been suggested, but currently there is no effective control strategy (Shanmuganathan et al., 2010).

Overall, it seems that no one form of pest control is without disadvantages, at least in the long term. An alternative approach is not to consider pest control as physical, chemical, or biological. Indeed, even each category of control need not be monolithic: this pesticide *or* that one; this biological augmentation *or* that natural enemy conservation. Taking a more pluralistic approach that combines different management strategies would seem to eliminate or reduce some of the problems that each approach has on its own, whilst also potentially providing a more effective solution. Such a philosophy is inherent within **Integrated Pest Management (IPM)** (Section 9.5).

FURTHER READING FOR THIS SECTION

An overview of the development of pesticides and their use and effects: Matthews, G.A. (2006) *Pesticides: Health Safety and the Environment.* Oxford: Blackwell Publishing Ltd.

An account of the development and use of herbicides, as well as further details on their modes of action: Cobb, A.H. & Reade J.P.H. (2011) *Herbicides and Plant Physiology*. Oxford: Wiley-Blackwell.

A general overview of biological control, and detail of some of the many applications: Bellows, T.S. & Fisher, T.W. (1999) *Handbook of Biological Control*. Millbrae, CA: Academic Press.

9.5 Integrated Pest Management.

IPM is an approach to pest management that seeks to combine different control measures in ways that are effective, but sympathetic to some of the problems of physical, chemical, and biological control methods. The IPM approach, at least in its ideal form, considers the system to be treated in a more sophisticated way and makes use of ecological theory and insight to control pests below the EIL.

IPM combines physical, cultural, chemical, and biological pest control methods in different ways to create a **multi-pronged** attack on pests. Cultural controls seek methods of agriculture that exploit aspects of pest ecology in ways that allow natural processes to keep pest levels below the EIL. An example of cultural control is the annual rotation of crops that suffer different pests. By changing the crop annually it may help to disrupt the population growth of potential pests, preventing them from causing economic injury (Radcliffe et al., 2009).

Biological pest control methods are a central component of IPM strategies. The biological aspect of IPM also includes managing the agricultural system and the surrounding environment so that pests are discouraged and beneficial organisms, such as pollinators or pest control agents, are encouraged. Genetic control methods can also be developed. Plant resistance through breeding or by genetic engineering (Hot Topic 9.2) can reduce the harm caused by pests. In some cases it may be that pest populations can be reduced by the introduction of sterile males or by manipulating female reproduction, for example, to produce only males. These genetic solutions can be used for plants and animals, and male sterility is a tactic being actively developed for mosquito control.

IPM is, as its name suggests, an integrated method, and chemical pest control can play an important role. Consequently, pesticides are used as part of some IPM operations. However, where they are used they can create problems if they affect biological control agents as well as target pest species, and so their use requires careful consideration. Pesticides also, of course, can lead to resistance and an aspect of IPM involves managing resistance. Using chemicals with restraint and changing their patterns of use focuses on the chemical side of the resistance equation, but the biological side, the problem of resistant pest organisms and the production of herbicide-resistant crop plants, also forms part of IPM's expanding portfolio of activities. Indeed it has been said that 'Good pesticide resistance management is just good IPM' (Hoy, 2009).

IPM strategies can create problems of their own:

- IPM approaches are frequently labour and knowledge intensive. They rely on accurate monitoring, for example, and they require a sound understanding of both general ecological principles, and of organisms and interactions specific to the crops, pests, and system being considered.
- An IPM strategy that works for one field may not work for a similar field in a different setting, simply because abiotic and biotic factors can differ between places, causing differences in ecological interactions. Thus, this adds to the burden of the first problem, since additional knowledge will be required to employ IPM strategies in different settings.
- IPM is not always robust enough to work entirely on its own through a full cycle of control. Combinations of biological, physical, cultural, and (limited) chemical pest management are often targeted at specific pests. It may be that over the course of the growing season further pests emerge that require recourse to broad-spectrum chemical pesticides for rapid and effective control.

IPM is underpinned by ecological theory and produces nuanced solutions to the complex ecological problem of pests. These solutions typically require more monitoring and more effort than chemical control, but reduce environmental harm and have more long-term potential. Some examples will serve to illustrate how IPM works in practice.

9.5.1 IPM control of purple loosestrife.

Purple loosestrife *Lythrum salicaria* is an attractive perennial plant that produces magenta-coloured flower spikes throughout the summer

Figure 9.10 Purple loosestrife *Lythrum salicaria* is an invasive species that can take over waterways and wetlands and is controlled using an IPM strategy.

Source: Photograph by Anne Goodenough.

Figure 9.11 Pocket gophers cause horseshoe-shaped mounds that can cause considerable damage to municipal lawns, gardens, golf courses and other human-built environments.

Source: Image courtesy of Becky Houtman/CC BY 2.0.

(as shown in Figure 9.10). It is native to central and southern Europe and parts of Asia, but is also established as a non-native plant in the USA, where it now occurs in every state except Florida (Swearingen, 2005). It thrives in wetlands, rivers, marshes, ponds, reservoirs, and ditches, and it quickly outcompetes and then replaces native grasses, sedges, and flowering plants. In short, it is an invasive species (Chapter 8) which can take over wetlands, lower plant biodiversity and reduce habitats for wildfowl; it is therefore an ecological pest that requires control.

Herbicides would certainly kill the plant, but spraying herbicides across wetland areas would also kill any remaining native plants and would likely harm aquatic fauna. Instead, an IPM approach is taken that combines physical, chemical, and biological control. The approach is more labour intensive than simple blanket spraying:

- **For small infestations:** remove by hand (physical control), or apply herbicide (chemical control) directly to the plant.
- **For larger infestations:** three insect species from Europe have been approved by the U.S. Department of Agriculture for use as biological control agents: a root-mining weevil *Hylobius transversovittatus*; and two leaf-feeding beetles, *Galerucella calmariensis* and *G. pusilla*.

9.5.2 IPM control of pocket gophers.

Pocket gophers Geomyidae are a group of small burrowing mammals that are native to Central and North America. They belong to six genera in the rodent family Geomyidae. Their burrowing activities are extensive and can lead to destabilization of soil. This can cause them to become pests if they burrow beneath farms, golf courses, gardens, or other human environments, as shown in Figure 9.11.

Gopher management takes an IPM approach that varies in strategy depending on the scale of the affected area. For small areas (<4 ha) in territories where it is legal, strychnine bait can be used. Strychnine is extremely toxic to all animals and considerable care must be taken if it is to be used. It must be applied below ground to prevent animals other than gophers consuming it (a **labour-intensive** process that is only possible for small areas), and spilt bait must be cleared up. This chemical control, undertaken in spring or autumn, should be combined with physical removal using specially designed gopher traps (a spring-loaded lethal device, not dissimilar to a conventional rat trap, but designed to sit within gopher tunnels).

For larger areas, individual gopher traps and poisoned bait are not efficient. Furthermore, increasing the area covered and therefore the amount of poison used increases the likelihood of non-target species (including humans) coming into contact with it. In these cases, a process called machine baiting is used, which uses rows of artificial burrow traps that are sunk into the soil (a process requiring specialist burrow building machinery) and baited to attract gophers (Salmon & Baldwin, 2009).

FURTHER READING FOR SECTION

IPM concepts and strategies: Radcliffe, E.B., Hutchison, W.D., & Cancelado, R. (Eds) (2009) *Integrated Pest Management: Concepts, Tactics, Strategies and Case studies.* Cambridge: Cambridge University Press.

For further case studies of IPM: http://www.fao.org/agriculture/crops/thematic-sitemap/theme/spi/scpi-home/managing-ecosystems/integrated-pest-management/ipm-cases/en/

9.6 Conclusions.

Pest control, at least in the developed world, has generally evolved from the wholesale application of chemical pesticides towards Integrated Pest Management practices that seek to find effective and sustainable methods to keep pest levels below the EIL and to maintain agricultural productivity. If there is one lesson of the history of pest control it is that sound practice greatly benefits from an understanding of ecological theory. IPM requires a particularly nuanced understanding of ecology and although this comes at a price, the advantages of an integrated approach, especially in the long term, are clear.

ONLINE ACTIVITY

Go to www.oxfordtextbooks.co.uk/orc/goodenough/ to download the activity that accompanies this chapter.

CHAPTER 9 AT A GLANCE: THE BIG PICTURE

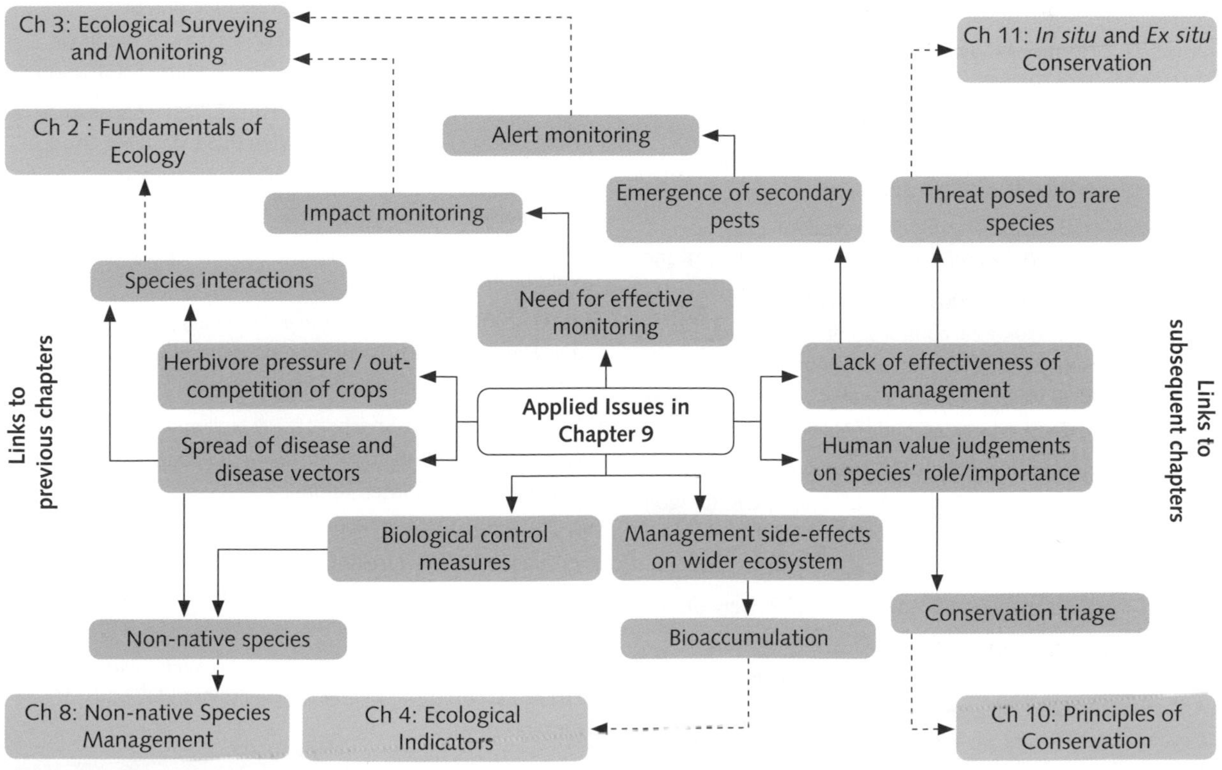

REFERENCES

Casida, J.E. & Quistad, G.B. (1998) Golden age of pesticide research: past, present or future? *Annual Review of Entomology*, Volume 43, 1–16.

Ceballos, A.I.O, Rivera, J.R.A, Arce, M.M.O. & Valdivia, C.P. (2012) Velvet bean (*Mucuna pruriens* var. *utilis*) a cover crop as bioherbicide to preserve the environmental services of soil. In: R. Alvarez-Fernandez (Ed.) *Herbicides—Environmental Impact Studies and Management Approaches*, pp. 167–184. Rijeka: Intech, Open Access Book.

Coats, J.R. (1982) *Insecticide Mode of Action*. New York, NY: Academic Press.

DeBach, P. & Rosen, D. (1991) *Biological Control by Natural Enemies*. Cambridge: Cambridge University Press.

Grewal, P.S., Ehlers, R., & Shapiro-Ilan, D.I. (2005) *Nematodes as Biocontrol Agents*. Wallingford: CABI International.

Hill, D. (1996) *The Economic Importance of Insects*. London: Chapman and Hall.

Hoy, C.W. (2009) Pesticide resistance management. In: E.B Radcliffe, W.D. Hutchison, & R. Cancelado (Eds), *Integrated Pest Management: Concepts, Tactics, Strategies and Case Studies*, pp. 192–204. Cambridge University Press, Cambridge, UK.

Insecticide Resistance Action Committee (2007) *Resistance Management for Sustainable Agriculture and Improved Public Health*. Brussels: Croplife International.

Invasive Species Compendium: *Buddleja davidii* (butterfly bush). Available at: http://www.cabi.org/isc/datasheet/10314

Krämer, W., Schirmer, U., Jeschke, P., & Witschel, M. (Eds) (2011) *Modern Crop Protection Compounds*. Weinheim: Wiley-VCH Verlag GmbH.

Kriticos, D.J. (2007) Buddleia weevil welcomed. *Landcare Research Newsletter*, Volume 39, 2.

Metcalfe, R.L. (1980) Changing role of insecticides in crop protection. *Annual Review of Entomology*, Volume 25, 219–256.

O'Neil, R.J. & Obrycki, J.J (2009) Introduction and augmentation of biological control agents In: E.B Radcliffe, W.D. Hutchison, & R. Cancelado (Eds) *Integrated Pest Management: Concepts, Tactics, Strategies and Case Studies*, pp. 192–204. Cambridge: Cambridge University Press.

Oregon Department of Agriculture (2014) *Noxious Weed Policy and Classification System 2014*. Salem, OR: Oregon Department of Agriculture Noxious Weed Control Program.

Pedigo, K.P., Hutchins S.H., & Higley, L.G. (1986) Economic Injury Levels in theory and practice. *Annual Review of Entomology*, Volume 31, 341–368.

Pimentel, D. (2009) Pesticides and pest control. In: R. Pehsin and A.K. Dhawan (Eds.) *Integrated Pest Management: Innovation-Development Process*, pp. 83–87. Netherlands: Springer.

Radcliffe, E.B., Hutchison, W.D. and Cancelado, R. (Eds) (2009) *Integrated Pest Management: Concepts, Tactics, Strategies and Case studies*. Cambridge: Cambridge University Press.

Richardson, B., Vanner, J., Ray, Davenhill, N., & Coker, G. (1996) Mechanisms of *Pinus radiata* growth suppression by some common forest weed species. *New Zealand Journal of Forestry Science*, Volume 26, 21–437.

Salmon, T.P. & Baldwin, R.A (2009) How to manage pests: pests in gardens and landscapes: pocket gophers. University of California Agriculture and Natural Resources publications 7433 UC Statewide IPM Program, Kearney Agricultural Center, Parlier. Available at: http://www.ipm.ucdavis.edu/PMG/PESTNOTES/pn7433.html

Sass, J.B. & Colangelo, A. (2006) European Union bans atrazine, while the United States negotiates continued use. *International Journal of Occupational and Environmental Health*, Volume 12, 260–267.

Shanmuganathan, T., Pallister, J., Doody, S., McCallum, H., Robinson, T., Sheppard, A., Hardy, C., Halliday, D., Venables, D., Voysey, R., Strive, T., Hinds, L., & Hyatt, A. (2010) Biological control of the cane toad in Australia: a review. *Animal Conservation*, Volume 13, 16–23.

Smith, J.W. (1998) Boll weevil eradication: area-wide pest management. *Annals of the Entomological Society of America*, Volume 91, 239–247.

Song, Y. (2014) Insight into the mode of action of 2,4-dichlorophenoxyacetic acid (2,4-D) as an herbicide. *Journal of Integrative Plant Biology*, Volume 56, 106–113.

Swearingen, J.M. (2005) *PCA Factsheet Purple Loosestrife*. Plant Conservation Alliance's Alien Plant Working Group. Available at: http://www.nps.gov/plants/alien/.

Tallent-Halsell, N.G. & Watt, M.S. (2009) The invasive *Buddleja davidii* (Butterfly Bush). *Botanical Review*, Volume 75, 292–325.

TEACH (2007) *Atrazine Chemical Summary US EPA, Toxicity and Exposure Assessment for Children's Health*. Available at: http://www.epa.gov/teach/

Whalon, M.E., Mota-Snachez, D., & Hollingworth, R.M. (2008) *Global Pesticide Resistance in Arthropods*. Wallingford: CAB International.

Williams, G.M., Kroes, R. & Munro, I.C. (2000) Safety evaluation and risk assessment of the herbicide Roundup and its active ingredient, glyphosate, for humans. *Regulatory Toxicology and Pharmacology*, Volume 31, 117–165.

World Health Organization (2013) *18th WHO Model List of Essential Medicines*. Geneva: World Health Organization.

PART 4

Conserving

Conservation is deliberate action to improve ecology, be that the extent, condition, or cohesion of habitat, or the presence, abundance, or reproductive success of species. Our exploration of this topic starts in **Chapter 10** by considering the underpinnings of conservation, as well as the scientific background, highlighting the concept that there is often no 'right' answer. It sets conservation in a theoretical context, but also debates why and under what circumstances species should be conserved and how conservation resources should be allocated.

Chapter 11 goes on to examine the conservation of species and the habitats that support them in the wild in more detail, while **Chapter 12** examines in detail the role of conservation in captivity. The section is rounded off by **Chapter 13**, which discusses the important concept of species reintroduction and looks to the future by examining the concept of habitat restoration at a landscape through the process of 'rewilding'.

10 **Principles of Conservation**

11 ***In Situ* Conservation**

12 ***Ex Situ* Conservation**

13 **Reintroduction and Rewilding**

Principles of Conservation

10.1 Introduction.

We are living in an age of **mass extinction**. Despite gaps in the fossil record that make calculation of natural background extinction rates complex, there are now some fairly robust estimates of extinction history that can be compared with current estimates. Such comparisons make for grim reading. Current rates are, on average, 1000 times higher than the background norm. In some taxa such as the cichlid fish of Africa's Lake Victoria, estimates are at least an order of magnitude higher at >10,000 times the background norm (Pimm et al., 2014).

Extinction rates, taken together with data from the International Union for Conservation of Nature (IUCN) that show 36% of the 62,549 currently-assessed species or subspecies are threatened by extinction, means that there is a widely held view that we are witnessing a mass extinction event (Barnosky et al., 2011). Of the 22,482 species listed as **threatened** by IUCN, 69 are already extinct in the wild; 4635 are listed as critically endangered; 6940 are endangered; 10,838 are vulnerable (2015 figures). Around 1.15 million known species have yet to be assessed and an estimated 7.5 million species are still to be identified (Mora et al., 2011).

Although the need for conservation has never been greater, there are also many **conservation success stories** that are object lessons in effective Applied Ecology. For example, the numbers of wild tigers *Panthera tigris*, in key areas of India and Thailand are increasing for the first time in many years, breeding success of the critically endangered hawksbill sea turtle *Eretmochelys imbricata* is improving due to conservation efforts, and Yarkon bream *Acanthobrama telavivensis*, once deemed extinct in the wild, occurs in self-supporting populations again following captive breeding and reintroduction.

In this chapter we will define conservation, examine what factors pose threats to species, and discuss extinction risk. We then provide an overview of conservation initiatives, including approaches in the wild (*in situ*) and in captivity (*ex situ*). The final section will debate **conservation prioritization** and discuss on what basis resources should be allocated, using concepts such as rarity, the role of keystone species, and public appeal. For *in situ* conservation, the wisdom of prioritizing areas with high species richness, endemism, or distinctive communities is considered (along with techniques for assessing such parameters) to lead into Chapter 11, while for *ex situ* conservation the likelihood of reintroduction and practicalities including husbandry knowledge, cost, and space are debated to lead into Chapters 12 and 13.

10.2 What is conservation and why is it necessary?

Conservation ecology is an important aspect of Applied Ecology that aims to safeguard species, habitats, ecosystems, and landscapes from destruction or degradation. The perceived importance of conservation is increasing as we deepen our understanding of the need for well-functioning ecosystems for global sustainability and the role of species and habitats within that, for example, through ecosystem services such as pollination. For many, there is also an ethical duty to conserve aspects of ecology that human activity is threatening.

The scale of conservation is perhaps best appreciated by examining statistics of the magnitude and geographical spread of current efforts, the need for further action, and the associated monetary costs, as shown in Figure 10.1.

10.2.1 Conservation versus preservation.

The terms 'conservation' and 'preservation' are often used interchangeably to mean the protection of a landscape, a habitat, or a species. However, these terms mean very different things:

- **Preservation:** to preserve a landscape and the biodiversity it contains in its 'natural' state without any interference from humans. This concept is a reaction to concerns that humans are encroaching on natural environments worldwide and the feeling that some areas should be protected completely from human activity—unused, uninhabited, and unmanaged. The term is used most often in the USA, having become part of common ecological vocabulary following the Wilderness Act of 1964.
- **Conservation:** to work within natural areas and manage the habitats and species they support. The founding principle of conservation is active management; this can vary considerably and can sometimes be very 'light touch', but it always exists.

Preservation tends to occur in countries with a large land area—for example, the USA, Canada, and

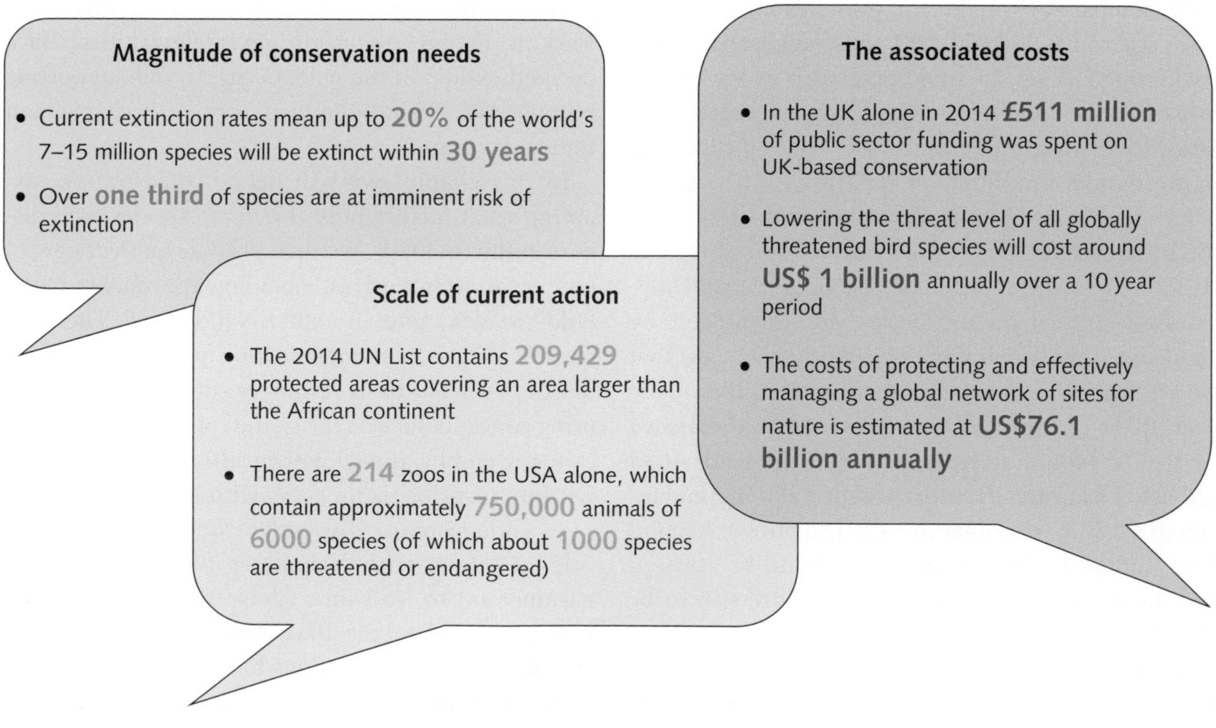

Magnitude of conservation needs

- Current extinction rates mean up to **20%** of the world's 7–15 million species will be extinct within **30 years**
- Over **one third** of species are at imminent risk of extinction

Scale of current action

- The 2014 UN List contains **209,429** protected areas covering an area larger than the African continent
- There are **214** zoos in the USA alone, which contain approximately **750,000** animals of **6000** species (of which about **1000** species are threatened or endangered)

The associated costs

- In the UK alone in 2014 **£511 million** of public sector funding was spent on UK-based conservation
- Lowering the threat level of all globally threatened bird species will cost around **US$ 1 billion** annually over a 10 year period
- The costs of protecting and effectively managing a global network of sites for nature is estimated at **US$76.1 billion annually**

Figure 10.1 Conservation statistics.

Source: Synthesis of information from WWF, DEFRA, IUCN and McCarthy et al. (2012) Financial Costs of Meeting Two Global Biodiversity Conservation Targets: Current Spending and Unmet Needs. *Science* 338(6109):946–9.

Australia—and where there are large areas of relatively undeveloped '**wilderness**' land. In large wilderness areas, natural systems are often self-regulating and external pressures can be buffered naturally. This partly stems from large areas having intact core and buffer zones rather than sites being dominated by edge effects (Section 4.2). Moreover, because of the size of such areas, they are also likely to be heterogeneous and support a range of different species naturally. The large scale also means that most species occur in large enough populations that short-term fluctuations in productivity or survival do not jeopardize their long-term viability. When combined, these factors mean that the most suitable strategy for large protected areas is protection rather than management.

In contrast, smaller or highly developed countries such as the UK often adopt a more **active** conservation approach to wildlife management. This is because small or highly fragmented habitats generally have less natural heterogeneity, such that active management is often needed to ensure species richness is high and population sizes remain viable. It should also be noted that a more active conservation approach is often seen as 'the norm' in countries that have had a history of working with landscapes (e.g. farming or active forestry involving planting and harvesting), such that social history can affect approaches to conservation.

There is some debate as to whether, in the 21st century, preservation is actually possible. Nowhere on earth is devoid of the influences of human activity, either through global impacts such as human-accelerated climate change or widespread changes such as atmospheric pollution. Even Antarctica—often regarded as the last great wilderness preserve—is subject to such impacts, with a warming climate causing ice sheet loss on the Antarctic Peninsula and DDT being recorded in Adelie penguins *Pygoscelis adeliae*. Moreover, all ecosystems change naturally through processes such as ecological succession (Section 2.8.2) and evolution (Section 2.7), such that the 'natural' state of an area is not constant. The combination of widespread human activity and natural change means that preservation can only ever protect areas from intense direct human impact and preserve their ability to change naturally: in other words, preservation can preserve only the 'natural' or 'pristine' status of an area to allow natural processes to happen, not a specific physical state.

In many ways, conservation is much more complex than preservation because it requires Applied Ecologists to make decisions about the best way to manage. This can involve **setting priorities** about which species are targets for conservation action (and which are not), deciding the best way of managing for target species, and debating what activities are permitted in conservation areas. This means that values and **subjective judgement** enter into the conservation arena, merging scientific objectivity with societal factors in ways that are not always mutually supportive.

As conservation and active management form the usual realm of Applied Ecologists, this chapter will focus largely on conservation rather than preservation, but it is important to remember that, in some instances, a **non-intervention policy** can be most appropriate.

10.2.2 The focus of conservation.

The most common driver for conservation initiatives is the protection of species, either for one focal species (single-species conservation) or for guilds or communities (multi-species conservation).

Species-focused conservation

Until fairly recently, species-orientated conservation primarily followed the **single-species conservation** approach. Such conservation is, conceptually at least, fairly straightforward, as there is one clear goal and often a reasonably clear set of steps that will give the best chance of that goal being realized. Single-species conservation can be highly effective, as shown in Table 10.1.

Single-species conservation undoubtedly still has its place, especially for very rare species or when *ex situ* (captivity-based) initiatives are important parts of overall conservation. At a site level, single-species approaches might also be particularly valuable for sites supporting endemic species or nationally/internationally important numbers of particular species. Increasingly though, species-orientated conservation is moving towards a **multi-species conservation** approach, especially within *in situ* (wild) contexts. For example, a heathland nature reserve, if managed

Table 10.1 Single-species conservation success stories (data from IUCN).

Species	Number of wild individuals before conservation	Number of wild individuals in 2015
Black-footed ferret *Mustela nigripes*	24	1000
Whooping crane *Grus Americana*	15	382
Mauritius kestrel *Falco punctatus* (Case Study 7.1)	4	400
California condor *Gymnogyps californianus* (Case Study 12.1)	15	231
Hawaiian goose *Branta sandvicensis*	20–30	2500

appropriately, will conserve multiple species. This often has more widespread benefits than focusing management on one focal species, and reduces the associated risks of doing so when there is incomplete knowledge of species' requirements.

The move to multi-species conservation is not without challenge, especially when multiple **priority species** have different habitat management requirements. For example, two grassland butterfly species in the UK, the Duke of Burgundy *Hamearis lucina* and the large blue *Phengaris (= Maculinea) arion* are both priority conservation species that have overlapping ranges. However, the species depend on different host plants for larval development. That, in itself, would not be a problem, but the complication arises because the local conditions and grazing management needed for those different host plants to flourish are completely different. As such, multiple sites could support one of the two species, but it is very difficult to create good habitat for both species (especially in a small site). This means that difficult decisions have to be made about the direction conservation action should take, what species or habitats should be prioritized, and how practicalities should be resolved. These issues are discussed further in Section 10.5.

Habitat-focused conservation

Species-focused conservation often necessitates management of landscapes (e.g. to ensure connectivity and buffer problems caused by fragmentation: Section 7.4) or management of habitats so that they are suitable for specific target species.

Habitat conservation for its own sake rather than for the species it supports is usually reserved for small habitats that are critically endangered, or that perform a useful ecosystem service. A habitat that meets both criteria is the peat bog, which is a priority habitat under EU Directive 92/43/EEC Conservation of Natural Habitats and of Wild Fauna and Flora (more commonly known as the Habitats Directive). The driving factors behind peat bog conservation are explored in Case Study 10.1.

10.2.3 Threats to populations and communities.

Numerous factors can cause species populations to decline, sometimes to the point of extinction. Factors can be grouped in three main categories:

1. Natural abiotic (e.g. storms, volcanic eruptions).
2. Natural biotic (e.g. disease, competitive exclusion).
3. Anthropogenic (e.g. hunting, poaching, pollution, habitat loss).

Several studies have tackled the daunting task of quantifying factors associated with the decline of threatened species to establish the main **extinction drivers**. In the first study of its kind, Wilcove et al. (1998) examined 1880 species in continental USA that were listed as endangered by the Nature Conservancy and/or covered by the Endangered Species Act. More recently, a similar study was undertaken by Evans et al. (2011) for 1700 Australian species listed under the Australian Commonwealth Environmental Protection & Biodiversity Conservation Act 1999. The percentage of species affected by all the different threats is shown in Table 10.2. It is noteworthy that both studies listed habitat change as the most common threat, with non-native species interactions in second place. For the USA analysis, habitat change was subdivided to show the relative importance of different types of habitat change on biodiversity; the results of that analysis are shown in Figure 10.2.

CASE STUDY 10.1

Habitat-focused conservation: why conserve peat bogs?

Peat bogs are increasingly under threat. One of the major threats is drainage-related degradation. However, increasingly the lack of management that used to be undertaken historically, such as shrub clearance, means that areas of this important habitat are undergoing ecological succession and many are becoming wooded.

Destruction and degradation of peat bogs is important because as well as supporting a range of specific species, including *Sphagnum* mosses, peat bogs have an important role within wider ecology and environmental regulation. Peatlands are thus an example of a habitat that is conserved less for the species it supports (important though that is) than for its ecosystem services and functionality:

- **Carbon storage:** peatlands have removed carbon from the atmosphere over thousands of years and have 'locked it up' in a carbon store. When peat bogs are drained, the peat starts to decompose and carbon and methane are released, which contribute to human-accelerated climate change.
- **Absorption of atmospheric pollutants:** well-functioning peatlands (particularly those near to industrial areas) can absorb airborne sulphur dioxide, nitrogen, and heavy metals.
- **Control of fire:** wet bogs can act as natural fire breaks, which is especially important in areas with a mosaic of peatlands and dry heathland.
- **Fluvial dynamics:** groundwater flow (the movement of water within the ground towards rivers after rain) is quicker through a dry degraded bog than a wet bog. This, in turn, means that the time that elapses between rainfall and peak river flow is reduced, which can increase flooding. From a drinking water perspective, draining of bogs can lead to a reduction in reservoir catchment capacity and can cause discolouration of drinking water.

FURTHER READING

Review of the effects of environmental management change on peat bogs: Holden, J., Shotbolt, L., Bonn, A., Burt, T., Chapman, P., Dougill, A., Fraser, E., Hubacek, K., Irvine, B., & Kirkby, M. (2007) Environmental change in moorland landscapes. *Earth-Science Reviews*, Volume 82, 75–100.

Hydrology and peat bogs: Evans, M.G, Burt, T.P, Holden, J., & Adamson J.K. (1999) Runoff generation and water table fluctuations in blanket peat: evidence from UK data spanning the dry summer of 1995. *Journal of Hydrology*, Volume 221, 141–160.

Table 10.2 The most important threats to American and Australian biodiversity based on the number of endangered species affected (categories are not mutually exclusive, so columns do not sum to 100). More details on habitat change in the USA (bold) are given in Figure 10.2.

	Continental USA (Wilcove et al., 1998)		Australia (Evans et al., 2011)	
	Plants (n = 641)	Birds (n = 56)	Plants (n = 975)	Birds (n = 104)
Natural abiotic	Negligible	Negligible	Negligible	Negligible
Natural biotic				
Natural species interactions	Not considered	Not considered	17%	24%
Natural biotic change	Not considered	Not considered	13%	23%
Disease	1%	4%	14%	30%
Anthropogenic				
Habitat loss/fragmentation	**90%**	**88%**	80%	88%
Non-native species	30%	48%	57%	81%
Over-exploitation	13%	39%	15%	49%
Pollution	12%	38%	11%	22%

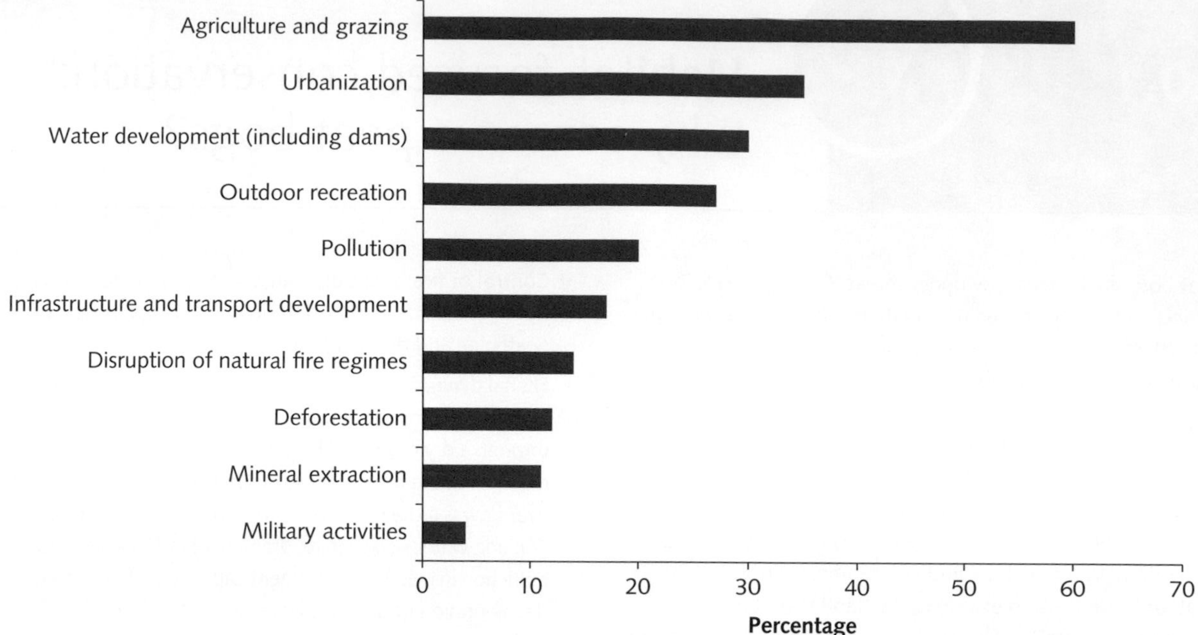

Figure 10.2 Habitat-related threats to US species listed as threatened shown as percentage of species affected (categories are not mutually exclusive, so data do not sum to 100).

Source: Data from Wilcove, D. S. et al. (1998) Quantifying threats to imperilled species in the United States. *BioScience*, 98, 607–615.

One key threat that was not covered by Wilcove et al. (1998) and Evans et al. (2011) is that of human-accelerated climate change. The effect of recent climate change on species' ranges was considered in Section 7.5.1. However, there are many other ways that climate change can affect species other than through spatial processes and range shift. For example, climate change can alter or decouple species interactions in ways that reduce breeding success and that, ultimately, could affect population dynamics to the point of extinction. Such processes are considered in the Online Case Study for Chapter 10, using birds as an example.

You can find the online case study at www.oxfordtextbooks.co.uk/orc/goodenough

FURTHER READING FOR SECTION

Review of key principles and challenges in conservation: Wallington, T.J., Hobbs, R.J., & Moore, S.A. (2005) Implications of current ecological thinking for biodiversity conservation: A review of the salient issues. *Ecology and Society*, Volume 10, 16.

Policy and legislation framework for habitat conservation in Europe: *EU Directive 92/43/EEC Conservation of Natural Habitats and of Wild Fauna and Flora*. Available at: http://ec.europa.eu/environment/nature/legislation/habitatsdirective/index_en.htm

Quantifying threats to species in Australia: Evans, M.C., Watson, J.E., Fuller, R.A., Venter, O., Bennett, S.C., Marsack, P.R., & Possingham, H.P. (2011) The spatial distribution of threats to species in Australia. *BioScience*, Volume 61, 281–289.

10.3 Extinction risk.

Before the different systems for studying and quantifying extinction risk are considered, it is first necessary to define exactly what is meant by extinction. Key terms include:

1. **Extinction**: complete loss of a species from the biosphere (e.g. dodo *Raphus cucullatus* and the woolly mammoth *Mammuthus primigenius*).

2. **Extirpation:** loss of a species from a specific site, province, or country (e.g. wolves *Canis lupus* used to occur throughout Europe including the UK; the UK population has been lost to overhunting).
3. **Extinction in the wild:** loss of a species in the wild; remaining individuals occur in captivity only (e.g. Guam rail *Gallirallus owstoni*, a bird that used to occur on the Island of Guam in the Pacific, and the scimitar oryx *Oryx dammah*, an antelope that used to occur in Chad).
4. **Functional extinction:** the species still exists in the wild and/or in captivity, but is not viable. Species can be declared functionally extinct when only one sex remains (for sexually reproducing species), when all individuals are too old to breed, or when remaining individuals are known to be infertile.

As noted in the introduction, the widely held view is that we are now at the start of a mass extinction event. This is a catastrophic loss of many species within, in evolutionary terms, a very short period (Leaky & Lewin, 1996). Analysis of the fossil record suggests that this is the sixth such **mass extinction event** (MEE), and it is important to place what we are currently witnessing into historical context, as shown in Figure 10.3. The last MEE was the extinction of the dinosaurs at the Cretaceous–Paleogene (K–Pg) boundary (also called the K–T or Cretaceous–Tertiary boundary) about 66 million years ago, possibly due to a meteor strike. The biggest MEE was the end-Permian event (251 million years ago), possibly due to volcanic eruptions, when up to 95% of species became extinct. The deficit in global biodiversity from the end-Permian event took an estimated 100 million years to be redressed through new species arising from speciation processes (Hallam & Wignall, 1997; Benton & Twitchett, 2003).

What makes the sixth mass extinction event unique is that it is the first to be bought about by the dominance of a single species: mankind. Indeed, some palaeontologists go so far as to say that we are in a new geological age, the **Anthropocene,** because our effect has been so great that it will be recorded in the fossil record. However, it is worth remembering that although attempting to avert extinction is the very essence of conservation, extinction itself is a natural process that ultimately befalls all species. It is the rule, not the exception: >99% of species

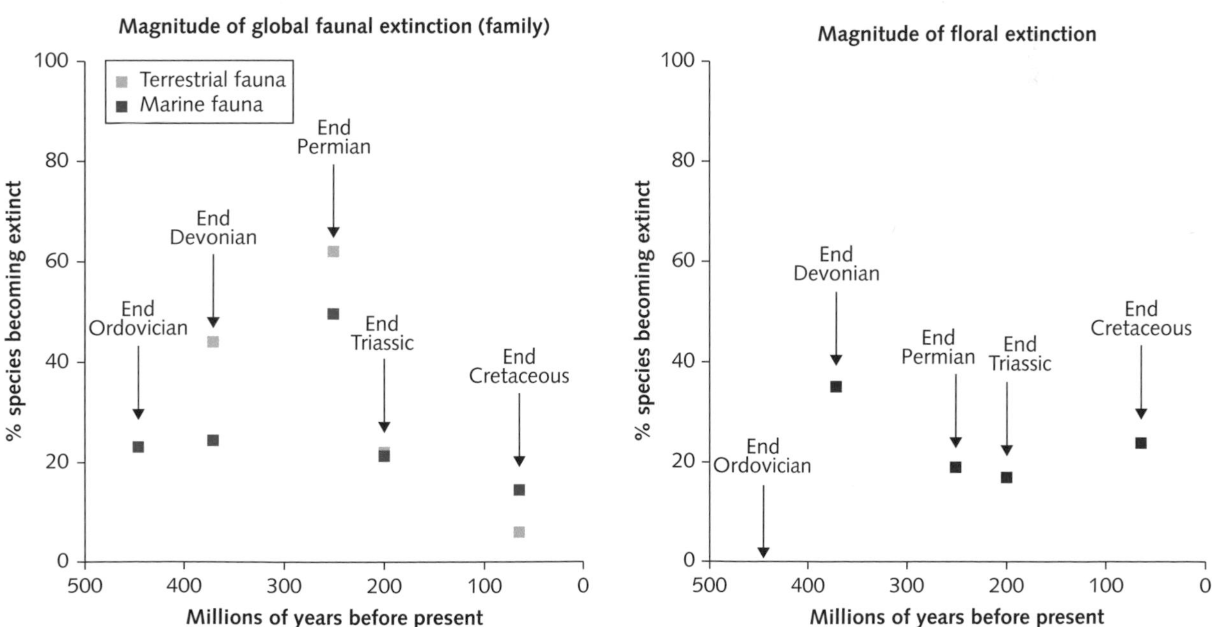

Figure 10.3 Past mass extinction events.

Source: Based on data from Table 1 of McElwain, J. C., & Punyasena, S. W. (2007). 'Mass extinction events and the plant fossil record'. *Trends in Ecology & Evolution*, 22(10), 548–557.

ever to have existed are now extinct, but the annual extinction rate has historically been very low. What conservation seeks to do, therefore, is not to prevent extinction, but rather to prevent extinction caused by human processes so that if and when extinction does take place, it does so naturally and at a similar rate to the background norm.

10.3.1 Studying extinction risk.

Extinction vortex

An extinction vortex is a series of linked processes leading to extinction. The concept was developed by Gilpin & Soulé (1986) and the term 'vortex' was used to signify a process that gathers speed in one inevitable direction, like water going down a plughole. There are four main extinction vortex models as shown in Figure 10.4; two of these are **allogeneic** (driven by external environmental factors and known as the R and D vortices) and two are **autogenic** (driven by internal genetic factors and known as the F and A vortices).

Chance extinctions can also occur. These are not driven by one particular causal factor, but rather occur through natural fluctuations in population productivity and mortality. This is termed **demographic stochasticity** and often occurs alongside fluctuations in the abiotic and biotic environment,

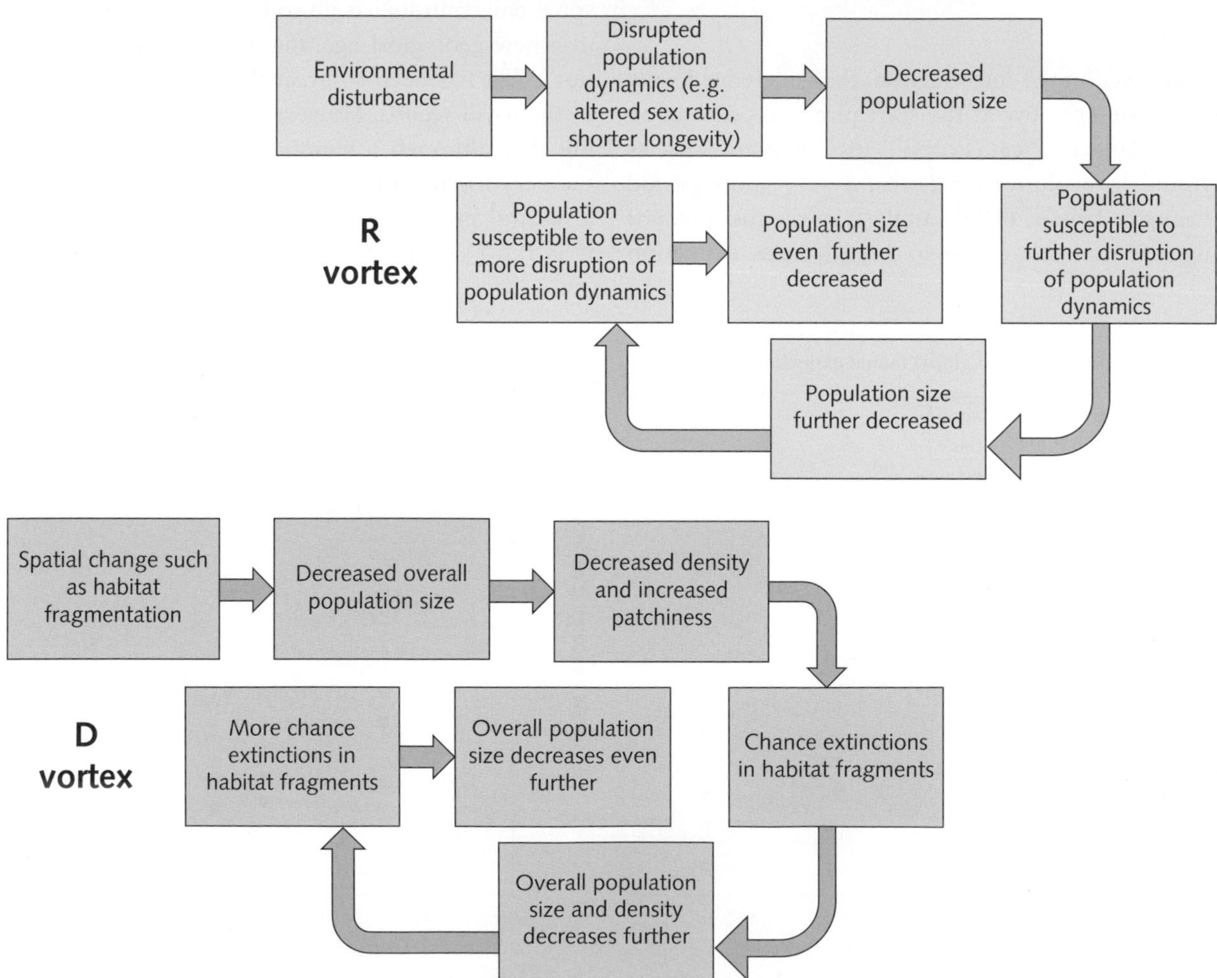

Figure 10.4 Extinction vortex models; R and D vortices are driven by change in the environment, and are therefore allogenic, whereas F and A vortices are driven by population genetics and are therefore autogenic.

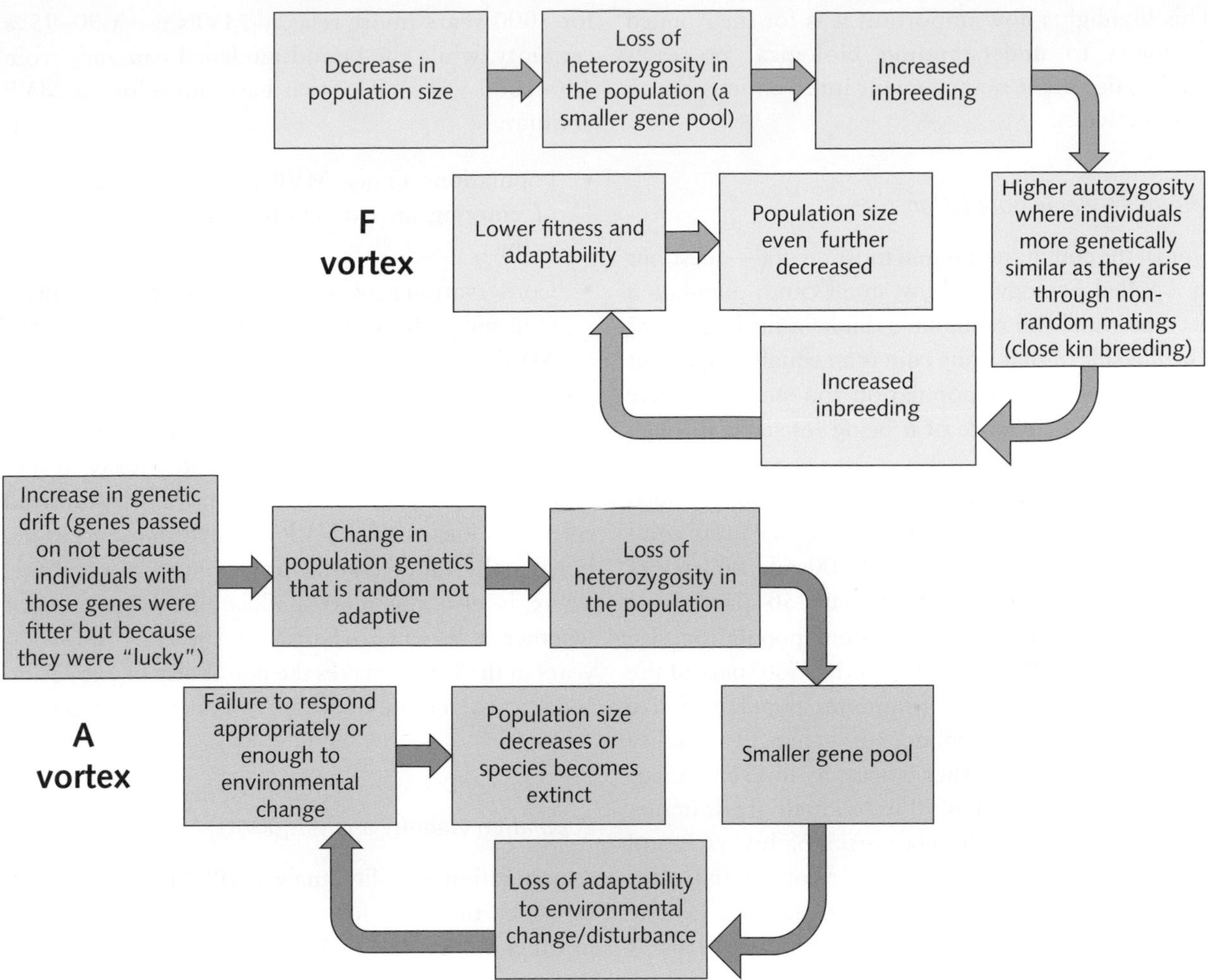

Figure 10.4 Continued.

which is termed **environmental stochasticity**. The risk of chance extinction due to stochastic processes is much higher for small populations as they are less able to buffer natural fluctuations compared to large populations. Imagine a population of just 10 individuals suffering an outbreak of disease or an unusually hard winter: if mortality increases to 50% just five individuals would survive. By chance, the surviving individuals could all be the same sex or too old to breed, thereby rendering the population functionally extinct. However, in a population of 100 individuals, that same 50% mortality would mean 50 individuals surviving, while in a population of 1000 individuals, 500 would remain: the risk of stochasticity-induced chance extinction decreases as the population increases.

One key lesson from extinction vortex modelling is that conservation aimed at addressing the original cause of decline might be insufficient to allow a species to recover if the vortex is too far advanced. An example is the greater prairie chicken *Tympanuchus cupido* in Illinois (Primack, 2006). This population declined to around 50 individuals due primarily to habitat loss and degradation. Contrary to expectation, however, simply restoring habitat had little effect because the vortex had already progressed to the inbreeding depression stage. The population only started to recover when it was outbred with individuals in other populations. In other words, the vortex had to be reversed from the 'inside' out as change aimed only at the 'outside' of the vortex (habitat) was not effective.

This highlights how important it is for the Applied Ecologist to understanding biological processes driving decline if management interventions are to be effective.

Minimum viable population

One of the commonest—and most pivotal—questions in Applied Ecology is: how small can a population become while still remaining viable in the long term? The flip side of that same coin is an equally important question: when is a population too small for there to be a realistic chance of it being rescuable through conservation action?

Early attempts to answer such questions tended to take a one-size-fits-all approach. The most common generalization was the **50-500 rule**, which was proposed by Franklin (1980). The '50' part of the rule was the perceived minimum population size for short-term persistence, and the '500' part of the rule was the perceived minimum population size for long-term maintenance of genetic variability and adaptability. In other words, by this rule, populations with fewer than 50 individuals are at immediate risk of extinction via vortex pathways R and D, while populations with fewer than 500 individuals are at long-term risk of inbreeding and genetic drift and extinction via vortex pathways F and A (Figure 10.4).

The 50-500 rule attracted criticism because the numbers were absolutes: no account was taken of a species' ecology, genetics, or life-history traits. In particular, there was no consideration of whether a species were r-strategists or a K-strategists. This is despite the population processes of a fruit fly *Drosophila melanogaster* for example, clearly being different to those of an African elephant *Loxodonta africana*, and it making little sense for them to have the same rule-of-thumb figures.

Despite its limitations, the 50-500 rule was successful in highlighting the need to formally consider threshold population sizes in conservation biology, and the need to do this on a species-by-species basis. Building on this, therefore, Shaffer (1978) proposed the concept of the **minimum viable population** (MVP). He defined the MVP being the minimum number of individuals to ensure that the population has a 99% chance of remaining in existence for 1000 years (more relaxed MVPs use a 90–95% certainty, while the period modelled can vary from 100–1000 years). The two basic ideas of an MVP are that:

- Populations under MVP are at significant risk of entering an extinction vortex and becoming extinct.
- Conservation programmes should only be considered successful if they raise the population above MVP.

The crucial difference between MVP and the 50-500 rule is that the former is computed for each species individually rather than being a standardized rule. This means that MVP is much more accurate, but equally takes considerably longer to compute. There is also the more philosophical question of whether a 95–99% chance of survival 100–1000 years in the future makes the possibility of extinction 'sufficiently remote or all too imminent' (Quammen, 1996).

Population viability analysis (PVA)

A **population viability analysis** (PVA) is the process by which the MVP for a given species is calculated. In other words, the PVA is the method; the MVP is the answer. PVA can be undertaken to answer several subtly different questions:

1. Is a particular population large enough to be viable?
2. Is a particular series of linked populations (a metapopulation) large enough to be viable?
3. At a global level, are there enough individuals of a species for the species to survive?
4. How many individuals need to be reintroduced into an area for the new population to be viable? (This will be considered in more detail in Section 13.4).
5. How many individuals can be removed from a given population (what level of harvesting is sustainable) via hunting, translocation to another site, or capture for captivity?

PVA is undertaken using simulation modelling software: the industry standard is the freeware package Vortex (Lacy & Pollak, 2015). All PVAs require model parameterization; this is the process through which species-specific ecological data are added to the model so PVA calculations—and the MVP result—are themselves species-specific. Generally information is needed on:

- **Population structure:** optimal sex ratio, age structure, breeding strategy (monogamy, polygyny, etc.).
- **Productivity:** number of broods per year, number of young per brood, proportion of males and females typically in the breeding pool, age at first reproduction, age at last reproduction.
- **Mortality:** longevity, survival rates per year for both sexes and each age class.
- **Genetic diversity:** previous history of population bottlenecks.
- **Spatial relationships between populations:** dispersal rate, dispersal frequency, direction of movement.
- **Carrying capacity:** habitat quality and suitability, size of habitat patch.

PVA models are generally run multiple times on the same data to add in random effects to reflect demographic stochasticity and catastrophes (abrupt decreases in population that occur at unpredictable intervals in response to storms, fire, and so on) and these are plotted as separate lines on the same graph. For a PVA to indicate that a population is viable, the population needs to remain extant in nine models out of 10 or 18–19 models out of 20 (i.e. 90–95% certainty of population/species persistence). This is shown graphically in Figure 10.5.

The first experimental use of PVA was by Mark Shaffer in 1978 for the Yellowstone grizzly bears *Ursus arctos*. The Yellowstone population was an isolated fragment of the main population in the wider Yellowstone to Yukon region (Case Study 7.3) and it was unclear whether it was large enough to be viable. Shaffer established that for a population of 30–70 bears occupying 2500–7400 km^2 (depending on habitat quality) there was <95% chance of survival for 100 years (Shaffer, 1978). Survival probabilities were most affected by changes in mean mortality, offspring sex ratio, and age of females at first reproduction.

Species-specific traits that determine extinction likelihood

There are a few, very general, rules-of-thumb that can be used to *estimate* relative extinction risk for different species when population size is consistent. These all utilize ecological or life-history traits:

- **Population strategy:** K-strategists tend to be more vulnerable than r-strategists due to lower annual productivity (Section 2.7.1). Although K-strategists can buffer decline for longer because adults are long-lived, such species are more vulnerable

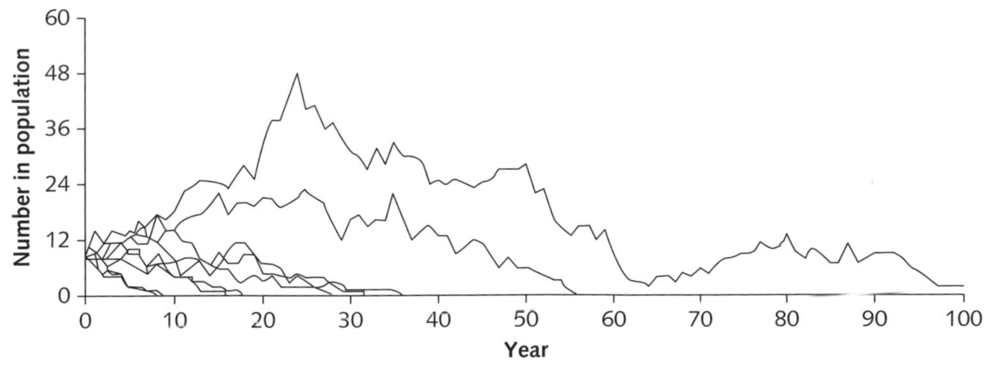

Figure 10.5 Adding demographic stochasticity and variable catastrophe scenarios into Population Viability Analysis means that of the 10 integrations of the same model generates variability in outcome, each shown by a separate line on the graph. Here, on only one occasion out of 10 does the population remain extant at year 100, so there is only a 10% likelihood of persistence after 100 years and that is at a very low level.

Source: Output here was generated using Vortex freeware [http://vortex10.org/Vortex10.aspx].

in the longer term. Small populations take a substantial time to recover from a population crash and are vulnerable to chance extinction while so doing.

- **Body size**: larger species tend to be more at risk than smaller ones because:
  - they generally have larger home ranges and so are more vulnerable to landscape change (Section 7.4.1);
  - they are often higher up the food chain and so are more vulnerable;
  - they are more likely to be K-strategists.
- **Specialism**: species that are highly specialized are more vulnerable than generalist species, particularly in an environment that is undergoing, or susceptible to, natural or anthropogenic change.
- **Adaptability**: there are two main ways that species can adapt to environmental change, either through **evolution** over successive generations (in which case the size of the gene pool is important; bigger = better) or via a process called **phenotypic plasticity**. Plasticity is the degree to which an individual can change in response to the environment in a non-genetic way—plants 'deciding' to flower early in a warm year can be a plastic response to local conditions. In the same way, species can change things such as the timing of reproduction according to local conditions, or adjust diet in response prey availability.
- **Endemic species**: these tend to be more at risk than cosmopolitan species. This is because if localized threats emerge that are detrimental to one population, this will have consequences for the entire species. This links to **range size** in a specific region too—species with smaller range sizes tend to be more vulnerable.
- **Dispersal and mobility:** species that are able to disperse rapidly over large distances over successive generations or that are highly mobile are less vulnerable to extinction as they can move away from threats.

It must be emphasized that these are general principles, but they start to pave the way for protocols to quantify species-specific extinction risk through PVA and species-at-risk classification systems.

10.3.2 International species-at-risk classification systems.

The two best-known international species-at-risk systems are the International Union for the Conservation of Nature IUCN red list, which uses an objective protocol developed in 1994 and revised substantially in 2001, and the NatureServe Imperilled list, which started in the 1980s. IUCN is entirely global in focus, whereas NatureServe operates at global, national, and sub-national scales (the G-scale being the global system).

IUCN and NatureServe share one common aim: to quantify species' **extinction risk.** This involves using proxies for extinction risk, such as population size and range size, and temporal trends in these parameters, while also considering species-specific traits that link to extinction likelihood (see above). Such schemes must be usable and objective for species in any taxonomic group in any part of the world. They must also be complex enough to be robust, and simple enough to be understood easily. Each system has seven categories of risk from 'extinct' through several levels of 'threatened' to 'secure', as shown in Figure 10.6. However, the mechanism of operation differs considerably:

- **IUCN**: uses a **rule-based approach** whereby a species is assigned to a threat category if it meets the quantitative threshold for at least one criterion—in this way it often takes the 'worst case scenario' approach. There are four broad criteria, each of which has sub-criteria, based around population size (and change therein) and range (and change therein). A PVA is also calculated at a global level to determine global MVP and overall percentage extinction risk, which is assessed against thresholds in a fifth criterion. One advantage of the rule-based method is that it is completely explicit about what parameter led to a species being listed as threatened (Mace et al., 2007, 2008).
- **NatureServe**: this uses a **point-scoring calculator** system, whereby a conservation status rank is assigned by assessing multiple factors (e.g. change in population and change in range) simultaneously. This method is less prone to outliers relative to IUCN, and the overall system is also less prone to inter-observer variability in assessors relative to IUCN (Regan et al., 2005).

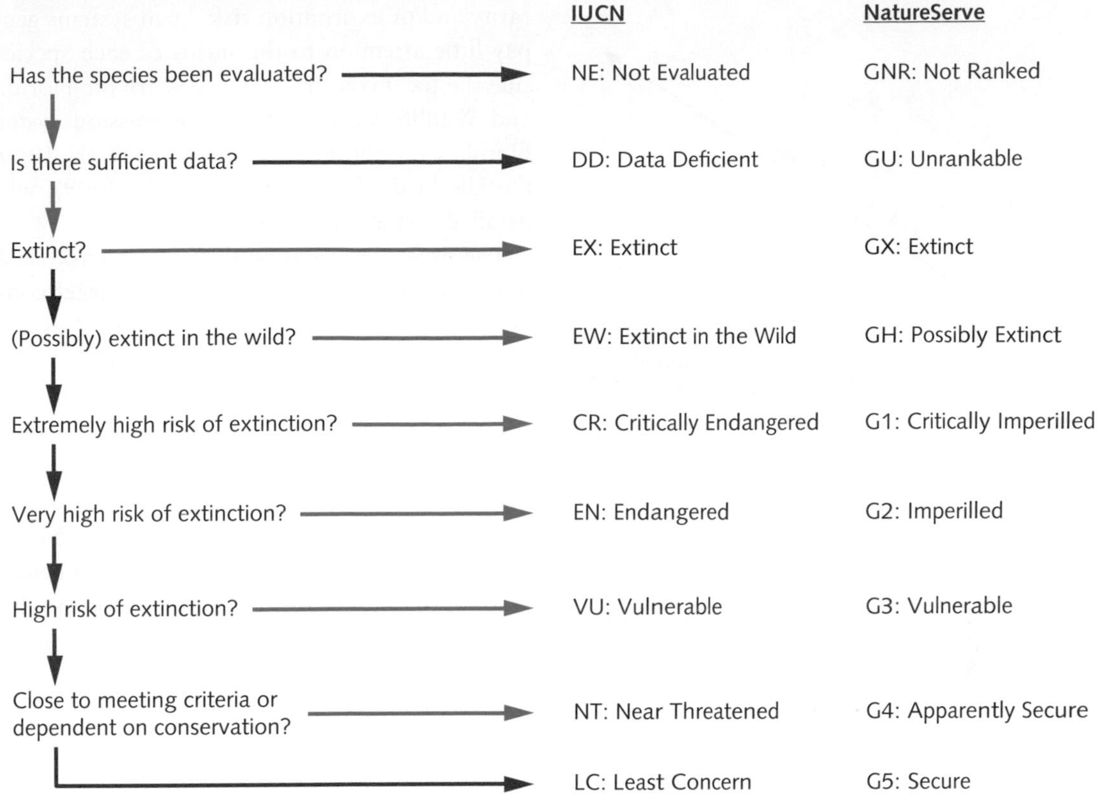

Figure 10.6 IUCN and NatureServe categories and decision-making process (red arrows denote 'No' and green arrows denote 'Yes').

In both systems, data are periodically reviewed and species are reclassified if necessary. **Reclassification** might be necessary because a species has become (more) threatened so extinction risk status needs to be increased, or to down-grade extinction risk due to natural population rebound or successful conservation intervention. Each year, new species are added to the system databases and are assessed for the first time, thereby moving from DD (data deficient) on IUCN, or moving from GU (unrankable) or GNR (not ranked) on NatureServe to an assessed category within that system. For some taxonomic groups, new classifications can mean substantial changes: in subsequent IUCN assessments, lemurs Lemuroidae changed from 76% of species being threatened (i.e. placed in critically endangered, endangered, or vulnerable categories) in 2010 to 95% of species being threatened in 2015. The majority of this change was due to data deficient species being classified (as threatened) for the first time.

Database expansion is a continual process and considerable progress has been made, as shown in Figure 10.7. However, species-at-risk systems are still far from comprehensive in coverage—just 5% of known species have been assessed by IUCN, for example. It should also be noted that a quirk in assessment procedures means that although the number of threatened species continues to increase with the increase in database coverage, the ratio is actually declining (compare the gradient of lines in Figure 10.7). In general, this is driven by initial assessments being biased towards species known to be at risk; an approach that is gradually being phased out.

Data from species-at-risk classification systems can, and should, provide valuable data to inform species management decisions. However, despite systems being widely used to inform legislative protection, and vital management decisions being made using their results, surprisingly little is known about their effectiveness.

10.3.3 National species-at-risk classification systems.

In addition to international systems for predicting extinction risk, there are national systems for classifying

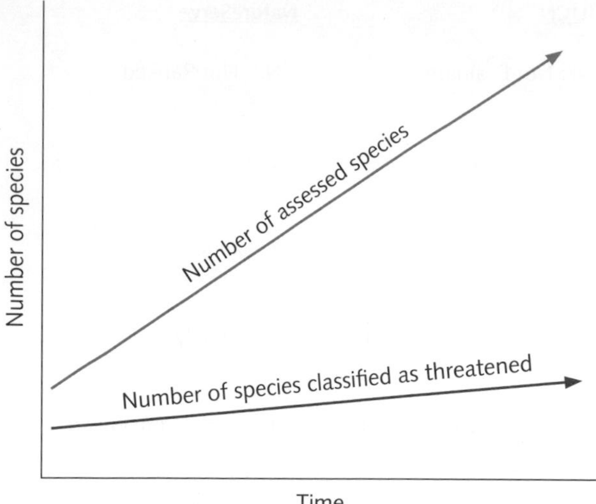

Figure 10.7 The number of species classified by species-at-risk schemes such as the IUCN as being threatened (i.e. classified as being Vulnerable, Endangered, or Critically Endangered under IUCN; red line) has increased less steeply over time than the number of species being assessed (green line).

rarity and/or **extirpation risk**. Such systems generally pay little attention to the status of each species outside the focal country. Examples are the Florida Fish and Wildlife Conservation Commission system for Florida's vertebrate fauna (Millsap et al., 1990) and the UK bird red list (Eaton et al., 2009), which is detailed further in Figure 10.8.

The presence of national species-at-risk classification systems means it is possible to **target conservation action** based on species' perceived extirpation risk. However, this does call into question the wisdom of using extirpation rather than extinction as a metric. In some cases, a species might be red-listed because it occurs in small numbers in one particular country, when in fact this is a localized population at the periphery of the geographic distribution and actually—globally—of very little consequence. The wisdom of conserving species in one area when the species exists in good numbers elsewhere is considered further in Hot Topic 10.1

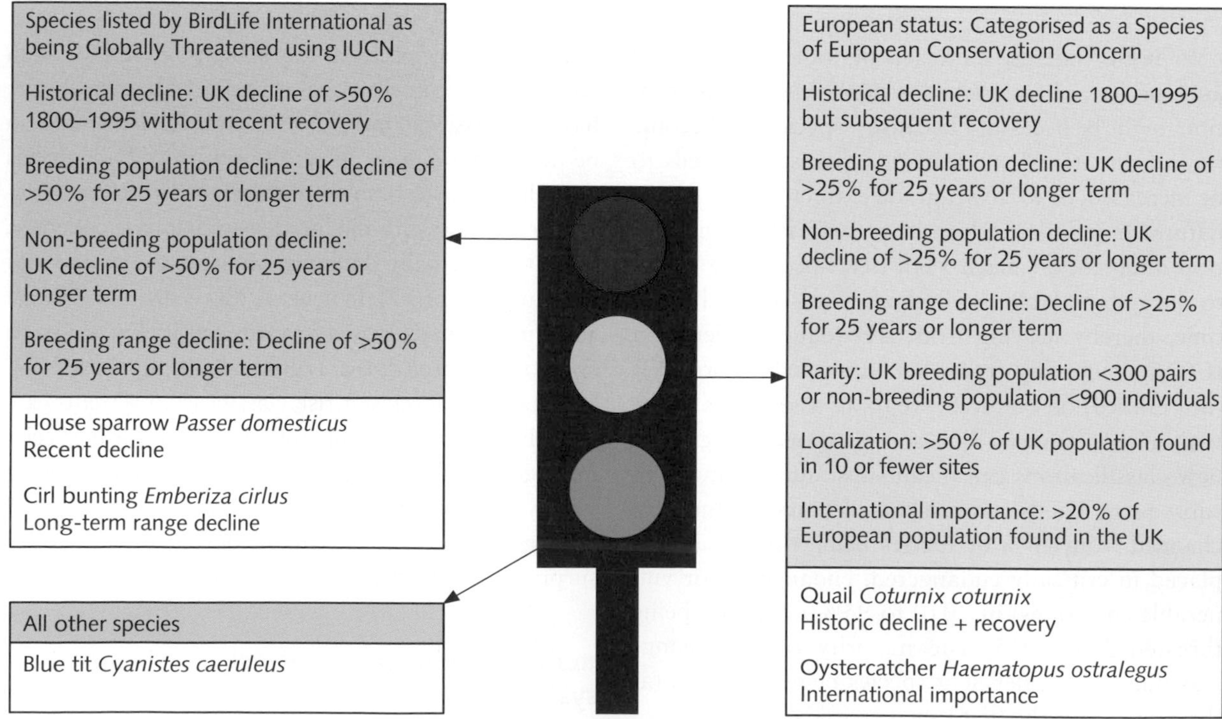

Figure 10.8 The UK bird national red list uses a traffic light system to show risk of extirpation. Species are classified as red (species of immediate conservation concern), amber (species of conservation concern), and green (species that are currently stable).

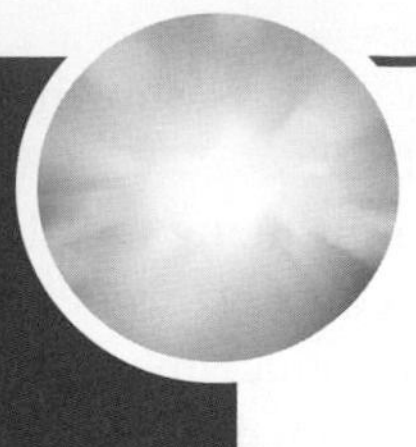

HOT TOPIC
10.1

Conserving species at the edge of their range

National conservation schemes often include initiatives for species that are declining in that specific country, but thriving elsewhere in their range. This frequently occurs when the country in question is at the edge of a species' range and gives rise to the question as to whether, in these circumstances, conservation action is sensible.

To some extent, the answer to this depends on the motivation for conservation. For example, if conservation of a specific species is 'just' to avoid its extinction, then the argument that conservation money would be better spent elsewhere is a strong one as the overall extinction of the species is unlikely. If, however, conservation is because of the role that a species has within an ecosystem (a keystone species, for example) or because it is being used as a focal flagship or umbrella species, it is the risk of extirpation that must be guarded against.

Subspecies also come into play here. For example, it has previously been debated whether or not the UK should conserve the swallowtail butterfly *Papilio machaon*, which is thriving in Continental Europe. However, the individuals in the UK are actually subspecies *britannicus* (Figure A) so extirpation of the *species* in the UK would mean extinction of the *subspecies*. It depends on personal views as to whether this makes a difference or not.

It should also be noted that just because a species is thriving elsewhere in its range at that point in time does not mean that the situation will continue. The passenger pigeon *Ectopistes migratorius* and the Rocky Mountain grasshopper *Melanoplus spretus* are both testament to the fact that very abundant species can be at risk of sudden extinction. Indeed, in some cases a species might first decline at the edge of its range, possibly because conditions there are more marginal, so any threats or stressors have more of an effect. In this way, decline at the periphery of a species' range could actually be an early warning of problems to come.

A slightly different approach is that of national responsibility. This involves considering the concepts of rarity and extirpation, but within the context of global patterns of distribution so that the priority for action focuses on both the small (national) and larger (global) pictures. For example, Schmeller et al. (2014) proposes a three-stage system whereby the assessment unit (e.g. species, subspecies, habitat, ecosystem) is defined, the current distribution is determined, and finally the importance of the distribution within a given focal area is determined.

Figure A Subspecies of swallowtail butterfly *Papilio machaon britannicus*, found only in England.

Source: Photograph by Anne Goodenough.

QUESTION

Do subspecies matter in conservation?

REFERENCE

Schmeller, D.S., Evans, D., Lin, Y.P., & Henle, K. (2014) The national responsibility approach to setting conservation priorities—recommendations for its use. *Journal for Nature Conservation*, Volume 22, 349–357.

10.3.4 Considering evolutionary distinctiveness.

In the species-at-risk schemes considered above, all species are considered as discrete 'genetic units' and the gap between those units is assumed to be consistent. In reality, this is not the case. Evolutionary distinctiveness is a measure of the distance between a given species in evolutionary terms from its nearest relative. From a genetic perspective, it can be argued that the loss of highly distinct species is especially important.

Imagine a situation with two closely-related species on one branch of a **phylogenetic tree** (the taxonomic equivalent of a human family tree). The loss of either one of those species would only mean a small reduction in overall genetic diversity. Conversely, if a species with no closely-related species and a uniquely divergent genome were to become extinct, that extinction would mean a bigger lost in terms of overall genetic diversity. This is shown in simple graphical form in Figure 10.9.

The concept of bringing together evolutionary distinctiveness and conservation is only possible in well understood taxonomic lineages, but is a challenge that has been adopted by the EDGE of Existence programme. EDGE stands for **Evolutionarily Distinct and Globally Endangered** and is discussed more in Case Study 10.2. Considering evolutionary distinctiveness in conservation fits with the second aim of the World Conservation Strategy: to preserve genetic diversity. This is often taken as being the diversity within species (i.e. preventing genetically driven extinction vortices), but it can equally apply to genetic diversity as a whole. There is an increasing appreciation of the need to move from solely species-based conservation to the conservation of genetic diversity at a level below species, including genetically distinct populations and subspecies.

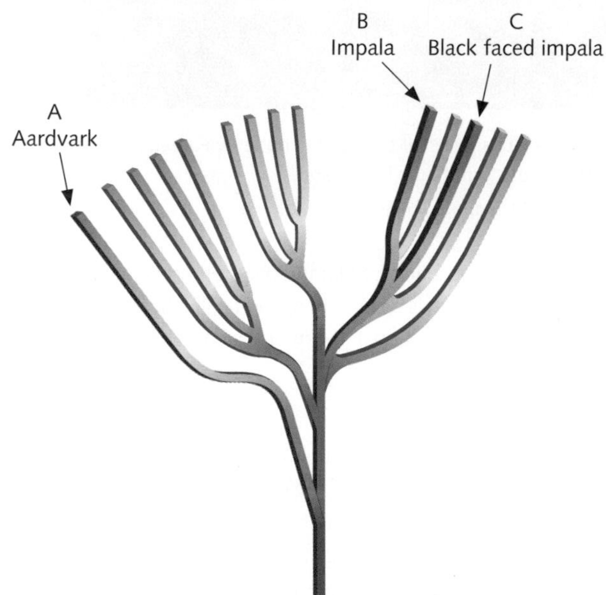

Figure 10.9 EDGE phylogenetic tree showing the evolutionary distinctiveness of different species. Here, species A (the aardvark *Orycteropus afer*) is more evolutionarily distinctive than species B and C, so its loss, in genetic terms, would be greater.

FURTHER READING FOR SECTION

Predicting extinction risk: Keith, D.A., McCarthy, M.A., Regan, H., Regan, T., Bowles, C., Drill, C., Corey, C., Pellow, B., Burgman, M.A., Master, L.L., Ruckelshaus, M., Mackenzie, B., Andelman, S.J., & Wade P.R. (2004) Protocols for listing threatened species can forecast extinction. *Ecology Letters*, Volume 7, 1101–1108.

IUCN red list: Overview available at: http://jr.iucnredlist.org/documents/redlist_cats_crit_en.pdf searchable database available at: http://www.iucnredlist.org/

NatureServe: Searchable database and information. Available at: http://www.natureserve.org/

Comparison of IUCN and NatureServe: Goodenough, A.E. (2012) Differences in two species-at-risk classification schemes for North American mammals. *Journal for Nature Conservation*, Volume 20, 117–124.

Seminal paper on PVA: Gilpin, M.E. & M.E. Soulé. (1986) Minimum viable populations: processes of species extinction. In: M.E. Soulé (Ed.) *Conservation Biology: The Science of Scarcity and Diversity*, pp. 19–34. Sunderland, MA: Sinauer Associates.

Recent example of PVA: Volampeno, M.S., Randriatahina, G.H., Kalle, R., Wilson, A.L., & Downs, C.T. (2015) A preliminary population viability analysis of the critically endangered blue-eyed black lemur (*Eulemur flavifrons*). *African Journal of Ecology*, Volume 53, 419–427.

Vortex software: Available at: http://vortex10.org/Vortex10.aspx

CASE STUDY 10.2

EDGE

The EDGE **Evolutionarily Distinct and Globally Endangered** programme is one approach to bringing phylogenetic considerations into species-at-risk classification systems (Section 10.3.4) and conservation prioritization (Section 10.5). The EDGE programme is overseen by the Zoological Society of London and aims to:

- Raise awareness of the world's most Evolutionarily Distinct and Globally Endangered (EDGE) species.
- Identify the current status of poorly known and possibly extinct EDGE species.
- Develop and implement conservation strategies for all EDGE species not currently being protected.
- Increase conservation capacity in the countries in which EDGE species occur, through supporting and training local scientists and conservation professionals to undertake research into focal EDGE species.
- Support all ongoing conservation activities for EDGE species.

So far, EDGE values have been calculated for mammals, amphibians, corals, and most recently, birds. In the case of birds, evolutionary distinctiveness was quantified for all species worldwide that had been described at the time of the study ($n = 9993$ species). Several parameters were quantified:

- **ED (evolutionary distinctiveness)**: the genetic distance between focal species and other species (mean = 6.2 million years; ranging from 0.8 million years to 72.8 million years for the oilbird *Steatornis caripensis*).
- **PD (phylogenetic distinctiveness)**: the overall distinctiveness of the taxonomic group (sum of ED value for all species in a taxonomic group).
- **GE (global endangerment)**: IUCN rank.
- **EDGE (evolutionary distinct and globally endangered)**: the ED value weighted by IUCN rank so distinctive species that are Critically Endangered score highest and species with low distinctiveness that are classified as Least Concern score lowest.

Evolutionary distinctiveness varied spatially, being highest in Australia, New Zealand, and Madagascar, especially for passerines (perching birds including all songbirds and crows), plus parts of Africa, the Middle East, and South America (especially for non-passerines such as seabirds, waterbirds, gamebirds, raptors, and owls) (Figure A).

The top 50 ED species include well-known 'oddities' such as the ostrich *Struthio camelus*, hoatzin *Opisthocomus hoazin*, and the oilbird *Steatornis caripensis*, which is the only nocturnal bird known to hunt using echolocation in a similar way to bats: all are rated as Least Concern by IUCN. Rarer examples include the shoebill *Balaeniceps rex* (Vulnerable), and the New Caledonian owlet-nightjar *Aegotheles savesi* (Critically Endangered). Once IUCN rating was factored in, the top 100 EDGE birds were identified; these are not necessarily the most distinct, but they are the most distinct *and* rare. The list is topped by the giant Ibis *Thaumatibis gigantean*, with the nightjar-owlet in second and the shoebill in 37th place.

The top 100 EDGE birds include species from 20 of the 29 living orders of birds. Other key facts are:

- More than 170 countries are represented on the ranges of the top 100 EDGE birds.
- India has the highest number of EDGE birds with 14 species.

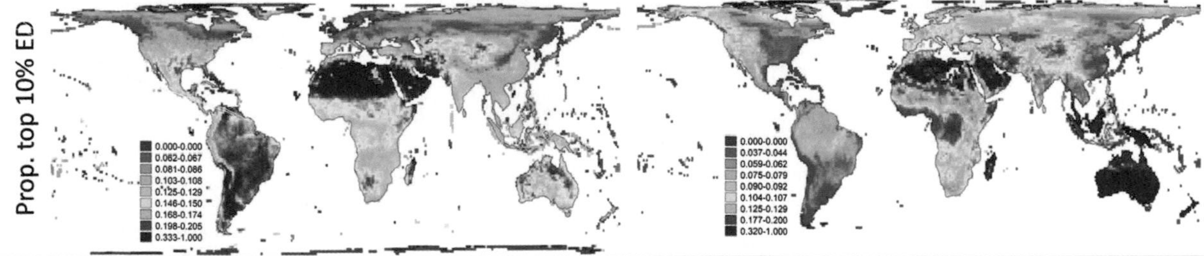

Figure A Relative evolutionary distinctiveness for birds across the globe (high = red, low = green).

Source: Reproduced from Jetz, W. et al. (2014) Global Distribution and Conservation of Evolutionary Distinctness in Birds. *Current Biology*, 24, pp. 919–930/CC BY.

- 62% of EDGE birds are endemic to one specific country.
- The Philippines has the highest number of endemic EDGE birds with nine species.

EDGE aims to promote conservation of ED species to avoid losing whole phylogenetic branches through extinction of one species. This very situation occurred through the extinction of the thylacine *Thylacinus cynocephalus* the last-known species of carnivorous marsupial that became extinct in Tasmania in the 1930s. Despite this, the lack of conservation for many EDGE species is striking.

FURTHER READING

EDGE online resources: Available at: http://www.edgeofexistence.org/index.php

Application of EDGE applied to birds: Jetz W., Thomas G.H., Joy J.B., Redding D.W., Hartmann K., & Mooers A.O. (2014) global distribution and conservation of evolutionary distinctness in birds. *Current Biology*, Volume 24, 919–930.

Application of EDGE applied to mammals: Isaac, N.J.B., Turvey, S.T., Collen, B., Waterman, C., & Baillie, J.E.M. (2007) Mammals on the EDGE: conservation priorities based on threat and phylogeny, *PLoS One*, Volume 2, 296.

10.4 Conservation strategies.

There are many different ways in which species and habitats can be conserved, and these will be explored in more detail in Chapters 11 and 12. In general terms, though, conservation action is often classified as being *in situ* or *ex situ*. *In situ* conservation involves conservation in the wild and encompasses all on-site techniques, whereas *ex situ* conservation involves conservation in captivity and encompasses all off-site techniques. A further subdivision is whether actions are single-site focused or more general in spatial focus (Table 10.3).

In situ initiatives are more sustainable than *ex situ* initiatives in the long term and provide **cascade benefits**—conservation of other species, protection of the wider environment, possibility for ecotourism, and so on. Accordingly, such approaches should be used where possible. However, if the overall population of a species is extremely small, *ex situ* conservation is often more likely to succeed in the short term and provide initial recovery from a critically low population size (or at least maintain the population size without further losses).

For some species, *in situ* and *ex situ* approaches are used at different times, in different places, or to meet different aims. For example, in the case of the Grand Cayman blue iguana *Cyclura lewisi*, and Jamaican iguana *Cyclura collei*, **insurance populations** were created *ex situ* while *in situ* habitat management and invasive species control was undertaken. Even a single conservation initiative can involve *in situ* and *ex situ* elements. For example, **species reintroduction** always involves *in situ* work—the actual reintroduction plus any pre-reintroduction habitat management and post-reintroduction monitoring—but can also have an *ex situ* element if release stock is from **captive breeding programmes** (Chapter 12). The steps involved in conservation are hierarchical in terms of desirability (and cost), as illustrated in Figure 10.10.

FURTHER READING FOR SECTION

Review of *ex situ* and *in situ* approaches: Pritchard, D.J., Fa, J.E., Oldfield, S., & Harrop, S.R. (2012) Bring the captive closer to the wild: redefining the role of ex situ conservation. *Oryx*, Volume 46, 18–23.

Important considerations when using *ex situ* conservation: Canessa, S., Converse, S.J., West, M., Clemann, N., Gillespie, G., McFadden, M., Silla, A.J., Parris, K.M., & McCarthy, M.A. (2015) Planning for ex situ conservation in the face of uncertainty. *Conservation Biology*, Volume 30(3), 509–609.

Discussion of conservation approaches for an example species; the critically endangered Vigors great Indian bustard: Dolman, P.M., Collar, N.J., Scotland, K.M., & Burnside, R. (2015) Ark or park: the need to predict relative effectiveness of *ex situ* and *in situ* conservation before attempting captive breeding. *Journal of Applied Ecology*, Volume 52, 841–850.

Table 10.3 Examples of conservation initiatives in relation to whether they are *in situ* or *ex situ* and single-site or multi-site focused.

	Single-site focused	General or multi-site focused
In situ	Designating protected areas to give specific sites legal protection (Section 11.4) Active management of nature reserves, including grazing, burning, fencing, planting, etc. (Section 11.3) Managing human–wildlife conflicts at specific sites through education and, in some cases, compensation for losses (e.g. crop destruction) Remediation of specific sites (Chapter 6)	Legislative protection at a national scale to protect vulnerable species (Section 11.4.1) or prevent/control introduction of non-native species (Section 8.6) Broad scale management such as agri-environment schemes where farmers are paid for initiatives that help biodiversity such as hedge planting (Section 11.5.3) Landscape ecology approach—protecting core areas, enhancing connectivity, creating buffer zones (Section 7.6)
Ex situ	Captive collections of animals—e.g. zoos, aquaria, and aviaries (Chapter 12) Captive collections of plants—e.g. botanical gardens, arboreta, and seed banks (Chapter 12)	Integrated management of multiple captive collections to facilitate captive breeding programmes—involves knowing genetic history of all captive individuals of a given species (listed in a studbook) and moving individuals between collections (Chapter 12)

In situ and *ex situ* conservation approaches each have their own advantages and challenges. Again these will be discussed further in the relevant chapter (Chapter 11 for *in situ*; Chapter 12 for *ex situ*), but Table 10.4 provides a useful introduction to give an initial general understanding of key concepts.

Table 10.4 Advantages and disadvantages of *in situ* and *ex situ* approaches to conservation.

	Advantages	Challenges
In situ	Cheaper than *ex situ* methods in general, especially long-term Always better to conserve species in the wild if possible—fewer ethical issues Conserving species conserves habitats too Engages local people—opportunities for helping people value local biodiversity	Protected areas cover just 15% of terrestrial land (ex. Antarctica) and 3% of marine areas (2014 figures from Deguignet et al., 2014) Virtual island problems—limited connectivity between core areas or reserves can prevent metapopulations occurring (Section 7.4) Lack of political support, especially in economically developing countries and where there are multiple demands on land Lack of financial resources or poor management Some threats such as poaching are hard to control
Ex situ	Usually maintain species' populations and generally population size grows Removes species from *in situ* threats Usually easier to provide medical attention; can be critical for very vulnerable species when every individual is vital for overall continuation of the species Can be used to provide individuals for reintroduction programmes (Chapter 13) Important for environmental education Usually self-funding and some funding often given to outreach *in situ* scheme at linked sites	High and ongoing cost (especially for large vertebrates)—maintaining some African species in zoos costs 50× more than conserving the same number in the wild (Leader-Williams, 1990). Multiple aims of captive collections (conservation, education, entertainment) means species most in need are not always protected Population sizes are generally small—can reduce genetic variability and cause population bottlenecks Disease outbreaks possible given high density Dependence on humans Restlessness/stereotypic behaviours Surplus stock—some species breed readily in captivity. Seeds of some plants don't store well; seed banks are inappropriate for species that reproduce vegetatively

Protect	Removing threats, reduce mortality
Enhance	Enhance natural reproduction, improve survival
Translocate	Move individuals away from threats or to create viable populations
Supplement	Short-term captive breeding programme to supplement wild populations
Reintroduce	Medium-term captive breeding programme aimed at reintroduction
Ark	Long-term captive breeding programme to keep species alive

Figure 10.10 Conservation hierarchy showing the desirability of different actions.

10.5 Conservation decisions.

10.5.1 The need for conservation triage.

The conservation hierarchy illustrated in Figure 10.10 suggests that if action at one stage is impossible or ineffective there is progression to the next step. However, **conservation resources** (money, time, space, and expertise) are not infinite, and demand is always likely to outstrip supply. Accordingly, Applied Ecologists need to develop criteria to establish where and for what species conservation action is most needed, and then apply these criteria within an objective decision-making process (see the Interview with an Applied Ecologist for this chapter).

Prioritizing conservation resources is effectively a system of triage: assessing the need for, and the likely benefits of, action in a given situation. Triage is derived from the French word *trier* meaning 'to sort into order'. The term is most frequently used in emergency and battlefield medicine to establish the order in which patients should be treated, and sometimes—however hard the decision might be—identify patients that are beyond reasonable hope of survival so much-needed time and resources can be redirected to patients with higher survival chances. Increasingly, the concept of triage is being used in conservation ecology to **prioritize** conservation targets. As Frederick the Great of Prussia once said: 'He who defends everything, defends nothing'.

One of the earliest calls for conservation triage was made by ecologist Norman Myers (1979). Myers loosely grouped species into one of three categories—critically endangered, threatened, and secure. He argued that conservation resources should be used to help species in the threatened category as those in the top category are beyond reasonable help, and those in the bottom one do not need conserving. In many ways this is a logical strategy to ensure that resources are used effectively and not wasted. However, almost 40 years on, this strategy still causes controversy. This controversy centres around four main themes:

1. **What constitutes 'reasonable' help?** If some species are ranked as being beyond reasonable help, there needs to be a robust definition of what 'reasonable help' actually is, which is far from straightforward. People might agree that spending a very substantial sum of money on a species with an extremely limited chance of long-term survival is not sensible, but then definitions are needed for 'very substantial sum', 'extremely limited', and 'long-term'.
2. **Arguments often look fine on paper, but ...** On paper, when dealing with anonymized data, it is possible to be objective. However, when people start considering the actual species involved, then things can get much more complicated as subjectivity starts creeping in, especially with charismatic or iconic species. Reallocating conservation resources from a species upon which vast sums are spent and that has little long-term chance of survival might sound fine until we know that

AN INTERVIEW WITH AN APPLIED ECOLOGIST 10

Name: Paul Butler
Organization: Rare (global conservation organization based in the USA: www.rare.org)
Role: Senior Vice President
Nationality and main countries worked: joint British and St Lucian; worked all over the world including 25 years in the Caribbean. Now working throughout Asia (especially China, Indonesia, and the Philippines) as well as Central/South America (especially Belize, Colombia, and Brazil).

What is your day-to-day job?

I work for the global conservation organization Rare. Rare has worked in 56 countries on over 300 locally-led behaviour change projects called Pride Campaigns. As Senior Vice President, my role is multifaceted. About a third of my time is spent on our new programme in China, which is encouraging rural farmers in Hubei Province to switch to growing organic crops and organic cotton to make agriculture more sustainable. A further third of my time at the moment is spent developing training programmes. The final third of my time is spent on the development side of Rare, especially visiting projects on the ground. A lot of my time is spent travelling!

You are involved with making conservation priority decisions. What's your approach?

Rare undertakes *in situ* conservation around the globe with local partners through its Pride Campaigns. We choose which Campaigns to support based on a number of criteria, which include:

1. Are the projects located in an area where there is high biodiversity under threat from human activities such as over-fishing, unsustainable farming practices, bush meat trade, and so on?
2. Is there a human dimension—for example food security or water resources?
3. Are threats primarily local in origin and is changing people's behaviour likely to make a difference?

In terms of the first, the rationale here is simple—if you conserve a hectare of habitat that supports a very high number of species, then, in theory at least, you get more 'bang for your buck' than if you conserve the same sized area with fewer species. That means much of our work is focused in the neotropics including Colombia, which is listed as one of the world's 'megadiverse' countries, hosting close to 10% of the planet's biodiversity. In terms of the second, our approach is fundamentally about the nexus of people and nature. For example, we work with fishers helping to establish functioning Marine Protected Areas that, through spill-over, both protect and preserve fish stocks, but also improve fisher catch per unit effort and thus improve livelihoods. As for the third, we are a behaviour change organization so it is vital that change is possible by working with local people at a local level. For example, there might be a low-lying coralline island in a very diverse area and where the human–nature links are very strong. However, if the main threat to the island was sea-level rise due to human-induced climate change, there would be little we could do with the people on that island to change anything: the problem is external to that site. That is not to say that the problem is not a very real one or that the people and ecology of that island don't need help, it's just that it would not be a situation where we could realistically hope to effect change by working at a grassroots level.

The key thing is, there is no one thing that can be used to prioritize action. We need to understand that not every intervention is going to succeed, or at least not every intervention is going to succeed to the level that we would like, but that does not mean that we should not try. The saying that really strikes a chord with me is 'it's better to light a single candle than curse the darkness'. Just because something is challenging does not mean that we should not consider it: challenges are made to be overcome. We just need to be as confident as we can be that we light the best candles we can with the resources we have got.

What is your most interesting recent project and why?

Oh there are so many to choose from! I'd probably say the Pride Campaigns that focus on near-shore fisheries and Marine Protected Areas, such as those in the Philippines. Our approach uses a concept often referred to as a TURF-reserve (Territorial User Rights in Fisheries coupled with reserves). The idea here is that there is a strict no-take zone, which is then surrounded by a buffer zone where fishers are given what we term privileged access (as distinct from a right, which is much harder to change or revoke if necessary). Studies of no-take areas have shown that they boost fish stocks by almost 450% (on average) inside the protected area and by over 200% (on average) in the area adjoining it. The problem with many traditional Marine Protected Areas is open access means that anyone and everyone can take the spill-over. It's like you carefully saving into a bank account, but your neighbour takes the interest (and even raids the capital as well). In this case, you have no incentive to save as it's the first person to get to the bank that wins. The TURF-reserve approach differs because of its concept of giving local fishers 'privileged access' to the area to ensure that no one person takes everything. This encourages people to take 'ownership' of their environment and zealously protect the no-take zone as they understand the benefits of doing so.

What do you see as the main challenges in your field and how can they be overcome?

The challenges of what we do are often logistic. We are working in some of the most pristine, but also some of the remotest, areas on the planet. Moreover, these areas also tend to be impoverished and there can be quite high rates of crime or corruption. Challenges can vary from not having electricity or basic transportation to much more serious issues such as physical hazards to staff and partners. So, it is challenging for sure, but if there were no challenges, there would be no need. It's easy to just say it is not worth bothering, but I don't believe in that. Going back to the candle analogy, while one candle might not seem very bright, if each and every one of us lights one think what a light that will then be cast—a real beacon of hope for others to follow.

Finally, how did you get into conservation ecology and what advice would you give to others?

In my final year at university, I was part of a group of people who solicited commercial sponsorship used to census the St Lucian parrot, a bird endemic to the Caribbean island of St Lucia. That was in 1977, and there were around 150 parrots remaining in the wild. I was invited back to the island by the St Lucian government when I graduated the following year to start to put some of our recommended actions into practice. A lot of the work I did there laid the foundations for the Pride approach that we have subsequently used in Rare all over the world. In terms of advice to others: seek out whatever experience you can and don't give up! Being an Applied Ecologist in the conservation realm is a wonderful career and one that you can take an enormous amount of pride in, so make sure you chase your dreams!

species is the Iberian lynx *Lynx pardinus*, or the Sumatran rhinoceros *Dicerorhinus sumatrensis*, for instance.

3. **How should species be assigned to a given category?** The wisdom or otherwise of this system depends not only on the categories used, but also on what basis species are assigned to those categories. One way to do this is using an international species-at-risk classification system such as IUCN or NatureServe (or, if evolutionary distinctiveness is important, the EDGE system outlined in Case Study 10.2). However:
   - no single system is universally accepted;
   - there is variability between systems and even between assessors using the same system;
   - lists are far from comprehensive (just 5% of known species have currently been assessed);
   - lists are biased towards terrestrial charismatic mammal and bird species and species that can be harvested or are of commercial value;
   - the validity of the assessed risk category depends on the accuracy of the data upon which the species has been assessed—inaccuracies can be magnified.
4. **This approach is over-simplistic:** this system only considers species rarity—other factors are not considered. This includes key ecological principles such as whether a species is a keystone species, as well as its public (and thus funding) appeal, and whether there are any benefits to humans.

These areas of controversy do not mean that the concept of conservation triage is not accepted. Indeed, most of the 600 scientists questioned by Rudd (2011) thought triage criteria should be established. The contentious issue is not whether triage should be used, but rather it is what the actual criteria should be. This also brings up debate about **conservation ethics**, which are discussed further in Hot Topic 10.2.

Prioritization of species based upon rarity and perceived threat has become standard practice for conservation scientists. Indeed, determining which species are thriving and which are not is probably the single most crucial factor in targeting conservation resources appropriately (Mace et al., 2008). However, many criteria other than rarity and threat can be—and should be—used within conservation triage. The sections below explore species-focused prioritization, site-focused prioritization, and the specific factors that are important in *ex situ* conservation.

HOT TOPIC 10.2

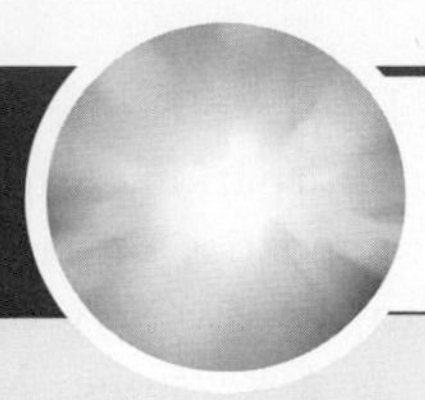

The ethics of conservation

In many ways, priority ranking is an integral part of conservation ecology. It is a pragmatic way of maximizing success and reducing the likelihood of wasted money, time, and effort that could have been more wisely invested. However, there are those that question whether we have the moral right to 'give up' on certain species: this is partly an ethical debate and can be particularly tricky when a species is endangered because of human action—do we then have an ethical responsibility to attempt conservation, whatever the cost? If so, how will this idealistic vision be funded? How do we get around the fact that extinction is ultimately likely to befall all species regardless of human action to 'save' them?

Because of resource limitations, sometimes it is necessary to divert resources from one species to another in the knowledge that this will cause extinction of the first. A parallel here is Hardin's (1974) lifeboat ethic, which was initially intended as a metaphor for resource distribution in 1974. The basic premise is a lifeboat with room for 10 people, but 100 people in need of rescue. Those that board are safe; those that do not will die. The ethics of that situation stem from the dilemma of whether (and under what circumstances) swimmers should be taken aboard the lifeboat. In conservation parlance, the lifeboat is conservation action, which has a finite capacity. The different swimmers are akin to different species, all of whom need rescue, but only some of whom can be 'saved', and the ethics here relate to which species should be conserved and why.

QUESTIONS

If we prioritize the conservation of some species over others, we are essentially 'giving up' on some species. How can we justify this approach?

Is the lifeboat ethic just a nice metaphor or is it helpful in framing conservation decisions?

REFERENCE

Hardin, G. (1974) Living on a lifeboat. *BioScience*, Volume 24, 561–568.

10.5.2 Species-focused priorities in conservation triage.

Examining species traits other than rarity is key in formulating conservation priorities. Such traits can be ecological such as prioritization of **keystone** species or the use of a focal **umbrella** species to protect multiple co-occurring species; others are centred around public support and funding through prioritization of **flagship** species.

Keystone species

A keystone species is one with an especially important role within an ecosystem. The concept was first outlined by Paine (1966) who used the term 'keystone' to link to the concept of architectural keystones—the stones at the apex of arches that are essential for structural integrity. The parallel with ecosystem structure is that although an ecosystem can involve many species, some are ecologically more important to that ecosystem than others, just as some stones are more important for the arch than others. For example, organisms that:

- **Control would-be dominant species** that, if otherwise unchecked, would dominate the ecosystem and exclude other species. An example is Paine's original study system in Mukkaw Bay, USA, where the seastar *Pisaster orhraceus* controls the would-be dominant mussel *Mytilus californianus*.
- **Are apex (top) predators** that prey upon other predatory species in lower trophic levels (mesopredators). This reduces the predation risk for their prey species in a process called **mesopredator release.**
- **Are ecosystem engineers** and modify the environment physically. This can result in new habitats being created as in the case of marram grass *Ammophila* which stabilizes sand

dunes. Alternatively ecosystem engineers can alter environments through the provision of new resources—for instance, plants such as clover *Trifolium* have a symbiotic relationship with nitrogen-fixing bacteria and thus increase the soil nitrogen levels by fixing atmospheric nitrogen.

- **Provide ecosystem services** that are ecologically-relevant (e.g. pollination by bumblebees, *Bombus*).

The conservation impacts of not conserving keystone species are neatly illustrated by two American examples.

1. The sea otter *Enhydra lutris*, is an endangered mesopredator that consumes *Strongylocentrotus* sea urchins. Sea otter numbers decreased around Alaska and the resultant lack of predation on sea urchins caused an eight-fold increase in sea urchin biomass and a resultant decrease in kelp from roughly 10 individuals per 0.25m^2 to only 1 per 0.25m^2 (Estes et al.,1998). Loss of kelp had a **cascade effect** on species that used it for shelter or food. One theory for sea otter decline is the start of, and rapid increase in, predation by orcas *Orcinus orca* (Estes et al., 1998). Although evidence is circumstantial, the IUCN has accepted this rare example of **predator switching** as a cause of decline. Sea otters are not traditional prey species, but following declines in other marine mammals, possibly due to starvation because of overfishing, orcas started to catch and kill sea otters from the early 1990s onwards.
2. In the terrestrial environment, grey wolves *Canis lupus* were extirpated from Yellowstone national park. This caused a dramatic rise in the population elk *Cervus canadensis*, their main prey, which led to overgrazing of quaking aspen *Populus tremuloides*. With the reintroduction of wolves in 1995 came the first elk predation for some 60 years and a resultant decrease in elk population. Aspen is now recovering in areas with high wolf (and thus low elk) populations (Ripple et al., 2001).

By their very definition, variation in abundance of keystone species has a greater impact than other species in an ecosystem. Thus, where keystone species can be identified, the case for prioritizing keystone species is clear in any (ecological) cost:benefit analysis. The problem is that identification of keystone species as being keystone is often only possible after their decline or extirpation. Attempts to characterize general characteristics of keystone species have foundered due to the diversity of ways in which a species can act as a keystone. Moreover, even for known keystone species, the effect of presence (and thus absence) on ecosystem dynamics are often context-dependent, as reviewed by Power et al. (1996). This highlights, once again, the inherent complexity of ecological systems.

Flagship species

Flagship species are high profile **charismatic** species with the ability to capture the imagination of the public. People can identify with cute-and-cuddly species such as koalas and lemurs, as well as more majestic species such as elephants and whales. The appeal of such species means that they are often used for conservation publicity and fund raising as shown in Figure 10.11. Perhaps the most famous is the World Wildlife Fund panda logo drawn by conservationist and artist Sir Peter Scott (who also started the IUCN red list in 1964).

There are examples of flagships in many taxa including insects (monarch butterflies *Danaus plexippus*), fish (clown fish *Amphiprion* for coral reefs and basking sharks *Cetorhinus maximus* for open water) and plants (giant redwoods *Sequoiadendron giganteum*; African baobab *Adansonia digitata*). However, there is a definite bias towards mammals and birds, which are over-represented relative to other taxa.

Species can also be held in special affection as emblems for specific areas. National examples include America's national bird, the bald eagle *Haliaeetus leucocephalus*, and Australia's national mammal, the red kangaroo *Macropus rufus*. Non-mammalian/avian examples include the national fish of South Africa (galjoen *Dichistius capensis*). Regional examples include the grey wolf *Canis lupus* in Yellowstone and the American alligator *Alligator mississipiensis* in the Florida Everglades.

There are two very different models for flagship conservation:

1. **Conservation of flagships** via organizations with a sole flagship focus (e.g. Space for Elephants in South Africa) or via high-profile initiatives

Figure 10.11 Flagship species used as 'poster boys' for conservation organizations.

Source: RSPB; Marine Conservation Society logos reproduced with permission. Fauna & Flora International logo © Fauna & Flora International.

run by general conservation organizations (e.g. Sumatran Tiger programme run by Flora and Fauna International). As shown in Figure 10.12, a considerable proportion of global conservation funding and attention is given to just a few flagship species.

2. **Use of flagship species to champion conservation** in general with flagship species becoming the 'shop window', rather than the sole focus for overarching charities, campaigns, and initiatives.

Opinion on using flagship species is divided. They can increase **public awareness** of, and funding for, conservation. From a purely economic perspective, flagships can be seen as a long-term investment as they are likely to bring a return on initial spending. From an educational standpoint, use of flagships can improve ecological knowledge, and not just of focal species. For example, Dietz et al. (1994) used lion tamarins *Leontopithecus*, to front a successful educational project in Brazil that improved knowledge of tamarins and ecology more generally.

Flagship species, particularly endemic species or those that are rare globally, but reasonably accessible in specific locales, can also be important for **ecotourism.** Walpole & Leader-Williams (2002) found that tourism revenue related to the Komodo dragon

Figure 10.12 Flagship-focused conservation schemes – four of the most popular species for conservation based on the number of Non-Governmental Organizations (NGOs) dedicated to them.

Source: Photographs by Anne Goodenough; data from Robert Smith, University of Kent, used with kind permission.

Varanus komodoensis resulted in the protection of biodiversity in the region and encouraged local positive attitudes towards conservation. Ecotourism based on golden snub-nosed monkey *Rhinopithecus roxellana* in China also increased tourist numbers and revenue. Moreover, after senior politicians visited, they changed their attitudes towards biological conservation and promised to transfer regular funds towards conservation (Xiang et al., 2011). Another excellent example is the mountain gorilla *Gorilla beringei beringei* ecotourism in Parc des Volcans in Rwanda, which will be considered in more detail in Section 11.6.

However, conservation can become over-focused on the charismatic few and not the species most in need for conservation action. An American study found that there was no significant difference in the amount of habitat and number of species protected in reserves containing flagship species versus those that did not (Andelman & Fagan, 2000). With the possible exception of tourism flagship species or those specifically used to front education or ecotourism, flagship species often have limited success at raising awareness.

Umbrella species

Umbrella species are species that are the focus of conservation initiatives that have far-reaching benefits. The idea is that by directing resources to *in situ* conservation of such species, the habitat in which that species occurs will also be protected, as will naturally co-occurring species. In this way, the focal species becomes a protective 'umbrella' for wider ecology. The aims of conservation schemes that use umbrella species vary. For example:

- **Planning nature reserve size/location:** capercaillie *Tetrao urogallus* have been used successfully in Switzerland to plan reserve location and management for a range of other birds (Suter et al., 2002).
- **Managing and restoring habitat:** the greater sage-grouse *Centrocercus urophasianus* has been proposed as an umbrella species to stop widespread degradation of the sagebrush ecosystem in the Western USA and restore former areas; the plan has not been implemented as most rare sagebrush-associated species would only benefit marginally from such a strategy (Rowland et al., 2006).
- **Reconnecting landscape patches:** three woodland carnivore species—European badgers *Meles meles*, weasels *Mustela nivalis*, and stone marten *Martes foina*—have been usefully used to identify woodland core areas and corridors in northern Italy (Bani et al., 2002).

The best umbrella species tend to be those that:

- Have large home ranges, or have a high MVP, but occur at low density, as this means that protected areas need to be large to be effective (**area umbrella**).
- Require specialist habitats so that rare or important habitat is conserved, thereby supporting other species with similar habitat requirements (**habitat umbrella**).
- Overlap in requirements for persistence with many other species; for example, similarity in patterns of movement through the landscape (**connectivity umbrella**) or responses to specific management actions (**management umbrella**).
- Overlap in range with many other species, thereby maximizing the number of other species that could potentially be conserved by proxy.

It can be especially useful if an umbrella species is also charismatic in some way and so can be used as a **flagship umbrella species**. A classic example is the Florida panther *Puma concolor coryi*, which occurs in pineland, woodland, and mixed-species swamp forest including the everglades. Conserving this species thus provides additional protection for two endangered habitats. Moreover, areas proposed specifically for panther conservation have at least 24 of 51 of the Florida endangered faunal species, including sandhill crane *Grus canadensis*, Audubon's crested caracara *Polyborus plancus audubonii*, and the gopher tortoise *Gopherus polyphemus*, as well as threatened plants like the night-scented orchid *Epidendrum nocturnum* (Cox et al., 1994). The panther is thus an ideal **tri-focus umbrella** (i.e. an umbrella spanning habitat, flora and fauna) (Figure 10.13).

However, despite some success stories, single-species umbrellas generally cannot ensure conservation of all co-occurring species (Roberge & Angelstam,

Figure 10.13 The Florida panther *Puma concolor coryi* protects rare habitats and species under its protective umbrella.
Source: Photographs clockwise from top by Ethan Oringel; James St. John/CC BY 2.0; and Anne Goodenough.

2004). A different approach is to align the flagship concept with the multi-species approach to conservation rather than single-species approach (Section 10.2) using Lambeck's idea of multiple umbrella species in the same location (Lambeck, 1997). Under this system, one species is typically identified as an area umbrella, another as a species that needs to disperse widely (connectivity umbrella) and a third with habitat-specific requirements (habitat umbrella). This has been used in Australia, where the hooded robin *Melanodryas cucullata* was identified as a good umbrella for area and habitat, while the eastern yellow robin *Eopsaltria australis* was a good connectivity umbrella (Watson et al., 2001).

There can also be useful overlap between umbrella species and **bioindicator** approaches (Chapter 4) with some focal species having joint roles. For example, the Siberian flying squirrel *Pteromys volans* has been used as both as an umbrella species and an indicator species for red-listed fungi, epiphytic lichens, and beetles in Finland. The significant association between flying squirrel presence and presence of 42 rare fungi, lichen, and beetle species that also need deadwood means that the species can be used both to indicate deadwood abundance and as an umbrella for multiple rare species (Hurme et al., 2008).

10.5.3 Site-focused priorities in conservation triage.

In addition to the species-focused prioritization of conservation resources, several other factors can be important for resource allocation within *in situ* conservation. These focus not on specific species, but on specific sites (and thus entire species communities and the threats to those communities).

Biodiversity hotspots and areas of high endemism

First proposed by Myers et al. (2000), **global biodiversity hotspots** support an extremely high number of species within a comparatively small area. As originally defined, there were 25 hotspots, all primarily terrestrial, but including the ocean between islands within archipelagos as appropriate. To be listed as a Myers hotspot, an area needed to meet two criteria:

1. Contain at least 0.5% or 1500 of the world's 300,000 **endemic** plant species.
2. Be subject to a high level of **threat**, having already lost >70% of primary vegetation.

Conservation of endemic species is, or at least should be, a conservation priority simply because the loss of such a species from the area where it is endemic is not extirpation, but extinction: local loss equals global loss. Of the original 25 hotspots, 10 contained >5000 endemic plants and a further five contained >2500 endemics, while 11 hotspots had already lost > 90% of their primary vegetation. Since 2000, the number of hotspots has increased to 35. Despite covering just over 2% of the earth's land surface, they contain 77% of the world's endemic plants, 43% of vertebrates (including 60% of threatened mammals/birds), and 80% of threatened amphibians.

Hotspots contain more islands than would be expected by chance, including Madagascar and New Zealand. The reason for this is two-fold. First, the isolation of islands tends to increase the number of endemic species because lack of gene flow between island–island or island–mainland communities causes speciation and subsequent endemism (Criterion 1), and secondly, because islands tend to be more vulnerable both to natural and anthropogenic change (Criterion 2). There are also a lot more hotspots in the tropics and subtropics than anywhere else because species richness is higher at the equator than at the poles: a concept known as a **latitudinal diversity gradient.** The tropical island of Madagascar is investigated in more detail in Case Study 10.3.

More recently, the hotspot idea was redeveloped for birds to allow for three different types of hotspot (Orme et al., 2005):

- Type 1: areas with **high species richness.**
- Type 2: area with **high threat level.**
- Type 3: areas with **high endemism.**

In all cases, the areas in the highest 2.5% of the range were declared hotspots. This led to 28 different hotspots as shown in Figure 10.14 (Richness = 2; Threat = 4; Endemism = 12; the remaining 10 were classified under at least two criteria). In total, these areas support 49% of bird species, including 1447 endemic species (59%). This demonstrates that there are many different types of 'hotspot'—in fact, in the last decade or so, even more hotspot templates have come out, including those that focus on '**crisis ecoregions**' and wilderness areas.

Areas with high species diversity

The hotspot principle can be applied at national or regional scales to prioritize sites with endemic or range-restricted species or with high species diversity. Diversity has two constituent parts:

1. **Richness:** a simple count of the number of species within a given area. This can be taxon-specific (e.g. floral richness) according to guild (e.g. farmland bird richness or woodland beetle richness) or total richness. Only species presence is considered not abundance; for instance:
   - **Site 1**—five reptile individuals each of a different species = reptile species richness of 5.

CASE STUDY 10.3

Madagascar

Madagascar, situated in the Indian Ocean to the East of South Africa and Mozambique, is one of the 'hottest' biodiversity hotspots. Although mainland Africa is the nearest continental landmass, Madagascar was actually most recently joined to India. Around 170 million years ago, the single supercontinent Pangaea had separated into Laurasia (containing most of what are now northern hemisphere landmasses) and Gondwanaland (containing most of what are now southern hemisphere landmasses). Within Gondwanaland, what is now Madagascar lay between the Africa–South America and Antarctica–India landmasses. About 160–135 million years ago the Antarctica–India landmass broke away from the Africa–South America landmass, taking Madagascar with it. Madagascar later split from Antarctica (130–80 million years ago) and finally India (about 88 million years ago).

This complex geological past has provided a complex biological present. Madagascan species largely evolved in isolation from common ancestors and responded to different local environments. This resulted in widespread allopatric speciation. This means that Madagascar has one of the highest rates of endemism of anywhere on earth: an estimated 85–90% of vascular plants, 86% of invertebrates, 90% of mammals, and 6% of birds are endemic. Endemism is not only evident at the species level, but also at higher taxonomic levels: Madagascar has an incredible eight plant families, five bird families, and five primate families that occur nowhere else on Earth. Probably the most iconic endemic species are the lemurs, of which there are around 100 known species.

In terms of overall species richness, Madagascar does not actually score particularly highly, but for endemic richness it is a very high conservation priority. Humans only colonized the island nation around 2000 years ago, but in that time, 90% of natural habitat has been destroyed. That habitat destruction means that numerous endemic species have gone extinct—for example the iconic elephant birds Aepyornithidae that stood over 3m tall. Of the species still present today, many are at a high risk of extinction, including many lemur species (Figure A) and, in many cases, loss of a species from Madagascar equals a global loss of that species (at least in the wild).

Figure A The Alaotran gentle lemur *Hapalemur alaotrensis* named after Lac Alaotra where it lives in Madagascar. It is classified as Critically Endangered on the IUCN red list.

Source: Photograph by Anne Goodenough.

FURTHER READING

Madagascar ecology: Goodman, S.M. & Benstead, J.P. (2005) Updated estimates of biotic diversity and endemism for Madagascar. *Oryx*, Volume 39, 73–77.

Species diversification in Madagascar: Vences, M., Wollenberg, K.C., Vieites, D.R., & Lees, D.C. (2009) Madagascar as a model region of species diversification. *Trends in Ecology & Evolution*, Volume 24, 456–465.

Lemur conservation: Schwitzer, C., Mittermeier, R.A., Davies, N., Johnson, S., Ratsimbazafy, J., Razafindramanana, J., Louis, E.E. Jr, & Rajaobelina, S. (2013) *Lemurs of Madagascar: a Strategy for their Conservation 2013–2016.* IUCN SSC Primate Specialist Group, Bristol, UK.

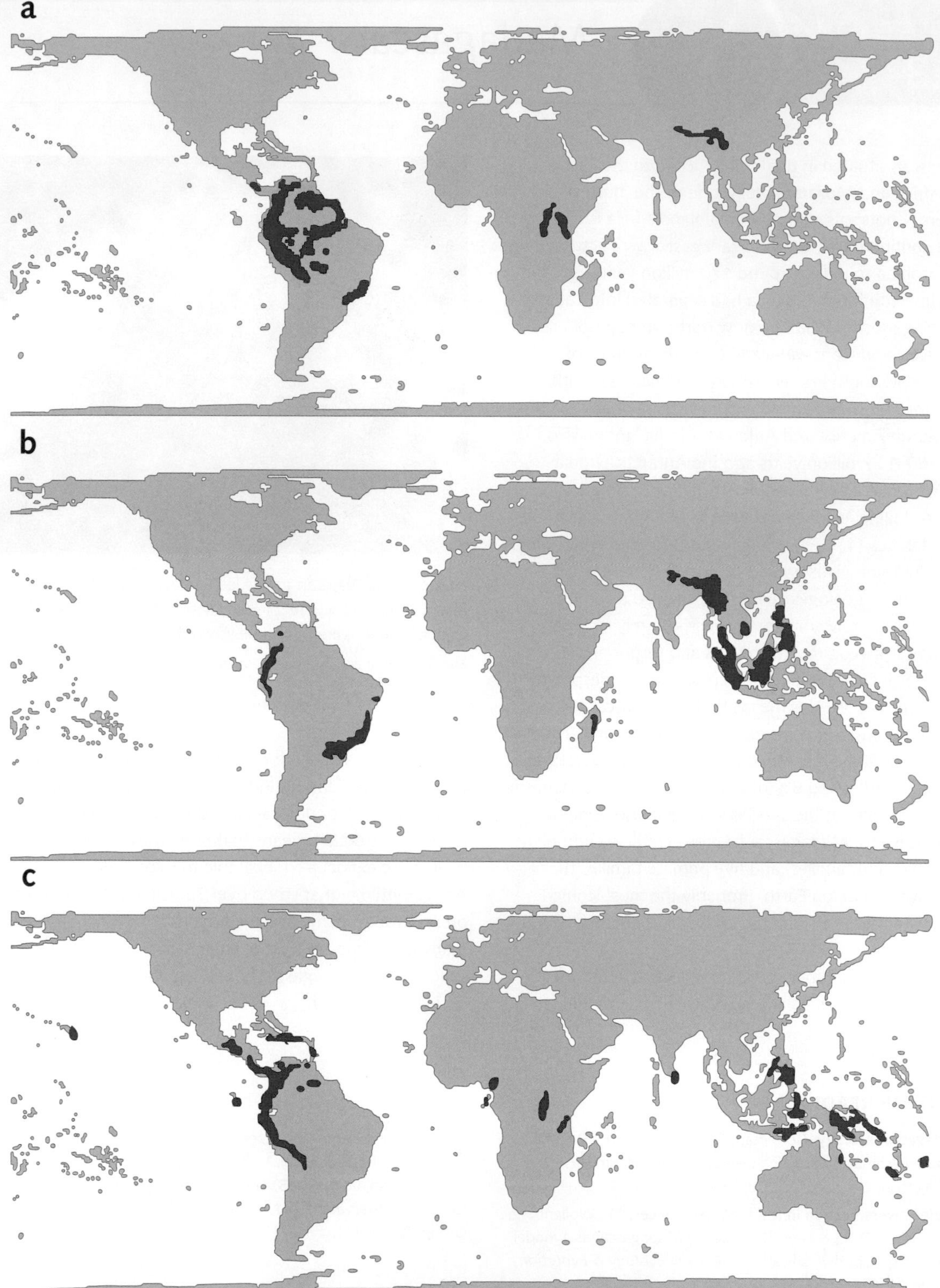

Figure 10.14 The locations of 28 global biodiversity hotspots for birds designated because they are: (a) hotspots of species richness, (b) hotspots of threatened species, or (c) hotspots of endemic species.Source: Reproduced from Orme, C.D.L. et al. (2005) 'Global hotspots of species richness are not congruent with endemism or threat'. *Nature*, 436(7053) pp. 1016–1019. Used here with permission from Nature Publishing Group.

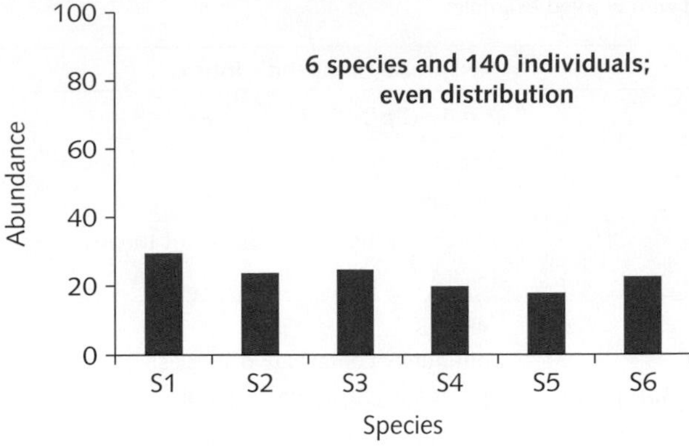

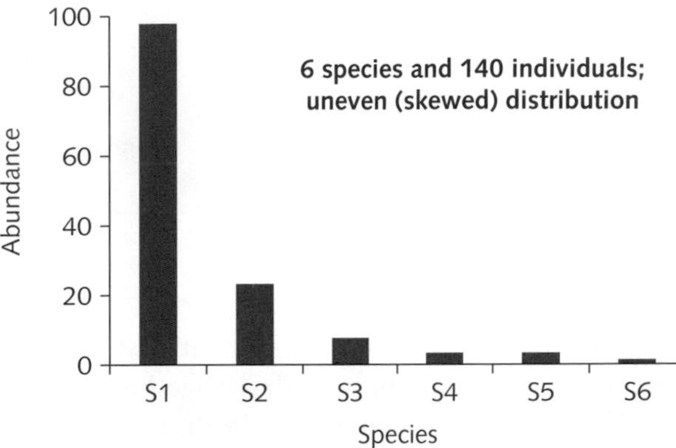

Figure 10.15 Graphical representation of different community structures in terms of the evenness of species' relative abundance. Here, the top community would score more highly than the bottom community in a diversity index and thus be regarded as more diverse as the species richness is the same but evenness is greater.

   - **Site 2**—500 reptile individuals, but only two species = reptile species richness of 2.
2. **Evenness:** this is a measure of community structure and can be understood most easily using a frequency histogram with each species being a separate bar. In an even community, each bar is approximately equal so the relative abundance of the different species is very similar. In contrast, in an uneven community, the bars differ because some species are dominant while others are rare. This is shown diagrammatically in Figure 10.15.

Species richness and community evenness can be considered independently of one another, or a **diversity index** can be used to combine these concepts into one metric. There are several indices, each with its own advantages and disadvantages. Two common indices are the Shannon–Wiener index (sometimes just Shannon or Shannon–Weaver) and Simpson's index, of which there are several different variants. These are based on slightly different concepts: Shannon–Wiener is based on evenness throughout the whole community whereas Simpson's is based on how dominant the dominant species are (i.e. the height of the tallest bar in Figure 10.15 relative to the others rather than the heights of all bars relative to each other). These two metrics are detailed in Table 10.5.

It is important to realize that diversity index values are ordinal so the gaps between values are unknown and potentially uneven. In other words, whilst 4 is certainly greater than 2, it would be wrong to say 4 is double 2. There are three other key rules for interpreting diversity index values:

1. Values make little sense alone; it is only in comparing values for multiple sites, or for the same site at different times, that the indices are useful.
2. Only values generated using the same index are comparable.
3. Values should only be compared when it makes logical (and ecological) sense to do so. It makes sense to compare butterfly diversity between two nearby sites with different management. It does not make sense to compare the diversity of marine fish in Australia with that of freshwater fish in Belgium, for example.

It should be noted that, while it is easy to calculate richness or diversity mathematically, obtaining field data on which to perform these calculations is much less straightforward. In many cases, species data simply do not exist. Even where they do, it is likely that they are for one taxonomic group not total biodiversity. For instance, beetle diversity is often used as a **proxy** of total diversity because of the ubiquity of this insect order and the comparative ease with which they can be sampled in comparison to other taxa (Section 4.5.1).

In terms of allocating conservation resources, it might make sense to prioritize conservation of areas with high richness/diversity over areas with low

Table 10.5 Shannon–Wiener and Simpson's diversity indices explained with worked examples.

	Shannon–Wiener index	(Reciprocal) Simpson's index
Equation	$H = -\Sigma p_i \ln p_i$	$D = 1-(\Sigma p_i^2)$
Elements	H = diversity value Σ = sum of p_i = the fractional abundance of the *i*th species ln = natural logarithm	D = diversity value Σ = sum of p_i = the fractional abundance of the *i*th species
Worked example	Data Community contains three species: S1 = 4 individuals; S2 = 9 individuals; S3 = 57 individuals Calculate fractions (S1 = 0.06, S2 = 0.13, S3 = 0.81) Take the ln of each (S1 = −2.81, S2 = −2.04, S3 = −0.21) H = −((0.06*−2.81*) + (0.13*−2.04) + (0.81*−0.21)) H = −(−0.59) H = 0.59 [note, might be 0.61 or 0.60 depending on rounding]	Data Community contains two species: S1 = 5 individuals; S2 = 7 individuals Calculate fractions (S1 = 0.417; S2 = 0.583) $D = 1 - ((0.417^2) + (0.583^2))$ D = 1 − (0.174 + 0.340) D = 1 − 0.514 D = 0.486
Interpretation	Higher values = more diverse community Values are 0-infinity, but normally run between 1.5 and 3.5	Higher values = more diverse community Values run between 0 and 1

species richness/diversity. That, of course, does not take any species-specific information into account. There might be times when it is better to prioritize an area with low diversity because species in that community are endemic, rare, or threatened. As such, richness and diversity values should not be used as a blunt tool, but rather as an added metric that might be useful. Diversity values can also be valuable for monitoring the effectiveness of conservation action on the overall community and thus complement monitoring of individual species.

Areas with distinct species communities

Another concept that can be useful for allocation of resources is **community similarity**. One of the most simple and intuitive metrics is the Jaccard Coefficient of Community Similarity (CC_j), which runs between 0 (no species in common) and 1 (all species are found in both communities):

$$CC_j = c/S$$

where S = overall number of species in two communities of interest, and c = number of species common to both communities of interest. If there are two focal communities that together contain 20 species with 15 common to both, the equation is 15/20 = 0.75. Thus the majority of species (75%) are found in both communities. The index is pairwise so to compare three communities, three tests are needed (1 and 2, 1 and 3, 2 and 3). This obviously rises exponentially.

The index can be useful when determining which of several possible sites would give most 'added benefit' relative to sites already conserved. Imagine that a species-rich grassland site was already protected by a conservation organization and that opportunity has arisen to buy and protect more grassland in the same area, but there are two possible sites. It might make sense to compare the community of both candidate sites in relation to that of the current site to identify, which is most dissimilar (conserving this site in addition to the current one would increase the total community, and thus the total number of species, protected). Alternatively, it might be better to target the most similar site if the aim is to manage a network of sites to establish metapopulations. Either approach is valid and would simply depend on the specifics of the sites and species involved, but both decisions would require knowledge of community similarity. This approach can also be used to evaluate restoration success (Section 11.3.1) by quantifying how similar the restored site is to a comparable pristine site.

Areas with historical legacy

Much of this chapter has been based upon the premise that all nature reserves have arisen by considering scientific methods and objective ways of examining and prioritizing species and habitats for conservation action. In other words, it has assumed that conservation is inherently based on evidence and sound decision making. Although conservation is gradually becoming evidence-informed (Section 11.7), politics and history have always played a part in shaping conservation and many of the nature reserves that exist today do so because of historical events. Often nature reserves, especially those owned by charities, occur because of legacies rather than having been selected because of their biodiversity. Other reserves may be purchased, but this is often driven by what areas are put forward for sale, cost, politics, the personal interests on those of charitable trusts, and the threats to specific areas at that point in time. Such issues are discussed further in Chapter 11.

10.5.4 Setting priorities within *ex situ* scenarios.

Within an *ex situ* scenario, the rationale for setting priorities can be subtly different, as conservation needs must be balanced with captive environments being centres for education and entertainment. Key considerations are:

- **Species with high reintroduction potential:** although *ex situ* collections are often viewed in isolation from *in situ* initiatives, from a conservation perspective *ex situ* work should only ever be part of a bigger picture, as will be made clear in Chapter 12. The main way that *ex situ* and *in situ* initiatives link is through reintroduction to supplement existing wild populations or to recreate extirpated populations. Thus, if the aim of captive collections is truly reintroduction (rather than being an ark—see the conservation hierarchy in Figure 10.10), the establishment of captive populations should be prioritized for species that need reintroduction and have a high chance of success.
- **Non-conservation priorities:** captive collections usually rely on entrance fees as a primary revenue stream. This means that many zoos and gardens prioritize flagship species—at least to some extent—to attract visitors and ensure the collection is financially sustainable. This is particularly important in the modern era when there are many attractions and activities competing for the public's time and money. Captive collections should also be centres of education, and this means that holding a range of different species from different taxa, with different life histories and from different parts of the world, is useful, as is holding species that showcase the range of threats to global biodiversity and the conservation schemes used to protect that biodiversity.
- **Practicalities:** captive collections—especially zoos—needs to carefully consider practicalities of space and cost, together with how much is known about husbandry requirements.

FURTHER READING FOR SECTION

Overview of conservation priorities and decision making: Mace, G., Possingham, H., & Leader-Williams, N. (2007) Prioritising choices in conservation. In: *Key Topics in Conservation Biology*, ed. D.W MacDonald and K. Service. Oxford: Blackwell Publishing, 17–34.

Use of flagship, umbrella and keystone species: Caro, T. (2010) *Conservation by Proxy: Indicator, Umbrella, Keystone, Flagship, and Other Surrogate Species*. Washington: Island Press.

Use of flagship, umbrella and keystone species: Smith, R.J., Veríssimo, D., Isaac, N.J., & Jones, K.E. (2012) Identifying Cinderella species: uncovering mammals with conservation flagship appeal. *Conservation Letters*, Volume 5, 205–212.

Harnessing the importance of charismatic mammals in Africa: Caro, T. & Riggio, J. (2013) The Big 5 and conservation. *Animal Conservation*, Volume 16, 261–262.

Use of biodiversity hotspots within conservation prioritisation: Bacchetta, G., Farris, E., & Pontecorvo, C. (2012) A new method to set conservation priorities in biodiversity hotspots. *Plant Biosystems—An International Journal Dealing with all Aspects of Plant Biology*, Volume 146, 638–648.

10.6 Conclusions.

Conservation ecology can be focused on species or habitats, it can have one focal outcome or many, and it can be *in situ* or *ex situ*. With the current high rate of extinction, which is many times higher than the background rate as the result of human activities, the need for conservation has never been greater. However, given that there is more demand for conservation action than there is resource to supply that action, tough decisions often have to be made about which species and habitats will be prioritized. Making these decisions is often far from objective, but relies on knowledge of change in habitats and extinction risk predictions for species. The notion that some species and sites are inherently more valuable than others is also important. Some species have keystone, flagship, or umbrella status while some habitats support a particularly high species richness or regulate ecosystem processes (such as hydrology). Non-ecological considerations of funding, national and local priorities, and stakeholder interests can also be important in conservation triage.

ONLINE ACTIVITY

Go to www.oxfordtextbooks.co.uk/orc/goodenough/ to download the activity that accompanies this chapter.

CHAPTER 10 AT A GLANCE: THE BIG PICTURE

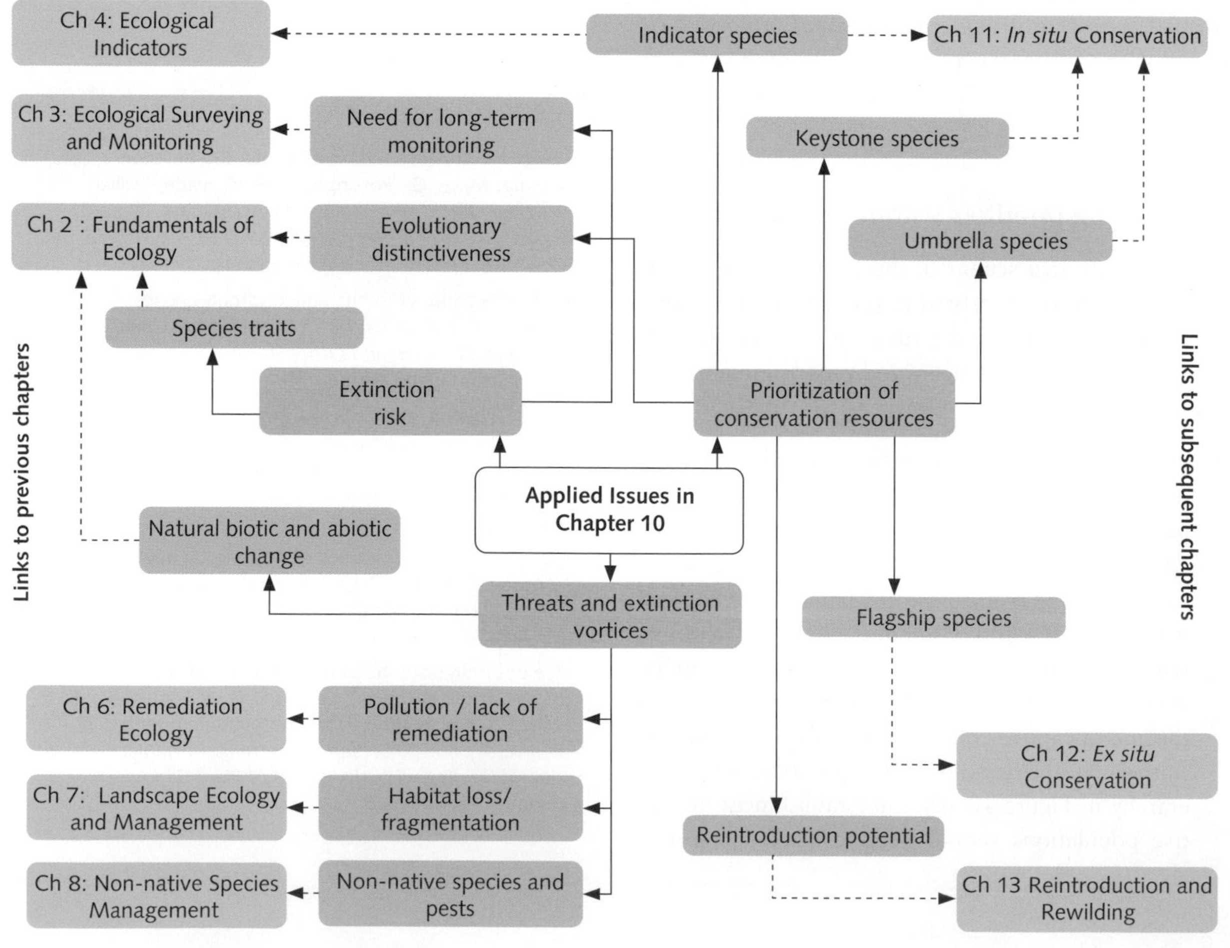

REFERENCES

Andelman, S.J. & Fagan, W.F. (2000) Umbrellas and flagships: efficient conservation surrogates or expensive mistakes? *Proceedings of the National Academy of Sciences of the United States of America*, Volume 97, 5954–5959.

Bani, L., Baietto, M., Bottoni, L., & Massa, R. (2002) The use of focal species in designing a habitat network for a lowland area of Lombardy, Italy. *Conservation Biology*, Volume 16, 826–831.

Barnosky D., Matzke N.S., Tomiya S., Wogan G., Swartz B., Quental T.B., Marshall C., J. McGuire L., Lindsey E.L., Maguire K.C., Mersey B., & Ferrer E.A. (2011) Has the Earth's sixth mass extinction already arrived? *Nature*, Volume 471, 51–57.

Benton, M.J. & Twitchett, R.J. (2003) How to kill (almost) all life: the end-Permian extinction event. *Trends In Ecology & Evolution*, Volume 18, 358–365.

Cox, J., Kautz, R., MacLaughlin, M., & Gilbert, T. (1994) *Closing the Gaps in Florida's Wildlife Habitat Conservation System*. Tallahassee, FL: Florida Game and Fresh Water Fish Commission.

Deguignet M., Juffe-Bignoli D., Harrison J., MacSharry B., Burgess N., & Kingston N. (2014) *2014 United Nations List of Protected Areas*. Cambridge: UNEP-WCMC.

Dietz, J.M., Dietz, L.A., & Nagagata, E.Y. (1994) The effective use of flagship species for conservation of biodiversity: the example of lion tamarins in Brazil. In: P.J. Olney, G. Mace, & A. Feistner (Eds). *Creative Conservation*, pp. 32–49. Springer, Netherlands.

Eaton, M.A., Brown, A.F., Noble, D.G., Musgrove, A.J., Hearn, R., Aebischer, N.J., Gibbons, D.W., Evans, A., & Gregory, R.D. (2009) Birds of Conservation Concern 3: the population status of birds in the United Kingdom, Channel Islands and the Isle of Man. *British Birds*, Volume 102, 296–341.

Estes, J.A., Tinker, M.T., Williams, T.M., & Doak, D.F. (1998) Killer whale predation on sea otters linking oceanic and nearshore ecosystems. *Science*, Volume 282, 473–476.

Evans, M.C., Watson, J.E., Fuller, R.A., Venter, O., Bennett, S.C., Marsack, P.R., & Possingham, H.P. (2011) The spatial distribution of threats to species in Australia. *BioScience*, Volume 61, 281–289.

Franklin, I.R. (1980) Evolutionary change in small populations. In: M.E. Soulé and B.A. Wilcox (Eds) *Conservation Biology: An Evolutionary Ecological Perspective*, pp. 135–140. Sunderland, MA: Sinauer Associates.

Gilpin, M.E. & M.E. Soulé. (1986) Minimum viable populations: processes of species extinction. In: M.E. Soulé (Ed.) *Conservation Biology: The Science of Scarcity and Diversity*, pp. 19–34. Sunderland, MA: Sinauer Associates.

Hallam A. & Wignall P.B. (1997) *Mass Extinctions and Their Aftermath*. Oxford: Oxford University Press.

Hurme, E., Mönkkönen, M., Sippola, A.L., Ylinen, H., & Pentinsaari, M. (2008) Role of the Siberian flying squirrel as an umbrella species for biodiversity in northern boreal forests. *Ecological Indicators*, Volume 8, 246–255.

Lacy, R.C. & Pollak, J.P. (2015) *Vortex: A Stochastic Simulation of the Extinction Process*. Brookfield, IL: Chicago Zoological Society.

Lambeck, R.J. (1997) Focal species: a multispecies umbrella for nature conservation. *Conservation Biology*, Volume 11, 849–856.

Leader-Williams, N. (1990) Black rhinos and African elephants: lessons for conservation funding. *Oryx*, Volume 24, 23–29.

Leaky, R. & Lewin, R. (1996) *The Sixth Extinction: Patterns of Life and the Future of Humankind*. New York, NY: Doubleday and Company.

Mace, G., Possingham, H., & Leader-Williams, N. (2007) Prioritising choices in conservation. In: D.W MacDonald and K. Service (Eds) *Key Topics in Conservation Biology*, pp. 17–34. Oxford: Blackwell Publishing.

Mace, G.M., Collar, N.J., Gaston, K.J., Hilton-Taylor, C., Akcakaya, H.R., Leader-Williams, N., Milner-Gulland, E.J., & Stuart, S.N. (2008) Quantification of extinction risk: IUCN's system for classifying threatened species. *Conservation Biology*, Volume 22, 1424–1442.

Millsap, B.A., Gore, J.A., Runde, D.E., & Cerulean, S.I. (1990) Setting priorities for the conservation of fish and wildlife species in Florida. *Wildlife Monographs*, Volume 111, 1–57.

Mora, C., Tittensor, D.P., Adl, S., Simpson, A.G., & Worm, B. (2011) How many species are there on Earth and in the ocean?. *PLOS Biology*, Volume 9, e1001127.

Myers, N. (1979) *The Sinking Ark: a New Look at the Problem of Disappearing Species*. Oxford: Pergamon Press.

Myers, N., Mittermeier, R.A., Mittermeier, C.G., da Fonseca, G.A.B., & Kent, J. (2000) Biodiversity hotspots for conservation priorities. *Nature*, Volume 403, 853–858.

Orme, C.D.L., Davies, R.G., Burgess, M., Eigenbrod, F., Pickup, N., Olson, V.A., Webster, A.J., Ding, T.-S., Rasmussen, P.C., Ridgely, R.S., Stattersfield, A.J., Bennett, P.M., Blackburn, T.M., Gaston, K.J., & Owens, I.P.F. (2005) Global hotspots of species richness are not congruent with endemism or threat. *Nature*, Volume 436, 1016–1019.

Paine, R.T. (1966) Foodweb complexity and species diversity. *American Naturalist*, Volume 100, 65–75.

Pimm, S.L., Jenkins, C.N., Abell, R., Brooks, T.M., Gittleman, J.L., Joppa, L.N., Raven, P.H., Roberts, C.M., & Sexton, J.O. (2014) The biodiversity of species and their rates of extinction, distribution, and protection. *Science*, Volume 344, 987–997.

Power, M.E., Tilman, D., Estes, J.E., Menge, M.A., Bond, W.J., Mills, L.S., Daily, G., Castilla, J.C., Lubchenco, J., & Paine, R.T. (1996) Challenges in the quest for keystones. *BioScience*, Volume 46, 609–620.

Primack, R.B. (2006) *Essentials of Conservation Ecology*. Sunderland, MA: Sinauer Associates.

Quammen, D. (1996) *Song of the Dodo: Island Biogeography in an Age of Extinctions*. New York, NY: Scribner.

Regan, H., Burgman, M., McCarthy, M.A., Master, L.L., Keith, D.A., Mace, G.M., & Andelman, S.J. (2005) The consistency of extinction risk classification protocols. *Conservation Biology*, Volume 19, 1969–1977.

Ripple, W.J., Larsen, E.J., Renkin, R.A., & Smith, D.W. (2001) Trophic cascades among wolves, elk and aspen on Yellowstone National Park's northern range. *Biological conservation*, Volume 102, 227–234.

Roberge, J.M. & Angelstam, P. (2004) Usefulness of the umbrella species concept as a conservation tool. *Conservation Biology*, Volume 18, 76–85.

Rowland, M.M., Wisdom, M.J., Suring, L.H., & Meinke, C.W. (2006) Greater sage-grouse as an umbrella species for sagebrush-associated vertebrates. *Biological Conservation*, Volume 129, 323–335.

Rudd, M.A. (2011) Scientists' opinions on the global status and management of biological diversity. *Conservation Biology*, Volume 25, 1165–1175.

Shaffer, M.L. (1978) *Determining Minimum Viable Population Sizes: A Case Study of the Grizzly Bear (*Ursus arctos *L.)*. PhD Thesis, Duke University, NC.

Suter, W., Graf, R.F., & Hess, R. (2002) Capercaillie (*Tetrao urogallus*) and avian biodiversity: testing the umbrella-species concept. *Conservation Biology*, Volume 16, 778–788.

Walpole, M.J. & Leader-Williams, N. (2002) Tourism and flagship species in conservation, *Biodiversity and Conservation*, Volume 11, 543–547.

Watson, J., Freudenberger, D., & Paull, D. (2001) An assessment of the focal-species approach for conserving birds in variegated landscapes in Southeastern Australia. *Conservation Biology*, Volume 15, 1364–1373.

Wilcove, D.S., Rothstein, D., Dubow, J., Phillips, A., & Losos, E. (1998) Quantifying threats to imperiled species in the United States. *BioScience*, Volume 48, 607–615.

Xiang, Z., Yu, Y., Yang, M., Yang, J.Y., Niao, M.Y., & Li, M. (2011) Does flagship species tourism benefit conservation? A case study of the golden snub-nosed monkey in Shennongjia National Nature Reserve. *Chinese Science Bulletin*, Volume 56, 2553–2558.

11 *In Situ* Conservation

11.1 Introduction.

Despite the undeniable importance of conserving species in captivity when they are at immediate risk of extinction in their natural environment (Chapter 12), ***in situ* conservation**—managing and protecting habitats and species in the wild—is still the mainstay of conservation ecology.

The term *in situ* conservation covers a wealth of **active management techniques** aimed at conserving organisms from different taxonomic groups, in different environments, and threatened in different ways. These include managing African grassland through rotational burning, providing supplementary food to increase the breeding success of the rare Mauritius kestrel *Falco punctatus*, translocating Tasmanian devils *Sarcophilus harrisii* to a remote island to create a disease-free population, deliberately scuttling ships, such as the *General Hoyt S. Vandenberg* off the Florida Keys, to create an artificial reef, and carefully pruning trees in Denmark to provide habitat for rare wood-boring beetles. Also included under the broad term of *in situ* conservation are **legislative and policy frameworks**, the creation of protected areas, and wider landscape-scale schemes, such as working with farmers to benefit biodiversity through agri-environment schemes. Working with the local community to educate and enthuse people about species and their interactions (the fundamental basis of ecology) is also vital.

In this chapter, we will consider some of the many types of *in situ* conservation using examples and case studies from around the world and from a range of environments. We will also consider some of the most contentious hot topics of conservation, including trophy hunting, poaching, and fishing quotas.

11.2 An overview of *in situ* management.

Mention the term '*in situ* management' in relation to ecology and the chances are that people will think of management undertaken specifically to conserve specific species and habitats. However, although a considerable amount of *in situ* management is undertaken for this reason (and is the primary focus of this chapter), it is important to recognize at the outset that management of natural environments is much broader than 'just' conservation and can be undertaken for many reasons. Broadly speaking, these can be placed in one of five groups as outlined in Figure 11.1, although there can be considerable overlap between them.

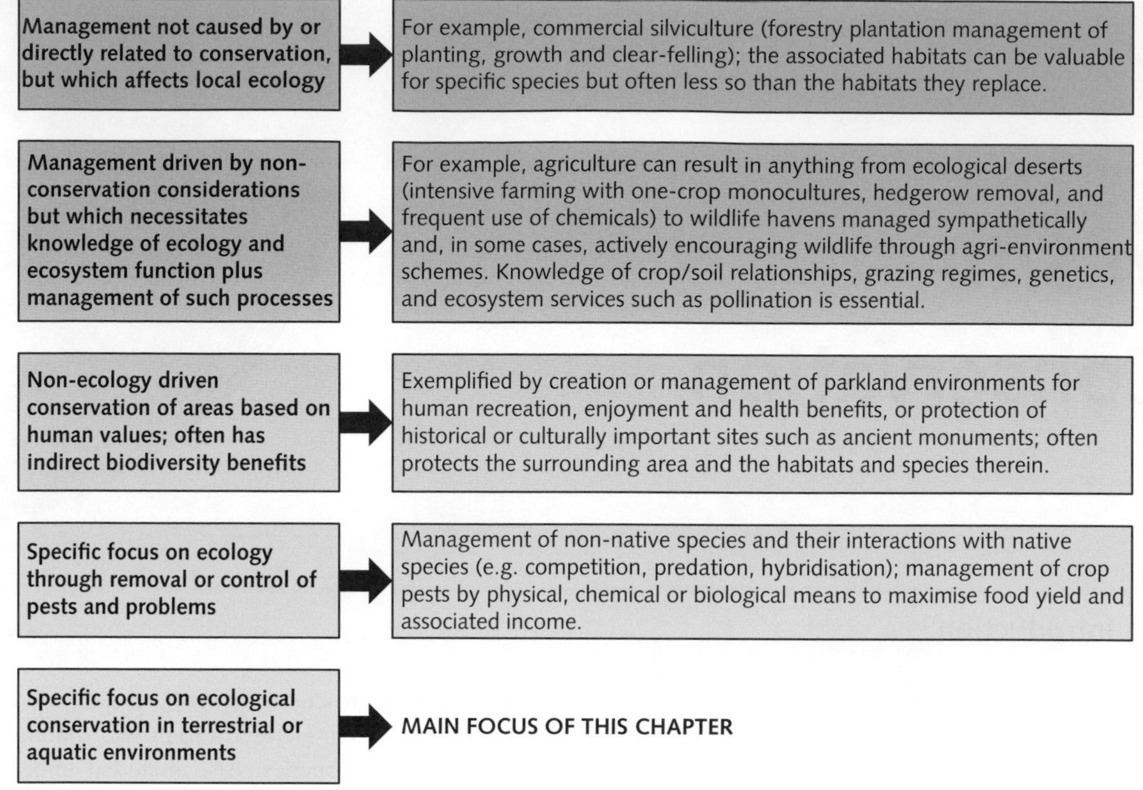

Figure 11.1 Reasons and motivation for *in situ* management of habitat and species.

11.2.1 Aims of *in situ* management for conservation.

There are six overarching aims for *in situ* conservation:

1. **To create:** establishing new areas of habitat or new protected areas might be undertaken to compensate for the loss of other similar sites (e.g. through development: Chapter 5), to enhance biodiversity of an area by adding additional habitat of high ecological value (**multi-species conservation**) or to create conditions that are suitable for specific species (**single-species conservation**) (Section 10.2.2).
2. **To restore:** recreating habitats that used to be supported at a site. This might be needed when the target habitat is a transitional one that has not been maintained due to lack of appropriate management or to reverse long-term degradation. Examples would be grassland that has not been routinely grazed or cut, and so, through the process of ecological succession (Section 2.7.2), has developed into scrub. Restoration might also involve the need to clean up a site through the process of remediation (Chapter 6).
3. **To protect:** protection of an area is possible either administratively by designating it as a legally protected area or via on-the-ground initiatives, such as putting up fences to prevent grazing animals accessing certain areas. Management of people, for example, asking visitors to remain on footpaths or not to fish in certain areas (Section 11.6.1), would also fall under this heading. This is perhaps the biggest priority for conservation, and is often the least resource-demanding and most cost-effective, especially relative to creation and restoration.
4. **To maintain:** once an appropriate management strategy has been devised and actioned, maintenance of a site can often be achieved simply by continuing this management. For example, if a species-rich grassland is in good condition having been managed through low-intensity grazing and control of fertilizer use, continuing this strategy

will probably maintain biodiversity effectively. The caveat here is that ecological systems are in flux, so it is important that the conservation strategy remains flexible enough to respond to changes or challenges. In the grassland example, the duration and/or intensity of management might need to be modified during a drought to avoid overgrazing. It would also be wise to undertake alert monitoring (Section 3.3.4), so that any immediate threats, such as invasion of non-native species, can be resolved quickly and effectively.

5. **To improve**: initiating management to improve the quality of a habitat or directly influence species' ecology (e.g. through supplemental feeding). The aim of improvement management is that such initiatives would be short term and then management could revert to more of a maintenance strategy. Improvement is subtly different from restoration since the latter is about reverting to a previous state, whereas improvement could be to try something new. Restoration is also typically a habitat approach to restore an overall community, whereas improvement might be very species-specific.
6. **To connect**: connecting areas of land and populations through improving connectivity. This can involve creating, restoring or improving corridors or stepping stone islands (Section 7.6.2).

11.2.2 Active versus custodial management.

In situ management can involve many different approaches and it is so it can be helpful to divide them into active management and custodial management:

- **Active management** involves a physical interaction with the environment—planting, digging, dredging, culling, and so on. There is a spectrum within active management as some actions demand more intense management over a longer period than others. For example, compare the management of a mature woodland that might only need very occasional intervention (e.g. to reduce herbivore numbers if woodland regeneration is too low) with that of a marginal wetland where intense scrub clearance might be needed each year to prevent the area reverting to scrub or woodland. In all cases, however, there will be a hands-on element that results in physical change to environments, habitats, species, or ecological processes, and active management is sometimes referred to as **manipulative management** for this reason.
- **Custodial management** is conservation through non-active means. This includes protection given to specific sites, habitats, and species through legislation, as well as specific directives and policy initiatives such as Biodiversity Action Plans (BAPs; Section 11.4.2). This form of management is sometimes called **passive management** because there is no active on-the-ground work, but this is rather misleading since it can imply it is easier than, or inferior to, active management. In fact, obtaining suitable legislative protection for a site or species is anything but easy (or passive!). Active management techniques are explored in Section 11.3, while custodial approaches are explored in Section 11.4.

FURTHER READING FOR SECTION

The basis of *in situ* conservation: **Convention on Biological Diversity Article 8:** Available at: https://www.cbd.int/convention/articles/default.shtml?a = cbd-08.

11.3 *In situ* conservation through active management.

Given the broad scope of active management and the number of habitats in which it can be used, it is perhaps unsurprising that there is a plethora of different active management techniques. These all require knowledge of species' ecology and ecological interactions if they are to be used effectively. Some of the most important techniques are highlighted in Table 11.1 and will be discussed in further detail throughout this chapter.

Table 11.1 Examples of *in situ* active management techniques.

Management focus	Examples
Managing habitat	• Maximizing habitat heterogeneity via grazing management; initiating controlled fires; suppressing fires; selective vegetation removal; regulation of water levels at wetland sites • Maximizing plant biodiversity by vegetation planting/seeding (also can increase faunal biodiversity though provision of new host plants) • Increasing structural complexity of vegetation by pollarding and coppicing
Managing species' competition	• Initiating regular disturbance, such as fire, grazing, or cutting to disrupt the ability of usually-dominant species to dominate (intermediate disturbance hypothesis—creating periodic disturbance to maximize biodiversity) • Selective targeting of dominant species to reduce population size (selective tree thinning or scrub clearance, culling, etc.)
Managing herbivory	• Preventing under-grazing: some areas need grazing to maintain a plagioclimax (e.g. to stop grassland becoming scrub) and to maintain diversity of grassland plants. This often requires arrangements with farmers and consideration of what animals to use, when, and for how long • Preventing over-grazing: creating exclosures by fencing areas; using plastic guards around saplings; pollarding rather than coppicing so new growth is above the browse line; culling of grazers/browsers • Management for grazers: using a fire regime to allow new growth; supplemental feeding during food shortage
Managing predation	• Predator-focus: supplemental feeding during food shortage; prey enhancement to increase prey availability; reducing disturbance; artificial nest/roost sites; monitoring key nests/breeding areas to manage persecution; creating water holes to attract prey • Prey-focus: creating habitat refuges by increasing concealment; using predator-proof fences; using predation-prevention plates on nest boxes or predator baffles; culling of predators especially non-native predators; live trapping and translocation; reducing predator reproductive success (e.g. by sterilizing)
Managing overpopulation	• Increasing the carrying capacity of a site to support more individuals (e.g. by increasing food supply) • Managing species' distribution: species translocation from overpopulated areas; facilitating natural movement along corridors from areas of overpopulation (Section 7.6.2) • Culling (could include trophy hunting—Hot Topic 11.1).
Managing eutrophication and pollution	• Reduce input: manage at source • Buffer effects or contain problem: reed beds in aquatic environments; trapping groundwater leachates (the liquid that drains or 'leaches' from landfill sites) • Remediation: removing nutrient stores (cutting and removing vegetation, clearing dead organic matter, removing topsoil in terrestrial environment, dredging in aquatic environment)
Managing pests, parasites and disease (Chapter 9)	• Chemical approaches (pesticides, insecticides, herbicides, fungicides) • Physical removal of pest species • Biological pest management

11.3.1 Managing habitat.

There are many different ways in which habitat can be managed with the aim of **creating** new habitat; **maintaining** existing conditions; **improving** habitat condition, value, or carrying capacity; and **restoring** an area to a previous (better) habitat type. These will in turn be discussed in the following sections.

Creating habitat

Creating new habitat can be an important aspect of *in situ* conservation management. This can be at a variety of spatial scales, from microhabitat provision to creating large-scale macrohabitat. An overview of this is given in Figure 11.2. Habitat creation can be undertaken to enhance biodiversity at a specific

Figure 11.2 Human-created habitats at different spatial scales, range from bird nest boxes and bee hives, to ponds and hedgerows, and finally forestry plantations and agricultural landscapes.

Source: Photographs by Anne Goodenough.

site, especially at micro- and meso-scale, or to compensate for the loss of similar habitat as a result of development.

Most often, habitat creation occurs at the **micro-scale**. For example, installing bird nest boxes can substantially increase the number of different species able to breed at a site and the size of the breeding population of each species, particularly in parkland or young woodlands where the main limiting factor is often lack of natural cavities (Newton, 1994). Nest boxes can increase the total number of nest sites (e.g. in managed cottonwood forests in the USA; Twedt & Henne-Kerr, 2001) or increase the number of high-quality sites in environments where nest quality, not quantity, is the limiting factor (e.g. aspen forests in Estonia; Lõhmus & Remm, 2004). Nest box provision can also increase reproductive success productivity since clutch size and fledging success are often higher in artificial nest boxes compared to natural cavities (Purcell et al., 1997). This is considered further in Case Study 11.1, which examines 60 years of nest box provision at Nagshead Nature Reserve, UK.

Other examples of micro-habitat provision include:

- **'Insect hotels'** can be created by placing hollow tubes, hollow stems of dead woody plants, split bamboo canes, drinking straws, and so on, in containers or simply drilling holes into blocks of wood. These are important for species such as solitary bees.

CASE STUDY 11.1

Provision of nest box microhabitats at Nagshead Nature Reserve, UK

Nagshead Nature Reserve, Gloucestershire, UK, is a site of major ornithological value that boasts a nationally important community of woodland birds. Most notably, the reserve supports several rare and declining migrant bird species that breed in some of the reserve's 450 nest boxes that together constitute one of the world's longest-running continually monitored nest box schemes.

Although the conservation value of the site and the nest box scheme is now recognized, both have rather unusual origins. The site was planted with English oak *Quercus robur* in 1814 to supply timber for the ship-building industry. However, the large number of mature trees formed monoculture stands that allowed Lepidopteran population explosions and, in the 1920s and 1930s, there were numerous caterpillar epidemics, primarily of oak roller moth *Tortrix viridana*. As a result, large areas of the oak woodland were defoliated each spring. This created a problem for the forest managers because without leaves in some years the trees could not photosynthesize, and their growth was both compromised and irregular.

Accordingly, nest boxes were erected in 1942 as part of a UK Forestry Commission scheme to prevent such defoliation. The aim was to attract breeding insectivorous woodland birds such as the common blue tit *Cyanistes caeruleus* as a form of **biological pest management** (Section 9.4.3). It is extremely doubtful how much impact could have been expected from the occupants of 84 boxes spread over 1 km^2, but before the effectiveness of the policy could be assessed, events took an unexpected turn. Around 20% of boxes were found to be occupied by a rare trans-Saharan migrant, the pied flycatcher *Ficedula hypoleuca*; this was the first confirmed breeding in the region. Over the years, the originally utilitarian nest box scheme has developed into a major woodland bird conservation scheme of over 450 nest boxes and the woodland becoming recognized as a Site of Special Scientific Interest, driven primarily by the breeding flycatcher population (Figure A).

Since 1990, the number of flycatchers at the site has declined dramatically, a pattern replicated across most other populations of this long-distance migrant across European breeding grounds. The focus has now shifted to undertaking research into population and productivity dynamics with the aim of making **evidence-informed**

Figure A Nagshead Nature Reserve and one of the many nest boxes on the reserve.

Source: Photograph by Anne Goodenough.

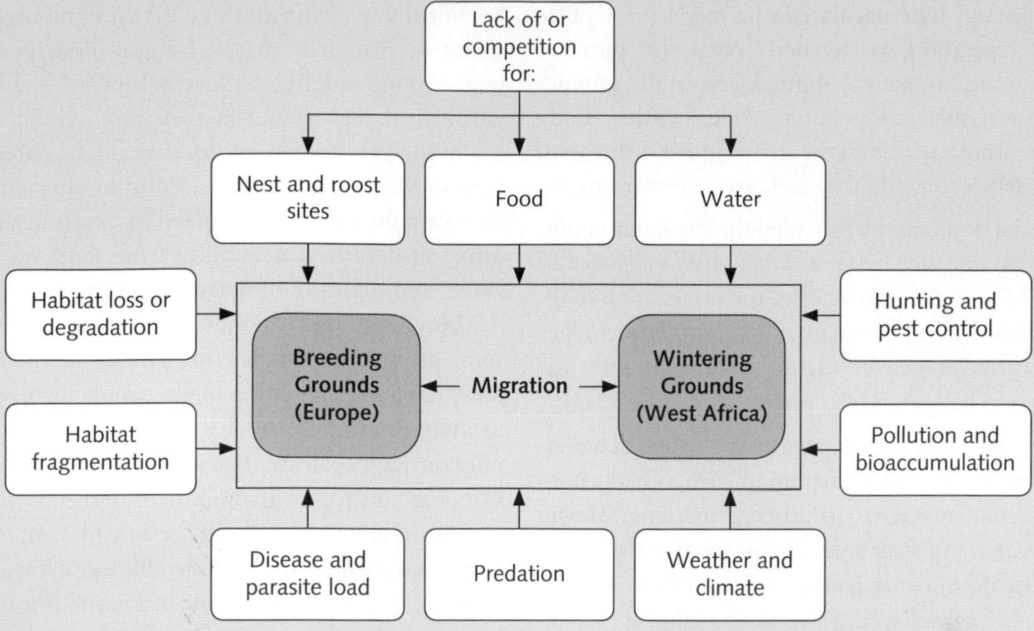

Figure B Pressures on pied flycatchers occur at breeding grounds, wintering grounds, and on migration.

conservation decisions to improve the species' prospects (Figure B).

Research has found that although some of the population decline is due to decreasing productivity, the greater part is occurring when birds are wintering in West Africa or on migration. In particular, it looks as though the amount of rainfall in the Mediterranean region (where birds stop on migration to 'refuel') has a major effect. This is controlled by the North Atlantic Oscillation, which is the northern hemisphere version of El Niño, and probably influences the number of flying insects upon which the birds rely (Goodenough et al., 2009).

Despite the larger part of the decline being driven by factors external to the reserve, maximizing productivity still has a role to play, first, to reverse the declining breeding success and, secondly, to try to increase success to help buffer other pressures. Research has found that birds using boxes facing south-southwest do less well than birds using boxes facing other directions, possibly because these boxes are warmer, wetter, and had higher levels of an allergenic fungus called *Epicoccum purpurescens* (Goodenough et al., 2008; Goodenough & Stallwood, 2012). Repositioning boxes is thus an easy way to improve success by a considerable amount—almost one chick per breeding attempt (~25% increase).

This case study illustrates how both reactivity and proactivity have a role to play in *in situ* conservation. The original conservation scheme was completely unintended and reactive to a rare species taking over nest boxes intended to be used by common species for pest control purposes, but the continuation of, and improvement in, conservation at this site is both planned and evidence based to ensure it is maximally effective.

REFERENCES

Goodenough, A.E., Elliot, S.L., & Hart, A.G. (2009) The challenges of conservation for declining migrants: are reserve-based initiatives during the breeding season appropriate for the Pied Flycatcher *Ficedula hypoleuca*? *Ibis*, Volume 151, 429–439.

Goodenough, A.E., Maitland, D.P., Hart, A.G., & Elliot, S.L. (2008) Nestbox orientation: a species-specific influence on occupation and breeding success in woodland passerines. *Bird Study*, Volume 55, 222–232.

Goodenough, A.E. & Stallwood, B. (2012) Differences in culturable microbial communities in bird nestboxes according to orientation and influences on offspring quality in great tits (*Parus major*). *Microbial Ecology*, Volume 63, 986–995.

- **Refugia** and **hibernacula** can be made for reptiles from vegetation, corrugated metal, or bitumen sheets approximately 2–4 m^2 placed on the ground. Because reptiles are poikilothermic (i.e. their body temperature varies with the environment rather than being self-regulated), they seek areas of warmth.
- **Artificial nests** can be provided for small mammals, such as dormice *Muscardinus avellanarius* and bats. Man-made earthworks for larger mammals, such as holts for European otters *Lutra lutra*, lodges for European beaver *Castor fiber*, and setts for European badgers *Meles meles* (see Figure 11.3).
- **Deadwood piles** are important for a range of woodland species, including mammals such as hedgehogs *Erinaceus europaeus* (Europe), martens *Martes americana* and deer mice *Peromyscus maniculatus* (America), and bog lemming *Synaptomys cooperi* (Canada), as well as deadwood-specializing invertebrates, mosses, liverworts, and lichens.

At a **meso-scale**, generally the most commonly created habitats are ponds and hedgerows (as shown in Figure 11.2). Such habitats might be created specifically with the aim of enhancing biodiversity or might be primarily driven by non-biodiversity need (e.g. a pond for livestock, stock-proof hedges). Regardless of the driving factors, any habitat creation is a biodiversity opportunity that can be enhanced by considering design features. Pond biodiversity value, for example, can be maximized by creating a range of different depths, accessibility from land via a gentle slope, and planting of native plant species.

Many examples of meso-habitat creation involve some element of 'engineering', but it is important to note that habitat creation might simply involve **ceasing to manage** an area to let new habitats develop naturally through ecological processes. Good examples are stopping cutting or grazing areas, either at the edges of fields (field margins—see Section 11.5.3), or within parks and gardens to create wildflower areas. Another simple (lack of) management technique is a reduction in cutting road verges and/or delaying cutting until after flowering plants have set seed (see Figure 11.4).

In the marine environment, artificial reefs can be created using human-made underwater structures—often ships that have been deliberately scuttled

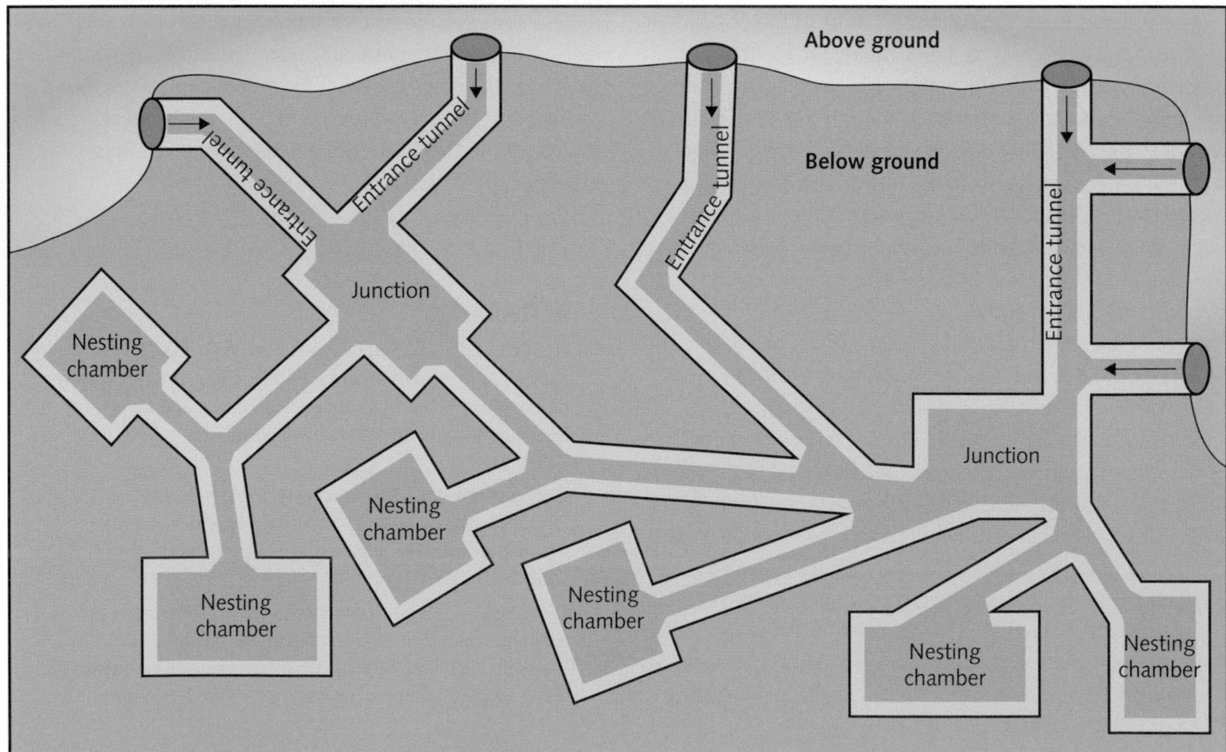

Figure 11.3 Artificial setts can be created for subterranean mammals, such as badgers, *Meles meles*, in Europe, but doing so is a hard engineering task. Such habitat creation is usually reserved for situations where the natural habitat of a legally protected species will be destroyed (under licence) at a development site.

Figure 11.4 'Creating' habitat can simply mean stopping a particular type of management. Where there is no impact on road safety through a reduction in visibility, stopping mowing the sides of roads at ecologically sensitive times of the year is becoming popular. From a council perspective this can also save money, which makes adoption of this more likely.

Source: Photograph by Anne Goodenough.

(e.g. *USS Yancey* was sunk as an artificial reef off Morehead City, North Carolina, in 1990, and *General Hoyt S. Vandenberg* off Key West, Florida, in 2009). Deliberate scuttling for coral reef creation needs to be very carefully planned so that sunken vessels do not pose a danger for current shipping. They also need to be carefully cleaned up to avoid pollution. Indeed, the danger of pollution is the reason almost 70% of the *Vandenberg*'s $8.4-million sinking budget went to clean-up efforts, including the removal of more than 10 tons of asbestos and almost 250,000 m of electrical wire (Harrigan, 2011). A bigger example is Cancun Underwater Museum in Mexico, which comprises more than 400 life-size sculptures made out of pH neutral clay and funded by the National Marine Park. Figure 11.5 shows how effective this strategy can be for creating thriving reefs, especially in areas with otherwise featureless sea beds.

Habitat creation at a **macro-scale** specifically for the purposes of conservation is comparatively rare. Most often, macro-habitat creation is driven by human need (e.g. plantations of trees for timber, creation of agricultural land, creation or rock exposures for aggregates, etc.). Sometimes, however, completely new sites are created as nature reserves, possibly to compensate for the loss of another site to development. Such situations provide a blank canvas in terms of habitat creation, as occurred at Newport

Figure 11.5 Artificial reefs can be created by scuttling ships, as in the case of the US Coast Guard Cutter *Duane*, which was sunk of Key Largo, Florida, and now supports a thriving ecosystem.

Source: Image courtesy of Matt Kieffer/CC BY-SA 2.0.

Wetlands in South Wales, which was constructed as a biodiversity compensation site following the loss of Cardiff Bay (Case Study 5.1). Habitat creation at Newport Wetlands is discussed in the Online Case Study for Chapter 11.

You can find the online case study at www.oxfordtextbooks.co.uk/orc/goodenough

Maintaining habitat by arresting ecological succession

A lot of habitat management involves keeping areas of land in a plagioclimax or subclimax state by preventing continuation of ecological succession (Figure 11.6). Put a little more simply, this means arresting the transition of, for example, grassland to woodland through the process of scrub encroachment or the transition of wetland to boggy grassland or wet woodland through the process of sedimentation.

Methods for arresting succession depend on the type of habitat involved, as well as logistical and financial constraints at a particular site, but all

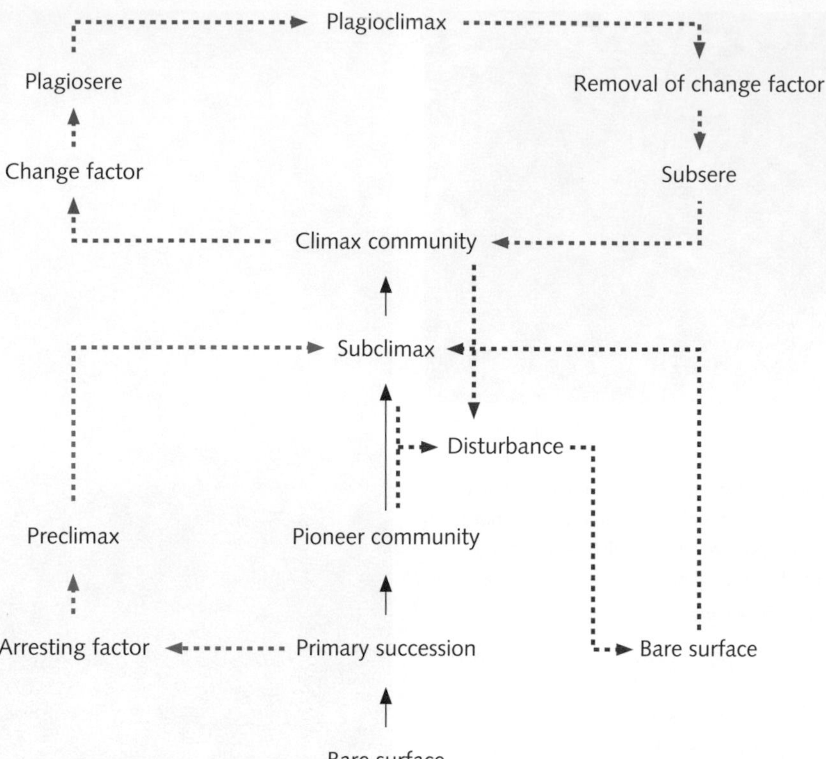

Figure 11.6 Ecological succession processes. Primary succession linear process shown using solid black lines – this should always occur. If, before the climax community is reached, an arresting factor such as grazing occurs, there can be a preclimax process as shown here with green dotted line. If a community reaches the climax community a change factor occurs, a plagioclimax process occurs as shown with the blue dotted line. If, at any point, the area is denuded (striped) of vegetation (e.g. by a hurricane), a secondary succession process occurs as shown by the red dotted line.

methods involve creating disturbance of some kind. In the terrestrial environment, such disturbance is usually grazing, cutting, or burning vegetation. In the aquatic environment, arresting succession might involve clearing vegetation at the edges of a wetland to prevent marginal creep and preventing too much sedimentation (e.g. by controlling soil erosion) as a maintenance management technique. If a restoration approach is needed, this would typically involve dredging to remove sediment.

Improving heterogeneity and structural complexity

Given that most species have a preference for, or perform better in, a specific habitat type, it generally follows that managing a site so it has multiple habitats (even if this is 'only' different types of one broad habitat type—different types of woodland, for example), will result in greater biodiversity, since the habitat needs of more species will be met. This links back to the concept of ecological niche explored in Sections 2.5 and 4.3.1. Although a niche is a theoretical construct, rather than a spatial place or particular habitat, it generally holds that the more habitats there are at a site, the more niches there are, and the more species that site can support.

As noted in Section 7.6.1, one of the ways of increasing species richness is to maintain natural heterogeneity (and the factors that cause it, shown in Figure 7.20) and/or artificially increase heterogeneity. Creating and maintaining a mosaic of different habitat types is, therefore, an important aspect of *in situ* conservation management. There are many ways in which high heterogeneity can be achieved. These can be loosely grouped into the four categories listed here, although there is considerable overlap between them. For simplicity, these approaches are discussed from the standpoint of managing a specific site (a nature reserve), but the concepts are equally applicable landscape-scale initiatives outside nature reserves and protected areas:

1. **Habitat zonation:** the overall reserve is divided into zones or patches, each with a different habitat. These habitats could differ substantially—for example, grassland and woodland—or might be fairly similar as in the case, for example, of subtly different types of heathland. In some cases, the different sections will be clearly demarcated on the ground, for example, by a fence, wall, hedge, or embankment. Usually, however, distinctions are blurred and ecotone transitional habitats occur (Section 7.2.4). Sometimes an area of land is already zoned when it becomes a nature reserve, whereas sometimes zones of different habitat types are created artificially; often it is a combination of the two, as in the case of the Newport Wetlands Reserve in South Wales (see the Online Case Study for Chapter 11). The zonation process can be extended to manage levels of human disturbance as discussed in Section 11.6.1.
2. **Within-patch differential management:** this is the process of enhancing heterogeneity within habitat areas. It is distinct from habitat zonation by its focus being within habitats, rather than between habitats and means that management practices are varied on a consistent basis in different parts of the same habitat patch. Examples would be selectively thinning some areas of woodland to create glades (clearings) with higher light levels, removing fish from some ponds to benefit amphibians, but retaining them in others, clearing scrub from some areas, but not others, and so on.
3. **Habitat grading:** whereas within-patch differential management focuses on creating differences to form a mosaic, habitat grading is about managing environmental gradients. A good example here is profiling of ponds or lakes to create a depth gradient (and thus light and temperature gradients), or artificially managing water levels within a wetland to create a mosaic of depth patterns (Figure 11.7 gives an example of how this can be achieved).
4. **Habitat rotational management:** in all of the above cases, management is consistent spatially—there might, for example, be one area of wet grassland and another area of marsh, and these two habitats are always managed accordingly. This contrasts with habitat management undertaken in rotation. Here, a habitat is divided into sections that are managed (usually by the removal of vegetation through clear-felling, grazing, or burning) in a preplanned rotating order. This creates a mosaic of different habitat types and ages, but the key point is that the role that each patch fulfils depends on where it is the rotation cycle. This means the same patch will, at different times, support different species. Rotations have been used in agriculture for many years as a good way to ensure, among other things, that the pests associated with one crop do not reach high levels (Chapter 9). In terms of conservation, good examples are rotational coppicing and rotational heath

Figure 11.7 Managing water levels within a wetland can be necessary to withstand periods of drought and flooding, as well as creating the ability to have a mosaic of lakes or lagoons with differing depths to increase species diversity. This can be done using either manual or automated sluice gates or pumps.

Source: Image courtesy of Beatrice Murch/CC BY 2.0.

burning, which will be discussed in more detail in Figure 11.8 and Section 11.3.2, respectively.

In addition to the concept that heterogeneity increases the number of species that can be supported at a given site, it can also be essential for species that have **complex habitat requirements**. Many insects, such as butterflies, have different requirements at larval and adult stages, and depend on sites that will support a full life cycle. Other animals use different habitats in close proximity to one another in the same landscape; heterogeneity is thus essential for many species' survival. In the European Alps, for example, chamois *Rupicapra rupicapra* require both meadows for grazing and rocky areas, where predation risk is lower, in which to sleep (Forman & Godron, 1986). Other species use different habitats in winter versus summer, such as black bears *Ursus americanus*, which use upland areas in the summer and move to lower elevations in winter.

Diverse habitat needs can be more subtle, as in the case of red grouse *Lagopus lagopus scotica* which need heathland in a range of different successional states within close proximity. Young heather (pioneer phase, <8 years) is a key food source; berry-producing plants (such as bilberry *Vaccinium myrtillus*) and insects in mid-successional heather (building phase, 8–20 years) are vital during breeding; and nests occur in mature stages (>20 years), where cover is greatest. Birds inhabit areas where all three heather ages intersect so it is essential to manage habitat so there are repeating patches of different stages by creating a cycle of cutting or burning heather patches.

There are two important trade-offs that have to be considered when management is directed at increasing habitat heterogeneity, particularly if this is being undertaken through a zonation approach:

- **Balancing the number of different habitats with the size of each habitat:** it is necessary to balance the 'more habitats = more species' concept with the fact that, as the size of each patch increases, so too does both species richness (species–area relationship: Figure 4.8) and the population size of each species. Accordingly, it is not sensible to create too many habitats in a small area. There is no magic formula for calculating how many habitats can, or should, be created in a specific land area since this depends on the aims of the reserve and its management, the site itself, and the types and distinctiveness of the habitats involved, but this is certainly a trade-off that needs to be considered. It is vital to ensure that each habitat patch is large enough to be meaningful and this will impose limitations on the total number of different habitats it is possible to have.
- **Balancing the number of different habitats and the amount of edge created:** as the number of habitats increases, so too does the amount of edge. Edge effects were discussed in detail in Section 7.4.2, where it was noted that small habitat patches can be unsuitable for interior-specialist species as the edge:interior ratio is skewed towards the non-favoured edge habitat. Conversely, multiple small patches with a lot of edge can benefit edge-specialist species. Again, this is something that should be factored into planning discussions, but there is no 'one size fits all' answer. It should be noted that sometimes management is designed specifically to create areas where different habitats intersect (necessarily at their edges), for example, managing heathland for game birds.

Restoring habitat by reviving traditional management

In many regions, land management has changed considerably over the last few hundred years. In addition to the ever-increasing intensification of agriculture and urban sprawl, some traditional land management practices have dwindled. This change has had a negative effect on species that have adapted to the opportunities provided by these practices and so, increasingly, nature reserve managers are reverting to forms of traditional management.

Within Europe, **coppicing** and **pollarding** both used to be practiced extensively. Coppicing is the selective cutting of woody plants, such as hazel *Corylus avellana*, to ground level to encourage regrowth in multiple small stems from a central stump, as shown in Figure 11.8. This management was undertaken for many centuries with a 10–20 year cycle to generate building materials and to make charcoal, but this also increased heterogeneity, especially in density of the shrub layer. As coppicing has dwindled, so too have coppice-adapted species, including heath fritillary butterflies *Mellicta athalia*. This species thrives in newly coppiced areas, but disappears quickly as soon as the areas become too shaded, so rotational management is essential (Warren et al., 1984).

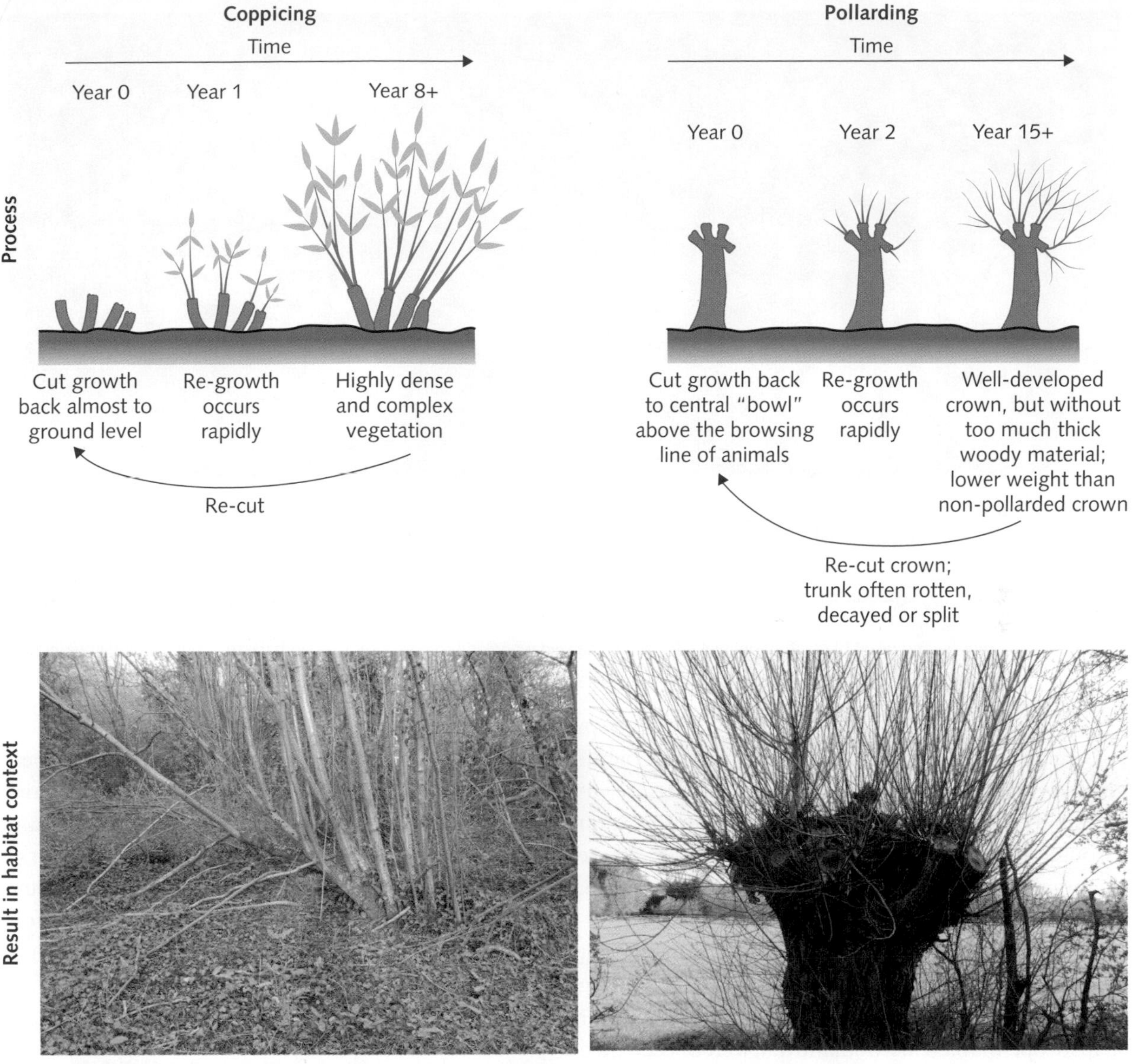

Figure 11.8 The process and result of coppicing and pollarding trees.
Source: Photographs by Anne Goodenough.

Pollarding, also shown in Figure 11.8, is a similar technique, but with growth cut back to around 2 m so that the regrowth occurs above browsing height to prevent animals eating new green shoots. Commonly pollarded species include willow *Salix*, and beech *Fagus*, but can include other broadleaved trees, such as oak *Quercus*, and lime *Tilia*, and, occasionally, conifers, especially yew *Taxus*. The process was undertaken to produce timber and animal feed, and thus was especially common in areas with comparatively few trees, such as Denmark and Scandinavia, or landscape dominated by pastoral farming. It is also used to control the size and form of trees in urban areas. Unlike coppicing, the biodiversity benefits of pollarding come less from a change in habitat structure complexity than from the effect on the tree itself. Periodic removal of the crown through pollarding acts to prevent the top of tree becoming too large and heavy for the trunk, and thus prolongs the life of the tree. The oldest part of the tree (the trunk) often has high biodiversity value. This wood, both dead and alive, supports many invertebrates, bryophytes, fungi, and lichen. It can be particularly valuable for saproxylic beetles and flies (those that eat dead and decaying wood), many of which specialize on pollarded trees.

Figure 11.9 The process and result of creating optimal rides in broadleaved woodland to create more open habitats for plants and invertebrates.

Given their habitat-modifying potential and biodiversity value, both coppicing and pollarding are increasingly being undertaken as part of conservation throughout Europe and, in some instances, where the practices have been used historically in America, Canada, Australia, and North Africa, too.

A final example of using traditional management techniques within woodland conservation is creating (or in some cases recreating) **woodland rides**—linear strips of open habitat in woodland that were once created for the purpose of horse riding. Creating or recreating such rides is often actively encouraged to increase flowering plant diversity, especially in mature woodlands. The ideal structure of woodland rides is shown in Figure 11.9.

Outside of woodland environments, traditional methods are also important. For example, the practice of ranchers creating cattle bomas (temporary enclosures) on Savannah grasslands in Kenya creates a patchwork of areas that are more fertile than others. This patchwork effect affects vegetation and the abundance of reptiles such as Parker's dwarf gecko *Lygodactylus keniensis* (Donihue et al., 2013). Another example is the management of coastal heathland of western Norway, which is a complex **habitat mosaic** of pasture, hay meadows, and both dry and wet heathland that is also periodically grazed. This is very much a cultural landscape and one that has very high biodiversity value (Webb, 1998).

It should be noted that it is not always possible to restore a habitat (or at least, not without huge effort) due to the way in which dominant species regenerate—for example, dipterocarp forests in Asia.

Restoring habitat through reversion techniques

Reversion involves changing the habitat of an area from its current state to a previous state. The most usual forms of reversion are creating species-rich grassland from arable fields (**arable reversion**) or woodland plantation (**forestry reversion**). Reversion is a long process and it can take many years before any benefits are seen. The best sites for reversion are those that:

- Have been subjected to the 'other' habitat use for a short amount of time (e.g. under 10 years for arable reversion; under 30 years for woodland reversion). This is because any soil changes will likely be minimal (soil fertility generally increases the longer a field is cultivated due to the application of fertilizers, while soil pH can change dramatically in woodland plantations, especially coniferous ones).

- Have been managed extensively (i.e. minimal fertilizer, low density of trees).
- Are bordered by the target habitat.

The general technique is to remove the existing vegetation and then either re-sow with a mixture of grasses and wildflowers, or allow **natural recolonization**. If sowing is undertaken, it is important that seeds come from a local source—this can sometimes be managed by cutting green hay from nearby fields and spreading this on the target fields so that seeds are spread naturally. Natural recolonization is a slower process and one that only really works if there is a good seed bank and/or close neighbouring fields. Simple techniques, such as bringing grazing animals or machinery from duties in neighbouring field onto the new site, will encourage transfer of seeds. Occasionally, if the soil is very fertile or the pH is suboptimal, the soil can be deep-turned so that the subsoil comes to the top and the topsoil is buried.

The evidence for the success of reversion is mixed. It is generally positive for flowering plants and insects, especially pollinators (review by Pywell et al., 2002). However, there can be disadvantages for other taxa. For example, a controlled before-and-after study in Sussex, UK, found that grey partridge *Perdix perdix* numbers declined much more rapidly on arable fields following their reversion to grassland compared with arable fields that were not reverted to grassland (Aebischer & Potts, 1998), while foraging brown hares *Lepus europaeus* generally avoided areas of farms that had been converted from arable crops to species-rich grasslands (Wakeham-Dawson, 1995). This demonstrates the complexity of *in situ* conservation decision making.

From restoration of habitats to rewilding

The ultimate form of *in situ* habitat-focused conservation is **rewilding**. This is the planned recreation of former habitats at a particular location through 'light touch' management using traditional techniques to recreate a wild landscape. Rewilding is usually applied to large areas using processes, such as naturalistic grazing or self-regulating hydrological regimes and will be considered in more detail in Section 13.6.

11.3.2 Managing species.

One of the main ways of managing specific focal species or communities *in situ* is through management and manipulation of habitat so that it becomes optimal for the target organisms. For this to occur, it is vital that species–habitat interactions (Section 11.7) are well understood. However, in addition to this indirect management, there are many other ways that species can be directly managed.

Translocations, insurance populations, and reintroductions

Species translocations—moving individuals from one area to another—is an important conservation strategy for both flora and fauna. It can be used in a variety of situations, for example:

- To move species away from a **threat** such as that brought about from site development and construction (see Hot Topic 5.1 for the use of translocation as mitigation in Ecological Impact Assessment).
- To create an **insurance population**: a subsidiary population at a site geographically removed from the main population(s). This strategy tends to be reserved for rare species with an imminent threat that is very hard to control, for example contagious disease. An example of where this has been used effectively is the *in situ* conservation of Tasmanian devil *Sarcophilus harrisii* (Case Study 11.2).
- To **supplement** a small population (especially one that is isolated and so unlikely to expand quickly through the process of immigration) or to expand the gene pool.
- To **reintroduce** a species to a former part of its range from which it became extirpated (Chapter 13).
- To reduce **overpopulation** at a particular site by moving some animals.

Supplemental feeding

Supplemental feeding is a high-maintenance and generally short-term form of *in situ* conservation, whereby additional food is provided to a population. This is usually done in periods of food shortage either to

CASE STUDY 11.2

Insurance populations in the case of the Tasmanian devil in Australia

Since the extinction of the thylacine (Case Study 10.2), the Tasmanian devil *Sarcophilus harrisii*, has had the distinction of being the largest extant marsupial carnivore (Figure A). Following its extirpation from mainland Australia around 3000 years ago, the Tasmanian devil is now found solely on the Australian island of Tasmania. The largely nocturnal and crepuscular opportunistic carnivore is considered to be Endangered by IUCN red list in 2015 due to a 60% decline in population size over 10 years (Hawkins et al., 2008).

The current populations have very low genetic diversity due to an island founder effect and several population bottlenecks (Jones et al., 2004). In more simple terms, this means that the population was greatly reduced during the initial colonization of Tasmania from the mainland (island founder effect) and, although it then grew, it has subsequently gone through several periods of very low numbers (population bottleneck), which caused inbreeding. As a result of this, the current populations on Tasmania have become very vulnerable to disease.

The primary cause of the decline is Devil Facial Tumour Disease (DFTD). DFTD is an unusual disease; it is not caused by a standard pathogen, but a non-viral transmissible cancer. The cancer cells are clonal, meaning that they are transferred between devils much like a tissue graft without provoking an immune response (Siddle et al., 2007). Low genetic diversity has led to a lack of variation in Major Histocompatibility Complex (MHC) genes. The MHC genes code for MHC molecules, which will bind to the surface of an antigen allowing recognition of the antigen and an immune response to be instigated. A lack of diversity in MHC genes is thought to have led to similar genes in both the tumour and the species genome, meaning that the devils' immune system is unable to recognize the tumours as foreign cells (Belov, 2011). DFTD is fatal to infected animals within six months.

One of the *in situ* strategies being used to tackle this problem is establishing an **insurance population**—a new population created in a disease-free environment and kept disease-free by isolation. This was instigated in 2012 and an insurance population was successfully established on Maria Island off the Tasmanian coast. This population is viewed as an ark so that if DFTD causes devils to become extirpated from Tasmania, there would be, at worst, a viable population elsewhere so that the species would not be lost and, at best, a source population for reintroduction (Section 13.4.2). Although this second plan is ambitious given the comparatively small size of the insurance population and the fact that red foxes *Vulpes vulpes* are predicted to move into large portions of the devil's range in its absence (intraguild competition has been shown to make reintroductions less likely to be successful) the insurance population at least keeps the possibility alive.

Figure A Tasmanian devil *Sarcophilus harrisii*.

Source: Image courtesy of James Stewart/CC-BY 2.0.

REFERENCES

Belov, K. (2011) The role of the Major Histocompatibility Complex in the spread of contagious cancers. *Mammalian Genome*, Volume 22, 83–90.

Hawkins, C.E., McCallum, H., Mooney, N., Jones, M., & Holdsworth, M. (2008) Sarcophilus harrisii*: IUCN Red List of Threatened Species*. Available at: http://www.iucnredlist.org/details/summary/40540/0

Jones, M.E., Paetkau, D., Geffen, E.L.I., & Moritz, C. (2004) Genetic diversity and population structure of Tasmanian devils, the largest marsupial carnivore. *Molecular Ecology*, Volume 13, 2197–2209.

Siddle, H.V., Kreiss, A., Eldridge, M.D., Noonan, E., Clarke, C.J., Pyecroft, S., Woods, G.M., & Belov, K. (2007) Transmission of a fatal clonal tumor by biting occurs due to depleted MHC diversity in a threatened carnivorous marsupial. *Proceedings of the National Academy of Sciences of the United States of America,* Volume 104, 16221–16226.

increase adult survival or reproductive success. A common example is providing food for garden birds during harsh winters when metabolic rate is high and both the availability of, and ability to find, food is low.

In more targeted cases, provision of supplemental food was a crucial part of *in situ* conservation for the Mauritius kestrel *Falco punctatus*, where it was used to improve chick survival (Case Study 7.1; Jones et al., 1995) and the critically endangered Iberian lynx *Lynx pardinus* in Spain, where it is used to improve adult survival by buffering the decline in the lynx's main prey, the European rabbit *Oryctolagus cuniculus* (López-Bao et al., 2010). It is also sometimes used for scavenging species, such as vultures (Case Study 11.3). Supplemental feeding is a frequent part of species reintroduction programmes (Section 11.3.2), when additional food is provided to translocated animals on their release into their new environment and gradually withdrawn.

Occasionally, providing additional food to predators is used to reduce pressure on prey, although this has mixed results. For example, predation rates on artificial nests containing chicken *Gallus gallus domesticus* eggs were almost halved on nature reserves in Texas, USA, where supplementary food was provided for predators such as striped skunks *Mephitis mephitis* (Vander Lee et al., 1999). However, a study by Redpath et al. (2001) found there was no difference in adult red grouse *Lagopus lagopus scotica* survival when local predatory hen harriers *Circus cyaneus* were given supplemental food.

Despite the advantages of supplemental feeding in some situations, there can be disadvantages. For example:

- Altering natural behaviour patterns.
- Dependency of animals on the human-provided food.
- Habituation to human contact.
- High intra- and inter-species aggression.
- Potential spread of disease in the high-density areas where individuals aggregate for food.
- Feeding non-target species, especially if these are pests such as rats *Rattus rattus*.

It should also be noted that supplemental feeding is an increasingly popular means by which tourism operators can facilitate close observation of wildlife. This can be done with visitors observing supplemental feeding that is undertaken as part of a conservation initiative, but is increasingly being used as a tourism hybrid between zoos and 'the wild', which, even without the disadvantages of supplemental feeding outlined above, has serious ethical implications as reviewed by Orams (2002).

Culling

It might, at first, seem odd to consider culling as a conservation strategy, but it can have two very important roles:

1. It can be necessary to cull a species that is posing a threat to another, endangered, species. This most often occurs with non-native species (Section 8.6.2), but can arise in other situations too, for example:
    - **Culling (native) predators:** fledging success increased at a common tern *Sterna hirundo* colony in eastern Canada following culling of native gulls *Larus* (Magella & Brousseau, 2001).
    - **Culling (native) competitors:** two studies from Australia have found increases in bird populations and species richness after the control of noisy miner birds *Manorina melanocephala*—a native but hyper-competitive species (Grey et al., 1997; Debus, 2008)
    - **Culling (native) disease hosts:** mountain hares, *Lepus timidus* (a **reservoir host** for ticks that carry louping ill virus), have been experimentally culled in the Scottish Highlands to benefit red grouse *Lagopus lagopus scotica*. Although there is some evidence that grouse breeding success increased at cull sites relative to non-cull control sites (Laurenson et al., 2003), important underlying differences in the pre-existing incidence of louping ill virus and habitat quality mean that the results of that study might not be reliable (Cope et al., 2004). This also illustrates how important it is that conservation decisions are based on sound science, a topic discussed further in Section 11.7.
2. On occasion, it can be necessary to cull individuals of a species that is the target of conservation action. This is most likely to occur when **overpopulation** occurs in a particular area, especially for large animals that are expensive or

CASE STUDY 11.3

Vulture restaurants

Vultures have the highly dubious distinction of being the most threatened group of birds in the world, with 65% listed as threatened or near-threatened on the IUCN red list in 2015. Decline in vultures is a conservation problem in its own right, but this situation is made worse by vultures' keystone status in terms of disposing of carrion and other organic refuse. Moreover, a decline in vultures can substantially increase populations of other scavengers, some of which are well-known **disease reservoirs** (e.g. in India dogs are the main source of rabies in humans, and their populations have increased substantially in parallel with the vulture decline).

The catastrophic decline of vulture populations in Asia (up to 97% in some areas) has been linked to poisoning by the drug diclofenac (Pain et al., 2003). Reasons for similar declines across Africa are less well understood, but have been linked to a reduction in carrion and a substantial proportional increase in the amount of carrion containing anti-inflammatory drugs (diclofenac in some countries or similar drugs in others); collisions with wind turbines and power line contact probably also have a role, as does killing for traditional medicine (Ogada et al., 2011). There have also been suggestions that poachers might, in some areas, be deliberately poisoning vultures to eradicate them from areas where they operate and thereby escape detection.

At supplementary feeding stations known as vulture restaurants, carcasses are made available to provide safe (non-poisoned) meat. Such vulture restaurants have become a widespread tool in conservation, with operation in a number of countries, including Spain, Nepal, India, Cambodia, Swaziland, and especially South Africa, where the Cape Griffon vulture is almost wholly dependent on vulture restaurants for their source of food (Figure A).

In many cases, this strategy is proving successful in safeguarding birds from poisoned meat (e.g. in Asia—Gilbert et al., 2007); however, the process is often reliant on the donation of dead livestock from commercial farms, and it is possible that these animals could have been treated with anti-inflammatory drugs prior to death. Recent research in South Africa also indicates that vulture restaurants have indirect benefit to other scavengers, notably brown hyena *Hyaena brunnea* and black-backed jackal *Canis mesomelas* (Yarnell et al., 2014).

Figure A Cape griffon vultures *Gyps coprotheres* feeding at a vulture restaurant in South Africa.

Source: Photograph by Anne Goodenough.

REFERENCES

Gilbert, M., Watson, R.T., Ahmed, S., Asim, M., & Johnson, J.A. (2007) Vulture restaurants and their role in reducing diclofenac exposure in Asian vultures. *Bird Conservation International*, Volume 17, 63–77.

Ogada, D.L., Keesing, F., & Virani, M.Z. (2011) Dropping dead: causes and consequences of vulture population declines worldwide. *Annals of the New York Academy of Sciences*, Volume 1249, 57–71.

Pain, D.J., Cunningham, A.A., Donald, P.F., Duckworth, J.W., Houston, D.C., Katzner, T., Parry-Jones, J., Poole, C., Prakash, V., Round, P., & Timmins, R. (2003) Causes and effects of temporospatial declines of Gyps vultures in Asia. *Conservation Biology*, Volume 17, 661–671.

Yarnell, R.W., Phipps, W.L., Dell, S., MacTavish, L.M., & Scott, D.M. (2014) Evidence that vulture restaurants increase the local abundance of mammalian carnivores in South Africa. *African Journal of Ecology*, Volume 53, 287–294.

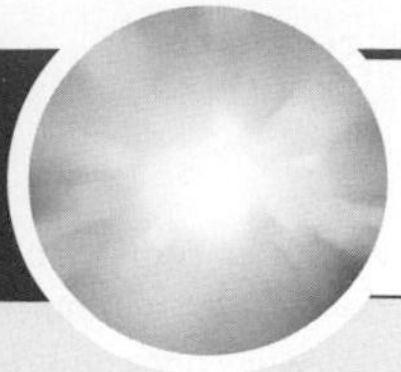

HOT TOPIC 11.1 Trophy hunting

The hunting of animals primarily for sport, rather than purely for subsistence, has a long history. However, it is the invention of modern firearms, the colonization of Africa and parts of Asia by Europeans, and the ability to travel the world with relative ease that has led to the activity that we now know as **trophy hunting**.

Trophy hunters typically pay large sums of money (that can easily amount to tens of thousands of US dollars) for the experience of hunting specific species or individual animals in order to secure a trophy, typically the skin, horns, tusks, or antlers that can be mounted for display. Trophy hunting is often portrayed in the media as an African activity undertaken by Americans, but in fact, such hunting occurs throughout the world and attracts various nationalities of hunters.

Trophy hunting is a controversial activity. Its detractors point to the detrimental effects that hunting can have on species populations and they also raise concerns over animal welfare, especially when hunted animals are not killed cleanly. Some hunting techniques, such as bow hunting and handgun hunting, compound welfare concerns as does the breeding of animals, especially lions *Panthera leo*, specifically to be hunted in small enclosures (an activity known as 'canned hunting'). Many detractors also object fundamentally to the notion of killing for sport, especially in the case of large and charismatic animals (Figure A). However, trophy hunting can play a role in conservation primarily through the income that it generates, which at least in principle can be used to fund conservation work.

Figure A The global outrage expressed by many over the killing of lions by trophy hunters leads to many debates on the ethics and value of recreational hunting.

Source: Photograph by Anne Goodenough.

Properly regulated trophy hunting can generate income and provide benefits (Lindsey *et al*., 2013). A lot of African trophy hunting occurs in South Africa, especially on privately owned game reserves that have been converted from cattle ranches. Such operations, if correctly managed, generate a surplus of wildlife, which in turn, can generate significant income if individuals are put 'on trophy'. Large, old males that are no longer breeding are the preferred choice of hunters so their removal will not necessarily threaten overall numbers, although group dynamics differ between species, so taking out old individuals is not a guarantee of effects being negligible.

Wildlife can be a nuisance or a hazard in places where people seek to expand grazing land, and to protect their crops, livestock, and families. In such regions, well-managed trophy hunting, where the income generated provides tangible community benefits, can place a value on wildlife and its continuing conservation. In addition, many such regions are not attractive to tourists, and hunting is the only way that wildlife can generate income (Hart, 2015).

As with many issues where humans, wildlife, economics, and ecology collide, trophy hunting is complex. There are success stories, where trophy hunting directly funds conservation and engages local communities (such as the Bubaye Valley Conservancy in Zimbabwe, where income from hunting funds lion and black rhino conservation (Hart, 2015)), but there are also clear problems. Overhunting of lions in Tanzania, for example, has contributed to their decline (Packer et al., 2011).

It is important to realize that trophy hunting is not a single activity, but involves different populations of different species living in different regions with differing regulations, political structures, land and wildlife ownership, and conservation priorities. For example, the hunting of a bull greater kudu *Tragelaphus strepsiceros* in a privately owned game reserve in South Africa is very different from the hunting of a lion in a poorly managed hunting concession in Tanzania. The media storm following the killing of Cecil the Lion has led to a simplification of the perception of trophy hunting, and its potential conservation value has tended to be lost in the ethical debate about the killing of animals for sport.

QUESTIONS

How might the group dynamics of some species (such as elephants and lions) lead to difficulties in the selection of suitable trophy animals?

It is argued by some that photo-tourism (non-consumptive utilization of wildlife) could replace trophy hunting (consumptive utilization of wildlife) as a source of income in many regions. What are the obstacles facing land managers in attracting tourists, rather than hunters?

REFERENCES

Hart, A.G. (2015) Viewpoint: uncomfortable realities of big game hunting. *BBC Science and Environment*, 1 September 2015. Available at: http://www.bbc.co.uk/news/science-environment-34116488

Lindsey, P.A., Balme, G.A., Funston, P., Henschel, P., Hunter, L., Madzikanda H, Midlane, N., & Nyirenda, V. (2013) The trophy hunting of African lions: scale, current management practices and factors undermining sustainability. *PLoS ONE*, Volume 8(9), e73,808.

Packer, C., Brink, H., Kissui, B.M., Maliti, H., Kushnir, H., and Caro, T. (2011), Effects of Trophy Hunting on Lion and Leopard Populations in Tanzania. *Conservation Biology*, Volume 25, 142–153.

impractical to translocate (or suffer high mortality or morbidity from translocation). It is vital to ensure that the 'right' animals are culled—for example, considering dominance hierarchy and family group structure. If animals *need* to be culled, it can sometimes make sense for managers to consider allowing trophy hunting. This can generate much-needed revenue that can, theoretically at least, be ploughed back into conservation, but is highly contentious, as discussed in Hot Topic 11.1.

Guarding and protection

Where poaching is a threat, one of the main management priorities can be guarding and protecting animals from being killed and/or detecting poachers. Actions can vary from relatively custodial (e.g. patrolling or placing motion-sensitive cameras at key sites such as water holes) to active (e.g. creating and managing fences or tagging animals with GPS tags that create an alert if the animals go beyond a specified geographical area).

In some situations, management involves altering habitat. For example, at Mankwe Wildlife Reserve, South Africa, rotational burning is used to create areas of new lush grass, and the distribution of animals is based on the burning pattern (highest at new-burn sites, lowest at old-burn sites). Due to poaching threats, the reserve now has a no-burn buffer strip close to fence lines to deter animals from being in these areas and lower the poaching risk. This works well for low-level subsistence poaching, but some poaching—such as killing of rhino for their horn—is a much harder battle to fight, as discussed in Hot Topic 11.2.

FURTHER READING FOR SECTION

Overview of habitat management techniques: Ausden, M. (2007) *Habitat Management for Conservation: A Handbook of Techniques*. Oxford: Oxford University Press.

Overview of habitat management techniques: Sutherland, W.J. & Hill, D.A. (Eds.) (1995) *Managing Habitats for Conservation*. Cambridge: Cambridge University Press.

Active management of woodlands using traditional techniques: Available at: http://www.forestry.gov.uk/pdf/Silviculture_Thinning_Guide_v1_Jan2011.pdf/$FILE/Silviculture_Thinning_Guide_v1_Jan2011.pdf

Arable reversion review: Pywell, R.F., Bullock, J.M., Hopkins, A., Walker, K.J., Sparks, T.H., Burke, M.J., & Peel, S. (2002) Restoration of species-rich grassland on arable land: assessing the limiting processes using a multi-site experiment. *Journal of Applied Ecology*, Volume 39, 294–309.

Successes and failures of species translocations: Griffith, B., Scott, J.M., Carpenter, J.W., & Reed, C. (1989) Translocation as a species conservation tool: status and strategy. *Science*, Volume 245, 477–480.

Tourism-driven supplemental feeding: Orams, M.B. (2002) Feeding wildlife as a tourism attraction: a review of issues and impacts. *Tourism Management*, Volume 23, 281–293.

HOT TOPIC **11.2**

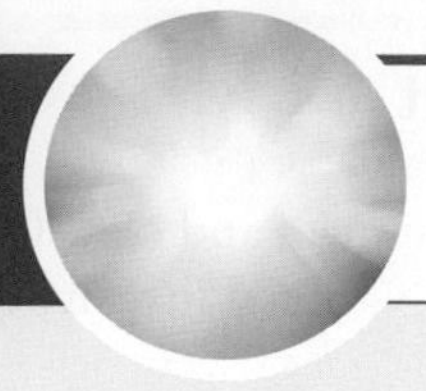

Rhino poaching

There are five species of rhinoceros: the Indian *Rhinoceros unicornis*, Sumatran *Dicerorhinus sumatrensis*, and Javan *Rhinoceros sondaicus* (Asia), and the white *Ceratotherium simum*, and black *Diceros bicornis*, (Africa). Depending on the species, rhinos grow one or more keratin horns at the front of their head. Historically, these horns have been used in dagger handles and traditional medicine, and have also been valued as hunting trophies (Hot Topic 11.1).

It is currently illegal to trade rhino horn, but despite this prohibition, rhino of all species are taken by poachers specifically for their horns. This is largely because of the use of horn in traditional Chinese medicine, which is growing in popularity, and the horn being viewed as a status symbol by wealthy Vietnamese. The relatively new use of rhino horn in Vietnam has helped to fuel demand and a concomitant rise in poaching over recent years. Since around 2008 the level of poaching of African species in particular has dramatically increased, with more than a 1000 taken annually from South Africa, where the majority of African rhino are now located (Emslie et al. 2013).

Combating rhino poaching is complex, expensive, and dangerous. Education in end-user countries can reduce demand, and determined law enforcement at ports and in end-user countries can certainly help, but none of these measures can negate the need for on-the-ground anti-poaching strategies.

In South Africa, most rhino (*c*. 75%) are found in Kruger National Park and this is also where most poaching occurs. Kruger is a large and topographically complex region with a long open border with Mozambique, both of which make poaching relatively easy and enforcement far more difficult. Military assistance, helicopter support, and surveillance using unmanned aerial vehicles (drones) are used, in conjunction with anti-poaching patrols, and both local and international intelligence to try to combat poaching. Similar tactics are being used in other areas to protect what rhino populations remain.

Around a quarter of rhino in South Africa are in private hands. These rhino owners are particularly stretched when it comes to defending their animals, since they receive little or no external support for their anti-poaching activities. Some owners are developing methods to contaminate or otherwise devalue horn, while others are dehorning their animals to make them less attractive to poachers.

Since rhino can be dehorned relatively safely in controlled procedures by trained individuals, and given that there are considerable stockpiles of rhino horn, some have called for the legalization of trade in rhino horn to prevent, or at least reduce, the current onslaught of poaching. Opponents of the trade argue that, among other things, it will stimulate demand (e.g. see Biggs et al. 2013)

The globalized trade in rhino horn continues despite being illegal and poaching continues despite huge efforts to combat it. High demand and high prices make rhino poaching a lucrative enterprise for those able to profit from it. Multifaceted solutions including end-user education, enforcement of existing laws, and additional anti-poaching surveillance to protect rhino, as well as detailed scientific consideration of more controversial measures, such as legalizing trade, are required if poaching is to be combated. However, such measures, especially on-the-ground protection and enforcement, are expensive and those protecting charismatic species of international significance need international financial assistance if they are to win what has rapidly escalated into a poaching war.

QUESTIONS

Rhino could be 'farmed' for their horn. Large numbers could be kept in relatively high densities and dehorned regularly to supply the market. If such an activity caused a large decrease in poaching and provided money for conservation could it be justified?

Rhino poachers are often highly organized, relatively well-resourced, and armed. Given that armed intruders on a reserve are unlikely to be anything other than poachers, is it ever justified to shoot on sight? In other words, is murdering poachers acceptable to save wildlife?

REFERENCES

Biggs, D., Courchamp, F., Martin, R., & Possingham, H.P. (2013) Legal trade of Africa's rhino horns. *Science*, Volume 339, 1038–1039.

Emslie, R.H., Milliken, T., & Talukdar, B. (2013) African and Asian Rhinoceroses—Status, Conservation and Trade. A report from the IUCN Species Survival Commission (IUCN/SSC) African and Asian Rhino Specialist Groups and TRAFFIC to the CITES Secretariat pursuant to Resolution Conf. 9.14 (Rev. CoP15). Gland: IUCN.

11.4 *In situ* conservation through custodial management.

Although active management techniques often have a greater share of the limelight, custodial management through legislation and policy frameworks is also a vital part of *in situ* conservation.

11.4.1 Legislation.

Legislation that pertains to *in situ* protection and management of ecology can be divided into **species-focused** legislation, **site-focused** legislation, and **activity-focused** legislation. In some instances, these different foci are covered in separate Acts; in others there is a single (main) overarching Act with multiple parts.

Species-focused legislation

Species-focused legislation usually operates at the level of the country through one or more Acts. In the UK, the primary legislation is the Wildlife and Countryside Act 1981, the USA has the Endangered Species Act 1973, and Australia has the Threatened Species Conservation Act 1995. These Acts tend to set out:

- **Protection for species**: in most countries national legislation sets out 'standard' protection for all species (or all species within specified taxonomic groups) and then details species that have additional special protection by virtue of their rarity, national importance, sensitivity to disturbance, etc. For example, in the Wildlife and Countryside Act 1981, (almost) all birds are protected from having their nests destroyed, but those listed on an additional Schedule—Schedule 1—have additional protection, whereby it is an offence even to approach or photograph a known nest.
- **Species exempt from protection**: this typically covers species that may be killed as pests and species that may be killed for sport or food (either generally or within a specified season). Such species might be explicitly noted (e.g. in the UK's Wildlife and Countryside Act) or exempt from being included in the relevant Act in the first place (e.g. in the USA's Endangered Species Act).
- **Species that need to be controlled**: this typically covers non-native species and can either be contained in the same Act as conservation measures (e.g. in the UK) or covered in separate legislation (e.g. the Substances and New Organisms Act (New Zealand) or National Invasive Species Act (USA)).

Failure to comply with legislation is typically punished either using fines or custodial sentences depending on the country, species, and nature of the offence. Most national species-focused legislation also has a system for authorizing actions that would otherwise be illegal if there is overwhelming need or if suitable mitigation can be put in place (Section 5.6 discusses this from a development perspective).

Site-focused legislation

Site-focused legislation covers the designation and management of protected areas. Most countries have several different types of statutory designation and thus several different types of protected areas. The difference between these can be in focus (e.g. designated for aesthetical value, designated for bird biodiversity, designated for roles in engaging local people with nature); importance of the ecology at the site (there is usually a scale within designation so sites that are unique have a higher level of protection than those that are important only at a local scale); and level of the designated authority (local, national, international, etc.).

Given the above differences between designations, it is both common and appropriate for single sites to have **multiple designations**. For example, in Costa Rica, the Tortuguero Conservation Area contains Tortuguero National Park and the Barra del Colorado Wildlife Reserve; the entire area is also listed as a Ramsar site as a wetland of international importance. The legislative structures behind designations can be both complex and convoluted, as demonstrated for the UK in Table 11.2.

Until comparatively recently, designation and management of protected areas was almost exclusively focused on terrestrial landscapes (including freshwater ecosystems), whereas the marine environment was largely ignored. Although protection of marine environments is still sadly lacking, things are slowly starting to change. Marine Protected Areas (MPAs) are defined by the IUCN as being 'clearly defined geographical space[s], recognized, dedicated, and managed through legal or other effective means, to achieve the long term conservation of nature with

Table 11.2 Protected area designations in the UK.

Level	Name	Details	Legislative framework
Local	Local Nature Reserves LNR (number depends on county)	Sites of local importance for biodiversity or that act to engage local people with nature	Designated and managed by local authorities after consultation with the relevant statutory nature conservation agency (e.g. Natural England in England)
National	National Park (n = 15 in 2015)	Generally notified for aesthetics, but give protection to some areas especially in terms of planning restrictions and development	Notified under National Parks and Access to Countryside Act 1949 (and amendments thereto in the Environment Act 1995 and Countryside and Rights of Way Act 2000) [Also National Parks (Scotland) Act 2000]
National	Area of Outstanding Natural Beauty (AONB; n = 46 in 2015)	As above	As above except National Parks (Scotland) Act 2000
National	National Nature Reserves (NNR; n = 224 in 2009)	Most important ecological sites in UK	National Parks and Access to the Countryside Act 1949; Wildlife and Countryside Act 1981 (as amended by Natural Environment and Rural Communities Act 2006)
National	Site of Special Scientific Interest (SSSI) [and Areas of Special Scientific Interest (ASSI) in Northern Ireland] (number fluctuates with registrations and deregistrations, but around 5000)	Listed for biological and/or geological features and value	Originally notified under National Parks and Access to the Countryside Act 1949. All SSSIs existing in 1981 were re-notified under Wildlife and Countryside Act 1981; sites designated since have been through the latter Act (as amended by Natural Environment and Rural Communities Act 2006). Improved provisions for management of SSSIs introduced by Countryside and Rights of Way Act 2000 (in England and Wales), Nature Conservation (Scotland) Act 2004 and Wildlife and Natural Environment (Scotland) Act 2010
European	Special Protection Area (SPA; n = 277 in the UK in 2015)	Found across Europe. Together with SACs (below) forms Natura 2000 European protected sites	Designated under EC Birds Directive (79/409/EEC)
European	Special Area of Conservation (SAC; n = 652 in the UK in 2014)	Found across Europe. Together with SPAs (above) forms Natura 2000 European protected sites	Designated under the EC Habitats Directive
International	Ramsar sites (n = 148 in the UK in 2015)	Wetlands of international importance	Designated under International Convention on Wetlands of International Importance (often called the Ramsar Convention as this was signed in Ramsar, Iran)
International	Biosphere reserves (n = 9 in UK)	Sites displaying a balanced relationship between people and natural environment; educating and inspiring the community to work together for a more sustainable future	UNESCO
International	World Heritage Sites (n = 29 in UK in 2015)	Usually applied to cultural heritage, but can be applied to unique landscapes (and thus habitats/species)	UNESCO (World Heritage Convention)

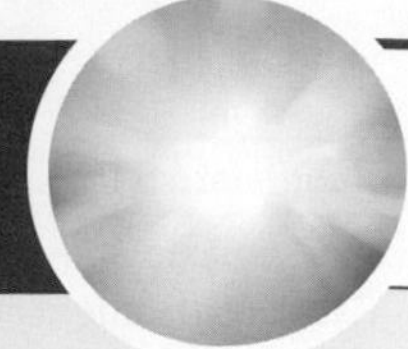

HOT TOPIC 11.3 Fishing quotas

Globally, 75% of fish stocks are fully- or over-exploited. Overfishing is thus a substantial and increasing problem, whereby off-take of fish occurs above the level at which fish stocks can replenish naturally. Where this occurs, populations of some fish species can be well below replacement level. Fishing pressure continues to threaten marine ecosystems, and the cultures and economies that depend on them (Griffith, 2008). The use of indicator marine species to highlight high fishing levels was discussed in Online Case Study Chapter 4. This Hot Topic looks at a more contentious issue: how to prevent, control, and police fishing levels.

There are several methods of reducing over-fishing, in theory at least:

- Limiting the total number or total weight harvested of a species (e.g. fishing quotas).
- Limiting where individuals can be harvested from (e.g. marine protected areas).
- Limiting or restricting harvesting of specific age/sex/size individuals within a harvested species (e.g. minimum mesh sizes in fishing nets to protect juveniles from exploitation).
- Limiting the number and/or equipment that can be used (e.g. limiting the number of fishermen and fishing boats).
- Limiting when individuals can be harvested (e.g protecting fish spawning periods).

To take the first possibility, fishing quotas can be based on 'total allowable catch' (TAC) for each species. However, as discussed by Baker (1999), if these are just issued with an annual (or seasonal) upper limit, each fisherman can be pressured to bring in the fish before the TAC is reached and the fishery is closed for the season. The fierce competition leads to investments in additional boats, making the industry highly inefficient and burdening already strapped local fishing economies. There are safety concerns as well because fishing boats go out in bad weather to get as much of the catch as possible. Regulations, such as per-trip catch limits, days-at-sea limits, and shortened seasons, can be used to slow the pace of fishing, but they often exacerbate the race and subsequent ecological damage (Leal et al. 2005).

Individual fishing quotas (IFQ) are often fairer and give each fisherman a stake in the fish stocks, and thus an interest in the long-term sustainability of fish populations. They also mean that fishermen no longer have to race to maximize their catch (Griffith, 2008). There is also evidence from New Zealand that IFQ fishermen are also more likely to fund ecological research as they recognize that better ecological data can guide decisions to improve the future stability and value of the fishery (e.g. Arbuckle & Metzger, 2000).

To take the second option, there has been considerable debate—much of it heated—about whether fishing should be allowed in Marine Protected Areas (MPAs), even if that fishing is part of an IFQ system. As of 2015, fishing is allowed in at least 94% of MPAs, and >99% of the ocean area; it is only complete banned in full Marine Reserves, which have to be no-take zones. It has been argued by Costello & Ballantine (2015) that this is illogical, given that almost any fishing will have some effect on biodiversity and natural ecosystem functioning. They argue that marine conservation should focus on Marine Reserves, rather than MPAs.

However, the situation is not that clear-cut. There are multiple reasons why MPAs might allow fishing:

- Lack of appreciation of trophic cascades in the ocean and the fact that baselines are shifting.
- Allowing some fishing may be a compromise to get some elements of biodiversity protected.
- A higher priority is given to terrestrial and freshwater biodiversity where extinction rates have been determined, even though recent assessments find more threatened marine than non-marine species (Webb & Mindel, 2015).

QUESTIONS

Do you think IFQs are suitable for use within the wider marine environment? Would anything else be better? Do you think fishing should be allowed within MPAs to any extent?

REFERENCES

Arbuckle, M. & Metzger, M. (2000) *Food for thought: a brief history of the future of fisheries management*. Nelson: Challenger Scallop Enhancement Company Ltd.

Baker, B. (1999) Individual fishing quotas—a complex and contentious issue. *BioScience*, Volume 49, 180–180.

Costello, M.J. & Ballantine, B. (2015) Biodiversity conservation should focus on no-take Marine Reserves. *Trends in Ecology & Evolution*, Volume 30, 507–509.

Griffith, D.R. (2008) The ecological implications of individual fishing quotas and harvest cooperatives. *Frontiers in Ecology and the Environment*, Volume 6, 191–198.

Leal, D.R., De Alessi, M., & Baker P. (2005) *The ecological role of IFQs in US fisheries: A Guide for Federal Policy Makers*. Available at: http://www.perc.org/sites/default/files/ifq_ecology.pdf

Webb, T.J. and Mindel, B.L. (2015) Global patterns of extinction risk in marine and non-marine systems. *Current Biology*, **25**, 506–511.

associated ecosystem services and cultural values'. Seven broad types are recognized, which are hierarchical in terms of level of protection:

- **Ia**: strict nature reserve—offers maximum protection and all resource removals are either strictly prohibited or severely limited.
- **Ib**: wilderness area—similar to a strict nature reserve, but generally larger and protected in a slightly less stringent manner.
- **II**: National park—prioritizes protection of the ecosystem, but allows light human use (usually recreational, sometimes including fishing in low-risk areas).
- **III**: natural monuments or features—established to protect historical sites, such as shipwrecks, and cultural sites, such as aboriginal fishing grounds.
- **IV**: habitat/species management area—established to protect a certain species or rare habitat. Might include spawning/nursing grounds for fish.
- **V**: protected seascape—accommodates a range of for-profit activities with the main objective being to safeguard regions that have distinct or valuable ecological, biological, cultural, or scenic character.
- **VI**: sustainable use of natural resources—allows a range of human uses of the area as long as they are deemed sustainable.

Activity-focused legislation

Activity-focused legislation covers livelihoods such as fishing and farming, as well as sports such as hunting and shooting. Control of fisheries via fishing quotas—and the contention around such legislation—is explored further in Hot Topic 11.3. It should also be noted that, in many countries, there will be additional constraints on activities within protected areas, which are covered under site-focused legislation.

11.4.2 Policy.

In addition to specific and formal legislation, *in situ* conservation is underpinned by policy. Policy can be thought of as a set of guiding principles or protocols on a particular topic. Policy can be aspirational, and thus characterized by targets, or more regulatory and characterized by thresholds. In the cases of conservation, some policy frameworks are indirectly relevant, such as reports by the Intergovernmental Panel on Climate Change (IPCC) and its implications.

Other policy frameworks are directly relevant, such as the international **Convention on Biological Diversity** (CBD) https://www.cbd.int/convention. This is a multilateral treaty that was devised in 1992 at the Earth Summit in Rio de Janeiro (often referred to simply as the Rio Convention) with the aim of developing national strategies for the conservation and sustainable use of biological diversity. One of the key Articles in this Protocol was that all signing parties (countries) should prepare a national biodiversity strategy to ensure actions were incorporated into the planning and activities of all those sectors whose activities can have an impact on biodiversity. One of the ways that some countries responded to this was to create plans for priority species and habitats referred to as Biodiversity Action plans or BAPs (e.g. UK, Australia, New Zealand, Tanzania, and Uzbekistan), Species Recovery Programmes (USA), or similar terms.

Biodiversity Action Plans (BAPs)

Biodiversity Action Plans are specific to an individual species or individual habitat, and outline its rarity, conservation status, threats, current conservation initiatives, and potential future conservation action. Their spatial scale can vary from national (normal for CBD-driven plans), regional, or site-specific (in which case they tend to be referred to as Local Biodiversity Action Plans, LBAPs). The BAP policy was an excellent first step in national-level biodiversity planning, but tended to ignore the complexities of management by focusing on species or habitats individually, rather than as part of an integrated ecosystem. BAPs still provide valuable information, however, and the process of creating LBAPS for small sites, such as school grounds, parks, and university campuses, is still conducted. In most countries, these have now been superseded by ecosystem-orientated plans following the 2010 CBD conversion, usually referred to as National Biodiversity Strategy Action Plans (NBSAPs) or sometimes as post-2010 Biodiversity Frameworks in the UK.

National Biodiversity Strategy Action Plans (NBSAPs)

NBSAPs have now largely replaced BAPs. They fulfil the same ideal—a national document outlining how biological diversity will be conserved, what targets have been set, and how these will be met and assessed. The main difference is that NBSAPs focus on the bigger picture by taking an ecosystems approach and considering ecology and environmental functioning in a broad, process-driven sense, rather than looking at each individual priority species and habitat individually and separately.

FURTHER READING FOR SECTION

UK Wildlife and Countryside Act (1981): Available at: http://www.legislation.gov.uk/ukpga/1981/69

Details on types of protected area in the UK: Available at: http://jncc.defra.gov.uk/page-4 (terrestrial) and http://jncc.defra.gov.uk/marineprotectedareas (marine)

UK Site of Special Scientific Interest database: Available at: https://designatedsites.naturalengland.org.uk/

USA Endangered Species Act: Available at: http://www.epa.gov/laws-regulations/summary-endangered-species-act

Worldwide National Biodiversity Strategy Action Plans: Available at: https://www.cbd.int/nbsap/about/latest/default.shtml

Worldwide Marine Protected Areas—challenges and solutions: IUCN. (2015) Challenges for Marine Protected Areas and examples for addressing them. Available at: https://www.iucn.org/about/work/programmes/marine/marine_our_work/marine_mpas/?21793/Challenges-for-Marine-Protected-Areas-and-examples-for-addressing-them

Original UK Biodiversity Action Plans (1994) via National Archives: Still a very valuable resource. Citing the UK Biodiversity Action Plan. Available at: http://tna.europarchive.org/20110303145238/http://www.ukbap.org.uk/cite.aspx

11.5 Reserve-based conservation versus wider countryside management.

Reserve-based conservation has traditionally been regarded as the mainstay of *in situ* conservation. However, due to the limitations of this approach, reserve-based initiatives often go hand-in-hand with conservation in the wider countryside.

11.5.1 Creation of protected areas and nature reserves.

Creation of protected areas is often a balancing act between the ideal scenario (as determined using ecological theory and modelling) and on-the-ground practical considerations. Two key considerations are: (1) the size versus number of reserves; and (2) their exact location.

SLOSS

In 1975, Diamond proposed a series of general principles for nature reserve design, which are described in Figure 11.10. Central to these was the debate of whether it was better to have a **Single Large** reserve or **Several Small** reserves, hence the acronym SLOSS. Diamond came down firmly on the Single Large side of the debate on the basis that there would likely be more species to start with (the species–area relationship, depicted in Figure 4.8) and the rate of extinction would be lower due to larger populations and because large areas tend to be more self-regulating (Section 7.4.1). However, if the conservation priority shifts slightly from minimizing extinction rates to maximizing species richness—not

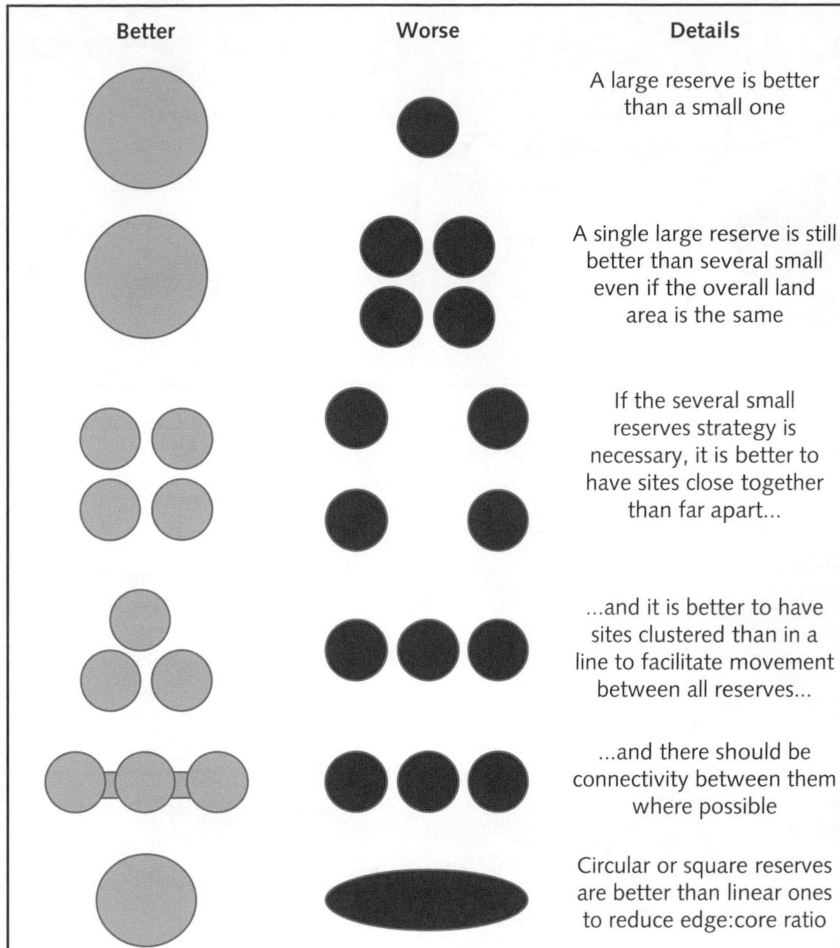

Figure 11.10 The six central ideas of nature reserve design by Jared Diamond encapsulated in the SLOSS (single large or several small) concept.

quite the flip side of the same coin—it can be possible to pack more species into a series of Several Small reserves than one big one, especially if those patches themselves differ slightly, thereby increasing habitat heterogeneity (Simberloff & Abele, 1976).

The SLOSS debate dates back to the mid-1970s, but is still a vital one for Applied Ecologists today, depending as it does both on conservation goals and the intricacies of specific species and sites. Evidence for the single large argument has been found for Caribbean spiny lobsters *Panulirus argus* in the Bahamas (Stockhausen & Lipcius, 2001), while evidence for the several small argument has been found both in terrestrial environments (e.g. butterflies in holm oak *Quercus rotundifolia* forests in Spain: Baz & Garcia-Boyero, 1996) and marine environments (fish and macroinvertebrates associated with seagrass *Zostera* in Australia (McNeill & Fairweather, 1993)).

There can also be a divide here between single-species and multi-species conservation (Section 10.2.2). Single species conservation might be better achieved in single large reserves due to the lower extinction rate and multi-species conservation might be better achieved in several small reserves with their higher species richness. However, even despite the extinction rate advantages, there can be disadvantages to single large reserves for single-species conservation; namely the concept of 'putting all our eggs in one basket'. In other words, if there is only one reserve and something happens to that reserve (fire, pollution incident, etc.) or a species protected within it (e.g. disease outbreak) the consequences could be severe and widespread. It is also important to consider where the species is actually located—this might already be in a series of small populations at some distance from one another, in which case the 'several small' scenario might be forced.

The other arguments outlined in Figure 11.10 are also subject to counter-arguments. For example, having reserves close together, clustered, or linked via

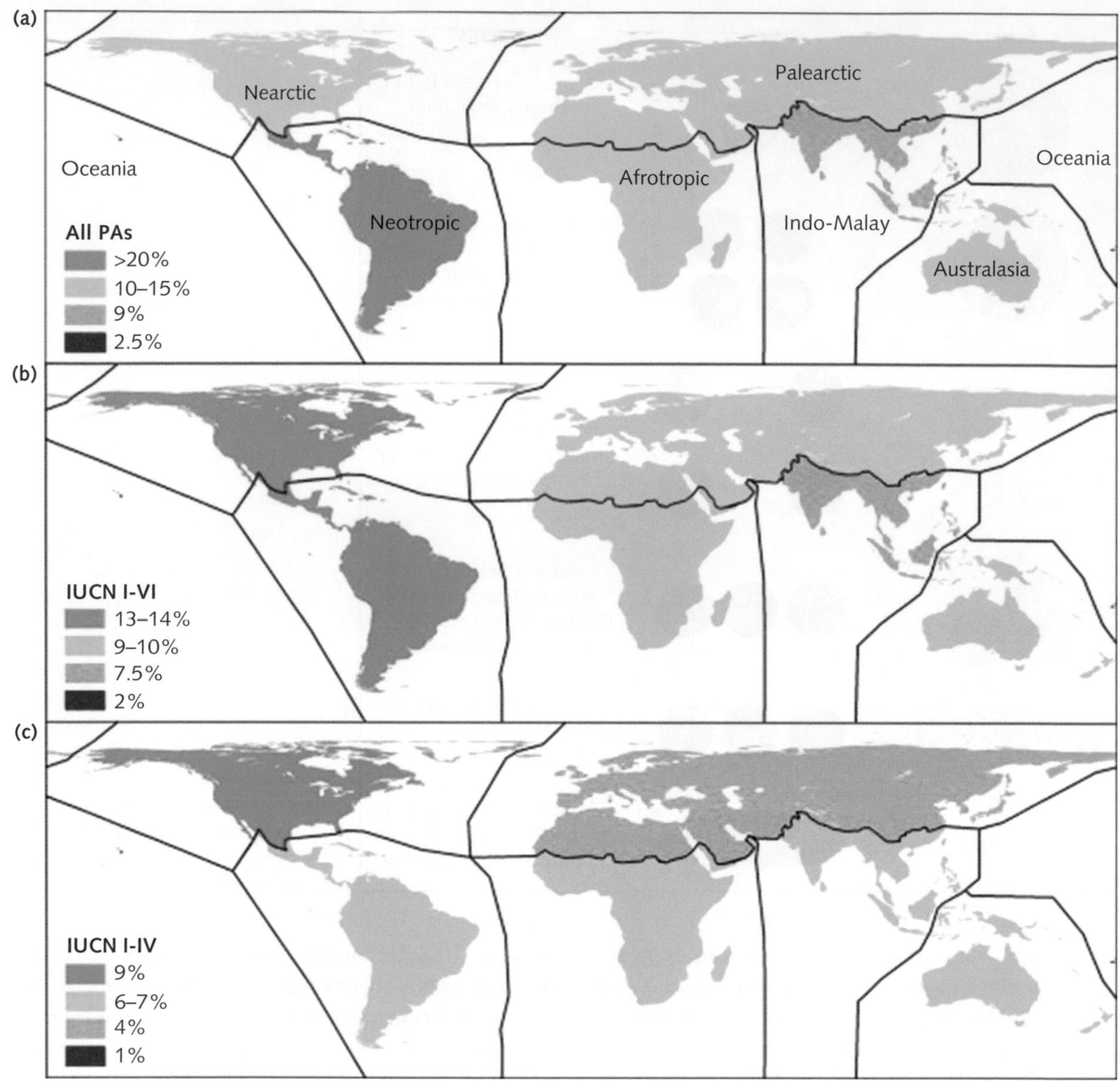

Figure 11.11 The relative percentage of land formally protected for biodiversity in different countries in (a) any form of protected area, (b) any IUCN-listed protected area, and (c) any strictly-protected IUCN-listed protected area.

Source: Reproduced from Jenkins, C. N., & Joppa, L. (2009) 'Expansion of the global terrestrial protected area system'. Biological Conservation, 142, 2166-2174, with permission from Elsevier.

corridors (Section 7.6.2), does have metapopulation advantages, but also means that problems such as pests, parasites, non-native species, and disease can also spread more quickly. Similarly, while minimizing the edge:core ratio might be sensible in many cases, it is not always possible (e.g. for linear habitats, such as rivers and hedgerows) and might not be desirable (e.g. if edge-loving species are the conservation target—Section 7.4.2). In many cases, the SLOSS guidelines are sound, but they are just that—guidelines—and should be subject to the critique of conservation ecologists, rather than being followed blindly.

11.5.2 Limitations of protected areas and nature reserves.

A lot of conservation management is focused on protected areas and nature reserves. The two entities

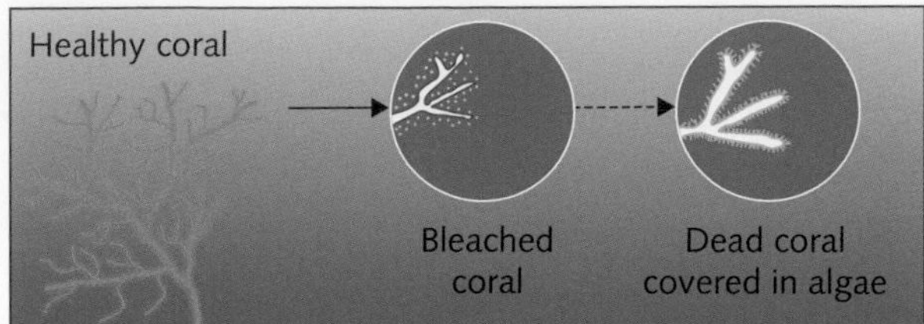

Figure 11.12 As sea temperatures rise, corals can become stressed and turn white through the process of bleaching. This happens when their symbiotic algae called zooxanthellae that are expelled, resulting in the death of those coral before their skeleton becomes covered by turf algae.

of protected areas and nature reserves can be one-and-the-same if an area is both given some form of legal protection (Section 11.4.1) and actively managed (Section 11.3), but equally it is possible to have a non-legally protected nature reserve or a protected area that is not actively managed.

The overarching aim of nature reserves and protected areas is to safeguard particular landscapes, protect specific habitat, and enhance populations of certain species. However, even if all nature reserves in a country have optimal management initiatives (which is rarely the case) and all management is 100% successful (virtually impossible), and all protected areas have full protection (also rarely the case), nature reserves and protected areas will still only be a minor percentage of total land mass, and usually are split into virtual islands.

Other than New Caledonia, a French territory comprising dozens of islands in the South Pacific, and Slovenia in Europe, no country had >50% of its land mass listed as protected in 2012 (World Bank, 2015). The average percentage listed as protected is 15.8%. However, 74 out of the 213 countries and territories have <10% protected, including some surprises given the value of their biodiversity (e.g. Madagascar = 5%; India = 5.2%), the presence of internationally renowned national parks (e.g. despite the presence of the vast and well-known Kruger National Park, South Africa only has 6.3% of land protected) and expressed commitment to conservation (e.g. Canada, with only 11.6%). Figure 11.11 shows the global differences in the percentage of land in protected areas.

Climate change

Alteration of species' range due to climate change was discussed in Section 7.5.1. The impact of this on the effectiveness of protected areas could be dramatic as target species' ranges shift out of the protected areas that are managed specifically for them. For some species, including those with substantial range change, with overlap between current and future ranges, and often simultaneous decreases in range size, current site-specific initiatives might not be suitable for long-term conservation.

Although protected areas are becoming increasingly less suited to species-specific conservation (e.g. Hannah et al., 2007), current distribution 'snapshots' are still used to inform site-based initiatives (Gaston et al., 2006). For example, for birds, the Important Bird Areas (IBA) programme designates areas based on the rarity of species currently present, while areas considered crucial for survival of rare species at a European level are often given Special Protection Area (SPA) status, and for wetland sites, international protection under the Ramsar Convention (Section 11.4.1). Given predicted range changes, creating additional protected areas by integrating future climate-informed distribution predictions into reserve selection algorithms (Araujo et al., 2004) might be a partial solution. Protected area supplementation already appears to be an effective strategy for plants in Europe, and a proactive approach (creating new protected areas using predictions) is less costly than a reactive one (trying to create new protected areas after species have moved), even given uncertainties of this approach.

In the marine environment, problems are, if anything, even greater. Temperature change in the oceans is generally less dramatic than on land, but oceanic ecosystems are very sensitive even to low-magnitude changes. Effects include coral bleaching (whitening due to loss of the symbiotic algae; Figure 11.12) and loss of important species, such as krill Euphausiidae, which play a key role in marine food chains. Climate-induced changes in sea level can also be a problem, especially for coastal protected areas. For example Cape Cross, on Namibia's Skeleton Coast, is one of

Figure 11.13 Flooded land at Coombe Hill, Gloucestershire. The ingress of water in such circumstances is outside the control of reserve managers.

Source: Photographs by Andy Jayne, used with kind permission.

the most important breeding grounds for the brown fur seal *Arctocephalus pusillus* (which, despites its vernacular name is actually a species of sea lion), but is threatened by rising sea levels.

Off-site pressures

How ever well a protected area is managed, it is always susceptible to external pressures. For example, forest **fires** that start outside a protected area can result in huge areas of nature reserve being damaged, as in the Ernesto Tornquist Provincial Park, Buenos Aires province, Argentina, in 2014. Flooding can also pose problems, especially for low-lying areas where the flood water can be contaminated. For example, Coombe Hill Reserve in Gloucestershire, UK, is managed for species-rich grasslands that depend on low inputs of nitrates. This site is in a Nitrate Vulnerable Zone (a UK designation for an area vulnerable to eutrophication and the effects thereof) and the reserve is carefully managed by taking an annual hay cut and removing the vegetation (and nutrients) from the site. The reserve's managers also have agreements with adjacent farmers to minimize fertilizer use to reduce nutrient input. However, the reserve is subject to periodic flooding and much of the flood water is now nitrate-rich. This can engulf large areas for several weeks (as shown in Figure 11.13) and dramatically increase nutrient levels. Such events are completely outside of the reserve's control and differ from the historical situation when flood water was much less nutrient-rich and thus not a problem in this regard.

The impact that such off-site pressures can have is exemplified by the fact that one of the most iconic American National Parks, Yellowstone, was placed on the World Heritage Site in Danger List in 1995 partly because of external pressures (National Parks Service, 2003). For example, there was ongoing contamination from runoff from previous gold and silver mining, and an active proposal to reopen mines strengthened these concerns. Although sources of contamination were outside the Park, those sources were upstream so that the effects were felt within the Park itself. Another off-site pressure was the fact that some Yellowstone bison were infected with the bacterium *Brucella abortus*, the causal agent of the cattle disease brucellosis. In harsh winters, bison herds frequently moved outside the Park to find food and there were concerns that they could transmit *Brucella* to farmers' livestock; this was viewed as a significant economic threat to the livestock industry that would revoke the brucellosis-free status of bordering states. In the harsh winter of 1997–98, up to 1000 bison that had wandered outside the Park boundary were killed because of the fear of brucellosis.

In addition to the specific off-site pressures that an individual area can face, there are also much broader-scale issues. Chief amongst these is climatic change and widespread pollution—for example, the pesticide DDT being recorded in Adelie penguins *Pygoscelis adeliae* in Antarctica (Hot Topic 4.1).

11.5.3 Wider countryside initiatives.

Because of the inherent limitations of protected areas (Section 11.5.2), wider countryside initiatives that aim to conserve species and habitat outside protected areas are extremely important. Wider countryside initiatives can include custodial measures, such as the protection afforded by species- and activity-focused legislation (Section 11.4.1). More active measures include working with non-conservationists and members of the public to encourage biodiversity-sensitive management of private land such as forested areas and agricultural land.

The importance of private land for species must not be underestimated. For example, the vast majority of the remaining population of the corncrake *Crex crex* lives on agricultural land, while 60% of endangered species in the USA occur on private forested land (Robles et al., 2008). Public land that is not managed specifically for biodiversity under normal conditions (e.g. military bases or training areas, state-owned parkland) also have considerable potential for *in situ* conservation schemes.

Agri-environment schemes

Agri-environment schemes generally involve farmers undertaking initiatives that will benefit biodiversity, despite the fact that these cost money or, more usually, result in decreased profits. Farmers are paid a subsidy to compensate them for loss of earning. Generally, the level of the subsidy is related to the extent of the environmental initiatives being undertaken and thus the financial penalty involved. Examples include:

- **Actions that decrease the amount of land being farmed:**
  - managing hedgerows;
  - creating field margins either through no-touch zones or deliberately seeding with native species;
  - digging ponds.
- **Actions that decrease yields:**
  - reduction in use of fertilizers or pesticides;
  - cutting crops at non-ideal times to allow flowers to set seed or nesting birds to fledge.

Urban gardens

Gardens are an important resource for wildlife and are frequented by many taxa including birds, mammals, reptiles, amphibians, and insects. In the UK, for example, it is estimated that the total area covered by UK gardens is in excess of 400,000 ha (Davies et al., 2009). Gardens may also act as wildlife corridors between larger areas and, in this way, may be important in a landscape context, as well as in their own right.

There is growing evidence that species that are suffering a decline in the wider countryside can be found in significant numbers in gardens, for example, the common frog *Rana temporaria* and the hedgehog *Erinaceus europaeus* (e.g. Gaston et al., 2005; Williams et al., 2014).

Extractive reserves

A different initiative is that of extractive reserves, which can be used in situations when living resources are extracted from (as opposed to grown or reared in) specific ecosystems. The idea is that although the natural resource—be it timber, fish, latex for rubber, or naturally-growing foodstuffs such as nuts—is still harvested, that harvesting is subject to regulation and control to ensure that it is sustainable. Such initiatives can be government-regulated or can be voluntary (e.g. Forest Stewardship Counsel; Rainforest Alliance). The concept of extractive reserves also applies to removal of animals through trophy hunting licencing (Hot Topic 11.1) and harvesting of fish stocks using offtake quotas (Hot Topic 11.3).

FURTHER READING FOR SECTION

SLOSS seminal paper: Diamond, J.M. (1975) The island dilemma: lessons of modern biogeographic studies for the design of natural reserves. *Biological Conservation*, Volume 7, 129–45.

Protected area limitations in the face of climate change: Hannah, L., Midgley, G., Andelman, S., Araújo, M., Hughes, G., Martinez-Mayer, E., Pearson, R., and Williams, P. (2007) Protected area needs in a changing climate. *Frontiers in Ecology and the Environment*, Volume 5, 131–138.

Agri-environment scheme effectiveness: Batáry, P., Dicks, L.V., Kleijn, D., and Sutherland, W.J. (2015) The role of agri-environment schemes in conservation and environmental management. *Conservation Biology*, Volume 29, 1006–1016.

11.6 The public and *in situ* conservation.

People and conservation interact in two very different ways: as custodians of land, habitat, and species, and as 'consumers' of conservation through visiting nature reserves and captive collections (Chapter 12). This section will explore these interactions in more depth, as well as the role of education and community engagement with the *in situ* conservation process.

11.6.1 Management of visitor pressure.

There is no denying that visitors can exert considerable—and often damaging—pressure upon some nature reserves and other protected areas. This can range from huge landscapes, such as the Masai Mara in Kenya, to very small local nature reserves. As such, the management of human use (and disturbance) is a key consideration.

It can be possible to routinely prevent access to important wildlife sites, especially if they are on private land. For example, in Gloucestershire, UK, the tiny 1 ha Badgeworth Nature Reserve and Site of Special Scientific Interest (once in the Guinness World Records for being the smallest nature reserve in the world) simply cannot support many visitors without damage to the rare adder's-tongue spearwort *Ranunculus ophioglossifolius*. As a result, the reserve is open to the public on a permit-only basis, and one or two publicized open days.

Figure 11.14 Management of the impact of people on ecology and conservation through zonation of impacts in Banff National Park, Canada.

Source: Photograph by Anne Goodenough.

In other cases, it might be appropriate to prevent access temporarily at specific times of the year—turtle nesting beaches being a prime example (Powell & Ham, 2008).

A rather more common strategy than denying access is to adopt a **zonation strategy**, whereby human disturbance is confined to one area. This might be using a 'carrot' approach by tempting people to an area (for example, by installing interpretation boards, picnic tables, a visitor centre, and so on) or using a 'stick' approach (for example, by stipulating that visitors should keep to the paths: Figure 11.14). The first of these approaches is effectively creating a honeypot site—the analogy being that people buzz around that area like bees around a honey pot. In such cases, damage to this area might be more substantial that would otherwise be the case, but other areas are left relatively undisturbed. As long as the honeypot is created away from areas with especially high ecological value or sensitivity, or in some cases core areas, this can be a very useful approach. This zonation approach has been used in Koh Chang National Marine Park, Thailand, to manage snorkelling activity across the reef (Roman et al., 2007).

There can also be ways of minimizing the impact of visitors without reducing numbers or managing distribution. A commonly used strategy is using boardwalks to prevent trampling of rare species, as shown in Figure 11.15. However, it should be noted that, although, in some cases, management of visitor pressure is necessary, it should not be the default strategy unless visitors are actually damaging the ecology. One of the most important ways that people engage with nature conservation is to visit nature reserves, and this should not be prevented or curtailed unless absolutely necessary.

11.6.2 Education and community engagement.

Ecological education can occur within formal curricula in schools, via science communication within the media, through targeted public awareness campaigns by statutory and non-statutory organizations, and, as noted previously, at specific sites through interpretation boards and nature reserve management leaflets. One of the most important aspects of ecological education is to enthuse people. Once people are enthused, they are much more likely to want to

Figure 11.15 Use of boardwalks at Meares Island, Canada, to minimize the impact of visitors on ground flora and soil erosion.
Source: Photograph by Anne Goodenough.

learn about ecological interactions, threats, conservation successes, and what they can do individually to make a difference, be that gardening in a wildlife-friendly way, joining a conservation organization, or volunteering at a nature reserve. This opens the door for community-linked conservation.

11.6.3 Community-linked conservation.

Community-linked conservation, as the name suggests, occurs when the local community is involved with, and invested in, a nature reserve and its management. This is sometimes referred to as community-based conservation (e.g. Berkes, 2004), but this presupposes that all, or at least most of, the conservation is community led and community managed. There are some examples of this—some conservation areas are entirely community-run, possibly with some government funding assistance. A good example is Anja Community Reserve in Madagascar, which is managed by the Anja Miray association, comprising members of the Betsileo tribe, for a range of species, including ring-tailed lemurs *Lemur catta*.

In many cases, however, the community is involved with and a fundamental part of conservation, but does not necessarily solely drive it. It might be, for example, that a nature reserve is run by a charitable organization (such as a Wildlife Trust) and there are regular volunteer working parties, reserve open days, wildlife explorer youth groups, public wildlife events, and so on (see the Interview with an Applied Ecologist for this chapter). Members of the local community might sit on reserve committees, and membership might be fee-paying to generate revenue, but there are some paid staff too. Such approaches can be very good ways of bringing conservation and community together, and fit more neatly into the concept of community-linked conservation, rather than community-based conservation.

Ecotourism

In theory, ecotourism provides an excellent way of bringing together people and nature. It can bring in much-needed revenue and jobs for local people, and enthuse, encourage, and inspire visitors to help conserve the species and habitats that they encounter. Evidence from the Galápagos Islands shows that well-designed and delivered interpretation during the ecotourism experience can not only increase knowledge and support of the host-protected area, but also general environmental behavioural intentions and philanthropic support of conservation (Powell & Ham, 2008). In short, it can be a way of making sure that species are worth more alive than dead, or that habitats are worth more intact than destroyed.

Of course, the realities are much more complex. In the first place, tourism that involves seeing ecology and ecological (eco) tourism are often very different. Going to see wildlife can often result in considerable **visitor pressure** (Section 11.6.1), especially in areas where infrastructure has to be specially constructed for tourists or where environments are very sensitive (e.g. coral reefs discussed in Section 11.6.1 or Greek beaches used by loggerhead turtles *Caretta caretta* to nest: Katselidis et al., 2013). Moreover, often very little of the money paid to experience these areas and their wildlife is actually seen by the local community, who, in some instances, are even excluded from native lands. In such circumstances 'eco' tourism can end up being destructive to both ecology and community. Some of the main reasons why ecotourism initiatives succeed and fail worldwide are listed in Table 11.3.

AN INTERVIEW WITH AN APPLIED ECOLOGIST 11

Name: Gareth Parry

Organization: Gloucestershire Wildlife Trust (local conservation charity that is part of a national network of similar charitable organizations based in the UK: http://www.gloucestershirewildlifetrust.co.uk)

Role: Head of Community Programmes

Nationality and main countries worked: British; worked primarily in the UK.

What is your day-to-day job?

I lead the strategic development of Gloucestershire Wildlife Trust's community work, leading a team that creates opportunities for people from a wide range of ages, abilities, and backgrounds to experience nature, and begin to appreciate the importance of conservation. This includes public events, activities, community development projects, education, and training programmes. We link the environment to key social and economic priorities, such as health and well-being, job skills, and social deprivation. We also help existing wildlife enthusiasts develop their knowledge and take action to support conservation, notably through volunteering. As an ecologist, I also act spokesperson on relevant wildlife topics and help to drive an evidence-led approach to conservation, particularly in relation to carnivores such as Eurasian otter *Lutra lutra*, which are my ecological passion.

What is your most interesting recent/upcoming project and why?

We recently began a community project at a country park on the edge of the city of Gloucester. Many of the surrounding communities suffer deprivation, with low incomes, and high levels of health and education inequality. The park is a fantastic free-to-access green space with some great wildlife habitats, but public use was both low and declining. The project aimed to encourage more local people to use the park, by providing opportunities and understanding the barriers that deterred them. Rather than focusing on activities that appeal to traditional wildlife audiences, such as species identification and guided walks, there was a focus on free family-focused activities that made being outdoors fun. In just two years over 10,000 people attended events and use of the park increased by over 40%.

Before this project people didn't feel 'entitled' to use the park, or know what to do there. It seems obvious now, but expecting the average person to begin their wildlife journey with bird watching or species identification was completely unreasonable. This goes to show how many people could be interested in wildlife if they are given the right opportunities and barriers are removed. Attitudes to nature can be complex because they are shaped by a complex combination of cultural values, knowledge, embedded behaviours, and personal incentives and disincentives, and it is our job to change these where necessary to encourage more people to value nature.

What's been best part of this particular project?

Before the project began, many local conservationists discounted it as a waste of time because they felt that our target urban communities 'were never going to be interested in wildlife'. Consultation undertaken during the project actually revealed that 'more places to see wildlife' was one of the top wishes of local residents. This just shows how out of touch Applied Ecologists can be with the way in which many people engage with nature and highlights the opportunities that exist to change this if barriers can be broken down.

What do you see as the main challenges in your field and how can they be overcome?

It is often thought that overcoming people–nature engagement challenges means improving the public's knowledge of wildlife issues. In reality, extensive research has shown that increasing knowledge has little impact on people's environmental attitudes or behaviours. Stimulating pro-environmental behaviour is affected by a complexity of factors, particularly embedded values and internal/external incentives or disincentives. For example, many people know about the catastrophic loss of traditional wildflower meadows in the UK, but continue to maintain their garden lawns with herbicides and regular mowing. Providing information is not enough; we have to help people to build an emotional connection by providing exciting, interactive and accessible experiences. Nature should be for everyone, but the conservation sector often engages with people who are already pro-wildlife—the 'easy wins'. This must change if we are to mainstream the natural environment as one of society's most valued assets and priorities.

Finally, how did you get into community-focused conservation and what advice would you give to others?

Like quite a few ecologists, I started off aspiring to traipse around the wilderness studying wildlife and publishing my research. I soon realized that, for all the great ecological research being undertaken, very little was filtering through to the front end of conservation. I kept seeing management plans and monitoring techniques that were years behind current ecological understanding. I became tired of hearing the standard disclaimers of 'we have always done it that way' and 'that's what the handbook says'. The handbook was out of date: I wanted to do something about this by working for a practical conservation organisation. I decided that engaging people was a key problem facing conservation and that was where I wanted to focus my efforts.

People get into community work through all manner of routes. There are a few key things that are crucial to working in this field. First, you have to be able to inspire people about wildlife, so knowing your stuff and being able to communicate to people of different ages, abilities, and backgrounds is vital. You have to understand what motivates people to support conservation and the best way to do this is to volunteer yourself. Many conservation professionals do not understand why engaging people, particularly those 'who aren't already interested', is important. Don't get deterred by that: remember that nature should be for everyone.

Table 11.3 Main reasons why ecotourism initiatives either become unsustainable (top) or sustainable (bottom).

Unsustainable case studies (*n* =70)	**Case studies in %**
Types of unsustainability	
Habitat alteration, soil erosion, pollution	45.6
Local community not involved, leads to consumptive land-use	25.0
Flag species affected, population decline, serious behaviour alteration	20.6
Not enough revenue creation for conservation. consumptive use practiced	8.8
Reasons for unsustainability	
Too many tourists	36.8
Local community not involved	27.9
Not enough control and management	14.7
Not enough local revenue creation	103
Protected area has priority over local people	7.4
Locals do not get environmental education	2.9
Sustainable case studies (*n* = 118)	
Effects of sustainable case studies	
More conservation (new areas, more effective)	44.1
Revenue creation increased for local communities, non-consumptive use	28.8
Increased revenue creation, regionally and nationally	21.2
Conservation attitude of local communities changed	5.9
Reasons for positive effects	
Local community involved at most stages	38.5
Effective planning and management	33.3
Ecotourism simply an economic advantage, locally and regionally	17.1
Flagship species alone	6.0
Differential pricing of entry fees	5.1

Source: Krüger, O. (2005) *Biodiversity & Conservation*, **14**(3), 579–600.

Where ecotourism does work, though, it can work very well. This occurs when the central concepts of ecotourism—celebrating biodiversity with people in such a way that it enhances conservation and community—are met. Good examples are the Anja Community Reserve, where all the guides are local people from the nearby community who are paid appropriately for their work. Another excellent example is the mountain gorilla *Gorilla beringei beringei* conservation project in Parc National des Volcan in Rwanda (Nielsen & Spenceley, 2011). Here, again, local people are employed as guides and rangers. Tourism is now a major source of income, but despite this, a limited visitor permit system is in operation. This prevents too many people being able to access (and spoil) the area, and generates substantial additional funds that are ploughed back directly into conservation *and* the community, including education and healthcare. However, even in well-managed ecotourism ventures that are sensitively managed, there can be problems. In the case of primate conservation, this can include tourists visiting with diseases such as the common cold or gastrointestinal complaints, which could potentially be transmitted to non-human primates (Muehlenbein et al., 2010).

FURTHER READING FOR SECTION

Zonation of human impact: Roman, G.S., Dearden, P., & Rollins, R. (2007) Application of zoning and 'limits of acceptable change' to manage snorkelling tourism. *Environmental Management*, Volume 39, 819–830.

Education for conservation: Hughes, C. (2012) Environmental education for conservation: considerations to achieve success. *Natural Areas Journal*, Volume 32, 218–219.

Community-based conservation (general): Berkes, F. (2004) Rethinking community-based conservation. *Conservation Biology*, Volume 18, 621–630.

Community-based conservation (specific example—Botswana): Twyman, C. (2000) Participatory conservation? Community-based natural resource management in Botswana. *Geographical Journal*, Volume 166, 323–335.

Ecotourism overviews: Brightsmith, D.J., Stronza, A., & Holle, K. (2008) Ecotourism, conservation biology, and volunteer tourism: a mutually beneficial triumvirate. *Biological Conservation*, Volume 141, 2832–2842

Ecotourism overviews: Krüger, O. (2005) The role of ecotourism in conservation: panacea or Pandora's box? *Biodiversity and Conservation*, Volume 14, 579–600.

11.7 The future: evidence-based initiatives.

A considerable amount of conservation practice is undertaken based on experience and anecdotal evidence. Experience-based conservation certainly has an important role in *in situ* conservation, but equally it is vital that, in addition to drawing on experience and traditional techniques, conservation becomes evidence-based.

Evidence-based conservation is about ensuring that management actions and policy are based on **empirical evidence**. Such evidence might be scientific peer-reviewed publications, practitioners' *tested* experiences, and independent expert assessment. In other words, the answer to a particular conservation question should not be 'we have always managed this situation by undertaking [*Insert management action here*] so we will do that again', but rather 'best practice guidelines drawn out from recent published research suggest that the optimal management strategy is [*Insert name here*] and we have previously found that to be effective through objective monitoring at another location so we will do that and monitor how things change'.

There have been increasing demands for practitioners to move more towards evidence-based conservation for some time (e.g. Cook et al., 2009). The main challenges here revolve around the importance of practitioners appreciating the need for this and then understanding the complexities of empirical testing, randomized designs, standardization of sampling procedure, and statistical significance to create evidence on which a decision can be based. The importance of evidence-based conservation is highlighted next for a frequent part of *in situ* conservation: managing habitat optimally for a specific focal species.

11.7.1 Researching species–habitat interactions.

For many species, it is vital that habitat is optimal if they are to be able to arrive, survive, and thrive at a site. The consequences of non-optimal habitat on species' ecology are explored in Figure 11.16. Because of the importance of optimal habitat, a large amount of the time and effort that goes into *in situ* conservation is spent managing habitat to create or maintain suitable conditions. This can involve substantive changes, such as creating saline lagoons from coastal grassland to attract wading birds (Online Case Study Chapter 11), or it can be very subtle, for instance, slightly altering grazing to create better conditions for flowering plants. Detailed and specific knowledge of species-specific habitat requirements is thus a vital part of evidence-based conservation.

Some very specialist species are found only in highly specific (and often highly range-restricted or complex) habitats. Examples include the koala *Phascolarctos cinereus*, which requires Eucalyptus forest in Australia, and marram grass *Ammophila*, a genus of plants found almost exclusively in sand dunes. Other species are more generalist, with species such as black rats *Rattus rattus* and raccoons *Procyon lotor*, exploiting a wide variety of different habitats. However, even when a generalist species is *able* to survive in a wide range of conditions because it has a wide tolerance range (Section 4.3.1), some habitats will likely be *preferred* over others.

As with creating new habitat (Section 11.3.1), it is important to consider requirements at several spatial scales when defining species–habitat interactions:

1. **Macro-scale:** this refers to broad habitat type—woodland versus grassland for instance—and tends to be based primarily on general species location. At this broad scale, it is important to ensure that the habitat in which a species is found does not become conflated with habitat *requirements*. The common dormouse *Muscardinus avellanarius*, a species of small mammal that occurs in much of Europe, provides a salutary lesson here. The traditional view was that this species occurred solely in woodlands and required hazel *Corylus avellana*, ideally after coppicing to allow individuals to move around arboreally (without touching the ground)—indeed, the alternative vernacular name is hazel dormouse and its specific name (*avellanarius*) relates to the specific name of European hazel *C. avellana*, itself derived from Avella, a town in Italy. This view perpetuated because ecologists then only looked for the species when habitat conditions were deemed 'suitable', thereby generating a self-reinforcing prophecy. In reality, the species is found in hedgerows and even areas of coastal scrub where hazel is completely absent.

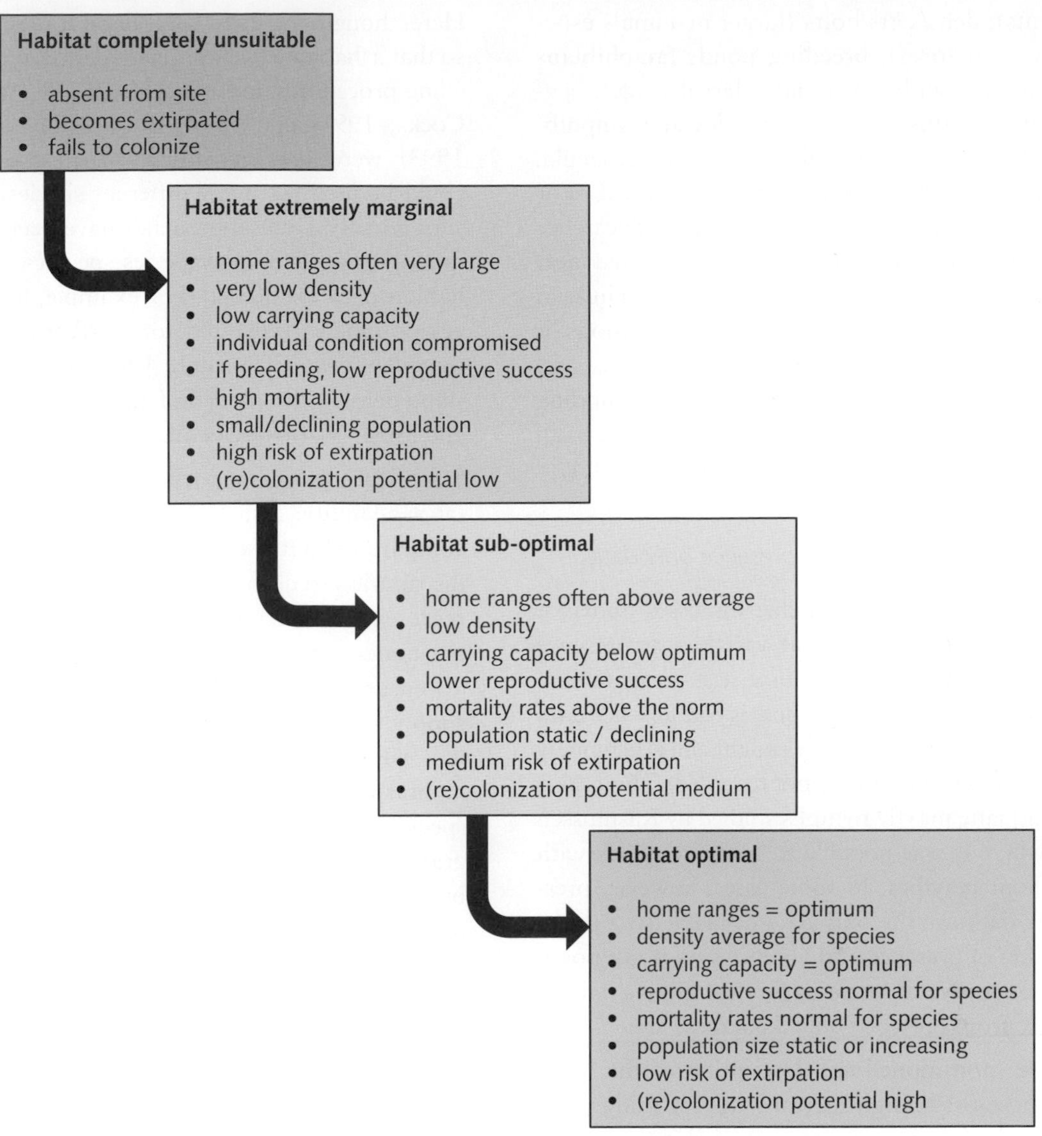

Figure 11.16 Effects of habitat suitability on species' ecology on a spectrum from completely unsuitable to optimal.

2. **Meso-scale:** this is a more specific level of habitat description that moves understanding of a species' habitat requirements from (say) woodland to considering what type of woodland (dominant species, underlying geology, proximity to landscape features such as water sources, and so on). Such analysis is often best done within a Geographical Information Systems framework, with species distribution overlying habitat information in a GIS layer stack (Figure 7.12). In many cases, the species layer will be a dot map showing occurrence than can be related to other habitat or landscape layers. This approach was used to establish black-tailed jackrabbit *Lepus californicus* in south-western Idaho, USA, occur in particularly high densities in areas with both scrub and cultivated crops (Knick & Dyer, 1997).
3. **Micro-scale**: very specific habitat requirements are usually investigated using a quantitative or modelling approach to unpick the intricacies of habitat requirements at a very local scale. For plants, this will be the area immediately surrounding specific individuals. For animals, this will be the specific habitat around nests (birds/

rodents), dens/setts/holts (larger mammals especially carnivores), breeding ponds (amphibians and insects with an aquatic larval stage), egg-laying sites (insects, some reptiles and amphibians, fish), roosts (bats, birds), and hibernacula. Sometimes, this is fairly straightforward. For example, some birds such as the pied flycatcher *Ficedula hypoleuca* nest in human-created nest boxes in preference to natural cavities. Thus, in order to study the detailed habitat interactions, it is possible to put nest boxes up in abundance and compare the habitat characteristics surrounding boxes that are used with those that are not. Such issues are discussed further in Case Study 11.1.

Quantifying interactions with presence-only data

In the case of the pied flycatcher discussed above, it is possible to compare habitat variables at presence versus absence sites because some nest boxes remain unoccupied. This means that there is true absence data as the species has been actively sought and is genuinely not there (rather than simply not recorded). This is also true for aquatic mayfly nymphs studied by Rasmussen (1988) where it was possible to compare ponds with and without mayflies. In some cases, however, presence-only data are the only data available so a direct comparison of presence and absence sites is not possible. There are several methods for determining habitat preference from presence-only data, these are:

1. **Simple (non-modelling) description of the habitat:** the most straightforward way of using presence-only data is to describe the habitat in which a species occurs. This is a relatively low-power approach but it can provide useful insights as in the case of Clymene dolphin *Stenella clymene* in Mexico, in relation to sea depth (Fertl et al., 2003). Despite, or possibly because of, the simplicity of this technique, it is commonly used.
2. **Habitat profiling:** the next level of complexity is using profile techniques to characterize environmental conditions associated with the presence records. Key among such approaches is habitat envelope modelling. This uses the same idea as climate envelope modelling (Section 7.5.1), which uses a species' current range to infer climatic requirements for that species (the 'climatic envelope'). Here, though, the focus is on habitat requirements so that a 'habitat envelope' is modelled. Early modelling procedures, including HABITAT (Walker & Cocks, 1991) and DOMAIN (Carpenter et al., 1993) were very generalist, which meant they could be used for many different species, but not very reliably. Later approaches have been broader in scope, but tend to be species-specific as detailed habitat data are needed. For example, habitat envelope models generated for northern goshawks *Accipiter gentilis* in Utah, USA, were based on stand density, elevation, and abundance of conifers and aspen (Zarnetske et al., 2007).
3. **Statistical model comparing habitat use to habitat availability:** a subtly different approach is to compare habitat use with the underlying habitat distribution to determine which habitats are over-represented (used more than expected) and which are under-represented (used less than expected). To ensure that any habitat 'preferences' are not simply reflecting an uneven habitat distribution it is vital to compare habitat percentage *use* to habitat percentage *availability*. For instance, if a species occurs more frequently in winter-grazed grassland compared with spring-grazed grassland, but the underlying habitat distribution is skewed towards winter-grazed grassland, the apparent 'preference' might only reflect habitat availability. This does not show what drives any differences, it just quantifies those differences.
4. **Statistical model comparing habitat use to habitat non-use:** to start to understand not only that differences occur, but what factors drive those differences, it is possible to compare habitat characteristics of sites where a species is present with the characteristics of locations in the same general area where the species is not (thought to be) present to create a comparison dataset. This is referred to as pseudo-absence data. For this to work, there must be a low chance that the species is actually present at pseudo-absence sites, and has remained undetected since that would compromise the comparison.
   - Generating pseudo-absence data is usually done in the field by creating new sample points and surveying them in the same way as if they supported the target species. A good example

of this approach is an analysis of barbastelle bats *Barbastella barbastellus* in central Italy (Russo et al., 2004).

- Alternatively, if a GIS approach is being used for a meso-scale analysis, random points can be computer-generated and compared with habitat information layers. This approach was used by Osborne et al. (2001) to analyse habitat requirements of the globally threatened great bustard *Otis tarda* in Spain; this showed a preference for homogenous grassland away from human activity.
- Usually field or computer-automated pseudo-absence points are determined randomly, but sometimes they are selected on a stratified random basis from the area(s) of a site with suitable baseline environmental conditions. This approach was used by Zaniewski et al. (2002) in a study of New Zealand ferns. It has the advantage of allowing the influence of more subtle habitat variables to be drawn out since the comparison is not swamped by substantial habitat differences. For instance, subtle differences in vegetation structure between fern-present and fern-absent sites might be masked if all fern-absent sites were, randomly, at a higher elevation than fern-present sites. The disadvantage is that more detailed information is needed about the species and the site to initially determine what areas could potentially be suitable.

5. **Exploring the intricacies of 'presence':** if 'presence' can be refined in some way, for example, to quantify abundance or density, it then becomes possible to compare sites where a species is abundant with areas where it is not, thereby circumventing the lack of absence data. This approach was used in Madagascar to compare the habitat characteristics of forest with a high density of blue-eyed black lemurs *Eulemur macaco flavifrons*, with areas with a low density (Schwitzer et al., 2007).

The statistics involved in comparing habitat use with availability are relatively simple goodness-of-fit tests, such as chi-squared tests or G-tests. For example, Jaquet & Gendron (2002) used this method to determine whether sperm whales *Physeter microcephalus* were uniformly distributed with respect to different marine habitats in California. When it comes to modelling the difference in habitat parameters between presence and pseudo-absence sites, analyses are rather more complicated. If there are only one or two habitat variables to compare, it is possible to use one-at-a-time using tests like *t*-tests that can compare the mean tree density (say) for sites with and without the focal species. If a whole range of habitat variables need to be examined, multivariate logistic regression is often best. This examines the effect of many habitat variables at the same time (multivariate literally means multiple variables) on a simple 0/1 dependent variable where 0 = the species is absent (or pseudo-absent) and 1 = the species is present.

There are two excellent papers that consider these issues and techniques in more detail specifically for species–habitat association modelling for Applied Ecologists: Pearce & Boyce (2006) provide a general review, while Redfern et al. (2006) provide a good overview for the marine environment with useful and easily accessible information on the statistics behind some of the approaches.

FURTHER READING FOR SECTION

The need for evidence-informed conservation: Cook, C.N., Hockings, M., & Carter, R.W. (2009) Conservation in the dark? The information used to support management decisions. *Frontiers in Ecology and the Environment*, Volume 8, 181–186.

The need for evidence-informed conservation: Arlettaz, R., Schaub, M., Fournier, J., Reichlin, T.S., Sierro, A., Watson, J.E.M., & Braunisch, V. (2010) From publications to public actions: when conservation biologists bridge the gap between research and implementation. *BioScience*, Volume 60, 835–842.

Importance of specialist practitioner advisory groups in evidence-informed conservation: Ewen, J.G., Adams, L., & Renwick, R. (2013) New Zealand Species Recovery Groups and their role in evidence-based conservation. *Journal of Applied Ecology*, Volume 50, 281–285.

Drawing on different types of evidence: Raymond, C.M., Fazey, I., Reed, M.S., Stringer, L.C., Robinson, G.M., & Evely, A.C. (2010) Integrating local and scientific knowledge for environmental management. *Journal of Environmental Management*, Volume 91, 1766–1777.

Excellent repository of information spanning all taxa all over the world: Available at: http://www.conservationevidence.com/. This is synthesized from scientific papers that have an applied conservation focus and is fully indexed.

11.8 Conclusions.

In situ conservation is a complex part of Applied Ecology that can involve active management and custodial approaches to safeguard habitats and protect species. For habitats, active approaches include creating new habitat, maintaining, or improving existing habitat, restoring previous habitat, or joining habitat together to increase connectivity. Active management of species can involve conservation translocations and reintroductions, supplemental feeding, and culling. Custodial measures include site-, species- and activity-focused legislation, as well as policy initiatives such as Biodiversity Action Plans. Wider countryside initiatives such as agri-environment schemes and living landscape approaches are increasingly being championed to move *in situ* conservation away from isolated patches protected as nature reserves and into a landscape context. The concept of *in situ* conservation being evidence-based with decisions being underpinned by empirical data is gradually becoming more widespread, but Applied Ecologists still have some way to go before this becomes the norm, especially at a local level.

ONLINE ACTIVITY

Go to www.oxfordtextbooks.co.uk/orc/goodenough/ to download the activity that accompanies this chapter.

CHAPTER 11 AT A GLANCE: THE BIG PICTURE

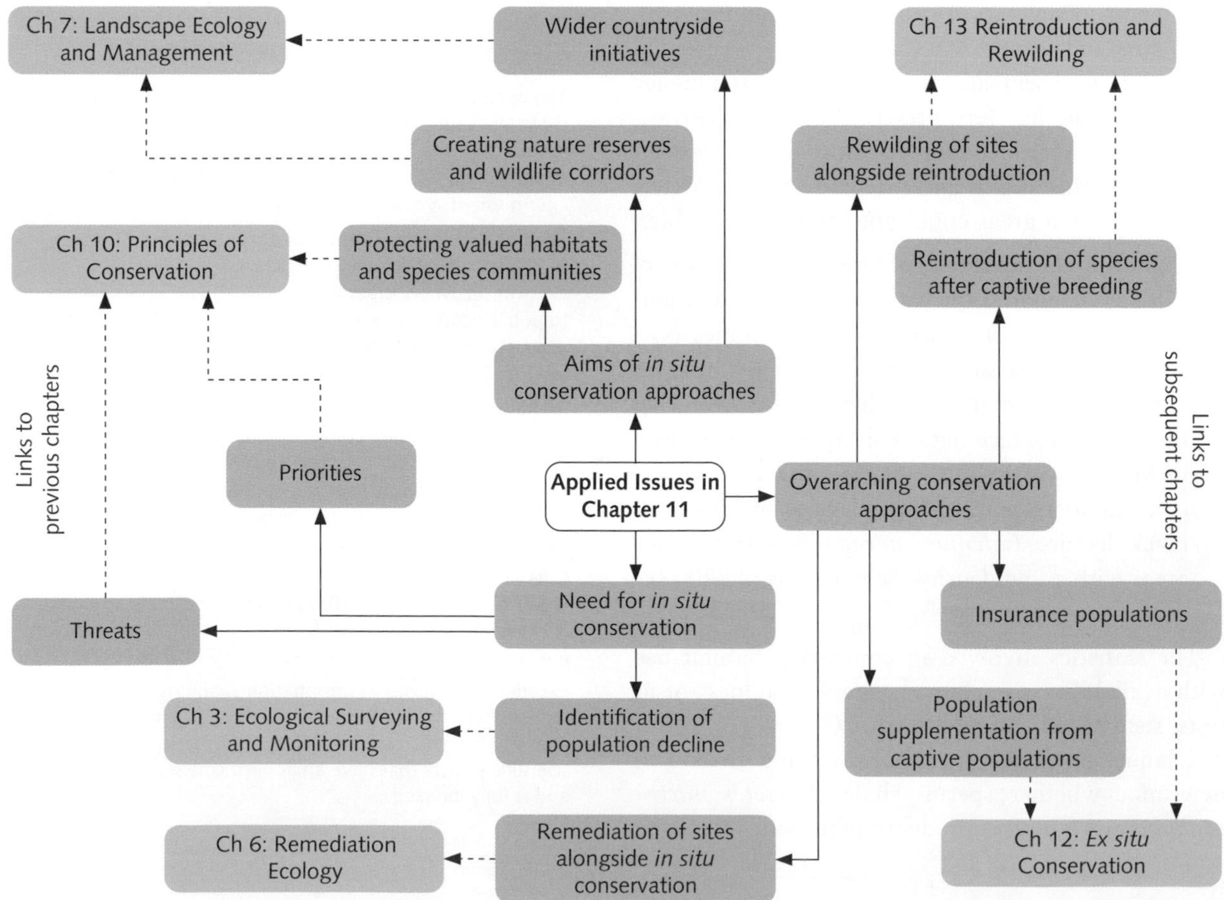

REFERENCES

Aebischer N.J. & Potts G.R. (1998) Spatial changes in grey partridge (*Perdix perdix*) distribution in relation to 25 years of changing agriculture in Sussex, U.K. Gibier Faune Sauvage. *Game Wildlife*, Volume 15, 293–308.

Araújo, M.B., Cabeza, M., Thuiller, W., Hannah, L., & Williams, P.H. (2004) Would climate change drive species out of reserves? An assessment of existing reserve selection methods. *Global Change Biology*, Volume 10, 1618–1626.

Baz, A. & Garcia-Boyero, A. (1996) The SLOSS dilemma: a butterfly case study. *Biodiversity and Conservation*, Volume 5, 493–502.

Berkes, F. (2004) Rethinking community-based conservation. *Conservation Biology*, Volume 18, 621–630.

Carpenter, G., Gillison, A.N., & Winter, J. (1993) DOMAIN: a flexible modelling procedure for mapping potential distributions of plants and animals. *Biodiversity and Conservation*, Volume 2, 667–680.

Cook, C.N., Hockings, M., & Carter, R.W. (2009) Conservation in the dark? The information used to support management decisions. *Frontiers in Ecology and the Environment*, Volume 8, 181–186.

Cope, D.R., Iason, G.R., & Gordon, I. J. (2004) Disease reservoirs in complex systems: a comment on recent work by Laurenson, M.K., Norman, R.A., Gilbert, L. Reid H.W. and Hudson, P.J. *Journal of Animal Ecology*, Volume 73, 807–810.

Davies, Z.G., Fuller, R.A., Loram, A., Irvine, K.N., Sims, V., & Gaston, K.J. (2009) A national scale inventory of resource provision for biodiversity within domestic gardens. *Biological Conservation*, Volume 142, 761–771.

Debus S.J.S. (2008) The effect of noisy miners on small bush birds: an unofficial cull and its outcome. *Pacific Conservation Biology*, Volume 14, 185–190.

Diamond, J.M. (1975) The island dilemma: lessons of modern biogeographic studies for the design of natural reserves. *Biological Conservation*, Volume 7, 129–145.

Donihue, C.M., Porensky, L.M., Foufopoulos, J., Riginos, C., & Pringle, R.M. (2013) Glade cascades: indirect legacy effects of pastoralism enhance the abundance and spatial structuring of arboreal fauna. *Ecology*, Volume 94, 827–837.

Fertl, D., Jefferson, T.A., Moreno, I.B., Zerbini, A.N., & Mullin, K.D. (2003) Distribution of the Clymene dolphin *Stenella clymene*. *Mammal Review*, Volume 33, 253–271.

Forman, R.T.T. and Godron, M. (1986) *Landscape Ecology*. Chichester: John Wiley and Sons.

Gaston, K.J., Warren, P.H., Thompson, K., & Smith, R.M. (2005) Urban domestic gardens (IV) The extent of the resource and its associated features. *Biodiversity and Conservation*, Volume 14, 3327–3349

Gaston, K.J., Charman, K., Jackson, S.F., Armsworth, P.R., Bonn, A., Briers, R.A., Callaghan, C.S.Q., Catchpole, R., Hopkins, J., Kunin, W.E., Latham, J., Opdam, P., Stoneman, R., Stroud, D.A., & Tratt, R. (2006) The ecological effectiveness of protected areas: the United Kingdom. *Biological Conservation*, Volume 132, 76–87.

Grey, M.J., Clarke, M.F., & Loyn, R.H. (1997) Initial changes in the avian communities of remnant eucalypt woodlands following reduction in the abundance of Noisy Miners *Manorina melanocephala*. *Wildlife Research*, Volume 24, 631–648.

Hannah, L., Midgley, G., Andelman, S., Araújo, M., Hughes, G., Martinez-Mayer, E., Pearson, R., & Williams, P. (2007) Protected area needs in a changing climate. *Frontiers in Ecology and the Environment*, Volume 5, 131–138.

Harrigan (2011) Relics to reefs: why fish can't resist sunken ships, tanks, and subway cars. *National Geographic*. Available at: http://ngm.nationalgeographic.com/2011/02/artificial-reefs/harrigan-text/1

Jaquet, N. & Gendron, D. (2002) Distribution and relative abundance of sperm whales in relation to key environmental features, squid landings and the distribution of other cetacean species in the Gulf of California, Mexico. *Marine Biology*, Volume 141, 591–601.

Jones, C.G., Heck, W., Lewis, R.E Mungroo, Y., Slade, G., & Cade, T. (1995) The restoration of the Mauritius Kestrel *Falco punctatus* population. *Ibis*, Volume 137, 173–180.

Katselidis, K.A., Schofield, G., Stamou, G., Dimopoulos, P., & Pantis, J.D. (2013) Evidence-based management to regulate the impact of tourism at a key marine turtle rookery on Zakynthos Island, Greece. *Oryx*, Volume 47, 584–594.

Knick, S.T. & Dyer, D.L. (1997) Distribution of black-tailed jackrabbit habitat determined by GIS in Southwestern Idaho. *Journal of Wildlife Management*, Volume 61, 75–85.

Krüger, O. (2005) The role of ecotourism in conservation: panacea or Pandora's box? *Biodiversity and Conservation*, Volume 14, 579–600.

Laurenson, M.K., Norman, R.A., Gilbert, L., Reid, H.W., & Hudson P.J. (2003) Identifying disease reservoirs in complex systems: mountain hares as reservoirs of ticks and louping-ill virus, pathogens of red grouse. *Journal of Animal Ecology*, Volume 72, 177–185.

Lõhmus, A. & Remm, J. (2004) Nest quality limits the number of hole-nesting passerines in their natural cavity-rich habitat. *Acta Oecologia*, Volume 27, 125–128.

López-Bao, J.V., Palomares, F., Rodríguez, A., & Delibes, M. (2010) Effects of food supplementation on home-range size, reproductive success, productivity and recruitment in a small population of Iberian lynx. *Animal Conservation*, Volume 13, 35–42.

Magella, G. & Brousseau, P. (2001) Does culling predatory gulls enhance the productivity of breeding common terns? *Journal of Applied Ecology*, Volume 38, 1–8.

McNeill, S.E. & Fairweather, P.G. (1993) Single large or several small marine reserves? An experimental approach with seagrass fauna. *Journal of Biogeography*, Volume 20, 429–440.

Muehlenbein, M.P., Martinez, L.A., Lemke, A.A., Ambu, L., Nathan, S., Alsisto, S., & Sakong, R. (2010) Unhealthy travelers present challenges to sustainable primate ecotourism. *Travel Medicine and Infectious Disease*, Volume 8, 169–175.

National Parks Service (2003) Yellowstone National Park: Site Progress Report to the World Heritage Committee. Available at: http://www.nps.gov/yell/planyourvisit/upload/whcreport10-03.pdf

Newton, I. (1994) The role of nest sites in limiting the numbers of hole-nesting birds: a review. *Biological Conservation*, Volume 70, 265–276.

Nielsen, H. & Spenceley, A. (2011) *The Success of Tourism in Rwanda: Gorillas and More*. Washington DC: World Bank.

Orams, M.B. (2002) Feeding wildlife as a tourism attraction: a review of issues and impacts. *Tourism Management*, Volume 23, 281–293.

Osborne, P.E., Alonso, J.C., & Bryant, R.G. (2001) Modelling landscape-scale habitat use using GIS and remote sensing: a case study with great bustards. *Journal of Applied Ecology*, Volume 38, 458–471.

Pearce, J.L. & Boyce, M.S. (2006), Modelling distribution and abundance with presence-only data. *Journal of Applied Ecology*, Volume 43, 405–412.

Powell, R.B. & Ham, S.H. (2008) Can ecotourism interpretation really lead to pro-conservation knowledge, attitudes and behaviour? Evidence from the Galapagos Islands. *Journal of Sustainable Tourism*, Volume 16, 467–489.

Purcell, K.L., Verner, J., & Oring, L.W. (1997) A comparison of the breeding ecology of birds nesting in boxes and tree cavities. *Auk*, Volume 114, 646–656.

Pywell, R.F., Bullock, J.M., Hopkins, A., Walker, K.J., Sparks, T.H., Burke, M.J., & Peel, S. (2002) Restoration of species-rich grassland on arable land: assessing the limiting processes using a multi-site experiment. *Journal of Applied Ecology*, Volume 39, 294–309.

Rasmussen, J.B. (1988) Habitat requirements of burrowing mayflies (Ephemeridae: Hexagenia) in lakes, with special reference to the effects of eutrophication. *Journal of the North American Benthological Society*, Volume 7, 51–64.

Redfern, J.V., Ferguson, M.C., Becker, E.A., Hyrenbach, K.D., Good, C., Barlow, J., Kaschner, K., Baumgartner, M.F., Forney, K.A., Balance, L.T., Fauchald, P.,Halpin, P., Hamazaki, T., Pershing, A.J., Qian, S.S., Read, A., Reilly, S.B., Torres, L., & Werner, F.E. (2006) Techniques for cetacean-habitat modeling. *Marine Ecology Progress Series*, Volume 310, 271–295

Redpath S.M., Thirgood S.J., & Leckie F.M. (2001) Does supplementary feeding reduce predation of red grouse by hen harriers? *Journal of Applied Ecology*, Volume 38, 1157–1168.

Robles, M.D., Flather, C.H., Stein, S.M., Nelson, M.D., & Cutko, A. (2008) The geography of private forests that support at-risk species in the conterminous United States. *Frontiers in Ecology and the Environment*, Volume 6, 301–307.

Roman, G.S., Dearden, P., & Rollins, R. (2007) Application of zoning and 'limits of acceptable change' to manage snorkelling tourism. *Environmental Management*, Volume 39, 819–830.

Russo, D., Cistrone, L., Jones, G., & Mazzoleni, S. (2004) Roost selection by barbastelle bats (*Barbastella barbastellus*, Chiroptera: Vespertilionidae) in beech woodlands of central Italy: consequences for conservation. *Biological Conservation*, Volume 117, 73–81.

Schwitzer, N., Randriatahina, G.H., Kaumanns, W., Hoffmeister, D., & Schwitzer, C. (2007) Habitat utilization of blue-eyed black lemurs, *Eulemur macaco flavifrons* (Gray, 1867), in primary and altered forest fragments. *Primate Conservation*, Volume 22, 79–87.

Simberloff, D. & Abele, L.G. (1976) Island biogeography theory and conservation practice. *Science*, Volume 191, 285–286.

Stockhausen, W.T., & Lipcius, R.N. (2001) Single large or several small marine reserves for the Caribbean spiny lobster? *Marine and Freshwater Research*, Volume 52, 1605–1614.

Twedt, D.J. & Henne-Kerr, J.L. (2001) Artificial cavities enhance breeding bird densities in managed cottonwood forests. *Wildlife Society Bulletin*, Volume 29, 680–687.

Vander Lee B.A., Lutz R.S., Hansen L.A., & Matthews N.E. (1999) Effects of supplemental prey, vegetation, and time on success of artificial nests. *Journal of Wildlife Management*, Volume 63, 1299–1305.

Wakeham-Dawson A. (1995) Hares and skylarks as indicators of environmentally sensitive farming on the South Downs. PhD thesis, The Open University, UK.

Walker, P.A. & Cocks, K.D. (1991) HABITAT: a procedure for modelling a disjoint environmental envelope for a plant or animal species. *Global Ecology and Biogeography Letters*, Volume 1, 108–118.

Warren, M.S., Thomas, C.D., & Thomas, J.A. (1984) The status of the heath fritillary *Mellicta athalia* Rott. in Britain. *Biological Conservation*, Volume 29, 287–305.

Webb, N.R. (1998) The traditional management of European heathlands. *Journal of Applied Ecology*, Volume 35, 987–990.

Williams, R.L., Stafford, R., & Goodenough, A.E. (2014) Biodiversity in urban gardens: assessing the accuracy of citizen science data on garden hedgehogs. *Urban Ecosystems*, Volume 18, 819–833.

World Bank (2015) Terrestrial Protected Areas. Available at: http://data.worldbank.org/indicator/ER.LND.PTLD.ZS

Zaniewski, A.E., Lehmann, A., & Overton, J.M. (2002) Predicting species spatial distributions using presence-only data: a case study of native New Zealand ferns. *Ecological Modelling*, Volume 157, 261–280.

Zarnetske, P.L., Edwards Jr, T.C., & Moisen, G.G. (2007) Habitat classification modeling with incomplete data: pushing the habitat envelope. *Ecological Applications*, Volume 17, 1714–1726.

Ex Situ Conservation

12.1 Introduction.

Human activities can degrade habitats and cause negative effects for species, including population decline and range contraction (Macintyre & Hobbs, 1999). Negative effects may reach a point at which active management strategies must be employed to reduce or reverse them. Strategies that conserve species and habitats are the tools of **conservation ecology**, a highly applied and increasingly essential component of modern ecology (Chapter 10).

In Chapter 10 the broad classification of conservation strategies as *in situ* or *ex situ* was introduced. *In situ*, or 'on site' conservation is very much focused on the site, or sites, in question. It might include legislative management, such as the establishment of specific legal protection, as well as active land management strategies, such as planting, burning, grazing, and culling. These are collectively designed to increase the **biodiversity** of sites or to conserve a specific species (or assemblage of species) of particular concern (Chapter 11). *In situ* approaches may also need to deal with **human-wildlife conflicts** and education (Chapter 10), as well as site **remediation** (Chapter 6) and **landscape ecology** (Chapter 7) in what becomes a complex multifactorial process that seeks to manage habitat, wildlife, humans, and the ecological interrelationships between them.

Ex situ means, literally, 'off site' and, in some respects, is conceptually more straightforward than *in situ* conservation (Engelmann & Engels, 2002). Individuals of species threatened in their natural habitats can be removed from those habitats (and, therefore, from the threat) and a breeding population can be established at a different, safe location eventually to provide individuals for reintroduction at the original location or a different site (for example, see Hot Topic 13.3). For some species this process may be quite straightforward, but for other species, such an approach may in practice be extremely difficult.

If the eventual goal of *ex situ* conservation is species **reintroduction** (Chapter 13) then a successful *ex situ* conservation programme will generally require *in situ* conservation techniques being employed in parallel (including legislative management, human–wildlife conflict resolution, and land management). It is worth noting though that in some cases animal species may have been put in captivity or and established captive breeding populations long before it was realized that natural populations were endangered. Likewise, some plant species have been successfully cultivated before any threats to their status had been identified. However, if this is not the case, and *ex situ* is being considered as a last resort, then organisms need to be taken from already fragile wild source populations that may be better conserved *in situ* (Dolman et al., 2015). In the future, the increasing combination of *in situ* and *ex situ* approaches may lead to the distinction between them becoming misleading or even obsolete (as discussed by Pritchard et al., 2012; Braverman, 2013). Currently, however, considering *in situ* and *ex situ* as separate but connected concepts is useful.

This chapter considers the need for *ex situ* conservation and some of the complexities of this approach and uses a case study of the Lord Howe Island stick insect *Dryococelus australis*, a particularly dramatic *ex situ* example, throughout the chapter. Additional examples will be used to illustrate specific points, but the Lord Howe Island stick insect usefully illustrates the need for *ex situ* conservation, the problems and issues that the approach must overcome, and the often necessarily close relationship between *in situ* and *ex situ* approaches. It also illustrates some of the problems of reintroduction, a topic that is explored more fully in Chapter 13.

12.2 *Ex situ* conservation: an overview.

Broadly, *ex situ* conservation can be subdivided into:

- **Bespoke and intensive captive breeding programmes:** these tend to be species-specific as part of an overall conservation action plan for that species, which is likely to be under urgent and considerable threat. The Panama Amphibian Rescue and Conservation Project (Figure 12.1), for example, was created in response to declines related to the invasive fungal pathogen *Batrachochytrium dendrobatidis*. It consists of two *ex situ* facilities in Panama that house populations of amphibians with the ultimate aim of creating 'insurance colonies' of 20 endangered amphibian species. These would have captive populations of a minimum effective population size of 500 individuals (Cikanek et al., 2014).
- **More general captive breeding programmes:** these are managed within multi-species collections, such as zoos, aquaria, and botanical gardens. Such programmes have to exist within a successful business, in many cases, and this can make the selection of captive-bred species difficult. Programmes need to balance conservation and economic priorities, with some species (large vertebrates) being more attractive to visitors but potentially harder to breed and reintroduce successfully and efficiently (Balmford et al., 1996).
- **'Gene banks':** genome resource banking is where seeds, pollen, living cells, tissues sperm, embryos, and DNA are preserved. This type of preservation is a long-term strategy and is exemplified by the Millennium Seed Bank Partnership (e.g. van Slageren, 2003).

Figure 12.1 The Panama Amphibian Rescue and Conservation Project is a bespoke and intensive facility for the captive breeding of endangered amphibians, including the golden frog *Atelopus zeteki*.

In practice there may be considerable overlap between all three subdivisions and especially between those that involve captive breeding. In their recent history, many zoos, aquaria, and botanical gardens and collections have become actively involved in international conservation efforts through captive breeding programmes. The expertise and specialized facilities of zoos and similar organizations mean that they are important *ex situ* locations for bespoke captive breeding programmes. In addition, such institutions raise both funds and awareness for conservation, which are significant activities in their own right.

The roles of an *ex situ* programme are considered to fall into four categories that also illustrate the considerable overlap with what might be regarded as *in situ* approaches. They are:

- addressing the causes of primary threats;
- offsetting the effects of threats;

- buying time;
- restoring wild populations (IUCN/SSC, 2014).

More specifically, these roles (taken from IUCN/SSC, 2014) might include:

- Developing an **insurance population** (to maintain a viable *ex situ* population of the species to prevent local, regional or complete extinction and maintain future conservation options).
- Effecting a **temporary rescue** to protect from catastrophes or predicted imminent threats (e.g. extreme weather, disease, oil spill, wildlife trade).
- **Maintaining a long-term *ex situ* population** after extinction of all known wild populations as a preparation for reintroduction or assisted colonization if and when feasible.
- **Manipulating demography,** for example, by temporarily removing individuals from the wild to reduce mortality during a specific life stage.
- Providing a **source for population restoration,** reintroducing the species into part of its former range or reinforcing an existing population.
- Providing a **source for ecological replacement,** re-establishing lost ecological functions. This approach may involve species that are not themselves threatened, but that contribute to the conservation of other taxa through their ecological role.
- Providing a **source for assisted colonization,** introducing a species outside of its indigenous range.
- Undertaking **research** that will directly benefit conservation of the species, or a similar species, in the wild.
- Acting as a basis for **education and awareness** that addresses specific threats to the conservation of the species or its habitat.

As with other aspects of Applied Ecology, *ex situ* conservation works best when it is built on a firm foundation of ecological theory. With *ex situ* conservation, that foundation includes insights gleaned from the entire ecological discipline, from knowledge of individual ecological requirements and their interactions with one another, to landscape ecology, habitat dynamics, and ecosystem functioning.

Conceptually, *ex situ* conservation undertaken as a last resort, rather than taking advantage of existing captive or cultivated populations involves at least three distinct phases regardless of the species involved:

1. Individuals must be **captured** or **collected** from the threatened population and then **transported** to a different, *ex situ*, location.
2. Animals must be kept alive and established in a **breeding programme** to produce more individuals, while plants must be cultivated with the aim of producing seeds. Note that animal sperm or other samples may be maintained as frozen samples and plant seeds may be kept in a dormant state. These strategies are discussed in Section 12.5.
3. Captive-bred or cultivated individuals, and potentially individuals that were removed from the threatened site, must then be transported to the original site (but only when it is ready to receive them), or to some other suitable habitat, and re-established. **Reintroduction** is the release of organisms into an area of their former range, from which they have been extirpated completely. If additional individuals are released to boost an existing population, this is termed **reinforcement** (unless additional individuals are being added to a population that was originally reintroduced, in which case it is **supplementation**) (Figure 13.1).

Clearly, the phases outlined above are risky (these risks are discussed in detail in Section 12.3 and 12.4) and there are many potential points of failure. Not only could *ex situ* conservation attempts fail, in extreme cases it is easy to see how failure could drive a species to extinction if a large proportion of the remaining individuals was taken. As well as being risky from a conservation and biodiversity perspective, *ex situ* conservation is expensive and requires considerable technical support for success.

Before undertaking *ex situ* conservation, it is first necessary to decide whether it is appropriate. The International Union for the Conservation of Nature (IUCN) has published guidelines for *ex situ* conservation, in which it states clearly when such an approach should be applied, specifically in cases where:

stakeholders can be confident that the expected positive impact on the conservation of that species will outweigh the potential risks or any negative impact (which could be to the local population, species, habitat or ecosystem), and that its use will be a wise application of the available resources. This requires an assessment of the potential net positive impact, weighted by how likely it is that this potential will be realized, given the expertise, level of difficulty or uncertainty, and available resources.

(IUCN/SSC, 2014)

To support stakeholders in making this decision, IUCN outline a five-step process. First, a threat analysis should be used to inform a detailed status review of the species in question (Dolman et al., 2015; Section 10.3). Secondly, the role that *ex situ* conservation will play in the overall conservation of the species should be clearly defined. Thirdly, the characteristics of the *ex situ* programmes should be determined with respect to the number and type of individuals involved and the duration of the programme, as well as the practical considerations of the proposal (these are discussed in detail in Sections 12.3 and 12.4). Fourthly, the expertise and resources required for success should be evaluated, and the feasibility and risks should be appraised (discussed in Section 12.4). Finally, a decision should be made that is informed (by making use of information gathered in the first four steps) and transparent; in other words so that it is clear how and why the decision was reached.

All stages of *ex situ* conservation have their own challenges, but overarching the entire endeavour is the requirement to obtain legal permission and ethical clearance for the project. In some cases, the *ex situ* programme will be undertaken in a legal jurisdiction different to the home site, either internally within a country or within a different country. For example, the Amur leopard *Panthera pardus orientalis* (Figure 12.2), has a total wild population estimated to be 30–35, but there are around 180 individuals acting as a captive and breeding *ex situ* insurance population in zoos all around the world (WAZA, 2015).

Licensing and permissions can be difficult, expensive, and time-consuming to obtain, especially if the collection site is in a different country to the captive breeding programme. Licenses and permissions may well be required to:

Figure 12.2 The captive population of Amur leopards *Panthera pardus orientalis* is more than five times the estimated wild population and acts as an *ex situ* insurance population.

Source: Image courtesy of Marie Hale/CC BY 2.0.

- Enter the country to conduct scientific research or to enter as a non-tourist.
- Enter a specific collection site.
- Collect and handle organisms within that site.
- Transport organisms within the country of collection.
- Export live organisms or seeds from the country of collection.
- Transport live organisms on a plane or ship, or across a land border.
- Have live organisms in transit through intervening countries.
- Import live organisms into the destination country.
- House and breed animals within a captive breeding facility or cultivate plants at an *ex situ* location.

Reintroduction programmes will probably require these permissions in reverse, as well as requiring permission to introduce organisms into the destination country. The legal work involved can be considerable, and might well include operating in a foreign language within an unfamiliar, and not necessarily stable, legal and political structure.

The ethics of *ex situ* programmes also require careful consideration. In essence, *ex situ* projects involve the removal of already threatened species from one area to be housed artificially in a captive breeding programme (in the case of animals) or cultivated within an artificial or alien environment (in the case of plants) with an uncertain chance of success. Animals and plants do not often respond well to being collected and transported, and any intervention that will possibly cause stress or potentially harm the organisms involved should be subjected to consideration by a suitably qualified **Ethics Panel**. When that intervention is also likely to cause a reduction in the wild population of a species then ethical consultation needs to be especially rigorous. For animals, it is also vital that wild-caught individuals and their captive-bred offspring are held in appropriate conditions and that any captivity-related stress or habituation are minimized. Such issues can be particularly complex for species that traditionally live in complex social groups (e.g. wolves *Canis lupus*) or that have large home ranges or nomadic lifestyles (e.g. Thomson's gazelle *Eudorcas thomsoni*, and whale sharks *Rhincodon typus*). The ethical considerations of cultivating plants may be reduced, but nonetheless it is necessary to ensure that plants are kept in conditions that, at the very least, do not cause harm to the individuals that have been collected, and ideally lead to the production of viable seeds or vegetative tissue.

There are also broader ethical considerations with *ex situ* conservation. First, the removal of individuals of a given species may have knock-on effects to other, potentially endangered, species involved in ecological interactions. For example, some plants have co-evolved with individual pollinator species (especially insects) and the removal of either partner from the natural habitat would be to the detriment of the other. Secondly, the eventual goal of such of programmes is to put captive-bred individuals back into the wild, and each step poses a risk to the individuals involved and, by extension, to an already fragile population.

Careful assessments of the current risks to the species *in situ* and of the risks to the individuals involved in *ex situ* projects, as well the likelihood of success, form the basis for an action plan, but are also essential in ensuring the project, is ethically sound. The considerable investment of money, time, and expertise required might be better placed in other conservation projects and deciding where best to place effort is far from straightforward. The final decision may be influenced by ecological insights but political considerations and subjective value judgements placed on the species concerned may also play a role (for example see Hot Topic 13.3).

FURTHER READING FOR SECTION

An overview of *ex-situ* conservation: Kasso, M. & Balakrishnan, M. (2013) *Ex situ* conservation of biodiversity with particular emphasis to Ethiopia. *ISRN Biodiversity*, Article ID: 985,037.

12.3 The Lord Howe Island stick insect: *ex situ* conservation in action.

The Lord Howe Island Group lies between Australia and New Zealand, about 600 km east of Port Macquarie, New South Wales. Twenty-eight islands, islets, and rocks make up the group, but the only large island is Lord Howe Island, which is crescent-shaped, about 10 km long and no more than 2 km wide. The island was once home to a thriving population of *Dryococelus australis*, a large, heavily built, flightless stick insect commonly known as the Lord Howe Island stick insect or, more descriptively, the tree- or land-lobster (Figure 12.3).

In 1918 a supply ship ran aground and as a consequence non-native black rats *Rattus rattus* were introduced to the island. As is a common occurrence with non-native species (Chapter 8), the rats found a plentiful supply of easily captured prey unused to such voracious predators and the stick insect population crashed. The last individual was sighted on the island in 1920, just two years after the rats' arrival. It seemed that the Lord Howe Island stick insect had undergone very rapid population decline and, ultimately, extinction. Moreover, because the species was **endemic** to the island, loss of the species in this one

Figure 12.3 The impressive Lord Howe Island stick insect *Dryococelus australis*, showing male on the right and female on the left.

Source: Used with kind permission from Rohan Cleave, Melbourne Zoo.

location was considered complete extinction, rather than local **extirpation**.

However, the story of this enigmatic insect does not end there. One of the islands in this group is a rocky sea stack called Ball's Pyramid, lying in the sea some 25 km from the main island. Thrusting more than 550 m out of the sea this remnant of a once active volcano attracted climbers in the 1960s keen to become the first to summit the pyramid of rock (Figure 12.4). In 1964, one such team of climbers discovered a dead Lord Howe Island stick insect.

Other climbers found a few more dead insects on later expeditions, but it took nearly 40 years for a living stick insect to be found. In 2001, two Australian scientists exploring the island found large insect droppings under a *Melaleuca howeana* shrub, a type of tea-tree endemic to the Lord Howe Island Group. One of the scientists, together with a local ranger, returned to the shrub after dark, a sojourn that can best be described as 'dedicated' given the terrain of Ball's Pyramid. Their dedication paid off, and that night they discovered two adult Lord Howe Island stick insects and one nymph (an immature individual), all female, living in the accumulated plant debris under that bush. A second expedition in 2002 located a population of 24 stick insects, 12 living beneath that solitary *Malaleuca* shrub and a further 12 dispersed among five nearby shrubs. The overall population was, at that time, estimated to be fewer than 40 individuals (Honan, 2008).

Figure 12.4 Ball's Pyramid is the dramatic location of the last surviving population of the Lord Howe Island stick insect.

Source: Photography by Rohan Cleave, Melbourne Zoo, used with kind permission

Quite how this remnant population made it to Ball's Pyramid is a mystery and their survival, in the plant debris nestled beneath a few shrubs on the precipitous cliffs, seems nothing short of miraculous. However, the fact they had survived offered conservation ecologists a unique opportunity: Lord Howe Island could be repopulated with these charismatic endemic insects.

It is clear that a population of 24 insects living under a few shrubs on what amounts to a small mountain jutting up from the open ocean, at best, is fragile. A rock falling from above, a storm, a hungry sea bird, or some pathogenic fungal spores borne on the wind could cause rapid and complete extinction. Other risks identified in the Department of the Environment Species Profile and Threats Database include the invasive vine Morning Glory *Ipomoea cairica*, which poses a threat to *Malaleuca*, the only known food plant of the stick insect, and the possible introduction of non-native invertebrates, such as the red imported fire ant *Solenopsis invicta* (Department of the Environment, 2015). Indeed, the potential danger caused by non-native species is amply demonstrated by the effect of the black rats introduced almost a century before.

Interestingly, another identified threat is poaching. Whilst insects might not seem like an obvious target of illegal hunting, rare insects are highly desirable to collectors both as living specimens and as dead, mounted exhibits. The remoteness of Ball's Pyramid makes illegal collecting unlikely, but if someone were to put a sufficiently high price on a specimen then poaching would become another threat that the Applied Ecologist must consider.

Whilst it is possible that other populations exist somewhere (although none have yet been found), conservation efforts in this case proceeded under the assumption that the few insects on Ball's Pyramid were the only individuals that remained. The Lord Howe Island Biodiversity Plan (Department of Environment and Climate Change (NSW), 2007) recognizes the need for priority recovery actions specifically for the Lord Howe Island stick insect, and it includes the need to maintain captive breeding programmes (note the plural), and reintroduce the stick insect back to the main island.

As both the captive breeding programmes and the reintroduction attempts occur at a site other than Ball's Pyramid, they are *ex situ* conservation. Other examples of *ex situ* approaches will be introduced throughout the chapter, but the Lord Howe Island stick insect conservation programme will act as an on-going case study to illustrate the steps of *ex situ* conservation and to explore some of the challenges the approach presents.

FURTHER READING FOR SECTION

Information on the Lord Howe Island stick insect, its rediscovery, and conservation priorities: Priddel, D., Carlile, N., Humphrey, M., Fellenberg, S., & Hiscox, D. (2003) Rediscovery of the 'extinct' Lord Howe Island stick-insect (*Dryococelus australis* (Montrouzier)) (Phasmatodea) and recommendations for its conservation. *Biodiversity and Conservation*, Volume 12, 1391–1403.

12.4 The stages of *ex situ* conservation.

Once an immediate and serious threat is clearly established and it has been decided that an *ex situ* approach is justified (as discussed in Section 12.2 and IUCN/SSC (2014)), carefully planning is required. For situations where *ex situ* conservation is being undertaken as a last resort, as was the case with the Lord Howe Island stick insect, then each of the three stages of collection, captive breeding, and reintroduction, reinforcement or supplementation need to be considered. Plants may present different challenges (for example transport of seeds is considerably more straightforward than transporting many animals) but conceptually the stages are similar. As discussed in Section 12.5 there may also be options for long-term storage of tissue and other material in gene banks, but this is considerably easier for plants (many seeds can be stored for long periods without substantially affecting their viability) than for animals.

12.4.1 Collection.

To establish a new captive breeding programme for animal species or a new *ex situ* cultivated population of plants, the conservation ecologist must first collect some individuals of the species at risk. It may be that the status review compiled in the initial planning stage has identified existing *ex situ* individuals that could be suitable for establishing a new programme, but in other cases it may be necessary to collect individuals from wild populations.

A number of important issues must be resolved prior to collection, and resolving them requires extensive consultation with different expert individuals and groups. Such consultation will take time but, in some cases, the threats to a species or its habitat may be considerable and immediate. In such situations there may not be much time to deliberate or consult and it may be necessary to prioritize action. However, in reality such cases are unusual, and wherever possible a well-informed and strategic approach should be the goal; after all, there may not be a second chance. Despite the obvious problems should things go wrong and clear guidelines provided by IUCN (IUCN/SSC, 2014), some conservationists (e.g. Dolman et al., 2015) are critical of the lack of rigorous evaluation taking place in many *ex situ* conservation programmes.

Key decisions in the initial collection phase include:

- How many individuals to collect.
- Which individuals to collect.
- When to collect.
- How to collect.

How many individuals to collect

With a small and vulnerable population, removing individuals to an off-site captive breeding programme whose success is uncertain, is a clear and considerable risk to the continued existence of the species. With very small populations the removal of even a few individuals can cause a proportionally high overall population reduction. In the case of Lord Howe Island stick insects, for example, captive breeding requires that a population that has survived unmolested for many years will be reduced by a minimum of 4% (the removal of one gravid female) and, more realistically, by a minimum of 8% (removing a male and female, although this assumes that both are fertile).

In some cases, it might be deemed necessary to remove all individuals to the relative safety of captivity. One such case is the California Condor *Gymnogyps californianus* and this is described in Case Study 12.1.

As well as the numerical effect that removing individuals has on a population, there may be more drastic, and ultimately catastrophic, effects. For instance, it is possible that some animal species aggregate at night or during the winter for warmth or to conserve water, as is the case with some invertebrates (Whitehouse, 2008), or perhaps they hunt collectively (as is seen in African wild dogs *Lycaon pictus*: Creel & Creel, 1995) as shown in Figure 12.5. In such cases, removal of a few individuals may cause knock-on effects for the survival of those that remain. Ideally, the ecology of the species in question will be well-characterized, and consultation with species experts has a good chance of revealing such issues. In those cases where knowledge is sparse, it may be necessary to undertake some field research or compare with similar species in similar habitats before deciding on the number of individuals to remove.

Ultimately deciding how many individuals to remove involves trading-off the continued survival and future reproductive success of the individuals if they remain at the home site against the chances of their survival and reproduction at the *ex situ* site. In many cases, it is a stated aim to establish captive-bred animals or cultivated plants back into the wild. The chances of this happening and the timescale in which it could occur also then play a part in balancing the trade-off between doing nothing, focusing solely on *in situ* conservation, or embarking on an *ex situ* programme in combination with *in situ* approaches. Knowledge is crucial in determining the course of action, but even for species where there is good understanding of their natural ecology and, ideally, of their behaviour and reproduction in captivity, there is no guarantee of success.

Which individuals to collect

The vast majority of animals reproduce sexually, and a captive breeding population will generally therefore require at least one male and one female to get established. Plants may be collected as seeds (and are certainly easier to transport that way), but germinating seeds, establishing seedlings, and growing

Figure 12.5 Animals may exhibit group behaviours that are crucial for survival. For example, groups might be important to prevent desiccation (e.g. littorinid snails), for warmth (e.g. ladybird aggregations), or for hunting (e.g. Africa wild dogs *Lycaon pictus*). Such behaviours should be taken into account when considering how many individuals to remove from a vulnerable population.

Source: Images (L-R) courtesy of Richard Stafford; Gilles San Martin; Lynne MacTavish/CC BY-SA 2.0.

CASE STUDY
12.1

The California condor—a captive breeding success story

The California condor *Gymnogyps californianus* is a New World vulture (Figure A). Weighing in at up to 12 kg and staying aloft thanks to a 3 m wing span, it is the largest North American land bird. A scavenger, the condor's numbers declined dramatically through the 20th century as a consequence of poaching, power cables, egg collecting, the feeding of trash to nestlings, habitat destruction, and lead poisoning. As a long-lived scavenger, condors are unusually vulnerable to the effects of the accumulation of lead in their bodies. Lead is a neurological toxin that can be fatal, and animals shot by hunters leave carcasses and spent lead-based ammunition that scavengers can ingest (Pattee et al., 1990). By 1987, only 22 condors remained in the wild (Snyder & Snyder, 2000).

In 1987 a radical *ex situ* conservation plan was put in place, based around a capture–breed–reintroduction model. In this case there was no need to select which individuals were to be captured because the California Condor Recovery plan, approved by the United States government, called for all remaining wild birds to be captured. This plan inevitably led to the species becoming extinct in the wild, and is an example of an unusually bold and brave conservation management decision. Were the captive breeding programme to fail then there would be no second chance for this iconic species.

Figure A The California condor *Gymnogyps californianus* is a New World Vulture.

Source: Image courtesy of Stacy Spensley/CC BY 2.0.

The captured birds were placed into captivity in the San Diego Zoo Safari Park and the Los Angeles Zoo, and put into a captive breeding programme with the twin aims of increasing the population and subsequently reintroducing birds back into the wild (Snyder & Snyder, 2000).

Condors do not lay many eggs, typically one every other year, but they can 'double clutch'. This means that if an egg is removed, the female will lay another to replace it. This allowed an initially slow captive breeding programme to gain momentum, since the initial egg could be removed and incubated artificially, with the resulting chick raised by humans. To prevent chicks from becoming habituated to human contact, keepers combined species-specific knowledge, husbandry techniques used in other species,

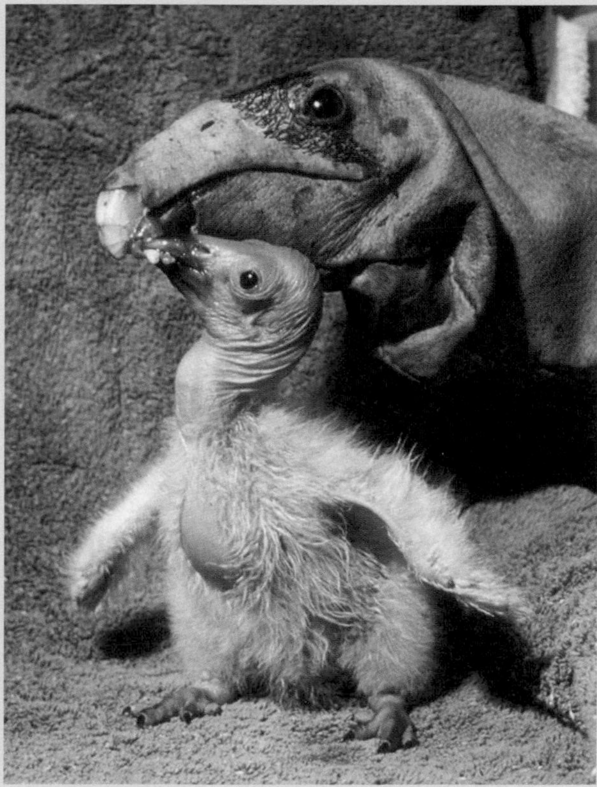

Figure B Hand-feeding a chick using a glove puppet.

Source: Image courtesy of San Diego Zoo, Ron Garrison.

and some imagination to construct realistic glove 'puppet parents' to feed the chicks (Figure B).

Reintroduction initially took a conservative, experimental approach. To determine protocols and to explore problems, captive-bred female Andean condors from South America were released in 1988 and following their successful recapture, captive-bred California condors were released in 1991 and 1992 in California, and 1996 in Arizona.

Although expensive (it is estimated to have cost more than 35 million USD), the programme has been successful, with the total population now exceeding 400 condors, more than half of which are living in the wild (U.S. Fish and Wildlife Service, 2014).

REFERENCE

Pattee, O.H., Bloom, P.H., Scott, J.M., & Smith, M.R. (1990) Lead hazards within the range of the California Condor. *Condor*, Volume 92, 931–937.

Snyder, N.F.R. & Snyder, H. (2000) *The California Condor: a Saga of Natural History and Conservation*. Princeton, NJ: Princeton University Press.

U.S. Fish and Wildlife Service (2014) *California Condor Gymnogyps californianus Recovery Program Population Size and Distribution*. Available at: http://www.fws.gov/cno/es/CalCondor/PDF_files/2014/Condor%20Program%20Monthly%20Status%20Report%202014-10-31.pdf

seedlings on to produce reproductive adult plants can be challenging. However, it may be that initially the aim is simply to place seeds within a seed bank facility (Section 12.5). Dioecious plants have male and female individuals, and to ensure a breeding population it will be necessary to rear both sexes from seeds. Furthermore, although some plants are monoecious, and have both male and female reproductive organs, many such plants are also self-incompatible. The number of seeds required will be dictated by the proportion of seeds likely to survive to reproduce and by the statistical likelihood of producing a favourable sex ratio for seed production.

A single gravid female or fertilized female plant might be sufficient to establish a population in some cases, but in only in animal species in which males are not essential for the survival of the young after hatching or birth. It is also possible that in species where low numbers of offspring or seeds are produced (so called **K-strategists**; Section 2.7.1), the offspring or seeds of a single female could all be of the same sex. In such cases, a single female may not be able to produce a sustainable captive population if a male is not also collected. It is worth noting that sexing some animals can be very difficult under field conditions, and might require the capture and careful, potentially invasive, examination of an individual. For example, the sex of smaller invertebrates like flies can often be distinguished, but only under a microscope. Capturing and handling animals has attendant risks both to the capturer and to the animal, and it should be kept to an absolute minimum. The services of somebody expert in the species who is able to identify the sex of an animal, ideally without capturing it, are invaluable to prevent unnecessary risk to an already threatened population.

For the Lord Howe Island stick insect, it was decided to remove four individuals in total, comprising two adult females and two adult males housed as two mixed sex pairs. These were collected from Ball's Pyramid in 2003, and although a small number in absolute terms it represented a *sixth of the entire known population* at that time.

For some animals it may be possible to remove eggs or juveniles rather than adults. This approach has the advantage of keeping breeding adults on-site, but it will not work for species where parental care is essential for juvenile survival. Of course, given that *ex situ* conservation is usually only undertaken when species are threatened *in situ* then leaving healthy breeding adults on site may not be the best course of action. Accordingly, the decision to remove adults or juveniles should be taken with careful reference to the initial threat assessment. It is also unwise to remove juveniles or eggs of species where juvenile ecology or incubation conditions are not well described. Juveniles are likely to have requirements that are quite different from adults (consider, for example, the feeding requirements of a leaf-eating caterpillar and the nectar-feeding adult butterfly) and the hatching success of eggs is notoriously sensitive to their incubation environment.

Invertebrates and amphibians are good candidates for egg-based removals, but parental care in these animals is not uncommon and egg husbandry can be far from straightforward. Amphibians are particularly difficult and the husbandry requirements of the vast majority of species are at best poorly known (Pough, 2007).

Other problems of focusing *ex situ* efforts on juveniles include:

- The difficulty of providing developing juveniles with suitable food.
- The microhabitat conditions at the natural rearing site may be difficult to measure and to replicate artificially—and might have profound effects, especially in species where the sex of offspring is temperature dependent as in crocodilians, for example.
- Rearing animals away from the environment in which they have evolved could potentially lead to problems establishing normal symbiotic bacterial flora in and on the animals, with knock-on effects for animal health (a factor that is poorly understood currently).
- Normal early life exposure to predators and prey, and to other natural stimuli (including social situations with members of same species), may be absent in captive-reared juveniles. For example, captive-bred predators are particularly prone to starvation and unsuccessful avoidance of competition with other predators (Jule et al., 2008). In some cases, captive-bred animals can be trained prerelease, for example, to avoid predators (Griffin et al., 2000).

Despite the obstacles, artificially rearing adult individuals from eggs has been successful in captive breeding programmes, notably in the California condor, *Gymnogyps californianus* (Case Study 12.1).

Collecting seeds of course is the plant equivalent of collecting eggs in animals. In many plants, seed production is prodigious and collecting seeds can be a safe and effective method of establishing *ex situ* captive populations or *ex situ* 'seed banks'. Indeed, the relative ease with which many plants can be transported and grown *ex situ*, the range of vegetative (asexual) propagation techniques, and the ability of seeds to be stored for extended periods of time help to make plants excellent candidates for *ex situ* conservation.

Once numerical and categorical decisions have been made (how many; females versus males; adults versus sub-adults or juveniles) it becomes necessary to consider the specific individuals to remove. It makes sense to establish the *ex situ* breeding population with healthy, vigorous individuals, but in practice it may not be possible to screen the threatened population for the best individuals to remove. For animals, it is likely that individuals of a certain age, or sexually mature individuals will be preferred, and as with sexing animals in the field it may require an experienced eye to determine the best candidates. To reduce the problems associated with in-breeding (Section 12.4.3) the founder individuals for the captive breeding programme would ideally be unrelated (or at least, not parent–offspring or siblings), and again knowledge and experience of the species, and particularly its social and dispersal ecology, is crucial for success.

When to collect

A pragmatic approach may be required when planning the best time to collect individuals. The optimal time for the species will be dictated by its ecology, but this must be balanced with human factors that will likely be decisive in determining the collection time. Some factors that may be important in deciding when to collect include:

- **The breeding ecology of the species:** it may be best to collect animals prior to the commencement of the breeding season or at a point during the breeding season. For seed collection, it will be necessary to collect after seeds have been set.
- **The seasonal ecology of the species:** it might be best to collect when the species is most active, least active, or even when it is dormant during the winter.
- **The availability of the *ex situ* captive breeding or seed banking facility:** there is little point collecting at the optimal time for the individuals in question if there is no facility ready to receive them.
- **Site-specific factors that may limit access or safety** at the collection site, including the weather, as well as logistical, social, and political issues that may limit access to certain sites.
- **The availability of key personnel and equipment** involved in collection and transportation.

How to collect

Capturing animals alive, without harm and minimizing stress, is a potentially dangerous and highly technical process often requiring great skill and expertise (e.g. using game capture bomas as shown in Figure 12.6). Removing plants can also be problematic especially if the root ball must be taken intact. The capture phase is further complicated by the need to select certain categories of individual, or perhaps even specific individuals.

Many animal capture methods are not random with respect to the individuals they are likely to capture and this should be accounted for when considering what methods to employ. For example, it may be that one sex is less inquisitive or bold, and is less likely to be trapped. For example, in many mammals, the male is the dispersing sex whereas in birds it is often the females that disperse (Liberg & von Schantz, 1985). Trapping techniques that capture animals likely to be dispersing away from nesting, sites could result in one sex being over-represented.

With so many factors to account for, the capturing part of the process requires a species- and location-specific strategy that should be planned in consultation with experts in the field collection of the species, or at least expert in dealing with related species in similar habitats.

Figure 12.6 Capturing animals alive can be technically demanding and involve highly specialized techniques. Medium and large antelopes and other species such as Burchell's zebra *Equus quagga burchellii* can be captured by erecting large curtains to funnel the animals towards a 'boma', often using a helicopter, where they can be moved onto vehicles.

Source: Photograph by Lynne MacTavish, used with kind permission.

Sometimes capture can be remarkably straightforward. Lord Howe Island stick insects are relatively easy to locate, can be readily sexed and aged (at least to life stage) in the field and can be caught and handled without risk to human or stick insect safety. Even seemingly straightforward cases, however, have problems that need to be solved. Finding, capturing, and handing stick insects is easy, but Ball's Pyramid is a remote and dangerous field location presenting transportation and on-site complications that, as with any collection site, need to be accounted for in the action plan and through careful, thorough risk assessment.

12.4.2 Transport.

By definition, *ex situ* conservation requires that individuals are moved from one location to another. Transporting plant seeds might be relatively simple, and even egg transport can be straightforward provided that favourable storage conditions can be maintained during transportation. Transporting living organisms, however, is often far from straightforward, and requires careful planning and specialized approaches. It is also subject to legislation. In the UK, for example, arrangements for transport must comply with the Welfare of Animals (Transport) (England) Order 2006, as well as the Convention on International Trade in Endangered Species of Flora and Fauna CITES (Hot Topic 3.2) and the Guidelines on Transport and the Regulations of the International Air Transport Association (IATA) (DEFRA, 2012).

The complexity of the task will be dictated by a number of factors including:

- **The size of the animal or plant:** larger organisms are more difficult and more expensive to transport.
- **Whether the species can be transported 'dry'** or if it must be transported in water (which is the case for most aquatic species).
- **How far the species is to be relocated and how long that relocation will take.**
- **The mode of transport:** time in transit might be traded off against comfort or cost, for example.
- **The specific husbandry requirements of the species when in transit:** plants may require light, but also need protection, making packing them especially difficult.

- **Whether an animal will be a 'good' species in transit**, or is prone to panic and stress.
- **The wider risks of escape:** transporting a large predator or a venomous animal, for example, would require careful 'incident planning' in case of escape. Likewise, care needs to be taken with potentially invasive plants (Chapter 8).
- The **logistics of loading and unloading.**
- **The potential for human-based problems:** such as transport worker strikes, political instability, or machinery breakdowns that could may cause fatal delays in the transport chain.

The Lord Howe Island stick insect is a relatively small terrestrial animal with no particular characteristics that make transit problematic. However, while many smaller invertebrates can be easily transported, the same cannot be said for other animals. Despite the obvious problems of transporting, for example, large marine fish or potentially dangerous predators, animals are transported regularly and successfully between aquaria and between zoos. In fact, the expertise exists to transport virtually any animal anywhere (Figure 12.7 shows a particularly dramatic example), but more complicated transportation requirements come with a higher cost and, generally, a higher risk.

12.4.3 Captive breeding.

In 2003, two adult pairs of Lord Howe Island stick insects were captured from Ball's Pyramid for captive breeding. They were transported to two sites on the mainland (Honan, 2008). Deciding on the number of *ex situ* locations is likely to be a compromise between what is ideal and what is practical. Ideally, a number of *ex situ* sites would be selected. A captive breeding site could be destroyed by a fire or some other accident, and having more than one site spreads out that risk. Realistically, it may not be possible to have multiple sites. For example, it may be that it is only possible financially to support one location or that only one location is available for some practical reason. It may also be that the number of collected individuals is simply not sufficient to establish multiple breeding groups. Such considerations are similar to the SLOSS (single large or several small) debate for the number and relative size of nature reserves for *in situ* conservation (Section 11.5.1).

Figure 12.7 With experience, care and the right resources virtually any animal can be safety transported virtually anywhere. Here, an antelope is being transported suspended upside down beneath a helicopter.
Source: Photograph by Lynne MacTavish, used with kind permission.

At the time of setting up the captive breeding programme, almost nothing was known of the biology or ecology of Lord Howe Island stick insects (Priddel et al., 2003). Clearly, such a situation is far from ideal, but it illustrates an important point—the threat to the remaining individuals was so great that setting up a captive breeding programme was deemed to be more urgent than spending time studying the insects in the field. In fact, the ease with which other stick insect species can be kept and bred in captivity (Frye, 1992) means that the risk of capturing and keeping them without much specific knowledge is considerably lessened. This is not always the case, and assuming that one species will be as easy to establish in captivity as similar species could be a dangerous assumption to make.

Captive breeding, plant cultivation, or gene banking (especially seed banking) programmes are often incorporated into existing facilities, and the specialized facilities and expertise already present in zoos, aquaria, and botanical gardens or collections make them obvious choices. For example, one pair of Lord Howe Island stick insects was taken to a private breeder in Sydney, Australia and the other pair taken to Melbourne Zoo (Honan, 2008). As well as the expertise, specialist support and facilities that are likely available in aquaria, zoos, and botanical gardens, such sites also provide an excellent opportunity to educate the public and to raise funds for conservation work. However, it should be noted that captive breeding in zoos cannot always be meaningfully equated with *ex situ* conservation (see the Interview with an Applied Ecologist for this chapter).

AN INTERVIEW WITH AN APPLIED ECOLOGIST 12

Name: Dr Tim Bray

Organisation: Bristol Zoo Gardens (UK zoological collection: www.bristolzoo.org.uk)

Role: Lecturer in Conservation Science

Nationality and main countries worked: British; worked in: UK, South Africa, Saudi Arabia, Czech Republic

What is your day-to-day job?

I work in the Conservation Science team at Bristol Zoo, which is diverse in its activities. There is a focus on the organizational and lecturing aspects of the job and project participation from other departments and institutions, which recently has included work on amphibian probiotics and reptile molecular marker development for assessing breeding programmes and lineage origins. We also update the primate assessments on the IUCN Red List and have our own research interests.

The zoo is involved in devising and running captive breeding programmes. Can you explain a bit more about how these work?

Captive breeding programmes are usually part of a broader European Endangered Species Programme, which considers all aspects of conservation for that species. One of the first and most important considerations is ensuring that there are enough individuals in the captive population, particularly initially. If you start off with only 10 founder individuals then all future generations are based the gene pool of those 10 individuals.

After a captive population has been founded, it is important to make sure that captive breeding is optimized to maintain as much genetic diversity over time as possible. There are two main considerations here. The first is the size of the population: small populations are more prone to genetic drift, whereas large populations are generally more stable. In larger populations the random changes in allele frequency that occur are less likely to result in fixation or loss of allelic diversity. The second consideration is avoiding inbreeding by allowing, and if necessary encouraging, matings between individuals that are not close kin. Keeping a studbook is one way to ensure that the genetic relatedness of each individual is known. In this way genetic diversity is maintained as far as possible by avoiding pairing individuals that are closely related wherever possible.

What is your most interesting recent project and why?

I haven't been working in a zoo environment very long, but already I am seeing what can (and needs to) be done in wildlife molecular genetics for breeding, identification, and uncovering geographical origins of organisms. I have been peripherally involved in a great project on this with the African pancake tortoise *Malacochersus tornieri*, to improve the way that individuals are added to breeding programmes when we have little or no prior information on their provenance—individuals that have been confiscated at airports, for example. Genetic tools are a perfect way of doing this. This is particularly relevant right now with the European Commission launching its Action Plan to 'crack down on wildlife trafficking'.

What do you see as the main challenges in your field and how can they be overcome?

Well, you first of all have the problem of defining species. This is obviously critically important when you are managing a species-specific captive breeding programme and can be much harder than it sounds, especially when individuals of different species look very similar to one another. Understanding hybridization is challenging too. Once a captive breeding programme is set up, one of the big challenges is the cost of moving animals between collections and the amount of paperwork necessary to do so in order to comply with legislation.

There are also problems resulting from the success of breeding programmes. Sometimes particular pairs are really good at breeding and produce lots of young. This might sound fantastic, but actually you don't want offspring of specific individuals to become too numerous and for the genes of those individuals to become over-represented in the population. There can also be the issue of what to do with any surplus animals, especially if reintroduction is complicated or not possible because the original threats have not been resolved. Space and resources are finite so this can be a real issue.

What's the most satisfying part of your work?

It might sound odd, given that I work in a zoo, but the most satisfying part of this role is the *in situ* aspect! A lot of the captive breeding work is done to supplement wild populations or reintroduce species, or at least to keep a viable species alive to allow reintroduction in the future if/when *in situ* problems have been resolved. Aside from these direct links, a good proportion of all revenue from people coming to visit Bristol Zoo is spent on *in situ* conservation initiatives. For example, we work on an *in situ* project in South Africa on African penguins. This species is very site-faithful and birds use the same breeding colonies year after year. This is usually an advantage, but fish stocks are declining globally and the distribution of good feeding areas is changing. Breeding success is decreasing massively in some colonies due to chick starvation as parents have to hunt for longer to find food. We work to collect chicks at the end of the season when parents cannot raise them or have abandoned them, and hand-rear them. Between 2006 and 2014, almost 4000 chicks were rescued to be hand-reared and, of these, 77% were released back into the wild.

Finally, how did you get into conservation ecology and what advice would you give to others?

My research career seemed to be moving away from conservation and I felt it becoming increasingly academic. I love the variety of amazing directions you can go with your research work, but I realized this disconnection with the practical side of biology was not what I wanted. I began to apply to more on-the-ground and conservation focused entities, and was fortunate enough to join the Bristol Zoo team. Keep your eyes open for opportunities, and make yourself available to help out on projects and give presentations—it may take years of applications and making the right connections (as it did in my case), but it is worth it to avoid compromising what you want to do.

Zoological and botanical collections that are open to the public, and are reliant on ticket sales to generate revenue, can only be financially sustainable if they can attract visitors and keep their running costs to an acceptable level. Thus, such collections must balance their entertainment value and their continued financial sustainability with any *ex situ* conservation aims. This inevitably influences which species are maintained. Not all collections practice captive breeding to the same extent, and while many do maintain captive breeding programmes, not all programmes can be said to be of conservation value. A zoo that successfully breeds fallow deer *Dama dama*, for example, is not breeding threatened animals in need of *ex situ* conservation and neither is it breeding animals with the ambition of establishing them back into the wild. Of course, the animals produced can be sold to other zoos or aquaria (raising valuable funds) or used to supplement the organization's own stock. This reduces the potential collection pressure on wild populations and might well serve to increase public awareness of such species.

Husbandry

Husbandry is the care and management required to keep animals and plants alive and healthy. Some animals and plants have requirements that are simple to provide and thus present few problems to those trying to keep them. On the other hand, others can be incredibly difficult to keep, requiring specific environmental conditions, and in the case of animals, unusual food, or specific forms of social contact with other members of the species.

Maintaining animals can be especially difficult and expensive. Larger animals and social animals may require very large enclosures, and zoos and aquaria need to provide the opportunity for the paying public to view animals, as well as providing the conditions those animals require. If species are held in a captive condition in a different climate zone from that which they would occur in the wild, this might also have to be recreated (setting up tropical houses with heaters and humidifiers for tropical birds, for example, or creating hot and dry enclosures for desert reptile species). Keeping animals alive is the most basic aim of a husbandry plan, but keeping those animals content is vital for their welfare and to encourage breeding. Environmental enrichment, such as adding climbing frames and toys to primate enclosures or hiding food in 'puzzle blocks' for big cats, provides a more natural captive experience and may help to enhance breeding success, as shown in Figure 12.8 (Young, 2003).

Captive animals can be reluctant to breed but with careful husbandry, good knowledge and, perhaps most crucially of all, plenty of hands-on experience, it often proves possible to breed in captivity even the most reluctant species. Sometimes interventions can be simple. Lesser flamingos *Phoeniconaias minor* have proved to be particularly difficult to get to breed in captivity, but can be encouraged to undertake mating displays and lay eggs by the installation of mirrors in their enclosures, creating the appearance of larger, more natural flocks (Scholefield, 2011).

Captive breeding need not take place in conditions that are normally associated with captivity. The Yangtze

Figure 12.8 Enriching the environment of captive animals can encourage natural behaviours, reduce stress, and encourage breeding. This can involve providing activities such as climbing frames for Schmidt's guenon *Cercopithecus ascanius schmidti* (left), or by concealing food in puzzle blocks for Sumatran tigers *Panthera tigris sumatrae* such that the animals have to manoeuvre the food out using thought process and manual dexterity (right).

Source: Images courtesy of David Wiley/CC BY 2.0 (left); and Rosie Wilkes from West Midlands Safari Park (right), used with kind permission.

finless porpoise, *Neophocaena phocaenoides asiaeorientalis*, is an endemic and endangered small cetacean (a group that comprises whales, dolphins, and porpoises) that only occurs in the middle and lower reaches of the Yangtze River, China. In the 1990s, the Tian-e-Zhou Oxbow, located in Shishou, Hubei, China, was approved by the central government as a National Natural Reserve for the Yangtze finless porpoise and another endemic freshwater cetacean the Baiji *Lipotes vexillifer*. This oxbow was formed naturally when it was cut off from the main channel of the Yangtze River in 1972 and is 21 km long and 1–1.5 km wide (Xia et al., 2005). As a contained freshwater body, with a controlled breeding programme of introduced wild-caught animals it is a captive breeding facility, but is neither an aquarium nor a zoo.

In the case of the Lord Howe Island stick insect, very little was known of its biology or ecology when the four founding individuals of the captive breeding programme were collected in 2003 (Honan, 2007). Consequently, the captive population was being simultaneously studied in detail, kept alive and encouraged to breed. Different food plants and egg-laying media were introduced and tested, together with different ways to house the insects. Initially, the adults were kept free-ranging in a glasshouse in the zoo and at night, when they were active, their behaviour was observed. Luckily, this behaviour included both mating and egg-laying, and expertise in keeping and breeding the stick insects was rapidly accumulated, resulting in around 1000 adults and thousands of eggs in the captive population in Melbourne Zoo (Crew, 2012).

Inbreeding and hybridization

Inbreeding occurs when closely-related individuals produce offspring. **Inbreeding depression** is the reduced survival and reproductive capacity (collectively termed biological fitness) that occurs in a population as a consequence of inbreeding (Charlesworth & Willis, 2009). It affects both animals and plants and is of great concern to those managing *ex situ* (and for that matter *in situ*) populations.

There is a genetic reason why breeding with close relatives (especially parents, offspring or siblings; so-called first-order relatives) can result in decreased fitness of individuals. Many deleterious conditions are genetically recessive, which means that both parents have to carry the deleterious copy of the gene in question for the offspring to be affected. Different forms of a gene are termed alleles, and if only one parent has passed on the recessive allele then the other parent's dominant copy of that gene will prevail and the offspring will be unaffected. The offspring will be a carrier of that recessive allele, however, and their offspring could be affected if the carrier were to breed with another carrier. Around one quarter of this pairing's offspring will inherit the recessive allele from both parents. These offspring are said to be **homozygous recessive,** and in these cases there are no dominant copies to mask the effects of the deleterious copy.

When close relatives breed, the likelihood of recessive alleles coming together in offspring is greater than when non-related individuals breed. This is because closer relatives are, by definition, more similar genetically, and so are more likely to have similar recessive alleles. The resulting offspring will likely have lowered fitness, which translates in reduced survival and reproductive output. In a worse-case scenario, the next generation may be unable to produce any viable offspring at all. Inbreeding can, potentially, result in a purging of recessive alleles from a population, and populations can survive being forced through what is termed a genetic bottleneck. For example, the northern elephant seal *Mirounga angustirostris* (see Figure 12.9), was hunted virtually to extinction by the end of the 19th century and the population likely fell to fewer than 20 individuals. Conservation efforts, tightly linked to legal protection measures, allowed the population to rebound and it now numbers more than 175,000 individuals (Weber et al., 2000). There is a very low genetic diversity within the current population as a consequence of this genetic bottleneck, and this could leave it vulnerable to disease or some other threat that some members of a more diverse population could survive. (See also details of extinction vortices (in Figure 10.4).

Captive-bred populations are typically very small within any one location and inbreeding depression is a very real threat. There is evidence that Lord Howe Island stick insects suffer from inbreeding depression within the captive-bred population at Melbourne Zoo over successive generations (Honan, 2008). Symptoms of inbreeding in this species included:

- Small egg size, which became more apparent with each successive generation.
- Low hatching rate.
- Small hatching size.
- Low nymph survival.
- Unusual abdomen development, especially in adults.

These inbreeding-related problems reduced the survival and breeding success of the captive population once they became apparent, and kept the population to around 20 individuals for the first three years or so of captive management. The problems were fixed by introducing new males into the captive-bred population, and within a year of the new males being introduced the population rose from 20 to 600.

Another genetic consideration pertinent to *ex situ* conservation, predominantly of plants, is **hybridization.** This is the process of interbreeding between individuals of different species (interspecific hybridization) or between genetically divergent individuals from different populations of the same species (intraspecific hybridization). Individuals so created may breed with one of the parent species, resulting in a movement of genes from one species to another in a process termed **introgression.** Plants hybridize much more often and much more successfully than animals. Consequently, both hybridization and introgression are primarily of concern to those maintaining *ex situ* collections of plants (as opposed to seed collections). In botanical collections where individuals from different species (especially members of the same genus) are maintained in close proximity to each other then it is possible for species maintained for the purposes of *ex situ* conservation to hybridize. Protocols to limit hybridization include covering flowers with bags and hand pollinating, physically separating plants likely to hybridize and managing the species present onsite to limit opportunities for hybridization.

Figure 12.9 The northern elephant seal *Mirounga angustirostris* has passed through a genetic bottleneck of approximately 20 individuals to more than 175,000 alive today. As a consequence, the population has low genetic diversity that could make it vulnerable to disease.

Source: Image courtesy of Franco Folini/CC BY-SA 2.0.

Studbooks

The introduction of new genetic stock into the captive population of Lord Howe Island stick insects and the subsequent rapid expansion of the population demonstrates both the extent to which inbreeding depression can constrain a population and the exceptional benefits of introducing genetic variation by introducing unrelated individuals, a process termed outbreeding. To keep captive-bred animal populations healthy, regardless of whether they are being maintained for conservation value and subsequent reintroduction, requires careful management of their genetic constitution, and this is achieved through **studbooks**.

A studbook is 'a compilation and source of genealogical data of individual animals which make up a particular population' (Biaza, 2015). The first official studbook (as opposed to informal verbal and written records) was the *General Studbook for Thoroughbred Horses* that was established in England in 1791. The first studbook for a wild animal in captivity did not appear until 1932 and was for the European bison *Bison bonasus*. By 1932 this animal was facing extinction and it was realized that captive populations (spread throughout Europe and across the world) would need to be managed in an active and cooperative way if the species was to survive (Biaza, 2015).

The establishment of a studbook is more than just a collection of data; it is an essential tool in the scientific management of captive-bred species and it allows for careful and well-planned breeding programmes to take place, that are coordinated between facilities across the world. The modern studbook is computer-based and its compilation and usefulness is greatly assisted by advances in animal identification (to individual level), sex determination, and the rise of computer software for record keeping and analysis.

Studbooks are managed at an international level by the World Association of Zoos and Aquaria (WAZA) and IUCN/SSC and a proposal for a new studbook must be approved by them before it can be officially established and accepted (WAZA, 2012a). The proposer must have the necessary credentials and experience, institutional support, and resources to maintain and publish the studbook. If accepted, then the official keeper of that studbook must then begin to compile detailed information on the individuals of that species kept throughout the world, including any known information on genealogy.

Once established, the studbook aims to maintain maximum genetic diversity and the characteristics of the species (including behavioural characteristics). By maintaining accurate and detailed records of individual animals it becomes possible to manage the genetic diversity of the entire captive population even though it may be dispersed across the globe and between many institutions. Analysis of individual genetic histories allows the studbook keeper (and other interested parties) to determine which pairings would be desirable, and which would not. It also allows the studbook keeper to manage the level of representation of some animals within the population. For example, if a pairing was particularly productive their offspring could quickly come to dominate the captive population and reduce the overall diversity. The studbook allows this fine scale genetic management to be achieved.

There are, however, still challenges with studbooks. The most important of these is that the charts that are created for each captive-held individual showing the proportion of its genes deriving from each of the original **founder individuals** are based on the assumption that each founder individual was genetically distinct from all other founder individuals. In other words, if two wild-caught founder individuals were actually siblings, or a parent/offspring pair, they would have had many more genes in common than would be expected using this model and this might affect how useful the genetic provenance graphs are for the whole species. Studbooks are, for many species, gradually reaching the stage that the majority of individuals in captivity are captive born. This means that age and parental history are well known. However, in some species, such as the Bornean orangutan *Pongo pygmaeus*, there are still a large number of wild-caught individuals whose profile is necessarily incomplete (WAZA, 2012b).

12.4.4 Reintroduction, supplementation, and reinforcement.

If a successful captive breeding programme is established then, with time, it may become possible to reintroduce a species into an area from whence it has been extirpated, add individuals to a reintroduced population through supplementation, or reinforce an existing population that is worryingly small or declining. Re-establishment of viable wild populations using one or more of these approaches is the stated goal of most captive breeding programmes for animals and plants of conservation concern. The actual release needs to be carefully planned. For some species, the release might be as simple as opening the transport container and allowing them to leave, but for others it might include a graded release. Animals may be introduced back into a 'contained wild', perhaps a fenced off area that can act as an acclimatization site and buffer between captivity and the wild. Plants may need to be fenced or protected from herbivores until they are established. These considerations and others are discussed in considerable detail in Chapter 13, which deals specifically with reintroductions.

A cautious approach has been adopted for the reintroduction of the Lord Howe Island stick insect. The biggest problem facing the species on Lord Howe Island itself remains the threat that led to its rapid near-extinction in the first place: non-native rats. These need to be completely eradicated, and although possible (it has been done on a number of islands), such eradication is practically difficult and potentially very expensive. The plans for Lord Howe Island are ambitious and involve releasing poison baits from the air while protecting native wildlife susceptible to the poison in rat-proof enclosures (dubbed 'arks') until the rats have been killed. However, to date, the proposals have not been carried out. There are Lord Howe Island stick insects on Lord Howe Island, but they are kept in rat-proof sheds and are captive breeding colonies, rather than successfully reintroduced wild animals (shown in Figure 12.10). If and when the species is reintroduced fully, is likely the reintroduction site will not be the same as the initial collection site to ensure that the captive-bred individuals have a realistic chance of survival and breeding.

It is instructive to consider that the obstacle preventing the reintroduction of Lord Howe Island stick insects is an *in situ* conservation problem resulting from global trade 100 years ago and a non-native species introduction (Chapter 8). It is also a conservation problem that will only be solved by the planned mass extirpation of a species. It is an excellent demonstration both of the complex and entangled nature of many Applied Ecology problems, and the problems that *ex situ* conservation can face even when captive breeding is, as in this case, highly successful. Another example of some of the difficulties even successful breeding programmes can encounter is outlined in Case Study 12.2. However, despite the challenges, *ex situ* conservation through captive breeding and reintroduction can be very successful. One such success story, with sand lizards *Lacerta agilis* in the UK and Europe, is detailed in the Online Case Study for Chapter 12.

You can find the online case study at www.oxfordtextbooks.co.uk/orc/goodenough

FURTHER READING FOR SECTION

Animal husbandry and breeding: Hosey, G., Melfi, V., & Pankhurst, S. (2013) *Zoo Animals: Behavior, Management and Welfare*. Oxford: Oxford Scientific Press.

The first attempt to reintroduce a wild population of a species when there were none left in the wild: Price, M.R.S. (2010) *Animal Reintroductions: The Arabian Oryx in Oman*. Cambridge: Cambridge University Press.

Figure 12.10 Lord Howe Island stick insects are back on Lord Howe Island, but are housed in enclosures like the one shown here (left = outside view; right = inside view). The rats that caused their extirpation are still present and would probably eat any stick insects that were released into open country.

Source: Photographs by Rohan Cleave, Melbourne Zoo, used with kind permission.

CASE STUDY 12.2

Unintended consequences of reintroduction—the case of the Mallorcan midwife toad

The Mediterranean island of Mallorca is home to a number of **endemic** species (species not found anywhere else), including the Mallorca midwife toad *Alytes muletensis* (Figure A). First described from fossil specimens in 1977, living toads were discovered in a remote gorge in the northern mountains of the island and described in 1981 (Mayol & Alcover, 1981). Its status as a rediscovered 'extinct' species makes the Mallorcan midwife toad an example of a **Lazarus species** (see also Section 3.6).

The toad is restricted to small streams deeply carved into the limestone mountains in the north of the island. With such restricted habitat and geographical ranges, and low population numbers (*c.* 500–1500 adults at the moment), the species is listed as 'Vulnerable' on the IUCN Red List of Threatened Species. However, populations are increasing because of intensive conservation efforts, which include a programme of **captive breeding** and **reintroduction** (EDGE, 2015).

The toads breed well in captivity and successful reintroductions have been taking place since the late 1980s. Indeed, it is estimated that more than 50% (Imperial College London, 2008) of the current wild population is from captive-bred stock. As a consequence of the success of this programme their status was changed from 'Critically threatened' to 'Vulnerable' in the Global Amphibian Assessment of 2004. However, despite the successes, in 1991 a theoretical risk of captive breeding became a reality in 1991.

Amphibian populations worldwide have been threatened by the fungus *Batrachochytrium dendrobatidis*, which among other symptoms causes a thickening of the skin, which interferes with normal respiratory and excretory functions (see also Figure 3.5). Captive breeding of Mallorcan midwife toads took place in a facility that also housed an endangered, and infected, species of frog from South Africa *Xenopus gilli*, before the fungus was known to science (Walker et al., 2008).

Captive-bred toads introduced into Mallorca in 1991 were infected with the fungus and that fungus has spread. However, in an odd twist, three of the four infected populations are doing well, suggesting that Mallorcan midwife toads may have some innate defence against a fungus that has devastated many amphibian populations across the world (Imperial College London, 2008). When *Batrachochytrium dendrobatidis* was discovered in the late 1990s, amphibian breeding programmes implemented screening, and captive breeding programmes are moving towards breeding threatened frogs in biosecure, quarantined facilities (Imperial College London, 2008).

Figure A The Mallorca midwife toad, *Alytes muletensis*.

Source: Image courtesy of tuurio and wallie/CC BY-SA 3.0.

As the authors of the study point out, the Mallorcan midwife toad 'provides a salutatory lesson of the need to ensure that breeding programmes are not hot-beds for cross-specific disease transmission, and that species are free of infectious agents prior to reintroduction (Walker et al., 2008).

REFERENCE

EDGE (Evolutionarily Distinct and Globally Endangered) (2015) *Mallorcan midwife toad* (Alytes muletensis). Available at: http://www.edgeofexistence.org/amphibians/species_info.php?id=600

Imperial College London (2008) *Captive Breeding Introduced Infectious Disease To Mallorcan Amphibians*. Available at: www.sciencedaily.com/releases/2008/09/080922122427.htm.

Mayol, J. and Alcover, J.A. (1981) Survival of Baleaphryne Sanchìz and Arover, 1979 (Amphibia: Anura: Discoglossidae) on Mallorca. *Amphibia-Reptilia*, Volume 1, 343–345.

Walker, S.F, Bosch, J., James, T.Y., Litvintseva, A.P., Valls, J.A.O., Piña, S., García, G., Rosa, G.A., Cunningham, A.A., Hole, S., Griffiths, R., & Fisher, M.C. (2008) Invasive pathogens threaten species recovery programs. *Current Biology*, Volume 18, R853–R854.

12.5 Gene banking.

Gene banking, or **genome resource banking**, is an *ex situ* approach that takes advantage of modern developments in storing biological material, freezing embryos, and artificial insemination to produce a biorepository of genetic material for future use. In a very real sense, gene banks are an insurance against the loss of genetic diversity or, in a worse-case scenario, against extinction (Eastwood et al., 2015).

Plants are, in many ways, much easier to conserve using *ex situ* gene banking approaches than are animals. Well-established vegetative propagation techniques allow single cuttings to be grown on into mature plants, and plants often require less complex and less intensive husbandry procedures. However, the feature of plant biology that makes them especially suitable for *ex situ* conservation is their production of long-lasting and easily stored seeds, often in great numbers. Seeds can be maintained in chilled seed banks for extended periods of time (potentially hundreds or even thousands of years). One of the best known of the seed banks is the Svarlbard Global Seed Vault, which acts as a repository for seeds that are significant for food production (shown in Figure 12.11). There are many other locations around the world storing seeds and plant material of economic, social, agricultural, and biodiversity significance of which The Millennium Seed Bank Partnership (MSBP), coordinated by Kew Gardens in the UK, is the largest (Eastwood et al., 2015).

The MSBP provides a global insurance against the extinction of plants in the wild by storing seeds. Plants and regions most at risk from changing land use and climate have been targeted through the 80 countries that form the partnership. In addition to research, education and public engagement, the project aims to collect seeds from 25% of all known flora globally (accounting for 75,000 species) by 2020 and to collect seeds from all of the UK's native seed-producing plants.

As with the captive breeding of animals or the cultivation of plants, seed collection for *ex situ* conservation requires multiple, sometimes complex, stages to be negotiated. These include:

- Getting permissions from land owners, governments, and organizations charged with protecting

Figure 12.11 An entrance leads down a tunnel deep inside the mountain that houses the Svalbard Global Seed Vault, halfway between mainland Norway and the North Pole. Permafrost acts as a natural deep freeze and the remote location helps to keep seeds safe from human interference.

Source: Photograph by Global Crop Diversity Trust, used with kind permission.

biodiversity in the territory of interest. Data sharing and other conditions may be imposed in addition to collecting constraints.

- Organizing expeditions to collect seeds, often in challenging locations. Such expeditions require enormous logistical and technical support both in the planning and in-country stages.
- Finding and positively identifying plants and seeds to ensure the correct species are being collected.
- Determining how many seeds to harvest and from which plants.
- Harvesting the seeds (up to 20,000 seeds from at least 50 individuals may be needed for long-term conservation).
- Sorting and caring for the seeds, and other plant material while in the field, especially ensuring the correct conditions of humidity and temperature for optimal storage.
- Transporting the seeds back to the seed bank via air or postal freight. Fruit, moist, or under-ripe seeds may need particularly careful handling, and freighting seeds may also require separate permissions.

Once collected, seeds are stored in large underground storage vaults at Kew's Millennium Seed bank at Wakehurst, UK, or in partner seed banks around the world. Here, they are subjected to further processing in a closed laboratory to prevent the escape of live insects and other organisms inadvertently collected with the seeds. Identifications are confirmed and seeds are assessed for storage needs before being cleaned, checked, X-rayed (to check for 'empty' or damaged seeds), weighed, measured, packed, dried, and sent for storage (Kew, 2015).

Seeds are stored at –20°C in sealed glass containers in which a constant 15% humidity is maintained. For species of high conservation priority, or those likely to be short-lived in storage, small samples are also stored in liquid nitrogen vapour at –196°C. Some seeds are expected to last for centuries or even millennia, but germination tests are carried to ensure the viability of the seeds, and should viability fall to unacceptable levels then the remaining seeds may be grown out to harvest fresh seed and to regenerate the collection.

Animals are less well suited for long-term storage. Their lack of rigid cell walls means that freezing is, except in highly unusual examples, fatal, and very few species lack any sort of long-term life stage equivalent to a seed (exceptions include the tardigrades). However, it is possible to store DNA in spermatozoa, eggs, embryos, and live tissue at low temperatures (cryopreservation), and such facilities have become known as 'frozen zoos', with San Diego Frozen Zoo being particularly well known (Watson, 2015).

Although whole, intact animals cannot be defrosted and reintroduced, embryos can be implanted, eggs fertilized, and DNA analysed. Spermatozoa are usually more accessible, and occur in larger numbers than eggs and embryos, and consequently are the primary cell type selected for preservation in many animal-based genome resource banking projects. Sperm can be obtained by the use of artificial vaginas, electroejaculation, and various massage techniques, and collection and storage protocols that have been developed for domestic animals serve as the starting point to developing bespoke techniques suitable for non-domestic species. As well as genome banking for insurance purposes, harvested wild sperm can be used to artificially inseminate individuals in captive or isolated populations to increase genetic diversity (Prieto et al., 2014). The usefulness of frozen animal genetic material relies entirely on the continued development of DNA technology and storage, as well as the improvement of methods for artificial insemination, in vitro fertilization, embryo implantation, embryo transfer, and cloning.

Animal gene banking is a technology-driven form of *ex situ* conservation that could also allow for future technology to resurrect currently extinct species for which there is stored genetic material. The possibly of 'de-extinction', as this 'resurrectionist' approach has been termed, has attracted considerable media attention (e.g. Roast, 2013). De-extinction, which might be thought of as ex tempo, as well as *ex situ*, is being pursued seriously by some scientists for animals that have recently become extinct and for whom viable DNA samples are available, but there are many problems with such an approach. De-extinction is discussed further in Hot Topic 12.1.

FURTHER READING FOR SECTION

De-extinction: Shaprio, B. (2015) *How to Clone a Mammoth: The Science of De-extinction*. Princeton, NJ: Princeton University Press.

HOT TOPIC
12.1

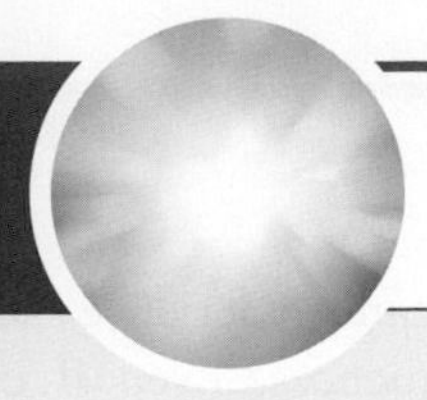

De-extinction

Extinction, the dying out of a species, is a natural event, and more than 99% of all species that have ever existed are now extinct. However, in many cases in recent history, human activities have led to extinctions and in some of these cases there is a desire to have these species back. Where genetic material exists (such as in preserved skins or in frozen samples) then modern DNA-based techniques give some hope that extinct species could be brought back to life. Using DNA from extinct organisms to create living representatives of those extinct species has become known as de-extinction (Sherkow & Greely, 2013).

De-extinction is far from straightforward and, to date, it has not proved possible to resurrect an individual from an extinct species for any meaningful period of time. The problems facing a de-extinction programme are numerous:

- Viable and complete DNA needs to be obtained, limiting de-extinction to recently extinct species for which suitably preserved material exists.
- DNA needs to be inserted into an egg and, in most cases, that egg will not exist. Thus, a closely-related species needs to act as an egg donor.
- The fertilized egg needs to divide and differentiate normally in vitro, and replicating the stimuli and environment needed to achieve this is far from straightforward.
- The resulting embryo needs to be transplanted into a suitable female 'mother' (an elephant for a mammoth, for example), to produce a viable egg (birds and reptiles) or a pregnancy (mammals).
- Once born or hatched the individual will lack the social cues and developmental environment of its species, and will rely on the 'donor' species or considerable human intervention for survival.
- To produce a breeding population will require multiple individuals surviving to reproductive age, then successfully reproducing in captivity.
- Introduction into a 'natural' environment will have the same problems experienced in more conventional *ex situ* approaches, with the added problem that the abiotic and biotic environments may have changed considerably during the extinction period.

Despite these problems, scientists have managed to bring, to date, one species back from extinction, albeit only for 10 minutes. The last Pyrenean ibex *Capra pyrenaica pyrenaica*, a female named Celia, died in 2000, marking the extinction of this species. Prior to her death, tissue samples were taken. Using cells from those samples and egg cells from a closely-related goat species, scientists were able, after many unsuccessful attempts, to produce a clone embryo of Celia that was transplanted to surrogate ibex and ibex–goat hybrid mothers. Only one clone was born alive from this project and this died after 10 minutes because of a lung defect (Folch et al., 2009). Of course, had this clone made it to adulthood a single female would not have been sufficient to establish a viable breeding population.

While some scientists are pursuing de-extinction approaches, others consider the possibility of bringing Tasmanian tigers *Thylacinus cynocephalus*, passenger pigeons *Ectopistes migratorius*, or mammoths *Mammuthus* back from extinction to be a distraction from the purpose of storing genetic material in order to bolster extant populations and to increase genetic diversity. However, it is only a matter of time before useable techniques to create a variety of extinct organisms from stored DNA move from science fiction to reality.

Supplementing small populations of conservation concern by breeding individuals from stored DNA is conceptually and practically simpler than de-extinction and in time will probably become part of the Applied Ecologist's conservation toolkit (Piña-Aguilar et al., 2009). The ethics and implications of this supplementation approach need to be carefully considered but, as with the practicalities, they are clearly simpler than the ethical considerations of bringing back an extinct species.

QUESTION

Even if we are able to resurrect a species, is it right that we should?

REFERENCES

Folch, J., Cocero, M.J., Chesné, P., Alabart, J.L., Domínguez, V., Cognié, Y., Roche, A., Fernández-Arias, A., Martí, J.I., Sánchez, P., Echegoyen, E., Beckers, J.F., Bonastre, A.S., & Vignon, X. (2009) First birth of an animal from an extinct subspecies (*Capra pyrenaica pyrenaica*) by cloning. *Theriogenology*, Volume 71, 1026–1034.

Piña-Aguilar, R.E., Lopez-Saucedo, J., Sheffield, R., Ruiz-Galaz, L.I. Barroso-Padilla J.D.J., & Gutiérrez-Gutiérrez, A. (2009) Revival of extinct species using nuclear transfer: hope for the mammoth, true for the Pyrenean ibex, but is it time for 'Conservation Cloning'? *Cloning and Stem Cells*, Volume 11, 341–346.

Sherkow, J.S. & Greely, H.T. (2013) What if extinction is not forever? *Science*, Volume 340, 32–33.

12.6 Conclusions.

Ex situ conservation can offer a lifeline to species that are likely to go extinct if such action is not taken. It can also be used to supplement populations that are under less severe but nonetheless real threat. However, removing individuals (in some cases every individual) to a safe location where they can reproduce and eventually be brought back to their home range is simple in principle, but can be exceptionally difficult in practice. There are a large number of issues that need to be considered, not least of which is the cost of implementing an *ex situ* conservation programme. Successful programmes (like the California condor example in Case Study 12.1) require an enormous amount of knowledge and experience, linked with substantial resourcing and, as highlighted in Case Study 12.2, even successful programmes can experience setbacks and unintended consequences. Even if captive breeding is successful and reintroduction or supplementation techniques are well established, the conservation threats that provoked the *ex situ* approach may still be present. The need for *in situ* and *ex situ* conservation to work together is well illustrated by the Lord Howe Island stick insect example used throughout this chapter. Captive-bred individuals are now back on Lord Howe Island, but must be housed in sheds within rat-proof enclosures to keep them safe. In the future, genetic techniques may offer hope for threatened species (and even for extinct ones), but until then the *ex situ* model of capture–breed–release remains the last resort for species for which *in situ* conservation alone is not sufficient.

ONLINE ACTIVITY

Go to www.oxfordtextbooks.co.uk/orc/goodenough/ to download the activity that accompanies this chapter.

CHAPTER 12 AT A GLANCE: THE BIG PICTURE

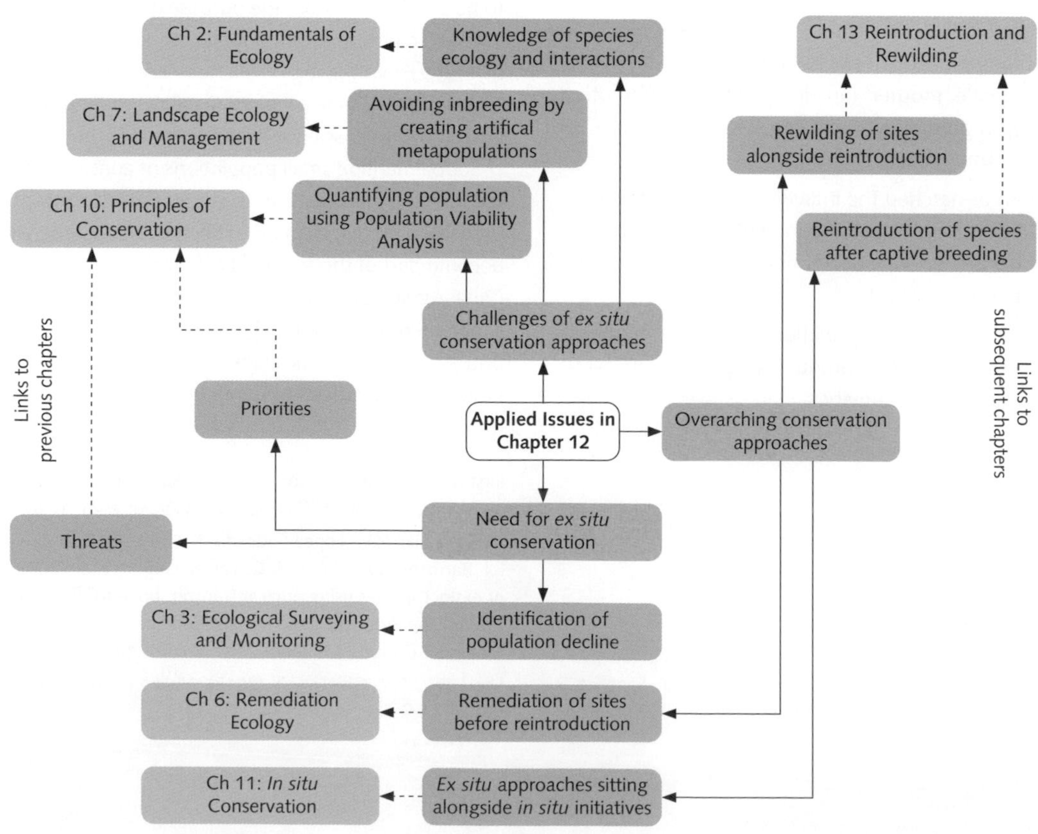

REFERENCES

Braverman, I. (2013) Conservation without nature: the trouble with *in situ* versus *ex situ* conservation. *Geoforum*, Volume 51, 47–57.

Balmford, A., Mace, G.M., & Leader-Williams, N. (1996) Designing the Ark: setting priorities for captive breeding. *Conservation Biology*, Volume 10, 719–727.

Biaza (British and Irish Association of Zoos and Aquariums) (2015) *Studbooks*. Available at: http://www.biaza.org.uk/resources/studbooks/.

Charlesworth, D. & Willis, J.H. (2009) The genetics of inbreeding depression. *Nature Reviews Genetics*, Volume 10, 783–796.

Cikanek, S.J., Nockold, S., Brown, J.L., Carpenter, J.W., Estrada, A., Guerrel, J., Hope, K., Ibáñez, R., Putman, S.B., & Gratwicke, B. (2014) Evaluating group housing strategies for the ex-situ conservation of harlequin frogs (*Atelopus* spp.) using behavioral and physiological indicators. *PLoS One*, Volume 9, e90218.

Creel, S. & Creel, N.M. (1995) Communal hunting and pack size in African wild dogs, Lycaon pictus. *Animal Behaviour*, Volume 50, 1325–1339.

Crew, B. (2012) Lord Howe Island stick insects are going home. *Scientific American*. Available at: http://blogs.scientificamerican.com/running-ponies/lord-howe-island-stick-insects-are-going-home/

DEFRA (2012) *Secretary of State's Standards of Modern Zoo Practice*. Available at: http://www.defra.gov.uk/wildlife-pets/zoos/

Department of the Environment (2015) *Dryococelus australis—Lord Howe Island Phasmid, Land Lobster.* Species Profile and Threats Database, Department of the Environment, Australian Government. Available at: https://www.environment.gov.au/cgi-bin/sprat/public/publicspecies.pl?taxon_id = 66752#summary

Department of Environment and Climate Change (NSW) (2007) *Lord Howe Island Biodiversity Management Plan*. Sydney: Department of Environment and Climate Change (NSW).

Dolman, P.M., Collar, N.J., Scotland, K.M., & Burnside, R.J. (2015) Ark or park: the need to predict relative effectiveness of *ex situ* and *in situ* conservation before attempting captive breeding. *Journal of Applied Ecology*, Volume 52, 841–850.

Eastwood, R.J., Cody, S., Westengen, O.T., & von Bothmer, R. (2015) Conservation roles of the Millenium Seed Bank and the Svalbard Global Seed Vault. In: R. Redden, S.S. Yadav, N. Maxted, M.E, Dulloo, L. Guarino, & P. Smith (Eds) *Crop Wild Relatives and Climate Change*, pp. 173–186. Oxford: Wiley-Blackwell.

Engelmann, F. & Engels, J.M.M. (2002) Technologies and strategies for ex-situ conservation. In: J.M.M.Engels, V. Ramanatha Rao, A.H.D Brown, & M.T. Jackson (Eds.) *Managing Plant Genetic Diversity*, pp. 89–103. Rome: International Plant Genetic Resources Institute (IPGRI).

Frye, F.L. (1992) *Captive Invertebrates: a Guide to their Biology and Husbandry*. Malabar, FL: Krieger Publishing Company.

Griffin, A.S., Blumstein, D.T., & Evans, C. (2000) Training captive-bred or translocated animals to avoid predators. *Conservation Biology*, Volume 14, 1317–1326.

Honan, P. (2007) The Lord Howe Island stick insect: an example of the benefits of captive management. *Victorian Naturalist*, Volume 124, 258–261.

Honan, P. (2008) Notes on the biology, captive management and conservation status of the Lord Howe Island stick insect (*Dryococelus australis*) (Phasmatodea). *Journal of Insect Conservation*, Volume 12, 399–413.

IUCN/SSC (2014). *Guidelines on the Use of Ex Situ Management for Species Conservation*, Version 2.0. Gland: IUCN Species Survival Commission.

Jule, K.R., Leaver, L.A., & Lea, S.E.G. (2008) The effects of captive experience in reintroduction survival in carnivores: a review and analysis. *Biological Conservation*, Volume 141, 355–363.

Kew (2015) *Millennium Seed Bank at Kew Royal Botanic Gardens*. Available at: http://www.kew.org/visit-wakehurst/explore/attractions/millennium-seed-bank.

Liberg, O. & von Schantz, T. (1985) Sex-biased philopatry and dispersal in birds and mammals: the Oedipus hypothesis. *American Naturalist*, Volume 126, 129–135.

Macintyre, S. & Hobbs, R. (1999) A framework for conceptualizing human effects on landscapes and its relevance to management and research models. *Conservation Biology*, Volume 13, 1282–1292.

Pough, F.H. (2007) Amphibian biology and husbandry. *ILAR Journal*, Volume 48, 203–213.

Priddel, D., Carlile, N., Humphrey, M., Fellenberg, S., & Hiscox, D. (2003) Rediscovery of the 'extinct' Lord Howe Island stick-insect (*Dryococelus australis* (Montrouzier)) (Phasmatodea) and recommendations for its conservation. *Biodiversity and Conservation*, Volume 12, 1391–1403.

Prieto, M.T., Sanchez-Calabuig, M.J., Hildebrandt, T.B., Santiago-Moreno, J., & Saragusty, J. (2014) Sperm cryopreservation in wild animals. *European Journal of Wildlife Research*, Volume 60, 851–864.

Pritchard, D.J., Fa, J.E., Oldfield, S., & Harrop, S.R. (2012) Bring the captive closer to the wild: redefining the role of ex situ conservation. *Oryx*, Volume 46, 18–23.

Roast, A. (2013) De-extinction: mammoth prospect or just woolly? *BBC News Science and Environment*, 20 August 2013. Available at: http://www.bbc.co.uk/news/science-environment-23602142

Scholefield, D. (2011) Flamingos breed at British zoo for the first time ever … After staff put up mirrors to help get them in the mood. *The Daily Mail*. Available at: www.dailymail.co.uk/sciencetech/article-2033604/Upon-reflection-40-years-peep-flamingos-wildlife-park-breed-supplied-mirrors.html#ixzz3p6gK8geR.

van Slageren, W. (2003) The Millennium Seed Bank: building partnerships in arid regions for the conservation of wild species. *Journal of Arid Environments*, Volume 54, 195–201.

Watson, J. (2015) *Survival for Some Endangered Species Hinges on 'Frozen Zoo'*. Available at: http://www.salon.com/2015/02/11/survival_for_some_endangered_species_hinges_on_frozen_zoo/

WAZA World Association of Zoos and Aquariums (2012a) *Resource Manual for International Studbook Keepers*. 65th WAZA Annual Conference, Cologne, Germany. Available at: http://www.waza.org/files/webcontent/1.public_site/5.conservation/international_studbooks/resource_manual/ISB%20Resource%20Manual_28Nov2012.pdf.

WAZA World Association of Zoos and Aquariums (2012b) *International Studbook of the Orangutan* (Pongo pygmaeus, Pongo albelii). Gland: WAZA World Association of Zoos and Aquariums International Studbook.

WAZA World Association of Zoos and Aquariums (2015) *Amur Leopard Conservation*. Available at: http://www.waza.org/en/site/conservation/waza-conservation-projects/amur-leopard-conservation

Weber, D.S., Stewart, B.S., Garza, J.C., & Lehman, N. (2000) An empirical genetic assessment of the severity of the northern elephant seal population bottleneck. *Current Biology*, Volume 10, 1287–1290.

Whitehouse, A.T. (2008) *Managing Aggregates Sites for Invertebrates: a Best Practice Guide*. Peterborough: Buglife.

Xia, J., Zheng, J., & Wang, D. (2005) Ex-situ conservation status of an endangered Yangtze finless porpoise population (*Neophocaena phocaenoides asiaeorientalis*) as measured from microsatellites and mtDNA diversity. *ICES Journal of Marine Science*, Volume 62, 1711–1716.

Young, R.J. (2003) *Environmental Enrichment for Captive Animals*. Oxford: Blackwell Science Ltd.

Reintroduction and Rewilding

13.1 Introduction.

Species can become extirpated (locally extinct) in part of their native range, or become completely extinct in the wild, for a variety of reasons. Sometimes the reasons for this are obvious; hunting (e.g. sea otters *Enhydra lutris* in Canada); habitat change (e.g. short-haired bumblebee *Bombus subterraneus* in UK: Case Study 13.1); or persecution (e.g. red kites *Milvus milvus*). In other cases, reasons for decline are unknown, which limits the effectiveness of management (e.g. large blue butterfly *Phengaris arion*: Case Study 8.2), or means that *in situ* conservation measures are simply ineffective or insufficient (e.g. California condor *Gymnogyps californianus*; Case Study 12.1).

It is sometimes possible—as in the case of the species listed above—to successfully **reintroduce** species to their former range, either by **translocating** wild-caught individuals or by releasing **captive-bred** individuals. Indeed, reintroduction is increasingly becoming a global phenomenon with other successful schemes including grey wolves *Canis lupus* in America, Eurasian beavers *Castor fiber* in Sweden, Père David's deer *Elaphurus davidianus* in China, Przewalski's horse *Equus przewalskii* in Mongolia, and giraffe *Giraffa camelopardalis* in Senegal.

Plant reintroductions also occur, although they are often less well studied. These typically involve translocating individuals (adults or seeds) directly or transplanting after **captive propagation**. Examples include reintroducing the critically endangered sand dune flowering plant *Dianthus morisianus* to Sardinia from captivity and translocating the juniper species *Juniperus sabina* from a nature reserve to new areas of the Russian steppe (Case Study 13.2).

While **restoration ecology** has a strong focus on species through reintroduction initiatives, it is also possible to focus on habitats and landscapes. This typically involves taking a managed, disturbed, or unnatural area and returning it to a previous, more natural, state through the process of **rewilding**. Rewilding schemes can occur separately from reintroduction, but where rewilding is undertaken reintroduction will often occur alongside (and vice-versa) to create 'wild' habitats inhabited by the species that would, historically, have been present. Despite success stories, reintroduction and rewilding are far from being cheap or straightforward and there is no ultimate guarantee of success, as is evident in the case of the Lord Howe Island stick insect (Section 12.4.4; Figure 12.12).

In this chapter we will consider these conservation strategies from a practical perspective and discuss legislation and policy frameworks. We also debate the ecological challenges and opportunities in reintroduction and rewilding and highlight links to *in situ* management (Chapter 11) and *ex situ* breeding programmes (Chapter 12). Finally, the chapter will critically debate the role of reintroduction and rewilding in modern conservation and discuss the dilemmas facing the Applied Ecologist involved with such schemes.

13.2 Basic principles of species reintroduction.

Reintroduction involves humans putting animals or plants back into part of their **former native range** from which they have been extirpated. Any species that is the subject of a *re*introduction must thus have existed previously in the area into which it is being released, even if this was a long time ago. If this is not the case, the release of a species is an introduction of non-native species, which often has profound ecological implications, and sometimes legal implications too (Chapter 8).

Species reintroductions are usually deliberate and carefully planned. However, there can be situations when **accidental reintroduction** occurs (or, more correctly, a reintroduction that occurs outside of a formal, planned, and legally licensed programme). An example is wild boar *Sus scrofa* re-establishment in a number of locations in the UK. Boar became extirpated as a wild species around the end of the 13th century (Yalden, 1999), and their reintroduction occurred through a series of escapes from farmed stock or non-permitted releases from captivity.

Different ecologists can use the term 'reintroduction' slightly differently. Some (e.g. Kleinman, 1989) use 'reintroduction' exclusively for captive-bred releases and use the term 'translocation' for situations where wild-caught individuals are moved. However, **translocation** is a much broader topic—for instance, individuals can be caught and released to move them away from a development site (a rescue translocation: Hot Topic 5.1) or to move individuals between isolated populations for genetic reasons (Fischer & Lindenmayer, 2000). Movement of species outside their native range is also considered to be translocation: see Section 8.3).

Throughout this chapter, the term reintroduction is used for the re-establishment of a species at a particular location either from captive stock or other wild populations as long as this fits with the general terms of reintroduction (former native range; extirpation; deliberate intention) as per the long-standing International Union for the Conservation of Nature guidelines on reintroduction nomenclature (IUCN, 1987). By this definition, reintroduction will always involve **conservation translocation**, which is defined by the IUCN as 'the intentional movement and release of a living organism where the primary objective is a conservation benefit: this will usually comprise improving the conservation status of the focal species locally or globally, and/or restoring natural ecosystem functions or processes' (IUCN/SSC, 2013).

A true reintroduction occurs only when a species has been totally absent from an area. If additional individuals are released to boost an existing population, this is termed **reinforcement** (unless additional individuals are being added to a population that was originally reintroduced, in which case it is **supplementation**) (Section 12.4.4). On rare occasions, introductions can be undertaken specifically for conservation through the processes of **assisted colonization** (i.e. intentional movement and release of an organism outside its native range to avoid extinction of the focal species) or **ecological replacement** (i.e. intentional movement and release of an organism outside its native range to perform a task such as pollination to re-establish an ecological function lost through loss of a different species). The different terms are outlined in Figure 13.1. The concepts of assisted colonization and ecological replacement are discussed in Hot Topic 13.1.

13.2.1 Aims of species reintroduction.

The principal aims of species reintroduction should be to re-establish a population that is:

- **Viable**: This means that population size must be sufficient to ensure the population is viable in the short term *and* have sufficient growth potential to be self-supporting in the long-term, ideally without any future population supplementation (or, if supplementation is planned, for a limited

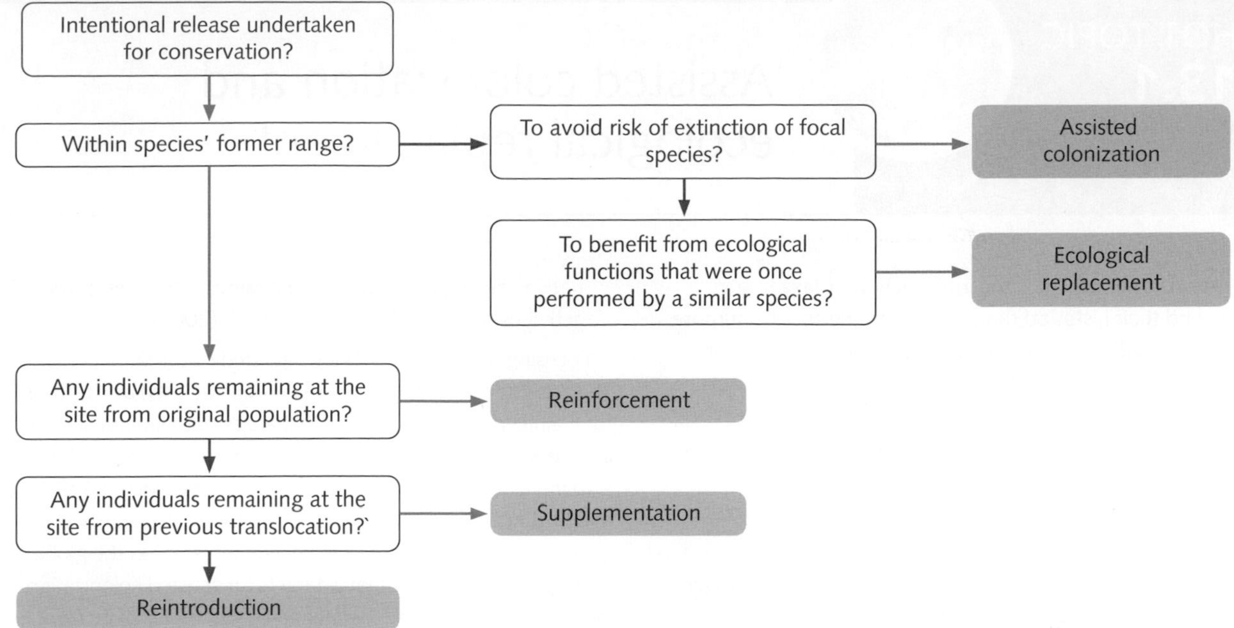

Figure 13.1 Schematic of reintroduction and translocation terminology. Green arrows indicate a 'yes' to the preceding question; red arrows indicate a 'no' to the preceding question.

time only). To achieve this, the number of individuals initially released (the founder population) must be at or above the short-term **minimum viable population (MVP)** (Section 10.3.1) and must have a suitable sex ratio and age structure to facilitate growth. It is also vital to ensure that the release site and adjoining land is big enough and has enough resources (for example food and nesting sites) to support a population at least as big as that needed to ensure the population is viable. In other words, the carrying capacity must not be lower than the MVP. These considerations are discussed in more detail in Section 13.4.3.

- **Self-supporting:** This encapsulates the notion that reintroductions should involve intensive effort only in the initial stages with minimal long-term management being required.
- **Free-ranging:** This underlines the importance of species being released into the wild, rather than in an enclosure, regardless of how large that might be. This was discussed for the Lord Howe Island stick insect *Dryococelus australis*, in Section 12.4.4. However, there are times when ecologists might consider reintroducing a species to form a population that can never be fully free ranging. For example, many African reserves such as Pilanesberg in South Africa and Etosha in Namibia are fenced, so a population reintroduced into these areas will never be truly free-ranging.

13.2.2 Reasons for reintroduction.

The global distribution of reintroductions is geographically biased, with most activity in developed regions (see map in Figure 13.2). There are many reasons for reintroducing a species to its native range. These can broadly be divided into three groups, which are not mutually exclusive:

1. **Species-focused motivations:** reintroductions undertaken as part of a conservation programme focused on a particular species (usually considered as the most common reason for reintroduction).
2. **Site-focused motivations:** reintroductions undertaken because of the role of a species within a particular ecosystem or at a particular site.
3. **Human-focused motivations:** reintroductions undertaken either because of the benefits or pleasure that humans derive from a species being present

HOT TOPIC 13.1

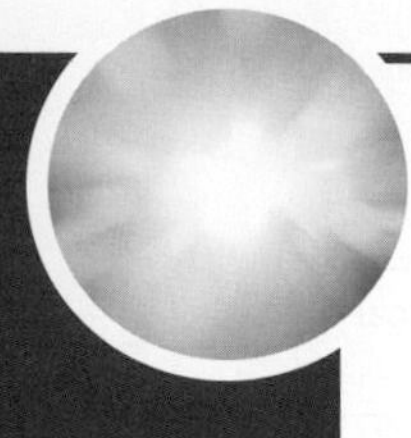

Assisted colonization and ecological replacement

Assisted colonization is the introduction of taxa to sites beyond their historical range, but with the specific purpose of conservation rather than aesthetics, agriculture, pest control, or any of the other reasons that typically motivate introduction of non-native species (Chapter 8). This concept has received considerable attention recently, and the risks and benefits have been vigorously debated.

There are three main reasons why assisted colonization might be considered:

1. To conserve a focal species through *in situ* conservation, but where there is no suitable site within the historic range because of habitat lost or **unresolvable threats** (e.g. inability to realistically control disease risk). This might follow a successful captive breeding programme and subsequent lack of reintroduction sites. Alternatively, it might be part of a rescue translocation where individuals of a species need to be collected from one site where they are under threat and moved somewhere else, but again when a suitable release site does not exist within that species' native range.
2. To buffer species against **climate change** by moving them to sites that have, or will have, suitable climate conditions. This might especially be done for poor-dispersing species, species that are strongly tied to a particular climate, or when the 'climatic envelope' (Section 7.5.1) moves a considerable distance or across barriers so that natural colonization is highly improbable. Such issues are discussed further in Gallagher et al. (2015).
3. To create or recreate **ecological function** through the process of ecological replacement. This involves the release of a species outside its historic range in order to fill an ecological niche left vacant by the extirpation of a native species, and is akin to the 'anticipatory restoration' activities proposed by Manning et al. (2009).

Focusing on climate-motivated assisted colonization, Thomas (2011) has suggested that Britain be used as a refuge for mainland European species that are climate-limited and will be at risk of extinction under climate change. He argues that 'expanding the dispersal of endangered species may represent the most effective climate change adaptation strategy available to conservationists to reduce extinction rates across the globe'. He highlights several potential targets for assisted colonization, including the world's most endangered cat, Iberian lynx *Lynx pardinus*, Spanish imperial eagle *Aquila heliacea adalberti*, and the Provence chalkhill blue butterfly *Polyommatus hispanus*. This last is currently restricted to northern Spain, southern France, and northern Italy and is at serious risk of extinction from climate change; southern England is predicted to become climatically suitable and the butterfly's host plant grows on calcareous grasslands in southern England.

At least one private group is already acting under the banner of 'citizen activism'. The Torreya Guardians have been planting seeds and seedlings across the eastern USA to expand the range of an endangered conifer *Torreya taxifolia*, whose modern distribution is confined to the Florida panhandle (Torreya Guardians, 2016).

However, the concept of assisted colonization has many critics. Ricciardi & Simberloff (2008) argue that 'even if preceded by careful risk assessment, such action is likely to produce myriad unintended and unpredictable consequences'. They point out the unpredictability of species introduction and the serious consequences that can occur as a result of introduction, as discussed in Chapter 8.

QUESTIONS

Given the ecologically sensible strategy of generally avoiding, and even preventing, non-native species introductions wherever possible, is it completely counter-intuitive to even consider assisted colonization?

What, if any, changes in legislation and policy would be needed for assisted colonization to become permissible and acceptable in different countries?

If assisted colonization is adopted, what safety measures would you put in place to minimize the risk of problems?

REFERENCES

Gallagher, R.V., Makinson, R.O., Hogbin, P.M., & Hancock, N. (2015). Assisted colonization as a climate change adaptation tool. *Austral Ecology*, Volume 40, 12–20.

Manning, A.D., Fischer, J., Felton, A., Newell, B., Steffen, W., & Lindenmayer, D.B. (2009) Landscape fluidity–a unifying perspective for understanding and adapting to global change. *Journal of Biogeography*, Volume 36, 193–199.

Ricciardi, A. & Simberloff, D. (2009) Assisted colonization is not a viable conservation strategy. *Trends in Ecology & Evolution*, Volume 24, 248–253.

Thomas, C.D. (2011) Translocation of species, climate change, and the end of trying to recreate past ecological communities. *Trends in Ecology & Evolution*, Volume 26, 216–221.

Torreya Guardians (2016) Available at: http://www.torreyaguardians.org

at a particular site, or the view that humans have an ethical obligation to reintroduce species extirpated through human activity.

Species-focused motivations for reintroducing species

As detailed in Chapter 12, captive breeding and propagation undertaken in zoological parks, aquaria, and botanical gardens can be undertaken to facilitate the continuation of those captive collections or as a form of 'ark': essentially a captive-dwelling **insurance population**. Alternatively, *ex situ* conservation can involve breeding programmes with the ultimate aim of reintroduction. In such cases, the *ex situ* conservation is one part of a much broader **species recovery plan** that likely also involves *in situ* management and eventual reintroduction. These schemes are generally internationally managed and usually involve studbooks (Section 12.4.3).

Taking individuals into captivity can be necessitated by severe and hard-to-manage threats in the wild that put the species at imminent risk of complete extinction. Alternatively, for species with a very small, but stable wild population, collection might be undertaken to increase the size (and thus the long-term viability) of the global population. Reproductive success and survival rates are often higher in captivity and captive-bred individuals can be used to supplement the wild population. The primary aim of reintroduction to the wild after successful captive breeding initiatives is to increase the chances of long-term species survival. This fits with one of the founding principles of conservation ecology: to ensure, wherever possible, that a species is not represented solely in captive collections.

Sometimes, reintroduction involves translocating individuals to a new site without an intermediate captive breeding step. This can fit with a species-focused motivation for reintroduction in one of two main ways:

1. **Increasing the number of wild populations:** this can mitigate the risk of something going wrong at any one site (disease outbreak, non-native predator being introduced, sea level rise, etc.). It also reduces the risk that loss of any one population, either due to threats or chance extinction, has a substantial effect on the survival of overall species. In an extreme case, with a species that exists at only one site, extirpation of that population equals extinction of the entire species.
2. **Increasing the total global population:** this might happen if the size of one population is being limited by carrying capacity (e.g. food resources are limiting the population growing). Translocating some individuals can then allow the initial population to expand up to the carrying capacity again while the new population can also grow, thereby increasing the total global population.

If enhancing species survival is the main motivation for reintroduction, as is commonly thought, it follows that most reintroductions should involve species listed as threatened on the IUCN red list (Section 10.3.2). However, this is often not the case. Seddon et al. (2005) found that in a sample of 138 reintroductions, 57% involved species listed as non-threatened (IUCN categories of least concern or near threatened), while only 21% involved endangered or critically endangered species. There were just three reintroductions (<1% of the total) that involved species extinct in the wild; including the California condor (Case Study 12.1). Similarly, only 50% of the avian and mammalian reintroductions examined by Fischer & Lindenmayer (2000) were undertaken for conservation reasons.

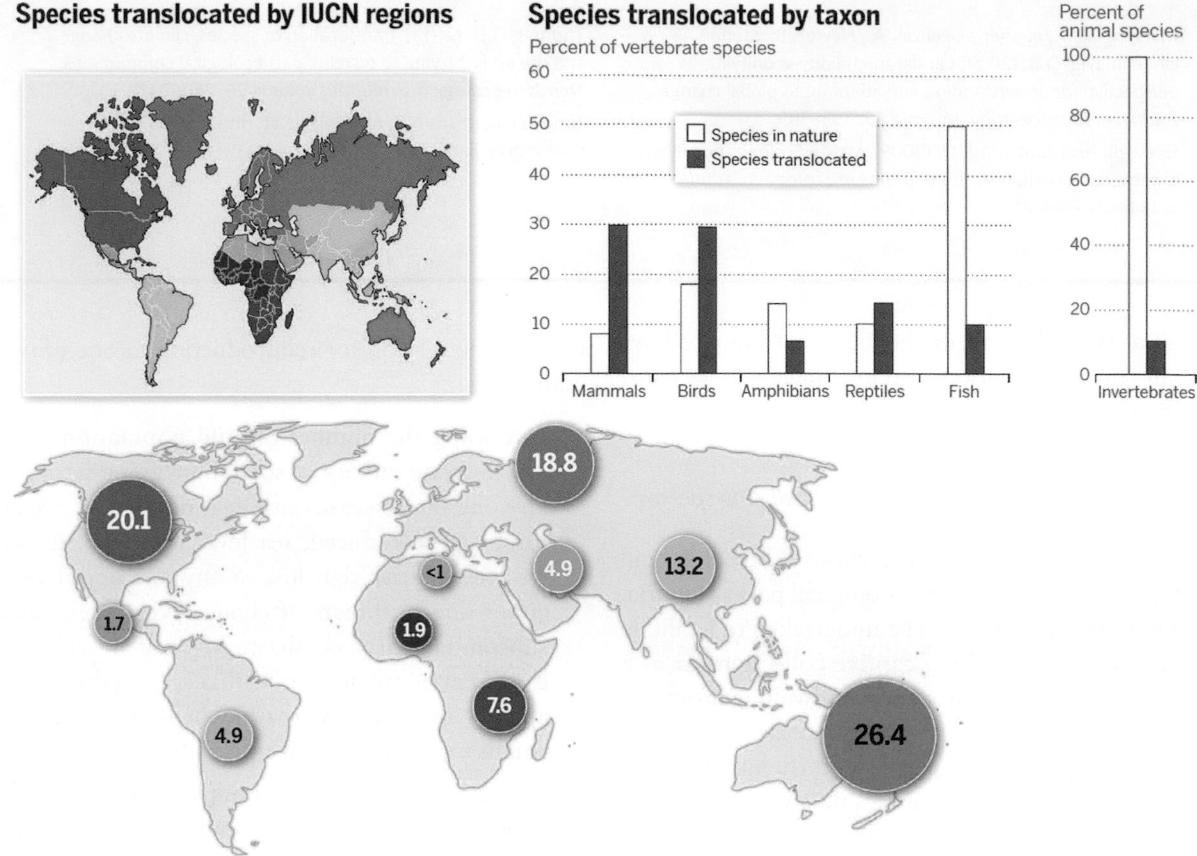

Figure 13.2 Global range of conservation translocations (reintroductions and reinforcements) by IUCN region. Small map shows the countries involved, main map shows the proportion of species reintroduced from a total of 303 reintroductions.

Source: Image reproduced from Seddon, P. J. (2014) 'Reversing defaunation: Restoring species in a changing world'. *Science*, 345(6195), 406–412, used with permission of The American Association for the Advancement of Science.

Occasionally a reintroduction scheme focuses on a **conservation surrogate**—a non-endangered species that is closely related taxonomically or ecologically (ideally both) to the eventual target species. This approach can identify problems that may affect the target species in later reintroductions. This piloting approach was used in Hawaii when a bird called the common Hawai'i 'amakihi *Hemignathus virens virens* was reintroduced as a test case for the reintroduction of endangered honeycreepers Carduelinae. This demonstrated that although suitable methods existed for captive breeding and release, reintroduction would be unlikely to be successful unless malaria-free reintroduction sites were available (Online Case Study Chapter 8) or strategies are developed to decrease mortality in honeycreepers exposed to disease after release (Kuehler et al., 1996). In this case, the pilot and target species were similar taxonomically (different subfamilies of the same family) and ecologically.

Site-focused motivations for reintroducing species

Some reintroductions can be motivated by the **ecosystem function** of a species and the importance of that function at a particular site, rather than the rarity of the species in question. Indeed, it has been argued that the primary goal of translocations should be to restore ecosystem function or balance rather than species composition (Armstrong & Seddon, 2008). Types of species particularly likely to have an ecosystem function include:

- **Ecosystem engineers**: species that change the physical nature of the habitat. For example, the

Eurasian beaver *Castor fiber*, which alters riparian habitat to create shallow pools used by dragonflies, has been reintroduced in a number of countries, including France, Germany, Spain, Belgium, and the UK. Beavers also increase the structural complexity of woodland by felling trees in a way that produces results similar to coppicing (Figure 11.7).

- **Apex predators:** species at the top of the food chain are important in ecosystem balance and thus another reintroduction target. A good example is the grey wolf *Canis lupus* which, following reintroduction into America's Yellowstone National Park, has reduced overgrazing by elk *Cervus canadensis* and allowed native plant communities to rebound as shown in Figure 13.3.
- **Keystone species:** species that have a pivotal role in an ecosystem (Section 10.5.2). For example, agoutis *Dasyprocta leporina* were reintroduced in Tijuca National Park, Brazil, partly due to the role of this South American rodent in seed dispersal (Cid et al., 2014).

Figure 13.3 Vegetation effects of wolf *Canis lupus* reintroduction to Yellowstone showing a lack of recent aspen recruitment (aspen <1 m tall) in an upland site (a; 2006) and the same site after reintroduction showing recent aspen recruitment with some aspen >2 m tall as a result of a reduction in grazing by elk *Cervus canadensis* (b; 2010). The dark, furrowed bark comprising approximately the lower 2 m of aspen boles represents long-term damage due to bark stripping by elk.

Source: Reproduced from Ripple, W. J., & Beschta, R. L. (2012) 'Trophic cascades in Yellowstone: The first 15 years after wolf reintroduction'. *Biological Conservation*, 145, 205–213, used with permission from Elsevier.

One of the challenges of the keystone species concept in general is that while it is intuitively easy to understand, identifying keystone species in real ecosystems is extremely challenging. Often species are identified as keystone only when they no longer exist in an ecosystem, which in turn makes them prime targets for reintroduction. The sea otter *Enhydra lutris* is a famous example. Sea otters were extirpated in British Columbia, Canada, leading to a population explosion in sea urchins *Echinoidea*, their main prey. This decreased the abundance of kelp Laminariales, on which the sea urchins fed, with concomitant effects on fish that relied on kelp forests for shelter (Estes & Palmisano, 1974). Sea otters were reintroduced to British Columbia directly from Alaska in the early 1970s (without a captive breeding intermediate step), which rebalanced the ecosystem as shown in Figure 13.4.

Human motivations for reintroducing species

There can be non-ecological reasons for reintroducing a species. For example, some people feel there is an **ethical responsibility** to reintroduce species that became extirpated due to human activity, especially direct activity such as hunting. Ethical obligations to reintroduce species might be considered to be particularly strong if the absence of that species had trickle down effects on other species or the wider ecosystem, as in the case of keystone species.

Alternatively, a species might be reintroduced because it provides a valuable **ecosystem service** such as pollination (Menz et al., 2011), or for

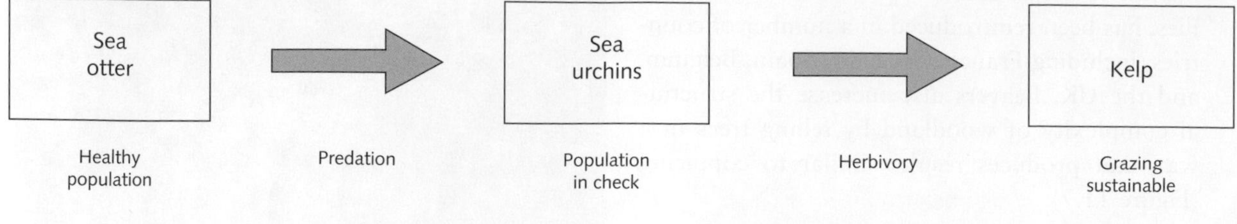

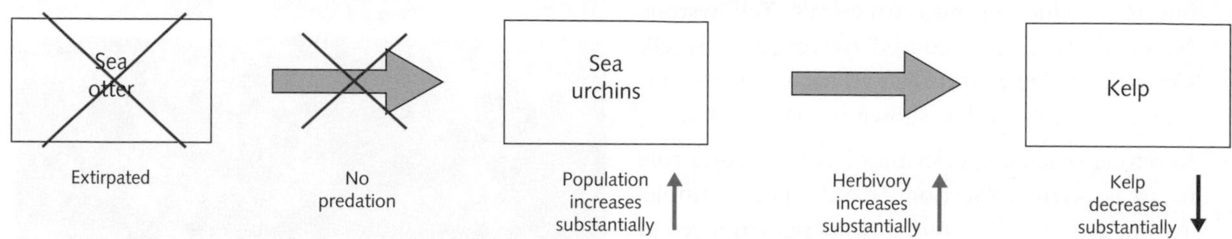

Figure 13.4 Ecosystem effects of extirpation of sea otters *Enhydra lutris* in British Columbia, Canada, which were reversed following their reintroduction.

aesthetic reasons. Proponents of reintroduction also highlight the potential of species reintroductions to enthuse and educate the public about ecological interactions, threats to biodiversity, extinction, and conservation (e.g. Dietz et al. (1994) for golden lion tamarins *Leontopithecus rosalia* in Brazil).

Another human motivation for reintroduction is **economic benefit**, especially potential tourism revenue. This can be seen, for example, in national parks and private reserves of Africa where rare species are key to attracting visitors. A good example here is Operation Genesis in Pilanesberg National Park, South Africa. This volcanic crater was a series of farms in the 1970s before the land was cleared, the crater was fenced, and around 6000 animals were reintroduced in the early 1980s in one of the biggest reintroduction projects ever seen (Motlhanke, 2005).

However, although the economic benefits of species reintroductions can be substantial, as in the case of Pilanesberg, this is neither guaranteed nor, in many cases, predictable. For example, before the European beaver *Castor fiber* was introduced into Scotland in a managed trial, it was estimated that the scheme could bring in over £2 million per year into the local economy, largely through tourism, with the 'pessimistic estimate' being a yield of ~£750,000 per year (Campbell et al., 2007). Ultimately, however, the value of 'wildlife experiences' was a much more modest £355,000–520,000 in total between 2008 and 2014 (Moran & Lewis, 2014).

13.2.3 Taxonomic bias in reintroduced species.

There tends to be a taxonomic bias in species that are the subject of reintroduction schemes. Seddon et al. (2005) examined the species involved in 699 reintroduction schemes listed in the IUCN Reintroduction Specialist Group database or otherwise described in the academic literature. The proportion of species in each taxonomic group was then compared to the taxonomic groupings of all known species. The findings suggest vertebrates are numerically overrepresented in reintroduction schemes relative to their abundance in nature, compared to plants and invertebrates, which were both underrepresented. Further analysis by Seddon et al. (2014) showed that this was largely driven by many more mammals and birds being reintroduced relative to their proportion in nature; the same was true, to a lesser extent, for reptiles. See Figure 13.2 for more details.

In addition to differential reintroduction effort between broad taxonomic groups, there is also bias within taxonomic groups. For example, mammals account for around 8% of all described vertebrate species, but over 40% of vertebrate reintroductions, while fish account for 50% of vertebrates, but fewer than 5% of vertebrate reintroductions. Within mammals there are further taxonomic differences, with carnivores such as big cats and ungulates such as antelope being overrepresented while rodents and bats are underrepresented.

Some of this taxonomic bias is likely to be driven by some species being more **charismatic** than others (flagship species: Section 10.5.2) and thus potentially more popular and fundable. It should, however, be noted that while it is certainly true that vertebrates, especially mammals and birds, are overrepresented numerically, there is often a strong case when deciding which species to reintroduce to seek to improve the overall structure of an ecological network by replacing species that have the greatest ecological function. Such species often tend to be larger vertebrates, so part of the reintroduction taxonomic bias might be deliberate and appropriate, rather than problematic.

FURTHER READING FOR SECTION

Review of reintroductions: Seddon, P.J., Griffiths, C.J., Soorae, P.S., & Armstrong, D.P. (2014) Reversing defaunation: restoring species in a changing world. *Science*, Volume 345, 406–412.

Ethics of reintroduction and de-extinction: Sandler, R. (2014) The ethics of reviving long extinct species. *Conservation Biology*, Volume 28, 354–360.

Taxonomic bias in reintroduced species: Seddon, P.J., Soorae, P.S., & Launay, F. (2005) Taxonomic bias in reintroduction projects. *Animal Conservation*, Volume 8, 51–58.

13.3 Reintroduction regulatory frameworks.

It is difficult to identify the first reintroduction scheme, but a likely contender is the release of 15 American bison *Bison bison* into the newly established Wichita National Forest and Game Preserve in Oklahoma in 1907 as shown in Figure 13.5 (Kleinman, 1989). The 1907 bison reintroduction was ground-breaking, not only in being such an early reintroduction, but also in anticipating the need for careful planning, pre-release assessment, strong community support and media backing (Beck, 2001).

The best practice adopted in 1907 has not always been upheld and although some reintroductions been very successful, others have been catastrophic failures. **Failed reintroductions** include those that have fallen short of recreating a viable and self-supporting population of the target species, as well as those that have created notable knock-on effects on the wider ecosystem. Such problems have often arisen because reintroductions have been ill-conceived or poorly planned. This checkered history, which has demonstrated both the potential of reintroductions and the consequences of failure, has led to the reintroduction policy and regulatory frameworks in place today.

13.3.1 Reintroduction policies.

In general, there is international support for conservation-oriented reintroductions. This is articulated most clearly in Article 9 of the **Convention on Biological Diversity.** This Article requires the signing parties (>160 countries) to adopt, as far as possible and appropriate, 'measures for the recovery and rehabilitation of threatened species and their reintroduction into their natural habitats under appropriate conditions'.

Figure 13.5 One of the first ever reintroductions: American bison *Bison bison*, Oklahoma, 1907.
Source: Photograph copyright Wildlife Conservation Society; reproduced by permission of the WCS Archives.

At a European level, the Convention on the Conservation of European Wildlife and Natural Habitats (the **Bern Convention; 1979**), to which >50 member states subscribe, also specifically encourages reintroductions. Article 11 requires the parties to 'encourage the reintroduction of native species of wild flora and fauna when this would contribute to the conservation of an endangered species, provided that a study is first made in the light of the experiences of other Contracting Parties to establish that such reintroduction would be effective and acceptable'.

13.3.2 Codes of practice for reintroduction.

The International Union for the Conservation of Nature (IUCN) Species Survival Commission (SSC) set up a Reintroduction Specialist Group (RSG) in 1988. Together with the Invasive Species Specialist Group (ISSG), the RSG created a set of **comprehensive guidelines** on reintroductions (IUCN/SSC, 2013). This guide to best practice is firmly rooted in the practicalities of reintroduction and covers every step of the progress from initial idea, through harvesting and release, to post-release monitoring and management.

The Association of Zoos and Aquaria has also devised guidelines focusing specifically on reintroductions that use captive-born individuals or those that have been in captivity for some time between capture and release (AZA, 1992). Although written some time ago, these guidelines are still relevant and useful.

13.3.3 Legislation for reintroduction.

The legislative framework for reintroductions is rather more complex than the logical, all-encompassing, codes of practice (Section 13.3.2). This is because the legislation with which reintroduction schemes must comply differs between countries and, in some cases, between states/provinces/counties within the same country. Moreover, there is often different legislation for different species and to regulate different parts of the reintroduction process (harvesting/collection, translocation, and release, as discussed in Section 12.4).

The legislative process is most complicated for reintroductions where collection of individuals takes place in a different country from the release site, especially if there is an intermediate captive breeding step. The legislative considerations for harvesting individuals from wild populations (for direct translocation to a new site or into captivity as part of a longer-term initiative) and moving those across country borders is covered in Section 12.4.2. Briefly, the necessary **permits** must be obtained for collection, transport of organisms within the country of collection, export of live organisms, transport of live organisms, and importation into the destination country. For reintroductions, additional permits are needed for release of organisms. Even the most legislatively straightforward reintroductions—when individuals from one population are harvested and translocated directly to a release site within the same country—usually still require permits covering collection, transport, and release.

The most important **international convention** pertaining to the regulation of reintroductions is the Convention on International Trade in Endangered Species of Wild Fauna and Flora (CITES) regulations. Reintroduction of species listed on Appendices 1, 2 or 3 (e.g. tiger *Panthera tigris*, Saint Lucia parrot *Amazona versicolor*, and most orchids including *Laelia lobata*) will need to comply with CITES requirements. Trade is prohibited, unless in exceptional circumstances, for species in Appendix 1, and is regulated for species in Appendices 2 and 3. Although reintroductions often constitute 'exceptional circumstances', the listing of species on CITES means that there are additional legislative hurdles to overcome. Export permits are required for species in Appendices 1 and 2; Appendix 1 species also require an import permit. The species in each Appendix are regularly revised.

Most countries have their own **national legislation** for what species can be collected from the wild and in what circumstances this is permissible, as well as transportation, import, and release. This tends to be undertaken through the principal legislative Act relating to wildlife—for example the Wildlife and Countryside Act 1981 (as amended) in the UK—usually together with nationally invoked import legislation and quarantine regulations.

Quarantine

In the case of reintroductions where animals are being translocated from a wild source population directly to a new release site in another country, individuals might need to be temporarily quarantined at a **quarantine facility.** This might be necessary to prevent, or at least reduce, the risk of translocating diseases and parasites with hosts.

The need for, and length of, quarantine depends on the import country's rules, the species concerned, and sometimes the export involved. Quarantine procedures are generally regulated through national legislation, but sometimes these relate to other statutes. If species need to be transported through intermediate countries (perhaps because overland travel is the only viable option) then there are likely additional legislative and quarantine obstacles to overcome.

Within Europe, Council Directive 92/65/EEC (**Balai Directive**) regulates the import and export of animals within Europe and between Europe and the rest the world. Zoos and other animal collections can seek approved status under the Directive by complying with strict veterinary protocols. **Approved centres** can then receive animals from other approved centres directly, without the need for quarantine. This facilitates the internationally coordinated captive breeding programmes (Chapter 12), including those leading towards reintroductions. Given that the veterinary protocols are agreed at a European level, release of individuals from approved centres into the wild as part of a planned reintroduction does not normally give rise to disease concerns, even if the approved centre and release site are not in the same EU country.

It is important to note that quarantine facilities and approved centres must be kept separate (even if on the same overall premises) as the entry of wild-caught individuals with no disease history or veterinary care into an approved centre would compromise the whole approved centre network. However, as Case Study 12.2 illustrated, in some cases unknown diseases can be accidentally introduced into wild populations despite all these precautions, as in the case of the Mallorca midwife toad *Alytes muletensis*.

FURTHER READING FOR SECTION

Reintroduction best practice: IUCN/SSC (2013) *Guidelines for Reintroductions and Other Conservation Translocations*. Gland: IUCN Species Survival Commission.

Guidelines on reintroduction of captive animals: AZA (1992) *Guidelines for Reintroduction of Animals Born or Held in Captivity*. Available at: https://www.aza.org/assets/2332/aza_guidelines_for_reintroduction_of_animals.pdf

CITES Appendices: Available at: https://www.cites.org/eng/app/appendices.php

Import/export of species to/from UK: Available at: https://www.gov.uk/government/collections/guidance-on-importing-and-exporting-live-animals-or-animal-products

13.4 The process of species reintroduction.

There are many stages of the species reintroduction process. An individual reintroduction scheme does not have to follow all stages in the exact order outlined below, but the basic structure is fairly typical.

13.4.1 Initial feasibility study.

The first stage in any reintroduction should be an initial feasibility study. This is the equivalent of a scoping study for Ecological Impact Assessment (Chapter 5): the basic idea is considered and a decision is made as to whether it could be desirable and realistic, and thus warrants further consideration, or not. The basic stages are:

1. Consideration of the **motivations** for reintroducing a given species. This will affect all aspects of the project, including possible benefits. Some of the reasons for reintroduction, and counter-arguments, are discussed in Hot Topic 13.2.
2. Drawing up of a **species profile**. If there is insufficient information on species biology and requirements, the viability of any reintroduction attempt must be questioned.
3. Assessment of the **original threats** that have caused the species to become extirpated. These either need to have been resolved or be capable of being managed at release sites. If the causes of the original extirpation are not understood or cannot be managed, any reintroduction attempt would be unlikely to succeed and should not be taken further.
4. Completion of a **generic risk assessment**. This should outline the potential benefits of reintroducing the species and discuss potential negative effects on ecology, environment, economy, and local community. If negative impacts are likely, detail should be provided on how serious these could be and how they could be mitigated. It should draw on current knowledge from both research and practitioner literature, with a specific focus on the focal species' interactions with co-occurring species and wider environment, as well as information on any previous reintroductions. If no previous reintroductions have been undertaken for the focal species in the same or similar habitat, it can be useful to examine past reintroductions of closely related or ecologically similar species. For example, when considering the feasibility of reintroducing marsh fritillary butterflies *Euphydryas aurinia* to the Netherlands, Jensen (2007) considered reintroduction of the same species in England *and* reintroduction of other butterflies such as the small pearl-bordered fritillary *Clossiana selene* to the Netherlands.

HOT TOPIC 13.2

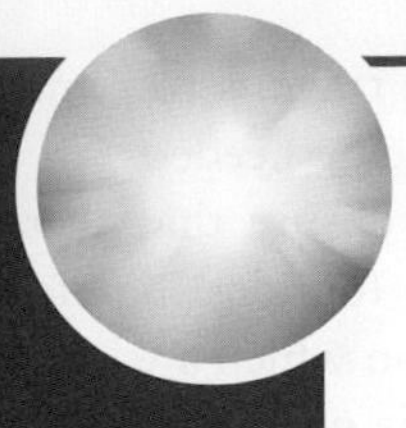

Arguments for and against reintroduction

Many Applied Ecologists have strong views as to whether reintroduction is an important part of conservation. There are plenty of arguments both for and against such initiatives, as outlined below, but it is vital to appreciate that there is no 'one size fits all' answer: relative advantages and disadvantages differ according to species, site, and motivations for reintroduction.

Arguments for...

- Can be important for species survival by increasing global numbers and number of populations to prevent human-accelerated extinction.
- Species belong in the wild, not in captivity, so the ultimate aim of *ex situ* conservation should be reintroduction.
- Can restore natural and self-regulating ecosystems and benefit multiple other species, especially why keystone or ecosystem engineers are involved.
- Can be an economic benefit (restoration of ecosystem services, ecotourism potential).
- Ethical responsibility.
- Has education potential and engaging people with nature and the natural world

Arguments against...

- Can be very expensive—the same money could be better spent conserving species already present in an area to prevent further decline (potentially bigger 'bang for the conservation buck').
- Relatively high failure rate.
- Can negatively affect other species, especially if the ecosystem has changed between original extirpation and reintroduction, and previously co-occurring species have adapted accordingly.
- Often focus on species doing well in the rest of their range (understandable if the motivation for conservation is motivated by what the species does; less so if it is motivated by species rarity in the focal area alone)

QUESTION

Having read the arguments above, are you for or against the reintroduction conservation approach?

FURTHER READING

Overview of the main reintroduction concepts and future directions: Armstrong, D.P. & Seddon, P.J. (2008) Directions in reintroduction biology. *Trends in Ecology & Evolution*, Volume 23, 20–25.

5. In rare cases, a **trial release** might be undertaken. For example, Sharifi & Vaissi (2014) released yellow-spotted mountain newts *Neurergus microspilotus* in western Iran, to establish whether juveniles were easy to translocate and survived their first winter, to inform a future, larger, reintroduction project.

Ultimately, the initial feasibility study is about weighing up the potential advantages and disadvantages of a reintroduction and assessing its chances of success. If the initial feasibility study is favourable, a detailed plan will be formulated to consider the following (outlined here and considered in more detail in Sections 13.4.2–13.4.5):

- **Founder individuals:** availability of source populations, genetic similarity and variability, morphology, and behaviour.
- **Release sites:** suitability in terms of habitat, carrying capacity, socioeconomic acceptability.
- **Methods:** translocation and release techniques.
- **Monitoring:** long-term monitoring and adaptive management.

13.4.2 Founder individuals.

The individuals released within a reintroduction scheme to start a new population are the founder individuals. As there is usually no (planned) supplementation of the population with additional individuals over time, the entire new population is founded on these individuals. Choosing founder individuals is thus of paramount importance.

Source and characteristics of founders

The first thing to consider is the source of the founder individuals. For animals, there are two basic options: using **wild-caught** individuals via a direct translocation or using **captive-bred** individuals. There are advantages and disadvantages to both scenarios, as highlighted in Table 13.1.

It is important that founder individuals are well-suited to their new environment. This is most likely when founders are the **same subspecies** as the extirpated population and similar morphologically and behaviourally. When it is not possible to determine this directly (e.g. if the characteristics of the extirpated population remain unknown) choosing a source population that is geographically close to the release site and from similar habitat can help increase the chances of the released individuals being well suited to the new environment. Close proximity can also reduce travel stress and reduces the chance of quarantine being needed.

It is vital to ensure that there is enough **genetic diversity** in the founder individuals to have a suitable gene pool on which a whole new population can be founded. This means selecting individuals known not to be inbred (captive-bred founders) or where inbreeding is unlikely because of the size of the source population and lack of history of **population bottlenecks** (wild-caught individuals).

It is sometimes necessary to harvest from more than one source population or use a **non-ideal source population**. For example, the ideal source of red kites for reintroduction to England was the red kite population still present in Wales. This was close geographically to the release sites (indeed the Welsh population was the remnant of the population that was once widespread over England and Scotland) and so the birds would have been close genetically to the extirpated populations. Moreover, harvesting birds from there would have allowed immediate release into the new areas as quarantine would have been unnecessary. However, the Welsh population was too small to withstand harvesting. As a result, the majority of individuals came from Sweden or Spain (via quarantine) where the populations were much larger and more buoyant.

The reintroduction of the short-haired bumblebee, which touches on many of the issues discussed in the preceding paragraphs, is discussed in Case Study 13.1.

Table 13.1 Advantages and disadvantages of using wild-caught and captive-born individuals in reintroduction.

Founders	Advantages	Disadvantages
Wild-caught individuals via direct translocation	• More likely to have natural behaviours • Less likely to fail due to lack of successful foraging/hunting • Less likely to be inbred	• More likely to translocate pests, parasites, or pathogens • Quarantine might be required; can exacerbate stress • Genetic history unknown • Can comprise source population if too many individuals are harvested
Captive-born individuals following captive breeding programme	• Genetic history known • Can be the only possibility if species is only represented in captivity or is the only source where harvesting a sufficient number of individuals is feasible	• Might not have a full suite of wild behaviours; could be more prone to poaching if humans not seen as a threat • Can be more likely to be involved in wildlife:human conflicts • Can be less likely to disperse from release area if used to enclosures • Might be less successful at foraging/hunting

CASE STUDY 13.1

Reintroduction of the short-haired bumblebee

The short-haired bumblebee *Bombus subterraneus* was once widespread across the south of England (Figure A). The species declined following the Second World War primarily due to changes in farming practices and habitat loss. It was last recorded at Dungeness in Kent in 1988 and was declared extirpated in 2000.

The species has recently been the subject of an especially complex reintroduction project, which aimed to:

1. **Reintroduce** the short-haired bumblebee in the UK. Meeting this primary aim involved creating a wild population that is viable and self-supporting.
2. Establish a **mosaic** of suitable bumblebee habitat linked by suitable **habitat corridors** throughout the southeast of England to facilitate the species to spread and **naturally colonize** new areas. This is a much longer-term goal than simply establishing a species at one new site and should be best practice.
3. Raise the profile of bumblebee conservation through **public outreach**. Another very sensible aim that shows the wider benefit that a reintroduction scheme can have, especially where other similar species are endangered through habitat loss and loss of native flowers, as is the case for other bumblebees.

Originally, New Zealand was going to be used as the source population. The reason that somewhere so far away was initially selected was that the New Zealand bees were directly descended from British bees exported some 120 years before to pollinate crops (Lye et al., 2010). However, there were substantial logistical issues with an antipodean source population, not least that the climate and seasons are completely opposite, so collecting bees in the New Zealand spring or summer would mean releasing them in the British autumn or winter. In addition, there were problems with captive rearing of queen bees in New Zealand in trials in 2009/10 and, more worryingly, genetic research showed high levels of inbreeding in the population, potentially reducing the fitness of any reintroduced stock.

Because of these issues, a European source population was decided upon. The most viable European population occurred in Sweden, which also has the benefit of having a

Figure A The short-haired bumblebee *Bombus subterraneus*.

Source: Image courtesy of Martin Andersson/CC BY-SA 3.0.

broadly similar climate to the UK. In 2011, Natural England and the project partners agreed to change the reintroduction source location to Skåne in Sweden.

In the springs of 2012 and 2013, bumblebee queens were released near Dungeness, the last site where they had been recorded some 24 years previously. Additional queens were released in spring 2014 and 2015 but, in this case, only after **disease screening** in Sweden to reduce the number being imported with disease. Worker bees have been seen in the Dungerness area, which means that at least one queen must have survived to create a new colony (Bumblebee Conservation Trust, 2015).

The key to a successful reintroduction of a species like this one, which is so completely dependent on its habitat, is the quality of those habitats. The project created bumblebee habitat prior to the reintroduction with over 800 hectares of flower-rich habitat within the release area of Dungeness and Romney Marsh **created**, **restored**, or **improved**. Likely as a result of these habitat improvements, England's rarest bumblebee, the shrill carder bee *Bombus sylvarum*, has returned to the area after a 25-year absence and the ruderal bumblebee *Bombus ruderatus* has come back after ten years (Bumblebee Conservation Trust, 2015). This demonstrates the trickle-down effects that are possible in reintroduction.

REFERENCE

Bumblebee Conservation Trust (2015) *The Short-haired Bumblebee Reintroduction*. Available at: http://bumblebeeconservation.org/about-us/our-projects/short-haired-bumblebee-reintroduction

Lye, G.C., Kaden, J.C., Park, K.J., & Goulson, D. (2010) Forage use and niche partitioning by non-native bumblebees in New Zealand: implications for the conservation of their populations of origin. *Journal of Insect Conservation*, Volume 14, 607–615.

For plant species, it is possible to transfer individuals from one site directly to another, or there can be an intermediate captive propagation step. For some reintroductions, a mixture of both methods is used, as was the case in the reintroduction of plant species to the Russian Steppe. This is discussed further in Case Study 13.2.

Population structure of founders

The number of individuals to release is one of the single most important decisions in a reintroduction plan. The ideal is to release as many individuals as possible since reintroduction success (Section 13.5) typically increases with the number of founders. There must be sufficient individuals released to allow positive population growth, buffer any mortality, and ensure that the founder population has sufficient genetic diversity. However, the source of individuals is always finite (and often very limited) and the costs increase with the number of founders involved, meaning that Applied Ecologists often have to find a difficult balance.

One tool to help determine the minimum number of founders is **population viability analysis** (PVA). This process was considered in Section 10.3.1 in terms of modelling whether a specific population had declined to the extent that it was no longer viable. In reintroduction, the idea is flipped to quantify the initial population size necessary for a particular species. This is a powerful approach and one that Seddon et al. (2007) have argued should be part of all reintroduction planning. The most common system is the freely available Vortex software (Armstrong & Reynolds, 2012). This has been used to model the number of beaver *Castor fiber* to release in Scotland (South et al., 2000) and to determine whether reintroducing New Zealand robins *Petroica longipes* to an island that was predator-free, but highly fragmented, would likely result in a viable population (Armstrong & Ewan, 2002).

In addition to providing a potential answer to the 'how many' question, PVA can also help determine the best composition of the founder population in terms of **sex ratio** and **age structure**. This is done after parameterizing the model with species-specific information such as longevity, age at first breeding, and mating strategy. For example, simulations for reintroduction of European bison *Bison bonasus* have shown that optimal herd growth rate and over all chances of population survival occur when around two-thirds to three-quarters of reintroduced individuals are female (Suchecka et al., 2014).

CASE STUDY 13.2

Reintroducting rare plants to the Russian Steppe

The Russian Steppe project run by the Botanical Garden of the Southern Federal University and the Botanical Garden of the Samara State University aims to recreate natural steppe communities. There are two parallel schemes using different methods.

1. Samara Region: planting individual species.

The natural plant community is **species-rich**, supporting more than 2500 species of vascular plants. Agriculture has led to the devastation of most of the steppe, but, in some areas, there has been a recent trend towards more extensive agriculture. This has led to secondary steppe communities developing through **secondary succession**. However, the replacement communities are **species-poor** and lack many of the valuable and rare species. Of particular note in the list of missing species are Red Data Book species, *Juniperus sabina* (considered **extirpated** in Samara Region for the last 50 years), *Paeonia tenuifolia* and *Iris pumila*.

Reintroduction sites were selected using the following criteria:

- Environmental conditions that meet the needs of the plants (soil, microclimate, etc.).
- Low intensity of anthropogenic pressure (cattle grazing, recreation, distance from settlements).
- Accessibility for initial translocation and monitoring.

Different methods were used for different species. In the case of *Iris pumila*, 150 new plants were **propagated** from captive stock in Russian botanical gardens. A similar strategy was used for *Paeonia tenuifolia* by splitting mother shrubs from captive stock to create 170 new plants. For *Juniperus sabina*, cuttings were obtained from three small populations in the Zhigulevskiy State Reserve. Reintroduction involved planting specimens to give the species an initial start, and also sowing seeds to create a long-term **seedbank**. Planted specimens were protected from the influence of adjacent plants by partial removal of **competitors**.

2. Rostov-on-Don Region: creating whole new communities using mass seeding.

This area has historically contained valuable plant communities supporting obligate steppe plants such as hermaphroditic grasses of the *Stipa* genus and rare tulips, *Tulipa* spp. Intense pressure from humans has almost completely destroyed the landscape, with almost 70% already decimated by agriculture and urbanization.

The specific goal for this phase of the project was to create sustainable steppe grasslands with *Stipa ucrainica* and *Stipa pulcherrima* as dominant species. These species were the focus not only because of their rarity (both are in the Russian Federation **Red Data Book**), but also because of their indicator status (*S. ucrainica* is an **indicator species** of dry steppe; *S. pulcherrima* is an indicator of stony steppe).

Seeding mixes were sown directly onto ploughed soil to create a large area of steppe on previously degraded land. The seeds were collected from relatively undisturbed natural steppe areas in the surrounding districted. In total, seeds of 98 typical steppe species from 28 families and 71 genera were collected to provide floristic diversity. Harvests were only made from sites with abundant *S. ucrainica* and/or *S. pulcherrima* as being indicators of good-quality natural steppe.

Summary.

These two projects demonstrate different approaches to reintroduction. The first is species-targeted, whereas the second is more community-focused, albeit using specific species as indicators. These case studies also showed the blurred distinctions that existed between **reintroduction** and rewilding (Section 13.6), with the second example tending towards a rewilding framework.

FURTHER READING

Shmaraeva, A. & Ruzaeva, I. (2009) Reintroduction of threatened plant species in Russia. *Botanic Gardens Conservation International*, Volume 6 , Art ID 0621.

Where founders are from a wild source population or populations, it is vital to ensure that the viability of the source population is not compromised by harvesting. Problems could occur if an unsustainable number of individuals are removed, or if harvesting affects the optimal sex ratio, age structure, or breaks up family groups (these and other issues are discussed in Section 12.4.1 in relation to collecting individuals for captive breeding). The PVA approach can thus also be used to determine whether harvesting a population as a source of individuals for reintroduction is likely to deplete that population to the extent that it is no longer viable. In the case of the bearded vulture *Gypaetus barbatus*, Vortex modelling demonstrated that the only captive population was not big enough to become source without substantial extinction risk (Bustamante, 1996).

Individual screening

Whether individuals come directly from the wild or a captive breeding scheme, the individuals being considered as potential reintroduction candidates will be assessed carefully. **Disease screening** will typically be undertaken (as for the short-haired bumblebee: Case Study 13.1), together with consideration of the individual's medical history for captive-bred individuals.

13.4.3 Choice of release site.

In some cases, the choice of release site drives the whole reintroduction. This generally happens when there are site-focused or human-focused motivations for reintroducing species such as ecosystem function or aesthetics (Section 13.2.2). In situations where species are the focus, or where the human motivation is general rather than tied to a specific site, the choice of release site becomes a key decision.

Habitat and carrying capacity

It is vital that the release site contains habitat appropriate for all life stages (adults and juveniles can have very different ecological needs) and throughout the year (summer and winter habitat needs, for example, can differ greatly). It should not be assumed that because a species used to occur at a site it will still be suitable, especially where a substantial time has passed since extirpation. In some cases, it is necessary to undertake **habitat restoration** or management before release.

It is important to consider the habitat's overall size to ensure that the carrying capacity of the area is at least as big as (and ideally substantially in excess of) the minimum viable population. Clearly, if the **carrying capacity**—the number of individuals that can be supported at a site—is below the minimum viable population, the release site is unsuitable. One of the world's most unusual release sites is the still-radioactive exclusion zone around the Chernobyl nuclear reactor in present-day Ukraine. The wisdom of using this release site is debated in Hot Topic 13.3.

Social factors

To be successful, reintroduction schemes need to have **public support.** Communities living in and around a release area will have legitimate interests in proposed reintroductions (IUCN/SSC, 2013). These can be complex, ranging from people worried about their own safety and potential impact on livestock—especially for carnivore reintroductions—to people concerned about the negative effects of possible ecotourism. Other people might be excited about a proposed release, perhaps because of the cultural or emotional value of having wildlife back in an area or because of possible economic benefits. Public perception of reintroductions, especially large carnivores, can often be either unrealistically positive or unrealistically negative.

It is vital that the local community is involved as a **stakeholder** in reintroduction plans as early as possible. Communication needs to be a two-way process. Those planning the reintroduction must maintain good, open, and honest communication with those affected by the proposal who must in turn be able to voice their concerns. Ensuring the stakeholders have the right information about a proposed reintroduction is vital, especially for carnivore species or species that are likely to modify an ecosystem.

Sometimes reintroductions are driven by public interest or demand, but more usually the onus is on the organizations involved in a potential

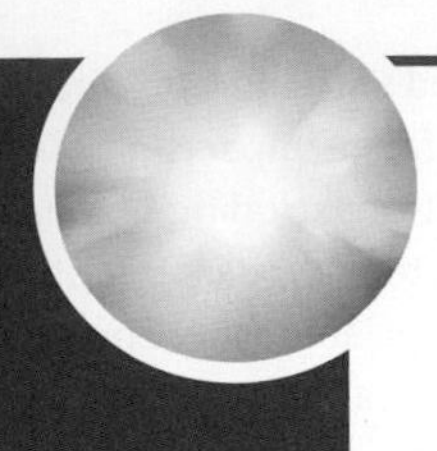

HOT TOPIC
13.3

Radioactive reintroductions: using Chernobyl's exclusion zone

On 26 April 1986, Reactor IV of the Chernobyl Nuclear Power Plant, in present-day Ukraine, went into meltdown. The worst nuclear disaster ever seen resulted in soil, water, and atmosphere being contaminated with radiation equivalent to 20 of the atom bombs that were dropped on Hiroshima and Nagasaki in World War 2.

Although the radioactive fallout into the atmosphere covered an extensive geographical area, the worst-affected area was immediately around the reactor. An initial 30-km exclusion zone was set up. This was quickly expanded to 4200 km^2 and all 116,000 people living in the zone were permanently evacuated (IAEA, 2006).

The initial ecological effects were substantial. One particularly highly irradiated area became known as the Red Forest because of the effects of radioactive dust. Over time, however, the exclusion zone has become a human-induced wilderness devoid of activities that normally threaten wildlife, such as agriculture, forestry, urbanization, and industry.

It appears that some species are booming. Helicopter survey data show rising numbers of elk, deer, and wild boar from one to ten years post-accident and no correlation between abundance and radiation level. Moreover, relative densities of these species within the exclusion zone are similar to those uncontaminated nature reserves; wolf abundance is actually more than 7× higher (Deryabina et al., 2015) (Figure A).

Somewhat counter-intuitively, European bison *Bison bonasus* has been reintroduced to the area and, in an even more unusual step, Przewalski's horses *Equus przewalskii* have also been released. In the case of the latter, this was a conservation introduction as the species was not historically present in the area. The decision to reintroduce Przewalski's horses came about because captive breeding had been very successful and space in captivity for the species was running out. The Chernobyl Przewalski reintroduction took place alongside a reintroduction to China, which started in 1985 and ended in 2005 (Xia et al., 2014).

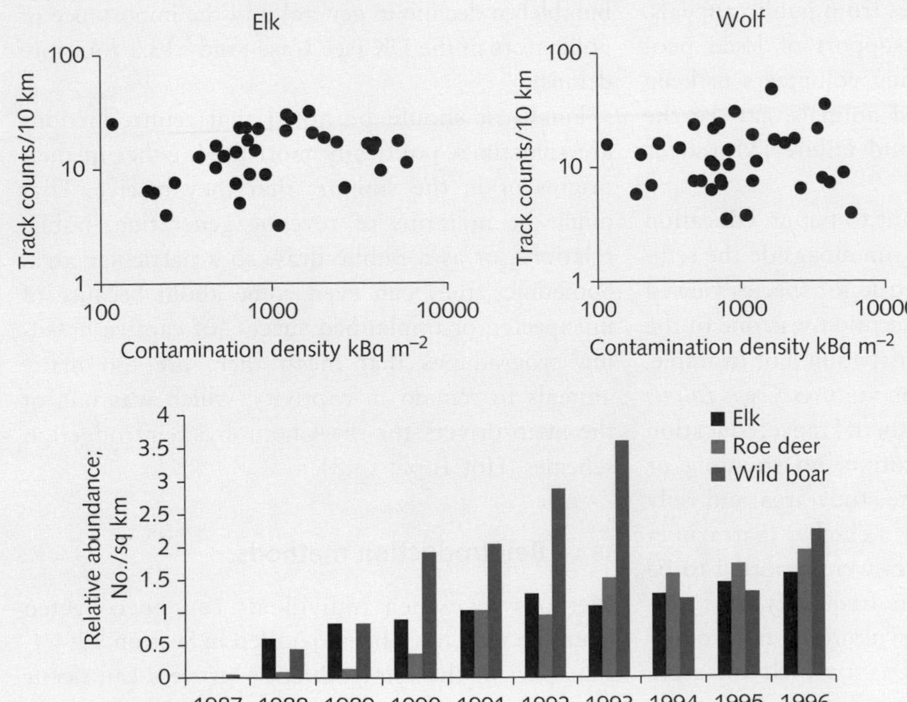

Figure A Mean number of track counts per 10 km in 2008–2010 for elk and wolf against mean radiation of each route (top) and change in relative abundance of three species in the 10 years after the Chernobyl accident (bottom).

Source: Reproduced from Deryabina, T. G. et al. (2015) 'Long-term census data reveal abundant wildlife populations at Chernobyl'. *Current Biology*, 25(19), R824–R826/CC BY 4.0.

Both (re)introductions have, so far, proven to be successful in establishing free-ranging and self-sustaining populations. The wisdom, or otherwise, of translocating species to an area known to be highly radioactive is uncertain: we simply don't know what the long-term effects will be on those individuals or on future generations in terms of genetic change or mutation. For now, though, translocated species seem to be thriving. As Baker & Chesser (2000) sum up: 'Radioactivity at the level associated with the Chernobyl meltdown does have discernible, negative impacts on plant and animal life. However, the benefit of excluding humans from this highly contaminated ecosystem appears to outweigh significantly any negative cost associated with Chernobyl radiation. Therein lies a paradoxical relationship.' The lessons that scientists can learn from this paradoxical relationship are outlined by the same researchers (Chesser & Baker, 2006).

QUESTIONS

It can be argued that reintroducing species to Chernobyl was a political, rather than ecological, decision. Do you agree and does this matter?

What could the long-term inter-generational effects of radiation be on Chernobyl's animals, both resident and reintroduced?

REFERENCES

Baker, R.J. & Chesser, K. (2000) The Chernobyl nuclear disaster and subsequent creation of a wildlife preserve. *Environmental Toxicology and Chemistry*, Volume 19, 1231–1232.

Chesser, R.K. & Baker, R.J. (2006) Growing up with Chernobyl. *American Scientist*, Volume 94, 542.

Deryabina, T.G., Kuchmel, S.V., Nagorskaya, L.L., Hinton, T.G., Beasley, J.C., Lerebours, A., & Smith, J.T. (2015) Long-term census data reveal abundant wildlife populations at Chernobyl. *Current Biology*, Volume 25, R824–R826.

IAEA (2006) *Environmental Consequences of the Chernobyl Accident and their Remediation: Twenty Years of Experience*. Vienna: International Atomic Energy Agency.

Xia, C., Cao, J., Zhang, H., Gao, X., Yang, W., & Blank, D. (2014) Reintroduction of Przewalski's horse (*Equus ferus przewalskii*) in Xinjiang, China: the status and experience. *Biological Conservation*, Volume 177, 142–147.

reintroduction scheme to garner public acceptance and support. Often a large part of the **funding** for reintroduction schemes comes from public appeals. In other cases, having the support of local people, for example simply asking volunteers to keep a watch on release sites and animals, can be the difference between success and failure (Moran & Lewis, 2014).

In some cases, it is important to put an **education programme** in place that can run alongside the reintroduction. This is especially true for species viewed as pests or where human persecution was one of the causal factors in the initial extirpation. For example, in the reintroduction of griffon vultures *Gyps fulvus* to river gorges in Aveyron, southern France, education was vital in reducing persecution: no shooting or poisoning was recorded in the study area and only a single nest was disturbed by a climber (Sarrazin et al., 1994). In addition, farmers were reported to be leaving carcasses in fields more frequently, thus providing a source of food for the vultures. A reintroduction project can also be used as a vehicle for more general education. For example, the reintroduction of the short-haired bumblebee was used to highlight bumblebee decline in general and the importance of pollinators in the UK (see Case Study 13.1 for more details).

Finally, it should be noted that reintroductions are sometimes politically motivated, either in their origins or in the support that they receive. That might be in terms of revenue generation, public relations, or as a public draw to a particular area. Sometimes, they can even come about because of unexpected or unplanned success of captive breeding programmes that mean there are too many animals to remain in captivity, which was one of the main drivers for the Chernobyl reintroduction schemes (Hot Topic 13.3).

13.4.4 Reintroduction methods.

Methods by which individuals can be collected from the wild have been detailed in Section 12.4.1. This part of the reintroduction process can occur a long time before the ultimate release where there is an intermediate captive breeding step: more than 60 years in the case of the Eagle Lake rainbow trout *Oncorhynchus mykiss aquilarum* recently reintroduced to California (Carmona-Catot et al., 2012).

Transporting animals to the release site, whether from the wild or captivity, can vary from being be a simple process to a complex logistical exercise (see the Interview with an Applied Ecologist for this chapter). Some species need climatically controlled environments or special containers; in some cases, blindfolds or earmuffs are used to reduce sensory input and transport stress. Some of the logistical challenges that Applied Ecologists can face when moving species, especially large mammals, are shown in Figure 13.6.

Release strategies

There are two basic types of release strategy: hard or soft:

- **Hard release**: individuals are taken to the release site and released straight into their new environment. This is the typical strategy for many taxa, including invertebrates, fish, and amphibians. For such species, there is indeed little that can be done to 'soften' the release experience.
- **Soft release**: individuals are acclimatized to their new environment rather more slowly, possibly using temporary enclosures at the release site, or installing refugia, nest sites, or hibernacula, or providing supplemental food for the initial post-release phase.

Figure 13.6 Capturing and transporting animals as part of reintroduction initiatives is not without challenge, especially for large African mammals where bomas (temporary cloth-based funnel enclosures into which game is driven), helicopters, cranes, and manpower can all be part of the process.

Source: All photographs by Lynne MacTavish, used with kind permission.

AN INTERVIEW WITH AN APPLIED ECOLOGIST 13

Name: Lynne MacTavish

Organisation: Mankwe Wildlife Reserve South Africa; http://www.mankwewildlifereserve.com)

Role: Reserve Operations Manager

Nationality and main countries worked: Zimbabwean; worked in South Africa

What is your day-to-day job?

I have worked on a Wildlife Reserve in South Africa for almost 20 years. I am the Reserve Operations Manager, which means I get involved with almost everything on the reserve and do a lot of the day-to-day running of the reserve. I run multiple research groups each year, and also coordinate university and college student groups. I am also involved with the Reserve's Anti-Poaching Unit (APU).

Explain how and why the Reserve first became involved in reintroductions?

Some of first animals that we reintroduced to the site, which was previously a buffer zone around an explosives factory and thus was fenced, were waterbuck *Kobus ellipsiprymnus*, tsessebe *Damaliscus lunatus lunatus*, and blesbok *Damaliscus pygargus phillipsi*. In the case of the blesbok, the animals were captured on other local reserves using a game capture boma. This is essentially a temporary horizontal funnel erected in the bush. It consists of a frame and canvas with one wide end tapering down to a holding crate or truck at the other end. Animals are chased into the wide end of the boma by a specialist helicopter pilot who has identified and separated a suitable group from the air. The pilots have to be very skilled and experienced as they have to be able to identify species from the air, and be able to sex and age them. For species with complex social structures, they have to make sure that the group structure is not compromised by splitting up a family group. The pilot also needs to know how far the species can be chased to make sure that the distance from separation to the boma is not too far. This is vital, as if individuals are over-chased they can get myopathy, which affects the muscles and can result in the heart being affected. Once the animals were captured, they were transported to Mankwe and released immediately in a hard release. In the case of the tsessebe, which don't respond well to boma capture, we used net capture, with individuals being chased into nets positioned in strategic positions in small numbers—ones and twos mainly.

What is your most interesting project and why?

We have reintroduced white rhinoceros *Ceratotherium simum* to our site. We originally reintroduced four individuals to the Reserve, but since then we have moved animals to and from the reserve to allow gene flow, or to manage the group's social structure, sex ratio, or group dynamics. Most recently, we exchanged Magoo, our sub-dominant male, for a new male called Brutus. The males were from different blood lines so this helps prevent inbreeding. Magoo went to a similar-sized reserve to be part of the breeding programme there. This involved darting both Magoo and Brutus in their respective Reserves by a vet from a helicopter. As soon as the animals were down, the ground crews moved in and blindfolded the animals while keeping them cool with cold water. When the animals came round, they were guided by means of a rope into a specially built rhino trailer, which was exactly the right size so they could not turn around and become stressed or injured. The animals were driven to their new reserves and released.

What do you see as the main challenges in species translocation and reintroduction?

Translocation is not an easy process and is not something to undertake lightly. It can be difficult and dangerous both to us and the individuals concerned, especially if you are dealing with large animals that are inherently dangerous. Animals can die from capture stress, injuries inflicted by other individuals in the confines of a boma, or failure to come around from the anaesthetic if they are darted. However, mortality is very low if the right precautions are taken and an experienced game capture team is employed. We rely on our own experience and that of the game capture organizations that we work with to ensure that risks are minimized as far as possible. We also employ a qualified veterinarian when doing game capture or darting to ensure that any injuries can be dealt with immediately, and anaesthetized animals can be monitored constantly and a reversal drug can be administered if the animal starts to show signs of distress.

What are the most satisfying parts of your job?

I am passionate about conserving wildlife, especially endangered species such as white rhinos. I strive to manage a healthy ecosystem where all species of fauna and flora are studied, researched, and protected. Future generations will have the task of managing and repairing the damage done to the environment by previous generations, and this can only be accomplished through education and awareness, so

I am passionate about showing people the beauty of wildlife, in the hope that they will gain a deeper understanding of how all species, no matter how big or small, are essential to sustaining a healthy ecosystem.

What's next for you, and why?
One of our main challenges here is the war on poaching, especially rhino poaching undertaken for horn. The reserve was hit by poachers in October 2014; two rhino were killed and one was orphaned and later sadly died, too. We ended up using the skills we had learnt for capturing rhino for moving them for a very different purpose: dehorning our remaining animals in an effort to keep them safe. Again, we used an aerial crew to capture, and a vet to dart all our animals in turn, and then the vet carefully dehorned each individual with the horn then being removed to a safe facility. We never expected to be using translocation methods in this way, but we will do whatever it takes to keep our population safe.

Finally, how did you get into wildlife conservation and what advice would you give to others?
Growing up in Zimbabwe I was surrounded by wildlife, and having a father that worked in National Parks enabled me to visit and be involved in conservation from a very young age. When my father started Mankwe it was my dream to work alongside him and turn Mankwe into a leading education and research centre. Working in conservation requires dedication and commitment—every day there are challenges and hard decisions to be made—but it is probably the most worthwhile job you can do.

Soft releases can increase initial survival by providing shelter, reducing predation, or decreasing foraging needs. They can be especially useful for captive-bred individuals where the transition to a new habitat is also the transition to a wild state or for species with a strong group structure (e.g. scarlet macaw *Ara macao cyanoptera*; see Figure 13.7 for details). One particularly interesting approach to reintroduction has been undertaken in the UK for common crane *Grus grus*, whereby juveniles had human foster carers during the rearing and soft release stages. This is considered in more detail in the Online Case Study for Chapter 13.

You can find the online case study at www.oxfordtextbooks.co.uk/orc/goodenough

Use of **temporary enclosures** within a soft release programme can make the difference between success

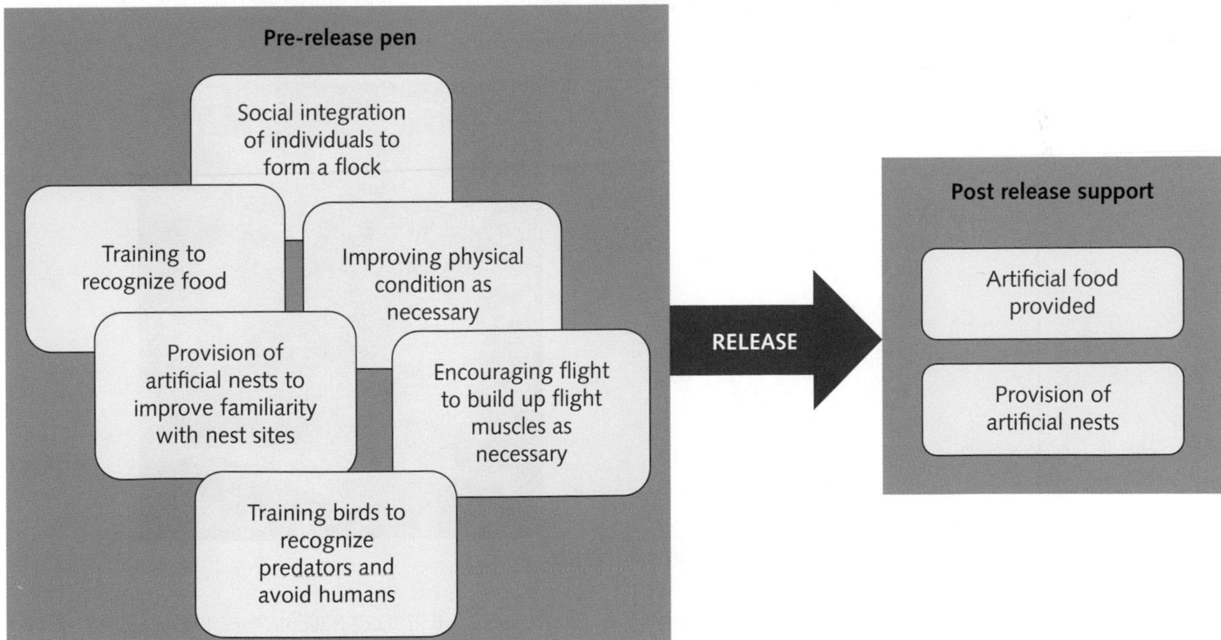

Figure 13.7 The soft release protocol for scarlet macaws *Ara macao cyanopterai* in Palenque, Mexico, involved all birds being held in a pre-release pen, while they formed into a cohesive group, were taught to forage, find nests, and display aversion to humans.

Source: Based on information from Estrada, A. (2014) 'Reintroduction of the scarlet macaw (*Ara macao cyanoptera*) in the tropical rainforests of Palenque, Mexico: project design and first year progress'. *Tropical Conservation Science*, 7, 342–364.

and failure. In some situations, however, there can be negative consequences of soft release if the new population over-relies on the additional resources and has subsequent problems integrating into the wild. For example, thick-billed parrots *Rhynchopsitta pachyrhyncha* reintroduced using a hard release approach in south-eastern Arizona, USA, had higher survival than birds reintroduced using a soft release approach (Snyder et al., 1994).

If **supplementary food is** provided, this can have impacts on individuals dispersing from the initial release site. In some cases this can be beneficial, as it creates a population core and prevents over-dispersal that can reduce the chances of individuals pairing up as a result of low encounter rates at low density. However, lack of **dispersal** can have long-term negative consequences. For example, the survival of burrowing owls, *Athene cunicularia* reintroduced to Minnesota, USA, was extremely high (97 birds out of 105), but no individuals dispersed from the immediate release site and none bred in subsequent years (Martell et al., 2001).

13.4.5 Post-release monitoring.

Post-release monitoring is a crucial part of reintroduction, both to allow management of that specific reintroduction (or effects arising from it) and to inform future reintroductions. Monitoring can focus upon:

- **Survival of released individuals:** assessed using repeated surveys to locate individuals that had been marked previously either individually or with their 'batch' details so that the release site and year can be ascertained. Ideally, marking is done so that individuals only need to be relocated, rather than retrapped to avoid disturbance (Sections 7.3.2 and 3.6.1 for more details on techniques used).
- **Movement:** undertaken at species-level by surveying over a wide area, possibly using a citizen science approach (Section 3.4.4), or at individual-level if individuals are fitted with satellite transmitters or radio tags (see Section 7.3.2 for more details on techniques and Figure 13.8 for how this was used to monitor movement of agoutis *Dasyprocta leporina* reintroduced into Tijuca National Park, Brazil). If baseline data are not known, monitoring might also be necessary for resident individuals at the same site (in the case of reinforcement) or at another site for comparison (e.g. Hermann's tortoises *Testudo hermanni hermanni* in the Mediterranean: Lepeigneul et al., 2014).

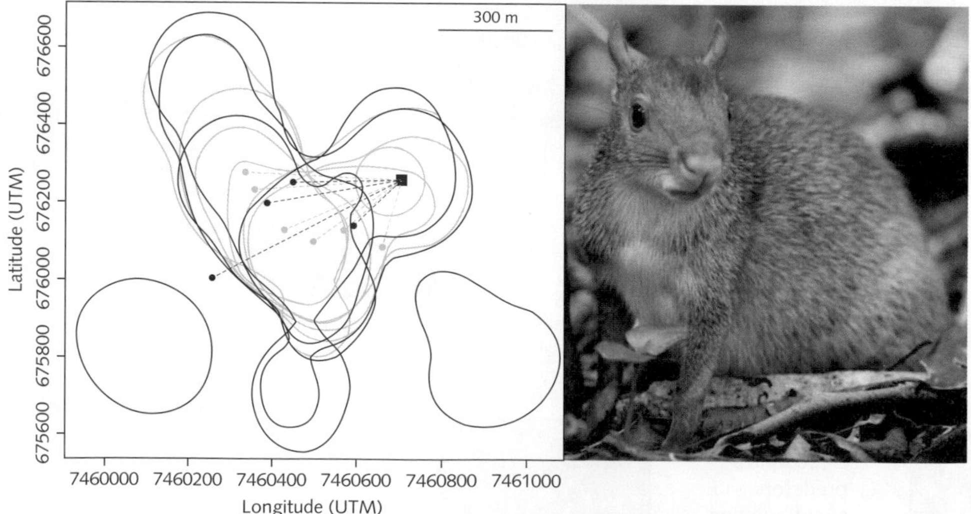

Figure 13.8 Home-range contours of individual agoutis *Dasyprocta leporine* reintroduced in Tijuca National Park, Rio de Janeiro, Brazil. The dark grey square represents the location of the acclimatization pen (release site). The dots indicate the centroid of the fixes used to estimate the home-range sizes for each individual. The dashed lines show the distance from the release site to the centroids. Males are in black and females in grey.

Source: Reproduced from Cid, B. et al. (2014) 'Short-term success in the reintroduction of the red-humped agouti *Dasyprocta leporina*, an important seed disperser, in a Brazilian Atlantic Forest reserve'. *Tropical Conservation Science*, 7(4), 796–813/CC BY 4.0.

- **Population change:** monitored by population census methods appropriate to the species. If population does not grow due to the founder population being of insufficient size, supplementation (adding additional individuals to the new population) might be appropriate.
- **Effects on other species and the wider ecosystem:** monitoring the effect of the reintroduction on other native species and their interactions is also important. Additional management might need to be put in place to resolve any issues resulting from reintroduction.

FURTHER READING FOR SECTION

Overview of the reintroduction planning process and associated risks: Morrison, S.A., Parker, K.A., Collins, P.W., Funk, W.C., & Sillett, T.S. (2014) Reintroduction of historically extirpated taxa on the California Channel Islands. *Monographs of the Western North American Naturalist*, Volume 7, 531–542.

Example species reintroduction plan: Hammer, M., Barnes, T., Piller, L., & Sortino, D. (2012) Reintroduction plan for the purple-spotted gudgeon in the southern Murray-Darling Basin. MDBA Publication. Available at: http://103.11.78.131/sites/default/files/pubs/PSG-final-corporate-style_v2.pdf

13.5 Factors affecting the success of reintroductions.

Research into reintroduction science is growing exponentially, as evidenced in Figure 13.9. Whether a reintroduction succeeds or fails depends on many interacting factors, but careful planning and implementation considerably increase the probability of success. The available data for the 1970s and 1980s suggest the majority of reintroductions failed to establish viable populations (Griffith et al., 1989; Wolf et al., 1996, 1998), but success has improved considerably in later years. In part this is because of better planning, but it is also a result of projects with a low chance of success not being progressed (Seddon et al., 2007).

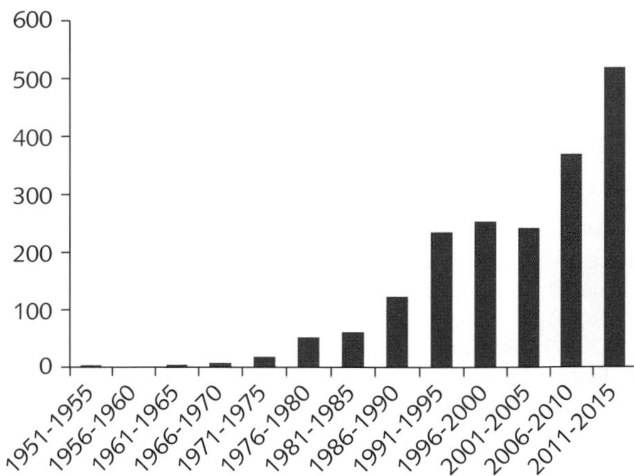

Figure 13.9 Number of reintroduction research papers published from 1951 to 2015 based on a Web of Science search for papers with 'reintroduction' or 're-introduction' in the title.

Although Applied Ecologists frequently talk about reintroduction success, there are many ways that 'success' can be defined, and currently there is no industry standard protocol. To address this, it has recently been suggested that reverse-engineering the **IUCN red-list criteria** could be a useful way forward. Under this scheme, reintroductions are considered successful if reintroduced populations do not meet red-list criteria (Robert et al., 2015).

In addition to the problem of a lack of agreement about what constitutes success, **publication bias** can also be an issue. Researchers are more likely to want to report, and have accepted for publication, accounts of reintroductions that were successful. Moreover, even when good follow-up monitoring is undertaken (Section 13.4.5), the timescale is often relatively short. This is an issue because reintroduction schemes that ultimately fail might appear to be doing well in the first few years. This is a similar issue to the **lag time** between non-native species' arrival and the impacts being evident (Section 8.5.2).

The above caveats aside, there are several factors that can help predict the chance of a reintroduction being successful. The most important of these is that the original threats have been controlled. Other factors have been identified by Wolf et al. (1996;

1998)—reanalysing a 1987 dataset compiled by Griffith et al. (1989)—and work by Fischer & Lindenmayer (2000) are:

- **The size of the initial founder population:** perhaps unsurprisingly there is an increased chance of success when more individuals are released. However, studies usually relate the number of founders with outcomes without allowing for differences in life history (e.g. Fischer & Lindenmayer, 2000). For example, suitable founder population size might be smaller for r-strategist species, which have short generation times and strong growth potential compared to K-strategist species (Section 2.7.1). Thus cross-taxonomic analyses of the ideal population size can be biased by underlying skew in the type of species being reintroduced. Much more valuable, albeit in a narrow sense, are comparisons of success for different reintroductions of the same species (e.g. black-faced impala *Aepyceros melampus petersi* in Namibia as shown in Figure 13.10).
- **Source population:** the success rate for reintroductions of wild-caught individuals is generally higher than reintroductions of captive-bred individuals. In the case of reintroductions undertaken for conservation, success rates were 29% and 15%, respectively (Fischer & Lindenmayer, 2000).
- **Diet:** species with broad diets are more likely to be reintroduced successfully than those with specialist diets.
- **Habitat quality:** this encompasses how well the abiotic and biotic conditions of a site match the ideal conditions for the target species.
- **Core range:** reintroductions into the core of the historical range are more likely to succeed than reintroductions to more marginal areas. This is because the habitat is more likely to be optimal and extirpation will likely have occurred more recently.
- **Location in the world:** successful outcomes are proportionally highest in Europe and North America and lowest in central/south America and Asia (shown graphically in Figure 13.11). This might reflect a difference in resources, more policy, and legislative support in some areas, a longer history of reintroduction and thus more local knowledge, or the types of species being reintroduced.

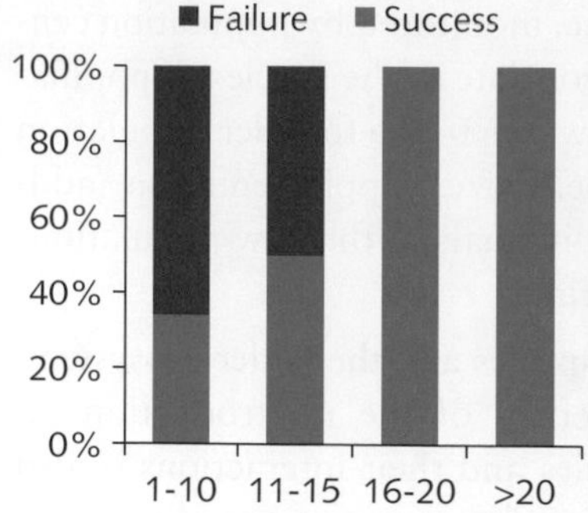

Figure 13.10 Effect of founder population size on ultimate success or failure of black-faced impala *Aepyceros melampus petersi* reintroductions to Namibian reserves: larger initial population sizes increased chances of success until 16 or more individuals were reintroduced, in which case all reintroductions were successful.

Source: Information from Matson, T. K. et al. (2004) 'Factors affecting the success of translocations of the black-faced impala in Namibia'. *Biological Conservation*, 116, 359–365. Photograph by Anne Goodenough.

FURTHER READING FOR SECTION

Defining reintroduction success criteria: Robert, A., Colas, B., Guigon, I., Kerbiriou, C., Mihoub, J.-B., Saint-Jalme, M., & Sarrazin, F. (2015) Defining reintroduction success using IUCN criteria for threatened species: a demographic assessment. *Animal Conservation*, Volume 18, 397–406.

Analysis of reintroduction success/failure for plants: Guerrant, E.O. & Kaye, T.N. (2007) Reintroduction of rare and endangered plants: common factors, questions and approaches. *Australian Journal of Botany*, Volume 55, 362–370.

Analysis of reintroduction success/failure for African carnivores: Hayward, M.W., Adendorff, J., O'Brien, J., Sholto-Douglas, A., Bissett, C., Moolman, L.C., Bean,P., Fogarty, A., Howarth, D., Slater, R., & Kerley, G.I.H. (2007) The reintroduction of large carnivores to the Eastern Cape, South Africa: an assessment. *Oryx*, Volume 41, 205–214.

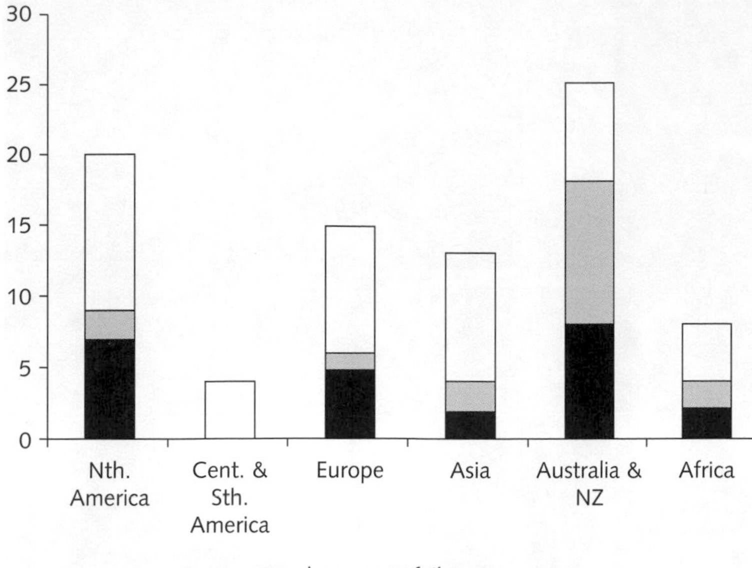

Figure 13.11 Global variability in reintroduction success (87 reintroductions conducted only for conservation purposes).

Source: Reproduced from Fischer, J., and D. B. Lindenmayer (2000) 'An assessment of the published results of animal relocations'. *Biological Conservation*, 96, 1–11, used with permission from Elsevier.

13.6 Rewilding.

In addition to their effect on individual species and extirpation processes, humans have also had a profound effect on the natural landscape. Urbanization, industrialization, and agriculture have substantial impact, removing or degrading habitats, and producing pollution. Human constructions such as highways can also fragment species' ranges and disrupt population dynamics, dispersal, and migration patterns (Chapter 7). Landscape-level activities compound activities at a smaller scale, such as hunting, fishing, and persecution, which affect species directly. As a consequence, there are very few **wildernesses** (uncultivated uninhabited areas with negligible human impact) left. Even where wilderness areas do exist, such as the Antarctic, some areas may be degraded by pollution and species may be threatened or extirpated.

While the 'pre-human ideal' is exactly that, - an ideal - large-scale restoration of wilderness could provide extensive areas of suitable habitat for species extirpated or greatly reduced by human activity. Small reserves often require intensive management, but much larger wilderness areas could, in theory, be **self-regulating**, with the absence of fences and human disturbance allowing animals to roam naturally in search of resources. During drier periods, for example, animals could follow resources and would not require water or supplementary food to be provided, as might be necessary in smaller reserves. Apex predators, which have long been extirpated from many regions, could be reintroduced to the restored wilderness. The wilderness would also support populations of other less-charismatic species associated with the habitats present.

Eventually, this approach could, in theory, be used to restore large areas to the original wilderness condition that existed before substantial human impact. Thus, to give an often used example, parts of the UK could be repopulated with European bison *Bison bonasus*, European beaver *Castor fiber*, and Eurasian lynx *Lynx lynx*, with wolves *Canis lupus* roaming in the highlands of Scotland and brown bears *Ursus arctos* hunting in restored 'wild wood' forests (Figure 13.12). Advocates of this approach, such as Monbiot (2013), say that it could be used to replace the current highly managed countryside with an unmanaged wilderness, teeming with native fauna and flora. The restoration of wilderness is the aim of a form of conservation that has become known as **rewilding**.

Figure 13.12 Some of the species that have been extirpated from the UK and which are proposed for reintroductions relating to rewilding projects. Clockwise from top left, Eurasian lynx *Lynx lynx*, the moose or European elk *Alces alces*, grey wolf *Canis lupus*, and white-tailed eagle *Haliaeetus albicilla*.

Source: Images courtesy of Tom Bech/CC BY 2.0; Alex Butterflied/CC BY-ND 2.0; Jacob Spinks/CC BY 2.0; Eric Kilby/CC BY-SA 2.0.

13.6.1 Is rewilding just 'big and ambitious' *in situ* conservation?

Terms like habitat restoration sound very much like the language of conventional *in situ* conservation. Indeed, the entire concept of rewilding sits within the **restoration ecology** framework already discussed in Section 11.3.1. The overall concept of restoration ecology can be thought of as any intentional activity that initiates or accelerates the recovery of an ecosystem to a previous state. Whereas many *in situ* conservation projects are looking to conserve what is already there and to make it work in conjunction with human pressures on the land, restoration involves changing the ecology of an area so it reassumes its initial, or pristine, condition. Small-scale restoration ecology projects can include reviving traditional forms of management, such as coppicing and pollarding (depicted in Figure 11.8), or more dramatic approaches to restore former habitats, for example through forestry or arable reversion. These techniques can be extremely useful on a fairly small scale. However, rewilding differs from most traditional conservation approaches, including many restoration approaches, in several key ways:

1. **Scale:** Rewilding is usually explicit in being a **landscape-level approach**. It seeks to establish core areas of habitat that can be linked by corridors wherever possible to form a 'wilderness network' (Section 7.6.2). This is in contrast to many conventional *in situ* conservation projects, which often deal with specific small-scale sites and in local contexts.

   It should be noted, however, that 'standard' *in situ* conservation does not need to be confined to small reserves and localized projects. Larger traditional conservation areas, such as the national parks of

East Africa or North America, are certainly landscapes in their own right, and many conservation organizations are now starting to see the benefit of living landscape and ecoregion approaches (Section 7.6.4). Indeed, connecting large conservation areas with wildlife corridors to accommodate migrating animals and wildlife relocating to find food and water is an established feature of conservation. Thus, whilst many proposed rewilding projects do tend to be large-scale enterprises, not all large-scale conservation efforts qualify as rewilding.

Similarly, the term rewilding is increasingly being used to describe smaller-scale projects. For example, in the UK, the Gloucestershire Wildlife Trust has realigned a section of the river Chelt to allow it to follow a more natural course. This restores the natural flood plain ecology of the habitat around the river and is described as rewilding (The Wildlife Trusts, 2015), although whether the term is any more useful than 'restoration' or even 'renaturalization' (which is also used) is debatable. Terminology and definitions can be difficult to establish, but when a debate becomes public and involves multiple stakeholders then clarity is vital. It would be perfectly possible for a local scheme restoring woodland to struggle to get public support if the term rewilding, with connotations of scale and apex predators, is used.

2. **The 'wild'**: Central to the rewilding approach is the restoration of the 'wild' coupled with an absence or extreme reduction of future management. Just as with reintroduction, the 're' is crucial: the approach is not just about letting nature run its course, but rather about restoring 'the wilderness' that existed previously at that location, if indeed that is possible and desirable. This part of rewilding is very straightforward conceptually, but as debated in Section 13.6.2, defining what the initial condition actually was can be problematic and is often more of an ideal than a reality that can be recreated.

3. **Reintroduction:** As noted in the earlier parts of this chapter and in Chapters 11 and 12, reintroduction is a well-established component of conservation, and indeed restoration ecology more generally (Section 11.3.2). Whether it is linked to *ex situ* captive breeding projects or to the direct translocation of individuals from other areas, the goal is always to establish sustainable populations at the target site. Rewilding can occur without species reintroductions, and species reintroductions can certainly occur without rewilding; however, the reintroduction will often occur alongside rewilding and vice-versa. This allows the (re)creation of 'wild' habitats inhabited by the species that would, historically, have been present. Indeed, in some cases, the reintroduction of large herbivores (often termed **mega-herbivores**) and apex predators is essential to the success of the rewilding project as a whole because these species regulate and manage the ecosystem in a natural way.

   Because of the importance of extirpated species within natural landscapes, rewilding projects typically have reintroduction at the forefront of the plan. For example, the European Bison Rewilding Plan has the goal of 'viable populations of free-ranging herds of bison *Bison bonasus* being established and restored across Europe and integrated into Europe's landscapes' (Vlasakker, 2014). Focusing rewilding efforts on an iconic and non-predatory species makes good sense. Restoring wilderness will have benefits for many species, but generating public support is likely to be easier when the species at the forefront of the proposal is perceived by most as non-threatening. Case Study 13.3 explores the European bison rewilding plan in more depth.

4. **No human intervention:** Conservation generally seeks to find ways for humans and the natural world to co-exist in some state defined by the aims of the conservation programme. Maintaining this status quo often requires human intervention, and so conservation is closely tied to the concept of **management**. The rewilding concept is different, at least on paper. By restoring areas to a natural state the resulting system is presumed to be self-regulating and so, after the initial restoration (and usually reintroduction) phases, human intervention should not be required and is, indeed, undesirable. Although this is an intuitive principle, there are issues with this approach and these are discussed in Section 13.6.2.

13.6.2 The challenges faced by rewilding schemes.

Rewilding is an attempt to solve the anthropogenic decline in species' abundance and biodiversity by setting aside relatively large, and potentially enormous, tracts of land that can be restored to an 'original' wilderness condition. It is a highly ambitious approach and rewilding schemes face many challenges in converting proposals into reality. These can be subdivided into challenges that are conceptual and those that are practical or logistic.

Conceptual problems: defining and developing the 'wild'

Rewilding, by definition, involves restoring the 'wild'. However, when areas have been subject to human interference, especially for a considerable period of time, it can be very difficult in practice to define what the initial state actually was.

A good example of this is debate over the nature of the primeval European forest and the role that large herbivores played in their ecology. The so-called Vera hypothesis maintains that the forests of ancient Europe had a relatively open structure and that these woodlands existed within a landscape mosaic of parkland and regenerating scrub. This led to the 'woodland-pasture' model, which proposes that large herbivores were the cause of the landscape-level wilderness (Birks, 2005). However, there are also studies suggesting that large herbivores are not required to maintain this landscape and that the sort of open woodland indicated by evidence of ancient pollen has only ever been maintained by humans (Mitchell, 2005). Knowing the exact nature of ancient forests is absolutely crucial for attempts to rewild parts of Europe: those involved in rewilding need to know what the target landscape looks like in order to have any hope of recreating it. It is vital that humans do not strive to create an idealized wilderness. At an even more pragmatic level, if the wild wood landscape turns out to be one that can only be maintained by human exploitation, then such a landscape is neither truly wild nor capable of maintaining itself without intervention.

If taken to its logical extent, restoring the wild not only requires reintroduction of species that used to be present, but also that **non-native species** introduced to the landscape after the target time should be eradicated. In practice, achieving non-native species eradication is usually extremely challenging. In the UK, for instance, rabbits *Oryctolagus cuniculus* were introduced by the Romans following their invasion in 43AD, but removing these highly abundant and ubiquitous animals would be expensive and very difficult (if not impossible) in practice. Removing introduced plants is also exceptionally difficult (Section 8.6.2), and indeed, both introduced plants and animals may have become important components of the ecosystem since their initial introduction. Rabbits, for example, provide food for a number of predators, such as foxes *Vulpes vulpes* and stoats *Mustela ermine*.

It should be noted that, in many cases, managed areas have a greater habitat heterogeneity, and thus often more species per unit area, than unmanaged ones. The concept of it being 'better' to have a hands-off approach thus depends on an individual's definition of 'better'. It might (possibly) be more natural, but in many cases it might actually, ironically, result in **biodiversity declines**. Moreover, expecting that an area can revert itself to its original state is a fallacy, at least in an acceptable timeframe. The seeds required to produce forest species, for example, will simply not be present in the seed banks of open farmland. Thus, although in the long term management effort might be low, the initial effort and expense involved in producing a wild and self-regulating landscape may, somewhat paradoxically, be considerable.

Logistic problems: land and public

In addition to conceptual challenges, there are also numerous practical and logistical hurdles to overcome in rewilding; for instance:

1. **Site availability and selection**: Rewilding requires that land be restored to wilderness and the first problem is finding sufficient land that is ecologically suitable. European bison *Bison bonasus*, for example, have a large home range and the minimum area required for a population to be considered a truly wild population is 10,000 hectares (Vlasakker, 2014). Any rewilding proposal focused on this species therefore requires a contiguous area of land approximately the size of the city of Paris. Given the pressure on land in many areas of conservation concern and the many competing parties with an interest in that land, locating potential sites is likely to be extremely difficult in many cases. This issue is discussed further in Case Study 13.3.
2. **Land ownership:** A significant problem facing rewilding projects is that land is owned. What is more, if rewilding is to succeed across large areas

then it is likely that multiple parcels of land will be required by the project to form core areas and corridors. Such parcels of land will likely have multiple owners. Whether owners are private individuals, mining companies, agricultural concerns, charities and trusts, community cooperatives, or nations, there will always be owners whose interests may not align with the interests of a rewilding project. Owners may be unwilling to sell the land or lease it in perpetuity. Depending on the geometry of the total tract of land required, relatively few landowners might have a disproportionate influence on the success or failure of the overall project. If, for example, the proposed final site has a 'pinch point' that connects multiple smaller areas together, then if the smaller areas are not viable individually, the owner of that crucial land effectively decides the fate of the entire project. It might be relatively straightforward (although potentially expensive) for smaller-scale rewilding projects to purchase the land required or gain permission to carry out rewilding. However, larger and more ambitious projects with multiple stakeholders unavoidably become more complex, more expensive, and potentially less tractable.

3. **Public perception of reintroductions:** As noted in Section 13.6.1, reintroductions, especially of large herbivores and apex predators, can be crucial within rewilding to recreate ecosystem functioning. However, such reintroductions can attract considerable and sometimes unfavourable media attention. Human–wildlife conflicts are especially pronounced with apex predators, and proposals such as the reintroduction of wolves *Canis lupus* to Scotland are sometimes met with understandable concern by local people who fear for their livelihood and safety (Munro, 2014).

13.6.3 Rewilding successes.

When judging the success or otherwise of rewilding schemes it is important that the terms 'reintroduction' and 'rewilding' do not get confused. The successful release of European beaver *Castor fiber* into rivers across Europe, for example, is not rewilding. Similarly, small-scale local habitat restoration is not rewilding, despite often being badged as such. For example, the organization Rewilding Britain lists 13 on-going projects, including the restoration of an English river and of montane scrubland to a Scottish valley. These projects are perhaps more usefully considered as examples of restoration ecology (Rewilding Britain, 2015). Likewise, it is also important not to confuse well-publicized proposals and ideas (such as rewilding large areas of North America (The Rewilding Institute, 2015) or restoring European bison (Case Study 13.3)) with realized projects.

Large-scale landscape-level rewilding has yet to be realized, but there are numerous small-scale success stories that could be viewed as **pilot studies** or as starting points for large-scale and ambitious projects such as Pleistocene Park in Siberia. This rewilding initiative seeks to restore the mammoth steppe ecosystem, (the dominant ecosystem of the Arctic in the late Pleistocene). The initiative's ambition requires the replacement of current ecosystems by grassland. However, currently Pleistocene Park is still largely conceptual, consisting of an enclosed area of just 16 square kilometres (Pleistocene Park, 2015).

The Oostvaardersplassen in The Netherlands is probably the most cited example of a successful rewilding project. It is a polder, an enclosed tract of low-lying land surrounded by raised embankments (dikes) that are characteristic of the North Sea coastal region of The Netherlands, as shown in Figure 13.13. It is a site of 56 square kilometres into which large grazing mammals such as koniks *Equus ferus caballus* (small semi-feral horses originally from Poland), and Heck cattle (a breed produced by attempts to breed back domestic cattle to their ancestral form) have been introduced. These animals, which are allowed to roam freely and without routine human intervention, are functional equivalents of the now extinct tarpan *Equus feras feras* (a wild horse), and aurochs *Bos primigenius* (a large wild ox). This is an example of assisted colonization (Section 13.2). Red deer *Cervus elaphus* and roe deer *Capreolus capreolus* have also been introduced into the area.

The result of the extensive grazing of animals in the Oostvaardersplassen is the creation of a grassland ecosystem that some think resembles the delta and riverbank ecosystems that would have been present in Europe before human intervention (ICMO2, 2010). Although often cited as a rewilding success

CASE STUDY 13.3

Rewilding the European bison

The European bison *Bison bonasus* (Figure A) was once native to most of lowland Europe, but was subjected to severe hunting pressure across its range. In 1927 the last wild bison was killed in Poland. The captive population of bison at that time was around 50 individuals, and the first studbook (see Section 12.4.3) for a non-domesticated species was started to manage the captive breeding of these animals. A number of reintroduction programmes were initiated, the first in 1929 in Poland, and as a consequence of a sustained conservation effort there are now c. 2700 free ranging bison in populations ranging from Ukraine to Spain (Pucek et al., 2004).

The organization Rewilding Europe aims to build on bison conservation successes to date by creating five new entirely free-roaming and completely wild-living bison populations of at least 100 individuals. They see the urbanization and intensification of farming and forestry in the most fertile areas of Europe as an opportunity. As extensive farming and pastoral livelihoods cease to be economically viable, people are abandoning rural areas and more than a million hectares across the EU are left fallow every year. There is therefore 'a space—ecological, social, and economic—for new initiatives to step in and create opportunities for conservation, restoration of land, and rebuilding of local economies.' (Vlaskker, 2014). By rewilding currently abandoned areas for bison they hope to provide space for an IUCN Red List Vulnerable species and opportunities for economic growth through ecotourism and related activities.

The Bison Rewilding Plan is an extensive document that details the background and strategy for the rewilding proposal (Vlaskker, 2014). The principal stages identified in the rewilding plan are:

Figure A The European bison *Bison bonasus*.

Source: Image courtesy of Francesco Carrani/CC BY 2.0.

1. **Habitat criteria and site selection**: bison have a number of ecological requirements that limit potential sites. They do not cope well with heat and struggle both to move and feed if faced with thick snow cover. They require natural meadows and deciduous forests with good and varied vegetation and they cannot cope with very wet conditions or handle steep slopes. Proximity to domestic cattle could cause disease transmission and this risk is greater in areas bordering non-EU Countries. Finally, the average home range for a free ranging bison population is 10,000 hectares.
2. **Selection of individuals**: the founder population should represent the maximum genetic variability possible and must have a suitable social structure with a mix of adult and juveniles.
3. **Public Perception**: bison are large herbivores and have the capacity to cause harm to people (cattle kill people even in well-regulated countries like the UK (McCarthy, 2015)) and to crops. However, reintroductions have the potential to provide economic benefits, and any rewilding scheme has to have clear and effective communication with all stakeholders.
4. **Soft release infrastructure development**: soft release, involving acclimatizing the animals in a facility prior to release into the wild, has a higher rate of success than hard release, but is more expensive. However, the soft release facility could become an attraction in its own right and a valuable revenue generator. This stage also needs careful consideration of logistical issues, such as transportation, tranquilization, veterinary care, condition monitoring, and supplemental feeding.

Reintroduction of bison to reserves across Europe has been achieved, but an approach based on the principles of rewilding, with reintroductions of truly wild herds across rewilded and connected landscapes, remains as a proposal, albeit a well-developed proposal, which builds on solid conservation successes. One of the main problems with rewilding is summed up by Rewilding Europe: 'A reintroduction project always deals with a species and a human component. The species component is generally the 'easy' component to assess and manage. The human component is generally much harder to manage' (Vlaskker, 2014).

REFERENCES

McCarthy, M. (2015) Hoofed and dangerous: Britain's killer cows. *The Independent* Wednesday 23rd September, 2015. Available at: http://www.independent.co.uk/environment/nature/hoofed-and-dangerous-britains-killer-cows-1776775.html

Pucek, Z., Belousova, I.P., Krasinska, M., Krasinski, Z.A., & Olech, W. (Eds) (2004) *European Bison. Status Survey and Conservation Action Plan*. IUCN/SSC Bison Specialist Group. Gland: IUCN.

Vlasakker, J. van de (2014) *Rewilding Europe Bison Rewilding Plan, 2014–2024*. Nijmegen: Rewilding Europe.

story there are some problems with the Oostvaarder-splassen as a rewilding story:

- The system lacks apex predators such as lynx *Lynx lynx* or wolves *Canis lupus*.
- Humans act as apex predators by culling within the reserve, which means that the area is neither self-regulating nor unmanaged (ICMO2, 2010).
- There are no currently corridors connecting this reserve with others.
- There can be ethical issues when the system breaks down, for example due to food shortage in particularly hard winters, when animals starve. Although this is, in many ways, a natural process, some people feel the area is effectively a wild zoo and this is thus unethical.

FURTHER READING FOR SECTION

The concept and scale of the rewilding approach: Foreman, D. (2004) *Rewilding North America: A Vision for Conservation in the 21st Century*. Washington: Island Press.

Pleistocene rewilding: Donlan, C. J., Berger, J., Bock, C.E., Bock, J.H., Burney, D.A., Estes, J.A., Foreman, D., Martin, P.S., Roemer, G.W., Smith, F.A., Soulé M.E., & Greene, H.W. (2006) Pleistocene rewilding: an optimistic agenda for twenty-first century conservation. *American Naturalist*, Volume 168, 660–681.

Figure 13.13 Oostvaardersplassen.

Source: Image courtesy of Peter Galvin/CC BY 2.0.

13.7 Conclusions.

Both reintroduction and rewilding are increasingly-debated concepts within both professional and public arenas. This debate has forced many to question, at a fundamental level, our attitude towards the natural world and what we should do about mediating our impacts upon it (e.g. Monbiot, 2013).

Reintroduction schemes can be undertaken for a plethora of different reasons, ranging from idealistic moral responsibility to restore a species in an area to highly pragmatic schemes designed to recreate key ecosystem processes. There have been many success stories, but success is far from guaranteed. Key ingredients in increasing the chances of success are maximizing the number of founder individuals and ensuring that the release site is within the core of the former range. Even more important is ensuring that the factor(s) that resulted in the original extirpation have been removed, controlled, or managed. Even then, there is no getting around the fact that reintroduction is a much more expensive form of conservation than most *in situ* initiatives.

If anything, rewilding faces even more difficulties in practice than reintroduction. This, together with its shorter history, means that extensive rewilding has struggled to become a reality. However, the conversations it has engendered are likely to lead to rewilding as an approach gaining momentum over the coming years. Whether that momentum will lead to successful large-scale rewilding schemes though, at the moment, is far from clear.

ONLINE ACTIVITY

Go to www.oxfordtextbooks.co.uk/orc/goodenough/ to download the activity that accompanies this chapter.

CHAPTER 13 AT A GLANCE: THE BIG PICTURE

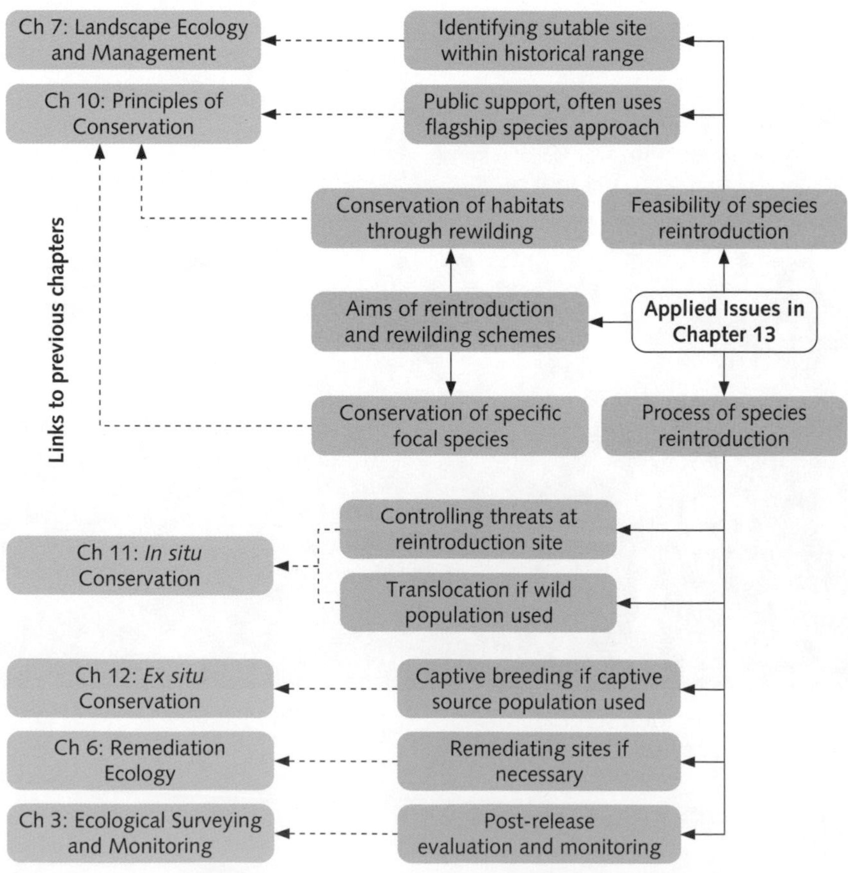

REFERENCES

Armstrong, D.P. & Seddon, P.J. (2008) Directions in reintroduction biology. *Trends in Ecology & Evolution*, Volume 23, 20–25.

Armstrong, D.P. & Ewen, J.G. (2002) Dynamics and viability of a New Zealand Robin population reintroduced to regenerating fragmented habitat. *Conservation Biology*, Volume 16, 1074–1085.

Armstrong, D.P. & Reynolds, M.H. (2012) Modelling reintroduced populations: the state of the art and future directions. In J.G. Ewen, D.P. Armstrong, K.A. Parker and P.J. Seddon (Eds) *Reintroduction Biology: Integrating Science and Management*, pp. 165–222. London: John Wiley and Sons.

AZA (1992) https://www.aza.org/assets/2332/aza_guidelines_for_reintroduction_of_animals.pdf

Beck, B.B. (2001) *A Vision for Reintroduction*. Silver Spring, MD, American Zoo and Aquarium Association.

Birks, H.J.B. (2005) Mind the gap: how open were European primeval forests? *Trends in Ecology & Evolution*, Volume 20, 154–156.

Bustamante, J. (1996) Population viability analysis of captive and released Bearded Vulture populations. *Conservation Biology*, Volume 10, 822–831.

Campbell, R.D., Dutton, A., & Hughes, J. (2007) *Economic Impacts of the Beaver*. Report for the Wild Britain Initiative. Available at: www.scottishbeavers.org.uk/docs/files/general/EconomicImpacts.pdf

Carmona-Catot, G., Moyle, P.B., & Simmons, R.E. (2012) Long-term captive breeding does not necessarily prevent reestablishment: lessons learned from Eagle Lake rainbow trout. *Reviews in Fish Biology and Fisheries*, Volume 22, 325–342.

Cid, B., Figueira, L., e Mello, A.F. D.T., Pires, A.S., & Fernandez, F.A. (2014) Short-term success in the reintroduction of the red-humped agouti *Dasyprocta leporina*, an important seed disperser, in a Brazilian Atlantic Forest reserve. *Tropical Conservation Science*, Volume 7, 796–810.

Dietz, J.M., Baker, A.J., & Miglioretti, D. (1994) Seasonal variation in reproduction, juvenile growth, and adult body mass in golden lion tamarins (*Leontopithecus rosalia*). *American Journal of Primatology*, Volume 34, 115–132.

Estes, J.A. & Palmisano, J.F. (1974) Sea otters: their role in structuring nearshore communities. *Science*, Volume 185, 1058–1060.

Fischer, J. & Lindenmayer, D.B. (2000) An assessment of the published results of animal relocations. *Biological Conservation*, Volume 96, 1–11

Griffith, B., Scott, J.M., Carpenter, J.W., & Reed, C. (1989) Translocation as a species conservation tool: status and strategy. *Science*, Volume 245, 477–480.

ICMO2 (2010) *Natural Processes, Animal Welfare, Moral Aspects and Management of the Oostvaardersplassen*. Report of the second International Commission on Management of the Oostvaardersplassen, The Hague/Wageningen, Netherlands.

IUCN (1987) IUCN *Position Statement on the Translocation of Living Organisms: Introductions, Re-introductions, and Re-stocking.* Gland: IUCN.

IUCN/SSC (2013) *Guidelines for Reintroductions and Other Conservation Translocations.* Gland: IUCN Species Survival Commission.

Jensen, B.H. (2007) *Feasibility study on reintroduction of Marsh Fritillary Euphydryas aurinia to Danish SACs*. Danish Ministry of the Environment. Available at: http://ec.europa.eu/environment/life/project/Projects/index.cfm?fuseaction=home.showFile&rep=file&fil=ASPEA_Feasibility.pdf

Kleinman, D. (1989) Reintroduction of captive mammals for conservation: guidelines for reintroducing endangered species into the wild. *BioScience*, Volume 39, 152–161.

Kuehler, C., Kuhn, M., Kuhn, J.E., Lieberman, A., Harvey, N., & Rideout, B. (1996) Artificial incubation, hand-rearing, behavior, and release of Common 'Amakihi (*Hemignathus virens virens*): surrogate research for restoration of endangered Hawaiian forest birds. *Zoo Biology*, Volume 15, 541–553.

Lepeigneul, O., Ballouard, J.M., Bonnet, X., Beck, E., Barbier, M., Ekori, A., Buisson, E., & Caron, S. (2014) Immediate response to translocation without acclimation from captivity to the wild in Hermann's tortoise. *European Journal of Wildlife Research*, Volume 60, 897–907.

Martell M.S., Schladweiler J., & Cuthbert F. (2001) Status and attempted reintroduction of Burrowing Owls in Minnesota, USA. *Journal of Raptor Research*, Volume 35, 331–336

Menz, M.H., Phillips, R.D., Winfree, R., Kremen, C., Aizen, M.A., Johnson, S.D., & Dixon, K.W. (2011) Reconnecting plants and pollinators: challenges in the restoration of pollination mutualisms. *Trends in Plant Science*, Volume 16, 4–12.

Mitchell, F.J.G. (2005) How open were European primeval forests? Hypothesis testing using palaeoecological data. *Journal of Ecology*, 93, 168–177.

Monbiot, G. (2013) *Feral: Rewilding the Land, the Sea and Human Life*. London: Penguin Books.

Moran, D. & Lewis, A.R. (2014) *The Scottish Beaver Trial: Socio-economic Monitoring, Final Report*. Scottish Natural Heritage Commissioned Report No. 799. Inverness: Scottish Natural Heritage.

Motlhanke, S.G. (2005) The Socio-Economic Impacts of Nature-based Tourism: The case study of Bakgatla ba-ga Kgafela in the Pilanesberg National Park. Doctoral thesis, University of Witwatersrand, Johannesburg.

Munro, A. (2014) Bring wolves back to highlands—John Muir Trust. *The Scotsman Monday* 19th May 2014. Available at: http://www.scotsman.com/news/environment/bring-back-wolves-to-highlands-john-muir-trust-1-3415318.

Pleistocene Park (2015) *Pleistocene Park and the North-East Scientific Station*. Available at: http://www.pleistocenepark.ru/en.

Rewilding Britain (2015) *Rewilding Projects*. Available at: http://www.rewildingbritain.org.uk/rewilding/rewilding-projects.

Robert, A., Colas, B., Guigon, I., Kerbiriou, C., Mihoub, J-B., Saint-Jalme, M., & Sarrazin, F. (2015) Defining reintroduction success using IUCN criteria for threatened species: a demographic assessment. *Animal Conservation*, Volume 18, 397–406.

Sarrazin F., Bagnolini C., Pinna J.L., Danchin E., & Clobert J. (1994) High survival estimates of griffon vultures (*Gyps fulvus fulvus*) in a reintroduced population. *Auk*, Volume 111, 853–862.

Seddon, P.J., Armstrong, D.P., & Maloney, R.F. (2007) Developing the science of reintroduction biology. *Conservation Biology*, Volume 21, 303–312.

Seddon, P. J., Griffiths, C. J., Soorae, P. S., & Armstrong, D. P. (2014). Reversing defaunation: Restoring species in a changing world. *Science*, Volume 345(6195), 406–412.

Seddon, P.J., Soorae, P.S., & Launay, F. (2005) Taxonomic bias in reintroduction projects. *Animal Conservation*, Volume 8, 51–58.

Sharifi, M. & Vaissi, S. (2014) Captive breeding and trial reintroduction of the endangered yellow-spotted mountain newt *Neurergus microspilotus* in western Iran. *Endangered Species Research*, Volume 23, 159–166.

Snyder N.F.R., Koenig S.E., Koschmann J., Snyder H.A., & Johnson T.B. (1994) Thick-billed parrot releases in Arizona. *Condor*, Volume 96, 845–862

South, A., Rushton, S., & Macdonald, D. (2000) Simulating the proposed reintroduction of the European beaver (*Castor fiber*) to Scotland. *Biological Conservation*, Volume 93, 103–116.

Suchecka, A., Olech, W., & Łopienska, M. (2014) Evaluation of the influence of demographic factors on the success of reintroduction of small herds of European bison. *Acta Scientiarum Polonorum. Zootechnica*, Volume 13, 67–80.

The Rewilding Institute (2015) *Our Mission.* Available at: http://rewilding.org/rewildit/about-tri/our-mission

The Wildlife Trusts (2015) Some Thoughts on Rewilding. Available at: http://www.wildlifetrusts.org/rewilding

Vlasakker, J. van de (2014) *Rewilding Europe Bison Rewilding Plan, 2014–2024*. Nijmegen: Rewilding Europe.

Wolf, C.M., Griffith, B., Reed, C., & Temple, S.A. (1996) Avian and mammalian translocations: update and reanalysis of 1987 survey data. *Conservation Biology*, Volume 10, 142–1154.

Wolf, C.M., Garland, T., & Griffith, B. (1998) Predictors of avian and mammalian translocation success: reanalysis with phylogenetically independent contrasts. *Biological conservation*, Volume 86, 243–255.

Yalden, D. (1999) *The History of British Mammals*. Calton: Poyser Natural History.

GLOSSARY

Key terms in Applied Ecology that feature in multiple chapters are defined below. Terms that are bold in the definition allow cross-referencing to other entries in the glossary.

Abiotic environment: the non-living environment and its physical characteristics, including temperature, light, water, soil, and geology.

Active management: management, often undertaken for conservation, which involves a practical interaction with the environment, such as planting, digging, dredging, and culling; differs from **custodial management**.

Allogeneic processes: processes driven by external environmental factors.

Anthropogenic: human-influenced or human-caused.

Applied Ecology: the subdiscipline of **ecology** that considers the application of ecology to real-world questions and challenges.

Assisted colonization: a **species introduction** undertaken specifically within a conservation framework to avoid extinction of that species.

Autogenic processes: processes driven by internal genetic factors.

Autotrophs: literally 'self-feeders'; organisms that are able to photosynthesize.

Background matrix: the dominant land type in a given area at a specific point in time.

Baseline ecological conditions: the habitats and species present at a site before development, management, conservation, or other intervention.

Bio-accumulation: the process by which non-degradable chemicals build up in biological tissues of an organism.

Bio-augmentation: adding naturally-occurring microorganisms that have been cultured in a laboratory to a focal site as part of a **bioremediation** programme.

Biodiversity hotspot: an area that supports an extremely high number of species, especially **endemic species**, within a comparatively small area (e.g. Madagascar).

Biodiversity indicators: using species presence or abundance to indicate biodiversity parameters, such as species richness or sustainability.

Biodiversity offsetting: a type of indirect **compensation** that is designed to give biodiversity benefits to compensate for losses as a result of human activity or development.

Biological indicators: using species presence or abundance to indicate biological or ecosystem processes, such as habitat age or grazing pressure.

Biological pest management: combines ecological processes like herbivory, predation, and parasitism with active human management to control pests.

Biological species concept: a method of defining whether different individuals are different species based on physical appearance and whether they interbreed regularly to produce viable and fertile offspring.

Bioremediation: taking advantage of naturally occurring biochemical pathways in organisms to remediate sites.

Biostimulation: the process of encouraging growth of microorganisms that occur naturally at a site as part of a **bioremediation** programme.

Biosurveys: using the presence of specific species or species communities in the real world to draw inferences about local conditions, generally in a non-standardized and qualitative way.

Biotic environment: the living environment including the interactions between species, such as competition, predation, and parasitism.

Biotic indices: standardized or semi-standardized methods that use species communities to answer key questions about an ecosystem, usually using a numerical framework.

Capture–mark–recapture (CMR): a method of ascertaining animal population sizes by capturing individuals, marking them, and then recapturing them to compare the ratio of marked to unmarked individuals (also called mark–release–recapture).

Carrying capacity: the number of individuals of a species that can be supported by the resources available at a specific site.

Citizen science: working with the public to collect scientific data.

Climate envelope modelling: the use of a species' current range to infer climatic requirements for that species (the 'climatic envelope') to predict the likely future range of that species under climate change scenarios.

Commensal relationship: a relationship between two species where one benefits and the other is unaffected.

Community similarity: the amount of overlap in the species community of two or more separate sites.

Compensation: measures taken to make up for the loss of, or permanent damage to, biological resources. Often involves the provision of replacement areas.

Competition: where supply of a resource, such as food or light, is lower than demand so individuals must vie with each other for those resources through an **intraspecific interaction** or **interspecific interaction**.

Conservation translocation: the intentional movement and release of organisms by humans where the primary objective is conservation.

Conservation triage: the process of deciding conservation priorities.

Conservation: a component of **ecological management**, conservation seeks to prevent the over-exploitation, degradation, and destruction of habitats and species through **active management** or **custodial management**.

Coppicing: selective cutting of woody plants to encourage regrowth in multiple small stems from a central stump; similar to **pollarding**, but occurs specifically at ground level.

Cosmopolitan species: species that occur in in many places across the world; opposite of **endemic species**.

Crepuscular species: species active at dawn and dusk.

Cryptogenic species: when there is uncertainty as to whether a specific species is **native** to an area or whether it is **non-native**.

Custodial management: management, often undertaken for conservation through non-active means, including legislative protection and policy initiatives; differs from **active management**.

Decomposition: the process by which dead organic matter, including dead organisms and faeces, is broken down.

Density-dependency: an effect or process where the magnitude increases with increasing **population density** (e.g. dispersal or reproduction).

Destructive survey methods: survey techniques that kill organisms; opposite of **non-destructive methods**.

Diurnal species: species active during the day; opposite of **nocturnal species**.

Ecological guild: a group of species that share the same environment and similar environmental resources.

Ecological Impact Assessment (EcIA): a formal process through which the ecological impacts of a proposed development are considered and minimized. Usually occurs as a discrete or semi-discrete part of an **Environmental Impact Assessment**.

Ecological indicators: the use of specific **indicator species**, multiple species within a taxonomic group, or even a whole community of species, to indicate something about the chemical, physical, and biological parameters of the environment in which they are found.

Ecological management: devising and implementing processes and procedures that enable us to manage the natural world for its benefit, for the benefit of humans, and in many cases for the benefit of both.

Ecological monitoring: studying change in species (individuals, populations, and communities), change in the interactions between species, or change in the **abiotic environment** over time to answer a specific question, check compliance with legislation or policy, or assess the effectiveness of management or conservation interventions.

Ecological replacement: a **species introduction** undertaken specifically within a conservation framework to re-establish an ecological function lost through the extinction of a different species.

Ecological resilience: the ability of a habitat or species to tolerate a change in conditions or recover from a major ecological disturbance.

Ecological succession: change in species community structure over time either on new land (primary succession) or following substantial disturbance (secondary succession).

Ecological surveying: the act of collecting primary ecological data.

Ecology: the study of interactions between species, or between species and their **abiotic environment**, either for its own sake (pure ecology) or to answer practical questions (**Applied Ecology**).

Economic Injury Level: threshold at which the amount of harm caused by a pest equals the cost of managing that pest.

Economic threshold: the point at which management action should be taken to prevent a pest population reaching the **Economic Injury Level**.

Ecoregions: geographically distinct areas with a common species community and that share similar environmental conditions and ecological dynamics.

Ecosystem engineers: species that modify the ecosystem in which they live and thus have a profound effect on their ecosystem.

Ecosystem services: an environmental function or process undertaken within an ecosystem that is of benefit to humans; for example, water filtration, flood regulation, pollination, food supply, or waste decomposition.

Ecotone: an intermediate habitat that has characteristics of the two adjacent habitats, such as an area of scrub between grassland and woodland.

Edge effects: change in species or habitat processes as a result of being close to the edge of a habitat or **protected area**.

Endemic species: species that occur in one restricted geographical location; opposite of **cosmopolitan species**.

Environmental Impact Assessment (EIA): a formal process through which the environmental impacts of a proposed development are considered and minimized.

Environmental indicators: using species presence or abundance to indicate environmental conditions, such as water nutrient content or soil moisture.

Environmental resources: any physical aspect of, or process in, the natural environment that is of use to humans. Includes non-renewable resources (e.g. coal, oil, minerals), renewable resources (e.g. wind, light, tidal power), and semi-renewable resources (e.g. timber).

Evolution: the change in gene frequency over time in response to selection pressures; can ultimately lead to the evolution of new species.

Evolutionary arms race: two or more species that evolve together and develop adaptations and counter-adaptations against each other to gain an advantage in an ecological tug-of-war; often seen in predator–prey or parasite–host relationships.

***Ex situ* conservation:** a form of **ecological conservation** that takes place in captivity.

Extinction in the wild: loss of a species in the wild; remaining individuals occur in captivity only.

Extinction vortex: a series of linked processes leading to **extinction.**

Extinction: complete loss of a species from the biosphere.

Extirpation: loss of a species from a specific site, province, or country.

Flagship species: a high-profile, charismatic, or evocative species with the ability to capture the imagination of the public.

Founder effect: the loss of genetic variation that occurs when a new population is established by a very small number of individuals.

Friction maps: a mapping technique that shows the ease with which a species can traverse a landscape; can help predict the impacts of fragmentation on species' dispersal.

Functional extinction: when a species still exists in the wild and/or in captivity, but is doomed to extinction, for instance, when only one sex remains, when all individuals are too old to breed, or when all individuals are known to be infertile.

Fundamental range: spatial area with suitable conditions that could theoretically support a species; sometimes called the potential range. Contrast with **fundamental range.**

Gene banking: creating a biorepository of genetic material for future use by storing biological material and freezing embryos.

Generalist species: a species with a wide **niche** that can tolerate a wide variety of environmental conditions; opposite of a **specialist species.**

Genetically engineered microorganisms: microbes that have been genetically engineered to improve their function; often used to improve performance within **bioremediation** contexts.

Genetically modified organism: a living organism whose genetic material has been artificially manipulated in a laboratory using genetic engineering.

Geographical Information Systems: digitization of ecological data into coordinates representing latitude and longitude so that they can be displayed graphically, together with other spatial information on factors such as geology or hydrology.

Habitat fragmentation: the process by which a landscape is 'interrupted' by human land use change and thus becomes increasingly patchy.

Herbivory: a feeding relationship whereby an animal consumes vegetation; animals that only eat plants are obligate herbivores, while those that incorporate plants as part of a wider diet are facultative herbivores.

Heterogeneous: mixed, diverse; opposite of **homogeneous.**

Heterotrophs: organisms that cannot fix carbon and other elements from the abiotic environment, and so must feed on organisms to gain the molecules they need for growth.

Home range: the area used by a specific individual of a species.

Homogeneous: uniform, similar; opposite of **heterogeneous.**

Inbreeding depression: reduced biological fitness in a given population as a result of **inbreeding.**

Inbreeding: when closely-related individuals breed with one-another; can lead to **inbreeding depression.**

Indicator species: species that can be used as an **ecological indicator.**

***In situ* conservation:** a form of **ecological conservation** that takes place in the wild.

Insurance population: an isolated population created or protected as insurance against the main population becoming **extirpated.**

Integrated Pest Management (IPM): a pest management approach that combines physical, chemical, and biological control measures.

Interspecific interaction: an interaction that occurs between individuals of different species; often applied to **competition.**

Intraspecific interaction: an interaction that occurs between individuals of a single species; often applied to **competition.**

Invasion pathways: the main routes by which **non-native species** are introduced to new environments, for example, trade routes.

Invasion risk: the probability of **non-native species** becoming established in a given area.

Invasive species: **non-native species** that spread and reproduce quickly in their new environment, and often come to dominate it to the point that they can become **pests**. Note that not all non-native species become invasive.

IUCN red-list: a formalized procedure to quantify species extinction risk developed by the International Union for Conservation of Nature (IUCN).

Keystone species: a species with a particularly important role within an ecosystem by virtue of its **population size**, its function as an **ecosystem engineer**, its interactions with other species, or the **ecosystem services** that it provides.

K-strategist species: organisms that have a 'live slow, live long' life history with slow growth, long lifespan, and few offspring at each breeding attempt; tend to predominate in more stable environments. 'K' comes from the German word *Kapazitätsgrenze*, which means **carrying capacity**; contrast with **r-strategist species**.

Lag effect: the time between an action (such as a **species introduction** or application of a **pesticide**) and its effect.

Landscape connectivity: the ease with which species can move from one patch to another, for example, via **wildlife corridors**.

Legacy pollutants: pollutants that exist in the environment as a consequence of former industrial processes that are no longer used or that have been superseded by cleaner, less polluting, alternatives.

Linear features: land use strips that link patches together; can be **wildlife corridors**.

Mass extinction event: the catastrophic loss of a very large number of species over a very large spatial scale within a time frame that is short in evolutionary terms.

Measuring: typically refers to quantifying an aspect of the physical environment using an established scale, often an SI unit or derivate of this, for instance, the depth of soil (cm) or flow rate of a stream (metres per second); can also be used for habitat measurements (area or length for linear habitats).

Metapopulation: a collection of semi-discrete populations with some movement of individuals between populations that provides partial gene-flow.

Minimum viable population: the minimum number of individuals needed to ensure that a population has a high chance of remaining in existence for at least 1000 years. Calculated by using a **population viability analysis**.

Mitigation: measures taken to avoid or reduce negative impacts before they occur.

Multi-species conservation: broadly-defined conservation actions aimed at conserving multiple species or an entire species community; differs from **single-species conservation**.

Mutualism: a relationship between two species where both benefit; this is referred to as obligate if neither species can survive without the other or facultative where the relationship is not essential to survival.

Native species: a species that occurs in an area as a result of natural processes.

Naturalized species: a **non-native species** that has become self-sustaining in the wild in its new environment.

Nature reserve: an area of land or water that is managed in some way for the benefit of biodiversity; many nature reserves are also **protected areas**, but the two terms are not synonymous.

Nearest neighbour: the distance from one patch of a habitat (or one individual of a species) to the nearest patch of the same habitat (or individual of the same species).

Niche: the sum of the ecological conditions a species needs to survive.

Nitrogen fixation: the ability of a species to convert atmospheric nitrogen to nitrogen that is available for plants to use.

Nocturnal species: species that are active at night; opposite of **diurnal species**.

Non-destructive survey methods: survey techniques that do not kill organisms; opposite of **destructive survey methods**.

Non-native species: a species that occurs in an area where it has not occurred historically, usually as a result of **species introduction** involving **translocation** by humans.

Palaeo-biomonitoring: use of historical records of species, usually fossils or pollen, to reconstruct habitat and ecological change.

Palynology: the study of pollen grains.

Parasitism: an ecological relationship where one species (the parasite) benefits at the expense of the other (the host).

Parasitoid: species that have a free-living (non-parasitic) adult stage, but females lay their eggs in close proximity to, on the surface of, or within the body of invertebrates and the larval form then develops inside the host, consuming it from within and eventually killing it.

Patches: areas that differ from, and are set within, the **background matrix**.

Pest management: control of **pests** to eradicate populations or reduce population size, usually using **pesticides** or **biological pest management**, sometimes via an **Integrated Pest Management** approach.

Pest: a species that is detrimental to human activity (e.g. reducing crop yields, spreading disease, damaging property). Often subject to control measures in the form of **pest management**.

Pesticide resistance: the process by which a **pest** species becomes tolerance of a **pesticide** so that the pesticide's effectiveness is reduced.

Pesticide: a chemical that kills a pest. Often subdivided into herbicides (plants), insecticides (insects and other arthropods), acarides (mites and ticks), nematocides (nematodes), fungicides (fungi), and so on.

Phenology: the timing of seasonal events such as leaf burst, bird breeding, amphibian spawning, etc.

Phenotypic plasticity: the degree to which an individual can change its response to the environment in a non-genetic way.

Phylogenetic species concept: a method of defining whether different individuals are different species based on molecular analysis and DNA.

Phylogenetic tree: a method of depicting how species are related to one another; the taxonomic equivalent of a family tree.

Phytoremediation: the use of plants in bioremediation.

Pollarding: selective cutting of trees to encourage regrowth in multiple small stems from a central stump; similar to **coppicing**, but occurs substantially above ground level.

Polymorphic species: the occurrence of different forms within a species.

Population bottleneck: a sharp reduction in the size of a population.

Population density: the number of individuals of a species per unit area (e.g. per km^2); differs from **population size.**

Population size: the number of individuals in a specific population or globally; differs from **population density**.

Population viability analysis: the process by which the **minimum viable population** for a species is calculated.

Predation: a feeding relationship whereby one species (predator) kills and consumes another (prey). Usually viewed as an animal–animal interaction, but it can be ecologically more useful to group with **herbivory**.

Predator release: a reduction in predation pressure, for example, due to the decrease in predators by natural processes, culling or **translocation.**

Preservation: the action of maintaining a landscape and the biodiversity it contains in its 'natural' state without much, if any, interference from humans.

Primary data: data collected directly by an Applied Ecologist or fieldworker to answer a specific question or for a specific project; differs from **secondary data.**

Protected area: an area of land or water that has got statutory protection that limits the actions that can take place there legally; many protected areas are also **nature reserves**, but the two terms are not synonymous.

Proxy data: data collected or used to infer information on a parameter of interest where it is not possible to measure that parameter of interest directly.

Ramsar site: a wetland of international importance.

Realized range: the spatial area in which a species actually occurs; contrast with **fundamental range.**

Receptors: environmental or ecological factors, such as air, water, habitats, and species that could be affected by a development and that are assessed in an **Environment Impact Assessment** or **Ecological Impact Assessment.**

Reinforcement: the release of individuals of a species to boost an existing population; contrast with **reintroduction** and **supplementation.**

Remediation: cleaning up **legacy pollution** at a site, often using specific organisms in a **bioremediation** framework.

Rescue translocation: movement of individuals away from a threat, usually a new development.

Restoration ecology: a branch of **Applied Ecology** that aims to recreate previous conditions; can include **remediation, reintroduction**, and **rewilding.**

Rewilding: the process of taking a managed, disturbed, or unnatural area, and returning it to a previous more natural state.

r-strategist species: organisms that have a 'live fast, die young' life history characterized by high growth rates (r), short lifespan, and numerous offspring per breeding attempt; tend to predominate in harsh or disturbed environments. Contrast with **K-strategist species.**

Scavengers: animal species that either specialize on carrion (obligate scavengers) or will take carrion if available (facultative scavengers).

Screening: the process by which a development proposal is assessed to determine whether an **Environment Impact Assessment** or **Ecological Impact Assessment** is necessary.

Secondary data: data collected before a specific ecological project, often for a different reason, but that is useful for that project; differs from **primary data.**

Single-species conservation: narrowly-defined conservation actions aimed at conserving just one species; differs from **multi-species conservation.**

Special Area of Conservation: an area designated at European level as being important for a priority habitat.

Special Protection Area: an area designated at European level as being important for birds.

Specialist species: a species with a highly constrained **niche** that can only tolerate a narrow set of environmental conditions; opposite of a **generalist species.**

Species community: all the species that co-occur within the same spatial area.

Species diversity: a measure of biodiversity that combines **species richness** and **species evenness**.

Species evenness: a measure of community structure that measures the relative abundance of each species; high evenness occurs if there are similar numbers of each species.

Species introduction: the **translocation** of a species by humans to a new environment.

Species population: a group of individuals of the same species.

Species recovery plan: a document describing protocols for protecting and conserving rare species.

Species reintroduction: the release of individuals of a species to an area of former range from which it had been extirpated. Contrast with **reinforcement** and **supplementation.**

Species richness: the number of species within a given area.

Statutory regulator: a legally established and/or government-appointed agency authorized to enact or enforce national or nationally devolved legislation on behalf of the relevant country or state, including that relating to the environment and ecology (sometimes referred to as the Competent Authority or Regulatory Authority).

Stochastic processes: random events that have a substantial and unpredictable effect of the overall population; can be either environmental or demographic.

Strategic Environmental Assessment: a holistic overview of a series of proposed developments or land use changes, rather than focusing on individual projects as per **Environmental Impact Assessment.**

Studbook: a compilation of genealogical data of individual animals; usually used to inform captive breeding programmes within ***ex situ* conservation.**

Supplementation: the release of individuals of a species to boost a population that was originally **reintroduced;** contrast with **reinforcement.**

Symbiosis: organisms that live together in close and long-term relationships, not necessarily beneficial ones.

Translocation: human-mediated movement of species from one area to another, possibly to avoid exposure to a threat (**rescue translocation**), to move individuals from their native range to their non-native range as a **species introduction**, or for conservation reasons as a **conservation translocation.**

Umbrella species: species that are often prioritized within **conservation triage** because protecting these species indirectly protects many other species and/or a specific habitat.

Vector: a species that spreads disease to another species without itself showing signs of illness.

Virtual island: an isolated habitat **patch** surrounded by a 'sea' of other land uses.

Wildlife corridors: **linear features** or landscape strips that species can use for moving around the landscape.

Zoonosis: a disease that can be transmitted from animals to humans either directly (e.g. Ebola) or via a **vector** that does not show signs of illness (e.g. a mosquito transmitting malaria or West Nile disease).

INDEX

Tables and figures are indicated by an italic *t* and *f* following the page number

aardvark *Orycteropus afer* 268
abiotic environment 393
  interactions 3, 8, 8*f*
abundance
  absolute 55
  common measures 85
  relative 55
  versus presence/prevalence 85
Acacia 242
ACFOR scale 85
acoustic surveying 62
active management 291, 393
  see also ex situ conservation; *in situ* conservation; management
adaptability, extinction risk relationship 264
adder's-tongue spearwort *Ranunculus ophioglossifolius* 320
Adelie penguin *Pygoscelis adeliae* 255, 318
Aedes aegypti mosquito 243
African baobab *Adansonia digitate* 276
African elephant *Loxodonta africana* 44, 47, 262
African pancake tortoise *Malacochersus tornieri* 344
African penguin *Spheniscus demersus* 209, 345
African tulip tree *Spathodea campanulata* 214
African wild dog *Lycaon pictus* 58, 338
agri-environment schemes 318–319
agouti *Dasyprocta leporina* 363, 380
air quality
  lichens as indicators 75–77, 79*f*, 80–81, 80*f*
  see also pollution
Alabama argillacea 244
Alaotran gentle lemur *Hapalemur alaotrensis* 281
Alcanivorax borkumensis 140
Alder *Alnus* 148
alert monitoring 32, 42, 43*f*
alfalfa *Medicago sativa* 233
alfalfa weevil *Hypera postica* case study 233, 233*f*
algal blooms 85
allogeneic processes 393
Alpine marmot *Marmota marmot* 163
American alligator *Alligator mississipiensis* 276
American bison *Bison bison* 365-6
American bullfrog *Lithobates catesbeianus* 61
American mink *Neovison vison* 200, 209*f*, 210, 215, 218
Amur leopard *Panthera pardus orientalis* 334
ancient woodland vascular plants (AWVPs) 90
anecdotal evidence 45
Anemone nemorosa wood anemone 111
animal sentinels 69, 70*f*
annual nettle *Urtica urens* 75
  see also dwarf nettle
Anopheles mosquito 226
Anthropocene 260
anthropogenic effects 3–4, 4*f*, 393
  monitoring 40
anthropogenic patches 158, 159*f*
ants, as biodiversity indicators 91–92
apex predators 11, 275
  reintroduction 363, 363*f*
aquatic species as indicators of organic pollution 81–83
  British Monitoring Working Party (BMWP) method 81, 82–83, 82*t*
  fully quantitative indices 83
  semi-quantitative indices 83
Arabidopsis thaliana 148
arable reversion 302
area umbrella 278
arithmetic mean 86
artificial nest creation 296, 296*f*
artificial reef creation 296–297, 297*f*
artificial selection 23
ash *Fraxinus* 94
  ash, European *Fraxinus excelsior* 111, 181-182
ash dieback case study 182–183, 182*f*
  see also Chalara fraxinea
Asian gypsy moth *Lymantria dispar* 194
Asian hornet *Vespa velutina* 214
Asian longhorned beetle *Anoplophora glabripennis* 199
Asian mongoose *Herpestes javanicus* 200
Asian tapir *Tapirus indicus* 59, 63
aspen *Populus* 94
assisted colonization 358, 360, 393
atrazine 238
Attini 16
Audubon's crested caracara *Polyborus plancus audubonii* 278
aurochs *Bos primigenius* 387
Australian pine *Casuarina equisetifolia* 208
autogenic processes 393
automated monitoring 47–48
Automated Water Quality Monitoring Network, North America 47
autotrophs 11, 393
average value calculation 86–87, 86*f*
aye-aye *Daubentonia madagascariensis* 58

Bacillus thuringiensis 240, 244
background matrix 157, 157*f*, 158, 393
badger, European *Meles meles* 30, 106, 278, 296
Baiji *Lipotes vexillifer* 346
bald eagle *Haliaeetus leucocephalus* 276
barbastelle bat *Barbastella barbastellus* 327
barley *Hordeum vulgare* 148
barrier methods for pest management 234, 234*f*
barriers 160
  permeability improvement 186
baseline conditions 393
baseline surveying 32, 33–34, 65, 107–110
basking shark *Cetorhinus maximus* 276
Batrachochytrium dendrobatidis 226
Bayesian network modelling 119
bearded vulture *Gypaetus barbatus* 374
Beauveria bassiana fungus 244
Bechstein's bat *Myotis bechsteinii* 30
beech *Fagus* 85, 301
  Fagus sylvatica 78
beetles, as biodiversity indicators 91
beneficial relationships 16–18

Bewick's swan *Cygnus columbianus* 166
bilberry *Vaccinium myrtillus* 300
bio-augmentation 142–143, 393
bioaccumulation 76, 393
  hyperaccumulators 138
  pesticides 244
  use in bioremediation 137–138
bioconversion factor 148
biodiversity
  audit 33
    case study 34, 34*t*
  hotspots 91, 207, 393
    conservation priority 280
    global biodiversity hotspots 280, 282*f*
  indicators *see* biodiversity indicators
  loss of 4*f*
  offsetting 127, 128, 393
  see also species diversity
Biodiversity Action Plans (BAPs) 313–314
biodiversity indicators 91–93, 393
  caveats 92–93
  composite index 92
  cross-taxonomic indicator 92
  population trends as sustainability indicator 92, 92*f*
  terrestrial invertebrates 91–92
  within-taxa indicator 92
biogeography 24–25
  geographical range 19, 162
  island biogeography, habitat fragmentation parallels 172–173, 173*f*
biological homogenization 197
biological indicators 88–90, 393
  marine animals as indicators of fishing levels 90
  plants as indicators of disturbance and grazing levels 88–89, 89*f*
  plants as indicators of habitat age 89–90, 89*f*
biological oxygen demand (BOD) 81
biological pest management 241–244, 242*f*, 393
  augmentation of natural enemies 242
  conservation of natural enemies 242
  introduced control 242
  non-native species 218
biological species concept 36, 393
biomes 51
bioreactors 145–147, 145*f*
  Leviathan Mine case study 146, 146*f*
bioremediation 136–153, 393
  genetically manipulated organism use 149
  hyperaccumulation 137–138
  marine oil spill case study 140, 140*f*
  metabolic breakdown 137
  using fungi 137
  using microorganisms 138–147
    bio-augmentation 142–143, 393
    stimulation of indigenous microorganism growth 139–142
    techniques 143–147
  using plants 137–138, 147–153
    see also phytoremediation
biosparging 144
biostimulation 143, 393
biosurfactants 142
biosurveys 393
  as ecological indicators 70
  see also surveying
biotic environment 393
  interactions 3, 8, 8*f*
biotic indices 70, 84, 393
  average value calculation 86–87, 86*f*
  fully-quantitative indices 83
  presence/prevalence versus abundance 85
  semi-quantitative indices 83
  see also biodiversity indicators; biological indicators; ecological indicators; environmental indicators
bioventing 144, 144*f*
birch *Betula* 80, 94, 148
birdcage plant *Oenothera deltoides* 175*t*
black bear *Ursus americanus* 124, 300
black rat *Rattus rattus* 205, 209, 216, 219, 305, 324, 335
black rhino (*see* black rhinoceros)
black rhinoceros *Diceros bicornis* 307, 309
black stork *Ciconia nigra* 44
black trout *Salvelinus fontalis* 200
black vine weevil *Otiorhynchus sulcatus* 242
black-backed jackal *Canis mesomelas* 306
black-faced impala *Aepyceros melampus petersi* 382
black-footed ferret *Mustela nigripes* 256*t*
black-tailed jackrabbit *Lepus californicus* 325
black-veined white *Aporia crataegi* 186
blackthorn *Prunus spinosa* 186
blesbok *Damaliscus pygargus phillipsi* 378
blowfly *Calliphora latifrons* 14
blowfly *Calliphora vomitoria* 19
blue tit *Cyanistes caeruleus* 294
blue whale *Balaenoptera musculus* 11
blue-eyed black lemur *Eulemur macaco flavifrons* 327
bobcat *Lynx rufus* 174, 229
body size, extinction risk relationship 264
bog lemming *Synaptomys cooperi* 296
Bornean orangutan *Pongo pygmaeus* 348
Bosavi woolly rat 37
boundaries 160–161, 161*f*
Bouteloua gracilis 211
bramble *Rubus fruticosus* 211
Braun-Blanquet scale 85
British Monitoring Working Party (BMWP) water oxygen monitoring method 81, 82–83, 82*t*
broad bodied chaser *Libellula depressa* 179
broad-leaved dock *Rumex obtusifolius* 78
broad spectrum pesticides 235
Brookesia micra 176
brown fur seal *Arctocephalus pusillus* 318
brown hairstreak *Thecla betulae* 186
brown hare *Lepus europaeus* 303
brown hyena *Hyaena brunnea* 14, 306
brown tree snake *Boiga irregularis* 195, 200, 209*f*
browsing 11–12
Brucella abortus 318
  see also brucellosis
brucellosis 318
brushtail possum *Trichosurus vulpecula* 211*f*
buckwheat *Eriogonum* spp. 210
Buddleia davidii 226-7
buff-tailed bumblebee *Bombus terrestris* 164
buffer zones 188
bumblebee *Bombus* 210
Burchell's zebra *Equus quagga burchellii* 342
Burmese python case study 229, 229*f*
burrowing owl *Athene cunicularia* 380
burying beetle *Nicrophorus* 15
butterflies, as biodiversity indicators 92

caddis fly Trichoptera spp. 210
California Condor *Gymnogyps californianus* 256*t*, 338, 339–40, 341, 357captive breeding case study 339–340, 339*f*
camera trapping 59–62, 63–64, 63*f*
Canadian lynx *Lynx canadensis* 10
Canadian pondweed *Elodia canadensis* 152
cane beetle *Dermolepida albohirtum* 246
cane toad *Rhinella marina* 193, 200, 211, 246
Cape griffon vulture *Gyps coprotheres* 19, 306
capercaillie *Tetrao urogallus* 278
captive breeding programmes 270, 332, 332*f*, 343–348
  case study 339–340, 339*f*
  husbandry 345–346, 346*f*
  hybridization problems 347
  inbreeding problems 346–347, 347*f*
  reintroduction 348–349, 349*f*
    case study 350–351, 350*f*
  studbooks 347–348
  see also ex situ conservation
capture–mark–recapture (CMR) technique 57–58, 58*f*, 393
  home range studies 166, 166*f*
caracal *Caracal caracal* 58
carbamates 236–237
Cardiff Bay Barrage case study 108–110, 108*f*, 109*t*, 110*f*
cardinal ladybird *Rodolia cardinalis* 242–3
Caribbean spiny lobster *Panulirus argus* 315
carrying capacity 20, 263, 393
  reintroduction release site 374
cascade benefits 270
cavity spot *Pythium sulcatum* 226
censusing 31
Chacoan peccary *Catagonus wagneri* 54*f*
Chalara fraxinea 181
  see also ash dieback
chamois *Rupicapra rupicapra* 300
charismatic species 276, 365
Cheltenham Biodiversity Audit case study 34, 34*t*
chequered skipper butterfly *Carterocephalus palaemon* 38
Chernobyl reintroduction programmes 375–376, 375*f*
chicken, domestic *Gallus gallus domesticus* 305
chickweed *Stellaria media* 225
chlorophleth maps 163, 164*f*
Christmas Bird Count, North America 39
Chrysanthemum 237
chytrid fungus *Batrachochytrium dendrobatidis* 41, 332
citizen science 49–50, 49*t*, 393
  non-native species surveying 212
climate change 4*f*
  effects on landscape ecology 181–183, 181*f*
  impact on protected area effectiveness 316–317, 317*f*
climate envelope modelling (CEM) 183, 393
climax community 25
clover *Trifolium* 276
clown fish *Amphiprion* 276
collared dove *Streptopelia decaocto* 195
Collision Risk Model (CRM) 120
commensalism 17–18, 393
common buzzard *Buteo buteo* 77
common crane *Grus grus* 379
common dormouse *Muscardinus avellanarius* 324
common frog *Rana temporaria* 198, 319
common periwinkle *Littorina littorea* 196
common pipistrelle *Pipistrellus pipistrellus* 77
common reed *Phragmites* 94, 152
Common Standards Monitoring (CSM) framework 52, 52–53, 53*f*
common tern *Sterna hirundo* 305
communities 20, 397
  climax community 25
  ecological indicators 73, 77
  evenness 283, 283*f*, 398
  interspecific competition 21
  seral communities 25
  threats to 256–258, 257*t*
community engagement 320–321
  reintroduction programmes 374–376
community similarity 284, 393
community structure succession 25
community-linked conservation 321–323
compensation 121, 122*f*, 127–128, 393
  mitigation and compensation monitoring 32–33, 45
competition 20–21, 393
  interspecific 21
    case study 22
  intraspecific 20
competitive exclusion principle 21
competitor release 231
compliance monitoring 32, 42–45, 66
composite index 92
composting, in bioremediation 144–145, 145*f*
connectivity 396
  gap analysis 189
  improvement 185–187, 186*f*
    barrier permeability improvement 185
    corridor creation 185
    hedgerows 186
    problems with 186–187, 187*f*
    stepping stone islands 185–186
  invasion risk relationship 186–187, 205
  maps 180
  umbrella 278
  see also corridors
conservation 253–255, 394
  cascade benefits 270
  decision-making 272–274
    limited resources 272
    triage system 272–274, 394
  edge-of-range species 267, 267*f*
  ethical aspects 274, 275, 335, 363
  habitat-focused 256, 257
  non-conservation priorities 285
  setting priorities 255, 285
  site-focused priorities 280–285
    areas with distinct communities 284–285
    areas with high endemism 280
    areas with high species diversity 280–284
    areas with historical legacy 285
    biodiversity hotspots 280
  species-focused priorities 255–256, 256*t*, 275–280
    flagship species 276–278, 277*f*
    keystone species 275–276
    species with high reintroduction potential 285
    umbrella species 278–280, 279*f*
  statistics 254*f*
  strategies 270, 271*t*
  versus preservation 254–255
  see also captive breeding programmes; endangered species; *ex situ* conservation; *in situ* conservation

conservation surrogate 362
contact insecticides 236
contamination *see* pollution
control action threshold (CAT) 232
Convention for the Protection of the Marine Environment of the North-East Atlantic 114, 114*f*
Convention of Biological Diversity (CBD) 313, 366
Convention on International Trade in Endangered Species of Wild Fauna and Flora (CITES) 44, 367
Convention on the Conservation of European Wildlife and Natural Habitats 366
Cook's petrel *Pterodroma cookii* 219
coppicing 299, 300*f*, 301, 394
core areas 188
cork oak *Quercus suber* 113
Corncrake *Crex crex* 319
corridors 160, 398
  creation 185
  disease spread 187
  fire spread 187
  hedgerows 186
  use by non-native species 186–187, 187*f*
  see also connectivity
Corynephorus canescens 52
cosmopolitan species 34
cotton boll weevil *Anthonomus grandis* 228, 244
cotton *Gossypium* 228
cottontail rabbit *Sylvilagus* 229
cottony cushion scale insect *Icerya purchasi* 242–243
cougar *Puma concolor* 124
coyote *Canis latrans* 174
coypu *Myocastor coypus* 218
Crassostrea gigas 210
cream milkvetch *Astragelus racemosus* 72
creeping lady's tresses *Goodyera repens* 90
creeping water primrose *Ludwigia peploides* 226
crepuscular species 58, 394
crested cow-wheat *Melampyrum cristatum* 37, 37*f*
crested wheat grass *Agropyron cristatum* 210
crisis ecoregions 280
cross-taxonomic biodiversity indicator 92
crowd sourcing 49
Cryptochaetum 243
cryptic species 36
cryptogenic species 196–197, 394
Culex quinquefasciatus mosquito 209*f*
culling 307–309
  non-native species 218
cumulative effects 116
cumulative impact assessment 117
custodial management 291, 394
  see also in situ conservation
Cuvier's beaked whale *Ziphius cavirostris* 187

DAFOR scale 85
dandelion *Taraxacium* 94
  Taraxacium officinale 200
data collection
  direct 55–58
    capture–mark–recapture (CMR) technique 57–58, 58*f*, 393
  indirect 58–59, 59*t*
  remote 59–62
data mining, from social networks 49
databases
  non-native species 214–215
  species-at-risk 264–265, 265*f*
dating methods 93
DDT (dichlorodiphenyltrichloroethane) 234–235
  bioaccumulation 76
de-extinction 353
deadwood piles 296
decomposition 14–16, 15*f*, 394
deer mouse *Peromyscus maniculatus* 210, 296
Deladenus siricidicola 244
demographic stochasticity 261
density-dependent effects 20, 394
dermestid beetle *Dermestes maculatus* 15
  see also hide beetle
Desmococcus viridis 79
desert tortoise *Gopherus agassizii* 167
development impact assessment 116–120
  impact magnitude and importance 118, 118*f*
  impact prediction 118–120
  potential for positive impacts 117–118, 118*f*
  types of impact 116–117, 116*f*
Devil's Hole pupfish *Cyprinodon diabolis* 219
Dianthus morisianus 357
dingo *Canis lupus dingo* 194
direct data collection and use 55–58
  basic species surveying methods 55, 56*t*
  capture–mark–recapture (CMR) technique 57–58, 58*f*, 393
  destructive methods 57, 394
    use of destructive indicator systems 76–77
  invasive methods 56–57
  non-destructive methods 55–56, 396
  observational methods 55–56
direct effects 116
disease
  ecological effects of 181
    ash dieback case study 182–183, 182*f*
  spread through corridors 187
disjunctions 162–163
dispersal, extinction risk relationship 264
distribution *see* species distribution
disturbance
  invasion risk relationship 204–205
  patches 159
diurnal species 394
diversity index 283, 284*t*
  see also biodiversity; species diversity
diversity–stability hypothesis 207
dodo *Raphus cucullatus* 258
dog rose *Rosa canina* 179
dog's mercury *Mercurialis perennis* 111
Dolania americana 24
DOMIN scale 85
dormouse *Muscardinus avellanarius* 56-7, 106, 126, 296
dot maps 163, 164*f*
downy brome *Bromus tectorum* 225
drift fencing 123, 123*f*
Duke of Burgundy *Hamearis lucina* 256
dunlin *Calidris alpine* 179
Dutch elm disease *Ophiostoma* 200
dwarf nettle *Urtica urens* 75
  see also annual nettle

Eagle Lake rainbow trout *Oncorhynchus mykiss aquilarum* 377
early warning 42

eastern yellow robin *Eopsaltria australis* 174, 279
ecdysis 16
ecological education 320–321
ecological enhancement 128
ecological guild 394
ecological impact assessment (EcIA) 45, 99, 104–112, 394
  baseline survey 33, 107–110
  comparative approach 105–107
  follow-up 111–112, 111*f*, 112*f*
  limitations and challenges 130–131, 130*t*
  outcomes 128–129
    post-development monitoring 129
    recommendations 128–129
  purpose of 104–105
  receptors 105–107, 107*t*, 397
  scoping 107–111, 111*f*
    national vegetation classification (NVC) systems 111
  see also development impact assessment; environmental impact assessment (EIA)
ecological indicators 69–73, 72*f*, 73*t*, 394
  advantages and disadvantages of 95*t*
  biosurveys 70
  biotic indices 70
  single- and multispecies approaches 71–72
ecological management *see* management
ecological replacement 358, 360, 394
ecological resilience 119, 394
ecological succession 24, 25, 39, 39*f*, 298*f*, 394
  arresting 297–298
  in decomposition 15
ecological value 112–116
  ecosystem service value 116
  habitat value 113–115, 115*f*
  individual organism value 115
  site value 112–113
  species value 115
ecology 3, 5, 394
  applied ecology 4–5, 8, 393
  definition 8
economic injury level (EIL) approach 228–231, 230*f*, 394
  problems with 231–232
    heterogeneity 231
    lag time 232
    quantification difficulty 231–232
economic threshold (ET) 232, 394
ecoregions 188, 189, 189*f*, 394
ecosystem engineers 115, 115*f*, 179, 275–276, 394
  non-native species 209
  population change effects 181
  reintroduction 362–363
ecosystem services 116, 363, 394
ecotones 160–161, 394
ecotourism 277–278, 321–323, 323*t*
edge effects 179, 394
  beneficial effects 184
edible frog *Pelophylax esculentus* 198
EDGE of Existence programme 268–270, 268*f*, 269*f*
education 320–321
  reintroduction and 376
elk *Cervus canadensis* 124, 276, 363
Ellenberg biotic index 75, 78
elm *Ulmus* 94
Elton, Charles Sutherland 10, 10*f*
Encarsia formosa 244
endangered species
  evolutionarily distinct species 268
    EDGE of Existence programme 268–270, 268*f*, 269*f*
  international species-at-risk classification systems 264–265, 265*f*, 266*f*
  minimum viable population 262, 359, 396
    population viability analysis (PVA) 262–263, 263*f*, 397
  national species-at-risk classification systems 265–266, 266*f*
  see also conservation; extinction
Endangered Species Act 1973, US 42
endemic species 394
  extinction risk 264
  islands 176
enemy release hypothesis 202–203
  competitor release 231
  predator release 40, 397
English oak *Quercus robur* 294
enhancement initiatives 128
Enterobacter agglomerans 149
Environment Protection and Biodiversity Conservation Act 1999, Australia 42
environmental DNA 60–61
environmental impact assessment (EIA) 100–103, 394
  challenges 130*t*
  developments requiring EIA 101–103, 102*t*
    responsibility for costs 102–103
    screening of proposals 101–102
  legislative framework 100–101
  process of 103, 104*f*
  purpose of 100
  receptors 103, 103*f*, 397
  strategic environmental assessment (SEA) 130–131
  see also Ecological Impact Assessment (EcIA)
environmental indicators 73–88, 394
  aquatic species as indicators of organic pollution 81–83, 82*t*
  average value calculation 86–87, 86*f*
  destructive indicator system use 76–77
  general lessons from 83–85
  lag time significance 84–85
  lichens as indicators of air quality 75–77, 79*f*, 80–81, 80*f*
  plants as indicators of soil properties 75
  species that make good indicators 85–88
environmental patches 158
environmental resources *see* resources
environmental stochasticity 261
Epicoccum purpurescens 295
equilibrium abundance 231
eradication of non-native species 215–217
  missed opportunities 216–217, 216*f*
ethical issues in conservation 274, 275, 335, 363
Eucalyptus 150
  Eucalyptus obliqua 55
Eurasian lynx *Lynx lynx* 383, 384
Eurasian otter *Lutra lutra* 180, 211, 296, 322
  see also European otter
Eurasian skylark *Alauda arvensis* 93
European (Eurasian) beaver *Castor fiber* 296, 357, 362, 364, 372, 383, 387
European bison *Bison bonasus* 348, 372, 383, 385, 386, 388–389
  European bison rewilding case study 388, 388*f*
European corn borer moth *Ostrinia nubilalis* 240
European directives 43, 100, 101
European eagle owl *Bubo bubo* 196
European elk *Alces alces* 40, 61, 124, 384
European mink *Mustela lutreola* 209*f*
European otter *Lutra lutra* 180, 211, 296, 322
  see also Eurasian otter
European perch *Perca fluviatilis* 194

European rabbit *Oryctolagus cuniculus* 58, 220, 225, 305
eutrophication, algal blooms as indicators 85
Evidence Reasoning model 119–120
evidence-based management 65, 324–327
evolution 21–24, 395
 definition 22–23
Evolutionarily Distinct and Globally Endangered (EDGE) programme 268–270, 268*f*, 269*f*
evolutionary arms race 12, 395
evolutionary distinctiveness 268, 269
ex situ conservation 270, 271*t*, 285, 331–335, 395
 case study 335–337, 336*f*
 ethical aspects 335
 gene banking 351–352, 351*f*
 guidelines 333–334
 permissions required 334
 stages of 337–349
  captive breeding 343–348, 346*f*, 347*f*
  collection 337–342, 338*f*, 342*f*
  reintroduction, supplementation and reinforcement 348–349, 349*f*
  transport 342–343, 343*f*, 377, 377*f*
 see also conservation; reintroduction
expansive native species 211
extinction 258, 395
 chance extinction 261
 de-extinction 353
 drivers of 256
 functional 259, 395
 in the wild 259, 395
 mass extinction events 253, 259–260, 259*f*
 rates 253
 see also endangered species
extinction risk 258–268
 extinction vortex 260–262, 260–261*f*, 395
 international species-at-risk classification systems 264–265, 265*f*, 266*f*
 minimum viable population 262, 359, 396
  population viability analysis (PVA) 262–263, 263*f*, 397
 national species-at-risk classification systems 265–266, 266*f*
 species-specific traits and 263–264
extirpation 259, 357, 395
 risk 266
extractive reserves 319

fallow deer *Dama dama* 345
false absence 54
feeding relationships 11–16
 decomposition 14–16, 15*f*
 herbivory 11–12, 11*f*, 395
 parasitism 12–14, 13*f*, 396
 predation 11, 12*f*, 397
fencing 187
 drift fences 123, 123*f*
 virtual fences 187
feral cat *Felis catus* 216, 219
feral goat *Capra hircus* 218
field maple *Acer campestre* 111
fire, spread through corridors 187
fishing pressure
 marine animals as indicators 90
 quotas 312
flagship species 276–278, 277*f*, 365, 395
Florida panther *Puma concolor coryi* 278–279
footprint traps 59
Forda formicaria 22
forestry reversion 302
founder effect 395
founder individuals *see* reintroduction
fox *Vulpes vulpes*, 55, 368, 386
fragmentation *see* habitat fragmentation
freshwater shrimp Gammaridae 83
friction maps 180, 395
fruit fly *Drosophila melanogaster* 262
functional extinction 259, 395
fundamental niche 19
fundamental range 162, 395
fungus-growing ants, tribe Attini 16

Galapagos giant tortoise *Chelonoidis nigra* 176, 200
Galerucella calmariensis 248
Galerucella pusilla 248
galjoen *Dichistius capensis* 276
gap analysis 189
gardens 319
gene banks 332, 351–352, 351*f*, 395
generalists 19, 395
genetic engineering 240
genetic swamping 218
genetically modified organisms (GMOs) 149, 240, 395
 microorganisms 395
 plant-incorporated protectants 238
 use in bioremediation 149
Geographical Information Systems (GIS) 169–171, 169*f*, 171*f*, 395
 habitat suitability mapping 171, 171*f*
geographical range 19, 162
Geoica 22
geolocators 167*t*
giant redwood *Sequoiadendron giganteum* 184, 276
giraffe *Giraffa camelopardalis* 357
global biodiversity hotspots 280, 282*f*
GLOBENET initiative 91
glyphosate 241
golden frog *Atelopus zeteki* 332
golden lion tamarin *Leontopithecus rosalia* 364
golden snub-nosed monkey *Rhinopithecus roxellana* 278
golden plover *Pluvialis apricaria* 179
gopher tortoise *Gopherus polyphemus* 278
Grand Cayman blue iguana *Cyclura lewisi* 270
grazing 11–12, 12*f*
 plants as indicators of grazing levels 88–89, 89*f*
great bustard *Otis tarda* 327
great crested newt *Triturus cristatus* 57, 106, 118, 123
great tit *Parus major* 93
greater kudu *Tragelaphus strepsiceros* 15, 307
greater prairie chicken *Tympanuchus cupido* 261
greater sage-grouse *Centrocercus urophasianu* 278
green bridges 123
 case study 124–125, 124*f*
green lacewing (order Neuroptera) 244
greenhouse whitefly *Trialeurodes vaporariorum* 242
grey partridge *Perdix perdix* 209*f*, 303
grey squirrel *Sciurus carolinensis* 193, 218, 225
grey wolf *Canis lupus* 40, 259, 276, 335, 363, 383, 384, 387
griffon vulture *Gyps fulvus* 376
grizzly bear *Ursus arctos* 59, 124, 263, 383
Guam rail *Gallirallus owstoni* 259
guinea worm *Dracunculus* 13
 guinea worm case study 13–14, 13*f*
gull *Larus* 209*f*, 225, 305

habitat 19
age, plants as indicators 89–90, *89f*
classification 50–51
condition assessment 51–53, *53f*
habitat-focused conservation 256, 257
heterogeneity 51
improvement of 184, *185f*, 298–301
management of *see* habitat management
monitoring 50–53
Phase One Habitat Surveying (P1HS) 51, *52f*
species–habitat interaction research 324–327, *325f*
translocation of 125, 126
value of *see* habitat value
habitat fragmentation 41, 123, 172, 395
direct effects on species 174–179
case study 177–178, *177f*, *178f*
extinction versus speciation 176
metapopulations 174–175, *178f*
severity and landscape resilience relationship 174
source-sink population dynamics 175–179
species vulnerability 174, *175t*
island biogeography parallels 172–173, *173f*
mapping 180
metrics 179–180
severity 172, *173f*
types of 172, *172f*
wider ecological effects 179
habitat loss 172
spatio-temporal monitoring 40–42
habitat management 292–303
arresting ecological succession 297–298
creating habitat 292–297, *293f*, *296f*, *297f*
habitat grading 299, *299f*
improving structural complexity 298–301
restoration by reversion techniques 302–303
restoration by reviving traditional management 301–302
rotational management 299–300
within-patch differential management 299
zonation 299
see also rewilding
habitat profiling 326
habitat suitability mapping 171, *171f*
habitat umbrella 278
habitat value 113–115
declining habitat 114
difficult to recreate habitat 114–115
habitat usefulness 115
highly specialized/restricted habitat 114
priority habitats 115
rare habitat 113–114, *114f*
vulnerable habitat 114, *114f*
Haekel, Ernst 9, *9f*
hair tubes 59
harbour porpoise *Phocoena phocoena* 61
hard rush *Juncus inflexus* 78
harebell *Campanula rotundifolia* 41–42
harlequin ladybird *Harmonia axyridis* 36, 212, 214
Hawai'i 'amakihi *Hemignathus virens virens* 362
Hawaiian goose *Branta sandvicensis* *256t*
hawksbill sea turtle *Eretmochelys imbricate* 253
hawthorn *Crataegus monogyna* 186
hazel *Corylus avellana* 300
hazel dormouse *Muscardinus avellanarius* 186
see also common dormouse
hazel, Eurpoean *Corylus avellana* 324
heat maps 163–165, *165f*
heath fritillary *Mellicta athalia* 300
Heck cattle 387
hedgehog, European *Erinaceus europaeus* 196, *209f*, 296, 319
hedgerows 186, 296
hen harrier *Circus cyaneus* 305
herbicides 238–241
herbivores 11–12, *11f*, 395
obligate 11
hermit crab *Pagurus* 209
Heterakis gallinarumi *209f*
heterogeneous environment 21
heterotrophs 11, 395
hibernacula 296
Himalayan balsam *Impatiens glandulifera* 121, 205, 210, 217
historic landscape reconstruction 93–94, *94f*
hitch-hiking 199, 201–202, *201f*
prevention 202
holm oak *Quercus rotundifolia* 315
home range 165, 395
capture-mark-resight studies 166, *166f*
natural variation use in studies 166, *166f*
technology use in studies 166–167, *167t*
stable isotope analysis 168–169, *168f*
honey bee *Apis mellifera* 194, 210, 211, 239
honey fungus *Armillaria* 182
honeysuckle *Lonicera periclymenum* 179
hooded crow *Corvus cornix* 179
hooded robin *Melanodryas cucullata* 279
Hubb's beaked whale *Mesoplodon carlhubbsi* 187
human activities *see* anthropogenic effects
husbandry 345–346, *346f*
hybridization 347
hydrocarbons 139
remediation 139
Hylobius transversovittatus 248
Hylocomium splendens 72
Hymenoscyphus fraxineus 199
hyperaccumulators 138
Hypogymnia physodes 72

Iberian lynx *Lynx pardinus* 114, 274, 305, 360
ice plant *Mesembryanthemum crystallinum* 209
ie'ie vine *Freycinetia arborea* 219
imidacloprid 238
impact assessment *see* Ecological Impact Assessment (EcIA); environmental impact assessment (EIA)
impact monitoring 32, 45, 66, 129
impact prediction 118–120
methods 119–120
reporting 120
imperative reasons of overriding public interest (IROPI) 129
in situ conservation 270, *271t*, 289, 331, 395
active management 291–309, *292t*
habitat management 292–303
species management 303–309
aims of 290–291

in situ conservation (*Cont.*)
community-linked conservation 321–323
custodial management 309–314
legislation 309–313
policy 313–314
education and community engagement 320–321
evidence-based initiatives 324–327
management approaches 291
motivation for 290*f*
protected area and reserve creation 314–318
limitations of 316–318, 316*f*, 317*f*, 318*f*
SLOSS debate 314–316, 316*f*
visitor pressure management 319–320, 320*f*
wider countryside initiatives 318–319
agri-environment schemes 318–319
extractive reserves 319
urban gardens 319
see also conservation
inbreeding 346–347, 347*f*, 395
captive populations 347
inbreeding depression 346, 395
Indian mustard *Brassica juncea* 148
Indian rhinoceros *Rhinoceros unicornis* 309
indicator species 72, 395
indicators *see* biodiversity indicators; biological indicators; ecological indicators; environmental indicators
indirect data collection and use 58–59, 59*t*
indirect effects 116
individual fishing quotas (IFQ) 312
insect hotels 293
insecticides 234–238, 236*f*
resistance 245–246, 245*f*
insurance populations 270, 303, 333, 334*f*, 361, 395
case study 304, 304*f*
integrated pest management (IPM) 247–249, 248*f*, 395
interference competition 20
intermediate disturbance hypothesis 88
International Union for the Conservation of Nature (IUCN)
red list 264–265, 265*f*, 266*f*, 381, 396
reintroduction guidelines 367
internationally important sites 113
interspecific interaction 395
competition 21
case study 22
mutualism 16–17, 17*f*, 396
parasitism 12–14, 13*f*, 396
symbiosis 17, 398
intraspecific interaction 395
competition 20
introduction 398
see also non-native species
introgression 347
invasion pathways 199–200, 200*f*, 395
corridor use 186–187, 187*f*
see also translocation
invasion risk 203–207, 204*f*, 395
environmental vulnerability 204–205
prevalence of translocation opportunities 203
propagule pressure 204
risk modelling 205–207
see also invasive species; non-native species
invasive species 202, 395–396
enemy release hypothesis 202–203
invasion meltdown 203
see also invasion risk; nativeness; non-native species
Iris pumila 373
islands
conservation priority 280
endemic species 176
habitat fragmentation parallels 172–173, 173*f*
island vulnerability to invasion 204
IUCN *see* International Union for the Conservation of Nature (IUCN)
ivy *Hedera helix* 225

Jaccard Coefficient of Community Similarity (CC_j) 284
jackal *Canis mesomelas* 17
Jamaican iguana *Cyclura collei* 270
Japanese beetle *Popillia japonica* 178
Japanese knotweed *Fallopia japonica* 36, 209*f*, 212, 212*f*, 216
Japanese white-eye bird *Zosterops japonicas* 219
Javan rhinoceros *Rhinoceros sondaicus* 309
juniper *Juniperus sabina* 357, 373

K-strategists 24, 396
extinction risk relationship 263–264
kelp Laminariales 363
keystone species 275–276, 396
reintroduction 363, 364*f*
kin selection theory 23
kiwi *Apteryx* 205
koala *Phascolarctos cinereus* 324
Komodo dragon *Varanus komodoensis* 277-278
konik *Equus ferus caballu* 387
krill Euphausiidae 317

ladybird beetles Coccinellidae 244
ladybird 338*f*
Laelia lobate 367
lag effect 84–85, 232, 396
non-native species impacts 212, 212*f*
lake trout *Salvelinus namaycush* 196
lamprey *Petromyzon marinus* 218
land farming 143–144, 144*f*
landscape 38, 156–162, 157*f*, 190
background matrix 157, 157*f*, 158, 393
boundaries 160–161, 161*f*
ecoregions 188, 189, 189*f*
historic landscape reconstruction 93–94, 94*f*
linear features 157, 157*f*, 160, 160*f*, 396
living landscapes 188
patches 157, 157*f*, 158–159, 159*f*
landscape connectivity *see* connectivity
landscape ecology 156
Geographical Information Systems (GIS) 169–171, 169*f*, 171*f*
habitat suitability mapping 171, 171*f*
management 184–189
connectivity 185–187, 186*f*
heterogeneity 184, 185*f*
species movement prevention 187, 188*f*
non-landscape process effects on 181–183
climate change 181–183, 181*f*
disease 181, 182*f*
ecosystem engineer population change 181
non-native species introduction 181

spatial patterns of individuals 165–167
home range studies 165–167, 166*f*
spatial patterns of species 162–165
mapping 163–165, 164*f*, 165*f*
species distribution 162–163, 163*f*
species range 162
large blue butterfly *Phengaris arion* 219–221, 256, 357
latitudinal diversity gradient 280
lavender *Lavandula spica* 55
Lazarus species 54, 54*f*
leafcutting ants
Acromyrmex 16, 17
Atta 16, 17
Atta cephalotes 179
Atta sexdens 179
leech *Hirudo medicinalis* 11
legacy pollutants 135–136, 396
legislation 42–45, 309–313
activity-focused 313
compliance monitoring 42–45
environmental impact assessment 100–101
non-native species management 218
reintroduction 367–368
site-focused 42, 310–313, 310–311*t*
species-focused 42, 115, 309–310
Leon Springs pupfish *Cyprinodon bovinus* 209*f*
Leopold matrix 120, 121*f*
Lessepsian migration 187, 187*f*
lesser flamingo *Phoeniconaias minor* 345
Leviathan Mine bioreactor case study 146, 146*f*
lichens as indicators of air quality 75–77, 79*f*
lichen zonation system 77
trunk and twig method 77, 80–81, 80*f*
life history theory 23–24
lime *Tilia* 301
limestone bedsraw *Galium sterneri* 78
Lincoln–Peterson index 57
linear features 157, 157*f*, 160, 160*f*, 396
lion *Panthera leo* 11, 307
lion tamarin *Leontopithecus* 277
littorinid snail 338
living landscapes 188
Local Biodiversity Action Plans (LBAPs) 313–314
local recorders 45–46
loggerhead turtle *Caretta caretta* 209, 321
long-finned pilot whale *Globicephala melas* 61
long-term species monitoring 39
see also temporal monitoring
Lord Howe Island stick insect *Dryococelus australis* 54, 332, 335-356, 342, 343, 346-349, 357, 359
conservation 335–337, 336*f*, 342, 343, 346, 349, 349*f*
Lotka–Volterra model 39–40, 40*f*

Madagascar case study 281, 281*f*
maize *Zea mays* 238
Malay civet *Viverra* 167*t*
Mallorcan midwife toad *Alytes muletensis* 350–351, 368
Mallorcan midwife toad case study 350–351, 350*f*
mammoth *Mammuthus* 353
management 97, 394
active management 291, 393
custodial management 291, 394
evidence-based 65, 324–327
landscape-scale 184
see also landscape ecology
monitoring relationship 65
rewilding and 385
site-level 184
species-level 184
see also ex situ conservation; *in situ* conservation; non-native species; pest management
management umbrella 278
manatee *Trichechus* 44
mapping 31, 34, 38–39, 38*f*, 41*f*, 56*t*
habitat fragmentation 180
habitat suitability mapping 171, 171*f*
species range and distribution 163–165
chlorophleth maps 163, 164*f*
dot maps 163, 164*f*
heat maps 163–165, 165*f*
Mariana's fruit bat *Pteropus mariannus* 174
marine aquarium alga *Culepra taxifolia* 200, 216
marine oil spill bioremediation case study 140, 140*f*
Marine Protected Areas (MPAs) 311, 312
marram grass *Ammophila* 275
marsh fritillary *Euphydryas aurinia* 368
marsh woundwort *Stachys palustris* 210
Marten, American *Martes americana* 296
masked palm civet *Paguma larvata* 44
mass extinction event (MEE) 253, 259–260, 259*f*, 396
Mauritius kestrel *Falco punctatus* 174–175, 256*t*, 289, 305
mayflies Ephemeroptera 24
measuring 31, 396
median 87
Melaleuca howeana 336
Messor 11
metaldehyde 241
metamorphosis 36
metapopulations 174–175, 178*f*, 396
methiocarb 241
midges Chironomidae 83
minimum viable population 262, 359, 396
minnow *Cyprinodon variegatus* 209*f*
mitigation 121–127, 396
focus of mitigation actions 125–127
development alteration 125
receiving environment alteration 125–127
hierarchy 122*f*
mitigation and compensation monitoring 32–33, 45, 129
spatial approaches 122–123, 123*f*
subcategories of 122*t*
temporal approaches 125
mobility, extinction risk relationship 264
mode 87
molecular analysis 60
monarch butterfly *Danaus plexippus* 50, 276
Moneses uniflora 90
monitoring 31, 31–32*t*, 394
alert monitoring 32, 42, 43*f*
approaches to 45–50
automated approaches 47–48
citizen science use 48–50
primary data use 46–48
proxy monitoring 48
secondary data use 45–46
semi-automated approachcs 48
compliance monitoring 32, 42–45, 66
habitats 50–53
impact monitoring 32, 45, 66, 129
management implications 65

monitoring (*Cont.*)
  mitigation and compensation monitoring 32–33, 45, 129
  non-native species 211–212
  post-development monitoring 119, 129
  purposes 32–33
  species 54–64
  temporal monitoring 32, 39–42, 65–66
    anthropogenic change 40
    natural change 39–40, 39*f*, 40*f*
    spatio-temporal monitoring 40–42
moose *Alces alces* 39, 61, 124, 384
  see also European elk
morning glory *Ipomoea cairica* 336
mosquito *Culex* 11
mosquito Culicidae 58
moulting 16
mountain gorilla *Gorilla beringei beringei* 278, 321
mountain hare *Lepus timidus* 305
movement prevention 187, 188*f*
  see also fencing
multi-species conservation 255–256, 396
multilateral agreements 188
mussel *Mytilus californianus* 275
mutualism 16–17, 17*f*, 396
mycoremediation 137
Myrmica sabuleti 220
Myrmica scabrinodis 220
Mytilicola orientalis 203
Myicola ostreae 203
myxomatosis case study 220–221, 220*f*
Myxomatosis cuniiculi 200

National Biodiversity Network (NBN), UK 46, 47*f*
National Biodiversity Strategy Action Plans (NBSAPs) 314
national vegetation classification (NVC) systems 111
national volunteer data 46
nationally important sites 113
native species 396
nativeness 194–197
  cryptogenic species 196–197
  ecological divergence 196
  spatial scale 195–196
  temporal scale 194
  terminology 195
  see also non-native species; translocation
natural change monitoring 39–40, 39*f*, 40*f*
  see also temporal monitoring
natural selection 23
naturalized species 194, 396
nature reserves 396
  creation 314–318
  limitations 316–318
    climate change 316–317, 317*f*
    off-site pressures 317–318, 318*f*
    SLOSS debate 314–316, 316*f*
NatureServe Imperilled list 264–265, 265*f*
nearest neighbour 180, 396
neonicotinoids 238, 239
Nesocodon mauritianus 219
nest box provision 293
  case study 294–295, 294*f*, 295*f*
new species discovery and classification 35–37, 37*f*
New Zealand robin *Petroica longipes* 372
niche 18–19, 73, 396
  competition relationship 20
  fundamental niche 74, 74*f*
  optimal range 74
  realized niche 74, 74*f*
  three-dimensional niche 74, 74*f*
  tolerance range 73–74, 73*f*
  two-dimensional niche 74, 74*f*
Nicotiana glauca 148
nicotine 237–238, 237*f*
night-scented orchid *Epidendrum nocturnum* 278
nine-spotted ladybird *Coccinella novemnotata* 50, 50*f*
nitrogen fixation 17, 396
no net loss principle 128
noctule *Nyctalus noctula* 77
nocturnal species 58, 396
noisy miner bird *Manorina melanocephala* 305
nominal data 55
non-native species 193–198
  co-existence with native species 208
  corridor use 186–187, 187*f*
  databases 214–215
  diversity 197
  impacts 4*f*, 197, 208–211
    buffering methods 218
    complexity of 210–211
    ecosystem engineering 209
    effects on landscape ecology 181
    genetic swamping 218
    lag effects 212, 212*f*
    negative impacts on native species 209*f*
    positive impacts on native species 210*f*
  invasion risk 203–207, 204*f*
    environmental vulnerability 204–205
    prevalence of translocation opportunities 203
    propagule pressure 204
    risk modelling 205–207
    see also invasive species
  likelihood of establishment 200–203, 203*f*
    enemy release hypothesis 202–203
    invasion meltdown 203
  management 197, 214–221, 215*f*
    biological control 218
    case study 198–199
    cast study 220–221, 220*f*
    counter-productive situations 218–219
    eradication 215–217
    legislation 218
    missed opportunities 216–217, 216*f*
    population control 218
    spatial containment 217–218
  monitoring importance 211–212
  native species responses to 211
  naturalization 194, 396
  prevalence 197, 197*f*
  prevention 197, 207–208
  rewilding programmes and 386
  secondary dispersal power 217
  spatial spread 217–218, 217*f*
  species value 115
  terminology 195
  threatened non-native species 219
  see also nativeness; translocation
northern elephant seal *Mirounga angustirostris* 347
northern fulmar *Fulmarus glacialis* 71
northern goshawk *Accipiter gentilis* 174, 177, 326
northern pool frog *Pelophylax lessonae* case study 198–199, 198*f*
nuthatch *Sitta europea* 41

oak *Quercus* 80, 85, 94, 301
oak roller moth *Tortrix viridana* 294
oil beetle *Meloe* 58
oil spill bioremediation case study 140, 140*f*

one-flowered wintergreen *Moneses uniflora* 90
Ophiostoma 199
optimal range 74
orange tip butterfly *Anthocharis cardamines* 186
orange-kneed tarantula *Brachypelma smithi* 14
orca *Orcinus orca* 276
ordinal data 55
organochlorides 236
organophosphates 236, 236*f*
otter, European *Lutra lutra* 180, 211, 296, 322
  see also Eurasian otter, European otter
overfishing 312
  fishing quotas 312
  marine animals as indicators 90

pacific oyster *Crassostrea gigas* 203
Paeonia tenuifolia 373
palaeo-biomonitoring 93, 396
palmate newt *Lissotriton helveticus* 55
palynology 93–94, 94*f*, 396
pangolin *Manis* 44
paraquat 238–241, 241*f*
parasitism 12–14, 13*f*, 396
  case study 13–14, 13*f*
parasitoids 14, 14*f*, 396
parathion 236, 236*f*
Parker's dwarf gecko *Lygodactylus keniensis* 302
Parmelia 77, 79
passenger pigeon *Ectopistes migratorius* 353
passive management 291
  see also management
patches 157, 157*f*, 158–159, 159*f*, 396
  isolated patch problem 174
  reconnection 278
  small patch problem 174
  stepping stone islands 185–186
peat bog conservation case study 257
Père David's deer *Elaphurus davidianus* 357
persistent organic pollutants (POPs) 236
pest management 226, 227–228, 232–246, 396
  biological management 241–244, 242*f*, 393
  case studies 229, 229*f*, 233, 233*f*
  chemical management 234–241, 236*f*, 241*f*
  integrated management (IPM) 247–249, 248*f*
  physical management 233–234, 234*f*
  problems with 244–246
    knock-on ecological effects 246
    pest resurgence 244
    resistance 245–246, 245*f*
    secondary pests 244–245
  theory of 228–232
    economic injury level (EIL) approach 228–232, 230*f*
pesticides 234–241, 397
  herbicides 238–241
  insecticides 234–238
  regulatory bodies 235
  resistance 245–246, 245*f*, 397
pests 224–227, 396
  definition 226
    problems with 226–227, 227*f*
  diversity of 224–226
  see also pest management
Phakopsora pachyrhizi (soybean rust) 42–3
Phase One Habitat Surveying (P1HS) 51, 52*f*
Phasmarhabditis hermaphrodita 244
pheasant *Phasianus colchicus* 209*f*
phenology 40, 397
  anthropogenic change 40
phenotypic plasticity 264, 397
phoresy 17–18
phylogenetic distinctiveness 268, 269
phylogenetic species concept 36, 397
phytoremediation 137–138, 147–153, 397
  phytodegradation 148–150
  phytoextraction 148
  phytostabilization 150
  phytovolatilization of atmospheric pollutants 150
  rhizofiltration 150–152, 152*f*
    case study 153, 153*f*
pied flycatcher *Ficedula hypoleuca* 93, 294, 326
pine marten *Martes martes* 179
Pistacia palaestina 22
PIT (Passive Integrated Transponder) tags 167*t*
plants
  as indicators of disturbance and grazing levels 88–89, 89*f*
  as indicators of habitat age 89–90, 89*f*
  as indicators of soil properties 75
  bioremediation role 137–138, 147–153
  genetically modified plants 238
  incorporated protectants 238
  secondary compounds 237–238, 237*f*
  use in bioremediation, *see also* phytoremediation
plasmids, bacterial 149
Plasmodium 12
Pleurodelinae 179
Poa sandbergii 210, 211
poaching 308, 309
pocket gopher Geomydiae 248
pocket gopher control case study 248–249, 248*f*
Pogonomyrmex 11
point sampling 56*t*
poison ivy *Toxicodendron radicans* 225
policy 313–314
  reintroduction 366
pollarding 300*f*, 301, 397
pollen analysis 93–94, 94*f*
pollutants
  legacy pollutants 135–136, 396
  persistent organic pollutants (POPs) 236
pollution 4*f*, 134–137
  aquatic species as indicators of organic pollution 81–83, 82*t*
  lichens as indicators of air quality 75–77, 79*f*, 80–81, 80*f*
  remediation 136–137
    see also bioremediation
  scale of 136
  sources of 135–136, 135*f*, 139*f*
  types of 135
polychlorinated biphenyls (PCBs) 139, 139*f*
polymerase chain reaction (PCR) 60, 61*f*
polymorphic species 36, 397
pond creation 296
po'o-uli, *Melamprosops phaeosoma* 37
pool frog *Pelophylax lessonae* 196–199
Poplar *Populus* 148, 149
population viability analysis (PVA) 262–263, 263*f*, 372–374, 397
populations 398
  bottleneck 397
  insurance populations 270, 303, 333, 334*f*, 361, 395
  metapopulations 174–175, 178*f*, 396

populations (*Cont.*)
minimum viable population 262, 359, 396
founder populations 372–374
population viability analysis (PVA) 262–263, 263*f*, 372–374, 397
size 397
source-sink dynamics 175–179
threats to 256–258, 257*t*, 258*f*
trends, as sustainability indicator 92, 92*f*
post-development monitoring 119, 129
powdery mildew 226
predators 11, 12*f*, 397
apex predators 11, 275
reintroduction 363, 363*f*
predator–prey population cycles 39–40, 40*f*
predator release 40, 397
preservation 254, 397
prickly parrot-pea *Dillwynia juniperina* 211
primary data use 46–48, 397
primary successions 25
productivity 263
propagule pressure 204
protected areas 316, 316*f*, 397
creation 314–318
limitations 316–318, 317*f*, 318*f*
SLOSS debate 314–316, 316*f*
protected species 115
legislation 309–310
Provence chalkhill blue butterfly *Polyommatus hispanus* 360
proxy monitoring 48, 397
Przewalski's horse *Equus przewalskii* 357, 375
pseudo-absence data 326–327
Pseudomonas fluorescens 149
Pseudomonas putida 149
Purple loosestrife *Lythrum salicaria* 247–248
purple loosestrife control case study 247–249, 248*f*
pyrethroids 327

quadrat sampling 56*t*
quaking aspen *Populus tremuloides* 276
quarantine 367–368

r-strategists 24, 397
as indicators of environmental change 85
extinction risk relationship 263–264
rabbit, European *Oryctolagus cuniculus* 305, 386
raccoon *Procyon lotor* 229, 324
radio telemetry 167*t*
radiocarbon dating 93
Ramsar site 397
rat, brown *Rattus norvegicus* 216
ratio data 55
realized niche 19
realized range 162, 397
receptors 103, 103*f*, 105–107, 107*t*, 397
red deer *Cervus elaphus* 23, 61, 387
red foxes *Vulpes vulpes* 179
red grouse *Lagopus lagopus scotica* 300, 305
red imported fire ant *Solenopsis invicta* 336
red kangaroo *Macropus rufus* 276
red kite *Milvus milvus* 357
red-bellied black snake *Pseudechis porphyriacus* 211
red-eared terrapin *Trechemys scripta elegans* 57, 59, 63
red-eyed tree frog *Agalychnis callidryas* 41
red-eyed wattle *Acacia cyclops* 209
red squirrel *Sciurus vulgaris* 225
redstart *Phoenicurus phoenicurus* 93
red-tailed bumblebee *Bombus lapidarius* 164
red-whiskered bulbul *Pycnonotus jocosus* 219
reed beds, rhizofiltration 150–152, 152*f*
refugia 296
regeneration patches 159
reinforcement 333, 348–349, 358, 397
reintroduction 36*f*, 270, 303, 333, 348–349, 349*f*, 357, 398
accidental 358
aims of 358–359
arguments for and against 369
basic principles 358–365
case studies 350–351, 350*f*, 371–372, 371*f*, 373, 375–376, 375*f*
factors influencing success 381–383, 382*f*, 383*f*
failed reintroductions 366
feasibility study 368–370
risk assessment 368–369
trial release 369
founder individuals 370–374
individual screening 374
population structure 372–374
source and characteristics of 370–372, 370*f*
funding 376
methods 377–380
hard release strategy 377
soft release strategy 379–380, 379*f*
supplemental feeding 380
temporary enclosures 379*f*, 380
post-release monitoring 380–381
ecological effects 381
movement 380–381, 380*f*
population change 381
survival 380
publication bias 381
reasons for 359–365
human motivations 363–365
site-focused 362–363, 363*f*, 364*f*
species-focused 361–362
regulatory frameworks 365–368
codes of practice 367
legislation 367–368
permissions required 334
policies 366
release site selection 374–376
habitat and carrying capacity 374
social factors 374–376
rewilding programmes 385
taxonomic bias 365
terminology 359*f*
relationships 8
beneficial 16–18
feeding 11–16
relative abundance 55
indicator species 72, 395
remediation 117, 136–137, 397
immobilization 136
pollutant removal 136
see also bioremediation
remnant patches 159
remote data collection and use 59–62
Renicola roscovita 210
rescue translocation 125, 126, 397
reserve creation *see* nature reserves
resources 395
competition for 20
increasing demands for 3–4
resource limitation 20
restoration energy 397
reversion techniques 302–303
rewilding 303, 357, 383–389, 384*f*, 397
case study 388, 388*f*
challenges 385–387
conceptual problems 386
land ownership 386–387
public perception 387
site availability 386
reintroduction and 385
scale 384–385
successes 387–389, 389*f*
rhinoceros poaching 308
rhizofiltration 150–152, 152*f*
case study 153, 153*f*

Rhododendron ponticum 200
Rhopilema nomadica 210
ring-tailed lemur *Lemur catta* 321
roe deer *Capreolus capreolus* 61, 387
rosemary *Rosmarinus officinalis* 55
rotenone 241
ruddy duck *Oxyura jamaicensis* 200, 218
ruderal bumblebee *Bombus ruderatus* 372
rush *Juncus* 75
Russian Steppe plant reintroduction case study 373

S

sable antelope *Hippotragus niger* 167
Saint Helena hoopoe *Upupa antaios* 205
St John's Wort *Hypericum perforatum* 200
Saint Lucia parrot *Amazona versicolor* 367
saltcedar tree *Tamarix*
sampling
  approaches 56*t*
  consistent sampling requirement 41–42, 64
sand lizard *Lacerta agilis* 349
sandhill crane *Grus canadensis* 278
satellite trackers 167*t*
savannah, plants as indicators of grazing levels 88–89, 89*f*
scavengers 14–15, 397
scenario analysis 120
scramble competition 20
scarlet macaw *Ara macao cyanoptera* 379
Schmidt's guenon *Cercopithecus ascanius schmidti* 346
scimitar oryx *Oryx dammah* 259
Scottish crossbill *Loxia scotica* 183
screening 397
sea otter *Enhydra lutris* 276, 357, 363
sea thrift *Armeria maritima* 71
sea urchin Echinoidea 363
sea urchin *Strongylocentrotus* 276
seastar *Pisaster orhraceus* 275
secondary data use 45–46, 397
secondary dispersal power 217
secondary successions 25
sedge *Carex* 93
seed banks 340, 341, 351–352, 351*f*
seed collection 339–340, 341, 351–352
selection 21
  artificial 23
  kin selection theory 23
  natural 23
  sexual 23
semi-automated monitoring 48
seres 25
sessile oak *Quercus petraea* 78
sexual selection 23
Shannon-Weiner index 283, 284*t*
short-haired bumblebee *Bombus subterraneus* 357, 370, 371–372
short-haired bumblebee reintroduction case study 371–372, 371*f*
Siberian flying squirrel *Pteromys volans* 279
sika deer *Cervus nippon* 218
silver birch *Betula pendula* 85
silver-washed fritillary *Argynnis paphia*, 179
Simpson's index 283, 284*t*
single-species conservation 255–256, 256*t*, 397
site value 112–113
  designations 112–113, 310–313, 310–311*t*
    multiple designations 310
  social value 113
  species richness relationship 113
Sites of Special Scientific Interest (SSSIs) 52
slender yellowtail kingfish *Alepes djedaba* 210
SLOSS (single large or several small) debate 314–316, 316*f*
small cranberry *Vaccinium microcarpum* 78
small Indian mongoose *Herpestes palustris* 55, 177
small pearl-bordered fritillary *Clossiana selene* 368
smooth newt *Lissotriton vulgaris* 55
snake's head fritillary *Fritillaria meleagris* 114
snowshoe hare *Lepus americanus* 10
soils, plants as indicators 75
Somerset hair grass *Koeleria vallesiana* 78
source-sink population dynamics 175–179
  case study 177–178, 177*f*, 178*f*
southern white rhino *Ceratotherium simum* 181
southwestern willow flycatcher *Empidonax traillii extimus* 219
Spanish imperial eagle *Aquila heliacea adalberti* 360
spatial surveying 32, 34–39, 38*f*, 65
spatial variation 24–25, 25*f*
spatio-temporal monitoring 40–42, 41*f*
  consistent sampling requirement 41–42, 64
Special Area of Conservation 397
Special Protection Area 397
specialists 19, 397
  extinction risk 264
speciation 23
species
  biological species concept 36, 393
  cryptic species 36
  monitoring 54–64
    absolute abundance 55
    likely absence 54
    prevalence 54–55
    relative abundance 55
    species presence 54
  new species discovery and classification 35–37, 37*f*
  phylogenetic species concept 36
  polymorphic species 36
  protected species 115, 309–310
  spatial patterns of 162–165
  species-focused conservation 255–256, 256*t*
  translocation 125, 126
  see also non-native species
species distribution 162–163
  continuous distribution 162, 163*f*
  discontinuous distribution 162, 163*f*
  disjunctions 162–163
  mapping 163–165, 164*f*, 165*f*
  modelling 205
species diversity 280–284, 398
  diversity index 283, 384*t*
  evenness 283, 283*f*, 398
  species richness 280–284, 398
    disturbance level relationship 88–89, 89*f*
    non-native species richness relationship 206–207, 206*f*
    site value relationship 113
  see also biodiversity
species introduction 398
species management 303–309
  culling 307–309
  guarding and protecting 309
  insurance populations 270, 303, 333, 334*f*, 361
  reinforcement 333, 348–349, 358
  supplemental feeding 303–307
  translocations 125, 126, 303
  see also captive breeding programmes; reintroduction
species range 19, 162
  mapping 163–165, 164*f*, 165*f*
species recovery plan 361–362, 398
species richness *see* species diversity
species turnover 207
species value 115

species–habitat interaction research 324–327, 325*f*
macro-scale 325–326
meso-scale 324–325
micro-scale 324
quantifying interactions with presence-only data 326–327
sperm whale *Physeter microcephalus* 327
spotted hyena *Crocuta crocuta* 58
spruce *Picea* 94
stable isotope analysis 168–169, 168*f*
starling *Sturnus vulgaris* 217
statutory regulator 398
Steinernema feltiae 244
Stephens Island wren *Xenicus lyalli* 205
stepping stone islands 185–186
sterile-male-release technique 218, 219*f*
stinging nettle *Urtica dioica* 75
Stipa 373
Stipa pulcherrima 373
Stipa ucrainica 373
stoat *Mustela ermine* 386
stochastic processes 40, 398
demographic stochasticity 261
environmental stochasticity 261
stone marten *Martes foina* 278
strategic environmental assessment (SEA) 130–131, 398
Streptanthus polygaloides 148
striped skunk *Mephitis mephitis* 305
strychnine 241
studbooks 347–348, 398
succession *see* ecological succession
sugar beet *Beta vulgaris* 148
sugarcane *Saccharum* 238
Sumatran rhinoceros *Dicerorhinus sumatrensis* 274, 309
Sumatran tiger *Panthera tigris sumatrae* 346
sunflower *Helianthus* 148
supplemental feeding 303–307
case study 305, 305*f*
reintroduced populations 380
supplementation 333, 348–349, 398
surveillance 31
surveying 31, 31–32*t*, 394
baseline surveying 32, 33–34, 65, 107–110
biosurveys as ecological indicators 70
direct data collection and use 55–58
environmental DNA utilization 60–61
indirect data collection and use 58–59, 59*t*
key principles 62–64
Phase One Habitat Surveying (P1HS) 51
purposes 32
remote data collection and use 59–62
spatial surveying 32, 34–39, 38*f*, 65
sustainability, population trends as indicator 92, 92*f*
Svalbard Global Seed Vault 351, 351*f*
Swallowtail *Papilio machaon britannicus* 267
swamp stonecrop *Crassula helmsii* 210
sycamore *Acer pseudopanatus* 148, 182, 196
symbiosis 17, 398

Tangalunga 167
tarantula hawk wasp Pompilidae 14
tarpan *Equus feras feras* 387
Tasmanian devil *Sarcophilus harrisii* 289, 303, 304
Tasmanian devil case study 304, 304*f*
Tasmanian tiger 353
see also Thylacine
taxonomy 35*f*
temporal monitoring 32, 39–42, 65–66
anthropogenic change 40
natural change 39–40, 39*f*, 40*f*
spatio-temporal monitoring 31*f*, 40–42
temporal variation 24–25, 25*f*
tens rule 200
tern *Sterna* spp. 209*f*
thick-billed parrot *Rhynchopsitta pachyrhyncha* 380
Thomson's gazelle *Eudorcas thomsoni* 335
thylacine *Thylacinus cynocephalus* 270, 304, 353
see also Tasmanian tiger
Thyroptera tricolor 175
tiger *Panthera tigris* 63, 174, 227, 253, 367
tipping point 116
toad *Bufo* 208
tobacco *Nicotiana* 149, 237
tolerance range 73–74, 73*f*
Torreya taxifolia 360
total allowable catch (TAC) 312
transect sampling 56*t*
translocation 199–208, 398
as a management technique 125, 126, 303, 358, 361*f*, 394
terminology 359*f*
invasion pathways 199–200, 200*f*
corridor use 186–187, 187*f*
hitch-hiking 199, 201–202, 201*f*
opportunities for 203
prevention 207–208
see also ex situ conservation; invasive species; non-native species; reintroduction
transposons 149
trapping, pest management 234
tree bumblebee *Bombus hypnorum* 194, 200
Tres Rios Wetland Project case study 153, 153*f*
Tricholoma virgatum 55
tri-focus umbrella 278
trophy hunting 306, 306*f*
tsessebe *Damaliscus lunatus lunatus* 378
Tulipa 373
twinflower *Linnaea borealis* 90

umbrella species 278–280, 279*f*, 398
urban gardens 319
Urophora affinis 210
Usnea 79
Usnea cornuta 72

vampire bat *Desmodus rotundus* 12
vector 398
Veldt Condition Index 89
velvet bean *Mucuna pruriens* 244
violet *Viola* 211
Virginia opossum *Didelphis virginiana* 229
virtual fences 187
virtual island 398
visitor pressure 321
management 319–320, 320*f*
volunteer participation 48–49
vulture supplemental feeding case study 305, 305*f*

walrus *Odobenus rosmarus* 44
Warming, Eugen 9, 9*f*

water vole *Arvicola amphibious* 118
waterbuck *Kobus ellipsiprymnus* 378
waterweed *Hydrilla verticillata* 78
weasel *Mustela nivalis* 278
weighted average 87
western mosquitofish *Gambusia affinis* 200
wetland rhizofiltration 150–152, 152*f*
  case study 153, 153*f*
whale shark *Rhincodon typus* 11, 335
white admiral *Limenitis camilla* 179
white-headed duck *Oxyura leucocephala* 218
white rhinoceros *Ceratotherium simum* 309, 378
white rust *Albugo candida* 226
white tailed deer *Odocoileus virginianus* 229
white-tailed eagle *Haliaeetus albicilla* 76, 384
whitethroat *Sylvia communis* 169
whooping crane *Grus americana* 256*t*
wild boar *Sus scrofa* 358wilderness areas 255, 383
Wildlife and Countryside Act 1981, UK 42, 309–310
wildlife crime 44
willow *Salix* 85, 148, 301
willow warbler *Phylloscopus trochilus* 169
within-taxa biodiversity indicator 92
Wolbachia 243
wood anemone *Anemone nemorosa* 111
wood thrush *Hylocichla mustelina* 167
woodcock *Scolopax rusticola* 58
woodlands
  plants as indicators of habitat age 89–90
  ride creation 184, 301–302, 302*f*
woodpeckers Picidae 225
woolly mammoth *Mammuthus primigenius* 258

xenobiotic compounds 139

Yangtze finless porpoise *Neophocaena phocaenoides asiaeorientalis* 345–6
Yarkon bream *Acanthobrama telavivensis* 253
yellow-spotted mountain newt *Neurergus microspilotus* 369
Yellowstone to Yukon (Y2Y) ecoregion case study 189, 189*f*
yew *Taxus* 301

zonation 299
  visitor disturbance management 320, 320*f*
zoonosis 398